CONSTRUCTIONAL STEEL DESIGN

World Developments

Proceedings of the First World Conference on Constructional Steel Design, held at Acapulco, Mexico, 6–9 December 1992, in association with the *Journal of Constructional Steel Research*

CONFERENCE CHAIRMEN

Professor P.J. Dowling, *Imperial College, London, UK*
Professor J.E. Harding, *University of Surrey, UK*
Professor R. Bjorhovde, *University of Pittsburgh, USA*
Professor E. Martinez-Romero, *EMRSA, Mexico*

TECHNICAL COMMITTEE

F.S.K. Bijlaard, *The Netherlands*
P.C. Birkemoe, *Canada*
J. Brozzetti, *France*
W.F. Chen, *USA*
A. Colson, *France*
B.L. Edlund, *Sweden*
H.R. Evans, *UK*
J.W. Fisher, *USA*
Y. Fukumoto, *Japan*
K. Ghavami, *Brazil*
M.R. Horne, *UK*
M. Ivanyi, *Hungary*
B. Kato, *Japan*
A.R. Kemp, *South Africa*

A.R. Lamas, *Portugal*
I. Lippi, *Brazil*
R. Maquoi, *Belgium*
C.D. Miller, *USA*
T. Moan, *Norway*
R. Narayanan, *UK*
D.A. Nethercot, *UK*
J. Odlund, *Norway*
M. Škaloud, *Czechoslovakia*
P.O. Thomasson, *Sweden*
N. Trahair, *Australia*
S. Vinnakota, *USA*
R. Zandonini, *Italy*

LOCAL ORGANISING COMMITTEE

Oscar de Buen y Lopez de Heredia
Ricardo Perez-Ruiz
Fernando Frias Beltran
Guillermo Wullf-Kerber

Fernando Guevara-Gonzalez
Joseph Zisman Cohen
Fernando Ibarra-Núñez

Sponsored by the Steel Construction Institute (UK), ICA Construction Group (Mexico), Cia Siderúrgica de Guadalajara (Mexico) and Robertson Mexicana (Mexico), and supported by the Mexican Society of Seismic Engineering and the Mexican Society of Structural Engineering.

CONSTRUCTIONAL STEEL DESIGN

World Developments

Edited by

P.J. DOWLING
Imperial College, London, UK

J.E. HARDING
University of Surrey, UK

R. BJORHOVDE
University of Pittsburgh, USA

E. MARTINEZ-ROMERO
EMRSA, Mexico

CRC Press
Taylor & Francis Group
Boca Raton London New York

CRC Press is an imprint of the
Taylor & Francis Group, an **informa** business
A TAYLOR & FRANCIS BOOK

CRC Press
Taylor & Francis Group
6000 Broken Sound Parkway NW, Suite 300
Boca Raton, FL 33487-2742

First issued in paperback 2019

WITH 104 TABLES AND 546 ILLUSTRATIONS

© 1992 by Taylor & Francis Group, LLC
© 1992 CROWN COPYRIGHT—pp. 225-239
CRC Press is an imprint of Taylor & Francis Group, an Informa business

No claim to original U.S. Government works

ISBN-13: 978-1-85166-953-0 (hbk)
ISBN-13: 978-0-367-86597-9 (pbk)

CIP Catalogue record for this book is available
from the British Library

Library of Congress CIP data applied for

Publisher's Note
The publisher has gone to great lengths to ensure the quality of this
reprint but points out that some imperfections in the original
may be apparent

Visit the Taylor & Francis Web site at
http://www.taylorandfrancis.com

and the CRC Press Web site at
http://www.crcpress.com

Preface

The idea for this volume was conceived during a lull in the proceedings of an international conference on advances in steel structures held by the Adriatic Sea. The first three editors discussed how they could further harness international collaboration amongst the community of professionals interested in constructional steel design. An outstanding programme of top quality papers from authors of international repute combined with a beautiful conference location had proved to be very successful and seemed a formula worth repeating to further progress international collaboration to a world stage. It was also of concern to us that we had failed to involve our Latin American colleagues working in the field of steel structures to any significant extent.

From these considerations emerged the idea to produce *Constructional Steel Design: An International Guide,* with chapters authored by leading international experts and which forms the companion volume to this one. It was further decided to introduce it to a world conference of steel specialists gathered in Acapulco, Mexico, to present recent developments in constructional steel design. This volume, *Constructional Steel Design: World Developments,* contains a collection of refereed papers chosen from some two hundred papers offered for presentation at the conference. The topics covered include aspects of frame elements, plate and box girders, connections, composite structures, dynamics and fatigue, seismic design, buildings, bridges and special structures.

The editors are grateful to all the authors for their contributions. They are also indebted to the Technical Committee which was formed by members of the Editorial Board of the international *Journal of Constructional Steel Research.* Special thanks are due to the chairmen of the various conference sessions, Yuhshi Fukumoto, Frans Bijlaard, Wilf Chen, André Colson, N. Hajdin, René Maquoi, Oscar de Buen, Geoff Kulak and Don Sherman, for giving so generously of their time during the conference.

If this volume contributes even in a small way to the growing collaboration and friendship between the individual members of the world-wide structural steel community, the editors will be well pleased.

PATRICK J. DOWLING
JOHN E. HARDING
REIDAR BJORHOVDE
ENRIQUE MARTINEZ-ROMERO

Contents

Composite Structures

Dynamics and Fatigue

Special Structures

Regional Developments

Late Submission

Frame Elements

DESIGN APPLICATIONS FOR BEAMS RESTRAINED BY ADJACENT MEMBERS

JOACHIM LINDNER
Univ.Prof. Dr.-Ing.
Technische Universität Berlin
Straße des 17. Juni 135, 1000 Berlin 12

ABSTRACT

The different types of restraints for different types of constructional detailing are discussed and methods to determine design values for these restraints are shown. Design formulae to take into account the restraining effects are explained. Numerical examples and parameter studies show the application of the proposed methods.

INTRODUCTION

In reality most beams in structures are at least partially restrained by adjacent members. The great effect of such restraints should therefore be accounted for in practical design.

DESIGN FOR LATERALLY UNSUPPORTED BEAMS

Beam design is affected significantly if lateral torsional buckling can occur and must be taken into account. For an easy treatment the beam is usually assumed as an unsupported beam. This means, that the beam is considered as an isolated member which can rotate freely about its longitudinal axis.

Most specifications define rules for unsupported beams. These rules vary in a wide range [1]. As examples the rules in the new German stability code DIN 18 800 part 2 [2] and the draft for Eurocode 3 [3] are specified by eq. (1).

$$M_x / (\kappa_M M_{px}) \leq 1 \tag{1}$$

where

M_x maximum bending moment about strong axis under factored loads

M_{px} full plastic moment

κ_M reduction factor

This reduction factor in defined differently in both codes. In DIN 18 800 [2] :

$$\kappa_M = (1 / (1 + \bar{\lambda}_M^{2n})^{1/n} \qquad\qquad \bar{\lambda}_M > 0{,}4 \qquad\qquad (2)$$

$\bar{\lambda}_M = \sqrt{M_{px} / M_e}$ non dimensional slenderness for lateral torsional buckling

M_e elastic critical lateral torsional buckling moment

$n =$ system factor, for distributed loads or moment gradient, = 2,5 rolled sections, = 2,0 welded sections
additional factor 0,8 for constant moment distribution

In EC 3 [3] :

$\kappa_M =$ reduction factor, flexural buckling curve a for rolled sections, flexural buckling curve c for welded sections - using $\bar{\lambda}_M$

Different statical systems like single span beams, continuous beam or cantilever can be taken into account by determination of the proper elastic critical moment M_e. Numerical values for that can be achieved by computer programs or simplified formulae which are available in literature or specifications like DIN 18 800 [2] or Appendix F to EC 3 [3]. For the time being very little information is available whether the design curve depends also significantly on the statical system or not. Therefore in all specifications the same design curve is used for different statical systems.
It is of importance for the practical design that for $\bar{\lambda}_M \leq 0.4$ the reduction factor $\kappa_M = 1.0$ is assumed.

INFLUENCE OF ADJACENT MEMBERS

In reality unsupported beams are very rare in practical design. All loads are introduced by neighboring seperate members and the stiffness of these members can be taken into account in much situations.

Adjacent members are normally present as individual members like cross beams or as floor elements like sheeting. The restraining effect of cross beams is mainly caused by its bending stiffness and to some extent by the joints. Contrary to this for sheeting commonly two restraining effects are present:
i)   the horzontal deflection of the upper cord of the beam is prevented by the shear stiffness R of the sheeting, if fasteners between sheeting and beam are present,
ii)   the bending resistance of the sheeting in combination with local deformations partly prevents the twisting of the beam.

EFFECT OF HORIZONTAL RESTRAINTS

If the entire compression flange of an I-beam is laterally fixed no lateral torsional buckling can occur. But in continuous beams the moment distribution is such that negative moments at the supports can lead to lateral torsional buckling

a)

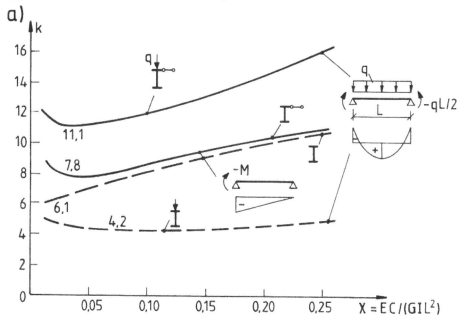

b)

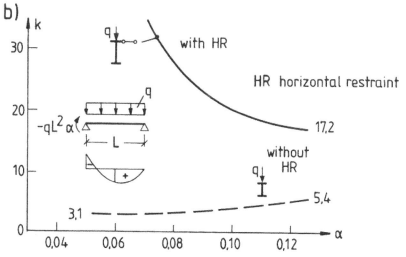

Fig 1. Lateral torsional buckling parameter k for beams with and without horizontal restraint

although the strength of such a beam is much higher than for a laterally unsupported beam. This effect can also be seen from figs. 1a and 1b, where values for the lateral torisonal buckling parameter k are shown which were calculated by computer programs. Using the factor k the elastic critical moment M_e can be determined from eq. (3).

$$M_e = (k / L) \sqrt{E I_y G I} \tag{3}$$

where

k	lateral torsional buckling parameter
L	beam length
I_y	moment of inertia weak axis
I	torsional constant

If floor elements are present the condition should be known when the upper flange of a beam can be assumed as horizontally fixed. In [2] a value for the neccessary shear stiffness R is given by eq. (4).

$$R \geq (EC \ \pi^2 / L^2 + GI + EI_y \ \pi^2 \ 0{,}25 \ h^2 / L^2) \ 70 / h^2 \tag{4}$$

where

C	warping constant
h	depth of the beam
70	medium value for different types of moment distribution

The effective shear stiffness R_e of trapezoidally corrugated sheeting can be calculated by the method of Bryan and must then be compared to eq. (4).

EFFECT OF TORSIONAL RESTRAINT

A simplified stability check requires a minimum stiffness of the torsional restraint coefficient c_ϑ (in EC 3 termed as spring stiffness K) by eq. (5).

$$c_\vartheta \geq (M_{px}^2 / EI_y) \ k_\vartheta \tag{5}$$

TABLE 1
Factor k_ϑ for different moment distributions

top flange	moment distribution				
free $\mathbf{I}$	4,0	3,5	3,5	1,6	1,0
restraint $\mathbf{I}$	0	0,12	0,23	1,0	0,7

The factor k_θ depends on the moment distribution and the fact whether the upper flange of the beam is horizontally restraint or not. Design values for k_θ can be seen from table 1, [2]. The values for c_θ were determined with regard to the demand that the load carrying capacity should reach 95 % of M_{px} if c_θ is taken into consideration. A further extension of the load carrying capacity would need an extensive enlargement of c_θ which seems not to be neccessary.

The effective torsional restraint coefficient c_θ depends on the rigidity of the joint between the beam which should be stabilized and the structural element which acts as an adjacent member. For trapezoidally sheeting three deformation components should be taken into account as shown by eq. (6). For cross beams as adjacent member the influence of $c_{\theta A}$ may be negligable.

$$1/c_\theta = 1/c_{\theta M} + 1/c_{\theta A} + 1/c_{\theta P} \qquad [kNm/m] \qquad (6)$$

where

$c_{\theta M}$ theoretical value with regard to the stiffness of the adjacent member only
        $= 4\,E\,I_a\,/L_a$ for continuous sheeting
        I_a, L_a effective moment of inertia, span of the adjacent member

$c_{\theta A}$ rigidity of connection, dependent on no. of screws, position of sheeting (inverted or not), sheet thickness, diameter of washers, thermal insulation

$c_{\theta P}$ distorsion of the beam investigated
        $= 5770/(d/t_w^3 + 0.5\,b_f/t_f^3)$, dimensions in [cm]

The rigidity of connection is most important especially for trapezoidally sheeting. To account for different width of beams, it is useful to calculate the effective value by eq. (7).

$$c_{\theta A} = c_{\theta A}' \, (b_f/100)^2 \qquad\qquad \text{if } b_f \leq 125 \text{ mm} \qquad (7a)$$

$$c_{\theta A} = c_{\theta A}' \, (b_f/100)1.25 \qquad\qquad \text{if } 125 < b_f \leq 200 \text{ mm} \qquad (7b)$$

Values for $c_{\theta A}'$ can be obtained by tests only. It was shown that a segment of a realistic constructed roof section may be used in tests for different types of loading and constructional details. Test configuration and conducting of the tests were described earlier, [5], [6]. Important parameters which influence the results significantly are:

-   type of the roofing skin (e.g. depth and width of the trapezoidally sheeting, plate thickness),
-   location and distance of the fasteners (at crest, in trough, fasteners in every ($b = b_r$) or alternate ($e = 2b_r$) crest/trough)
-   type of fasteners (normally selftapping screws diameter 6.3 mm)
-   roofing construction (with or without intervening thermal insulation, type of thermal insulation)
-   type of loading (gravity load, uplift loading by windload)
-   magnitude of loading.

An example for one type of construction which was investigated in tests is shown in fig. 2, an example for a torsional moment-rotation curve measured during the test is shown in fig. 3. From fig. 3 it can be seen that the stiffness related to an small value ϑ can be much higher than for the value $\vartheta = 0.1$ which was taken as an unfavourable reference value. All design values given in [2] and [4] are based on this unfavourable assumption of $\vartheta = 0.1$.

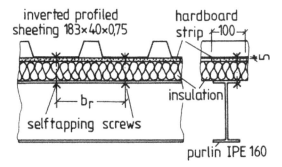

Fig. 2. Roof construction with thermal insulation

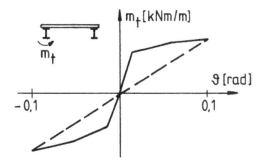

Fig. 3. Example for the torsional moment rotation curves in tests

Design values are given in DIN 18800 [2] and in literature, e.g. [4]. Values in [5] are not evaluated on the basis of eq. (7). Appendix A to Eurocode 3 [3] mentions the torsional restraint coefficient too but offers a sligthly different test configuration and gives the same design values as DIN 18 800 part 2. Some values from [2] and [4] are given in table 2.

If the simplified stability check of eq. (5) is not used, the positive influence of the torsional restraint can be accounted for in another way too. In this situation the elastic critical moment M_c can be calculated taking into account c_θ by calculating a torsional constant I^* instead of I using eq. (8).

$$I^* = I + c_\theta \, L^2/(G \, \pi^2) \tag{8}$$

This value I^* must then be introduced in a suitable formula, e.g eq. (3), or in a computer program.

TABLE 2

Torsional restraint coefficients for trapezoidally corrugated inverted sheeting, fasteners selftapping screws diameter 6.3 mm, gravity load

type of connection		c_{0A}'
without thermal insulation:		
bottom chord, $e = b_r$,	diam 22 washer	3.1
bottom chord, $e = 2b_r$,	diam.22 washer	2.0
top chord, $\quad e = b_r$,	saddle washer steel $t \geq 0,75$ mm	10.0
top chord, $\quad e = 2b_r$,	saddle washer steel $t \geq 0,75$ mm	5.2
with thermal insulation 60 mm Styrodur 3000S:		
bottom chord, $e = b_r$,	diam. 22 washer	4.7
bottom chord, $e = 2b_r$,	diam. 22 washer	2.9
top chord, $\quad e = b_r$,	saddle washer steel $t \geq 0,75$ mm	5.0
top chord, $\quad e = 2b_r$,	saddle washer steel $t \geq 0,75$ mm	3.2

NUMERICAL EXAMPLE

A roof is investigated with purlins of IPE 240 section as continuous beams, span L = 11 m, loaded by uniformly distributed load. The trapezoidally sheeting 40x183x0.75 mm (I_{ef} = 21.6 cm^4/m) is fastened at the bottom chord, $e = b_r$.

$$R = (21000 \cdot 0.0003739 \ \pi^2/11^2 + 8100 \cdot 0.00129 +$$

$$21000 \cdot 0.0284 \ \pi^2 \ 0.25 \cdot 0.24^2/11^2) \ 70/0.24^2 \qquad = 14300 \ kN$$

It must be shown additionally that the effective value R_e is greater than R, [7] or DIN 18 807.

$c_{0M} \quad = 4 \cdot 21000 \cdot 0.00216 / 3.0 \qquad\qquad = 60.5$ kNm/m

$c_{0P} \quad = 5770/(24/0.62^3 + 0.5 \cdot 12/0.98^3) \qquad = 53.9$ kNm/m

from table 2 : $\qquad\qquad\qquad\qquad\qquad c_{0A}' \quad = 3.1$ kNm/m

eq. (7a): $\qquad\qquad c_{0A} \quad = 3.1 \ (120/100)^2 \qquad = 4.46$ kNm/m

eq. (6) :

$1/c_0 \quad = 1/60.5 + 1/53.9 + 1/4.46 \qquad c_0 \quad = 3.86$ kNm/m

eq. (5) and table 1:

$c_0 \quad \geq (88.0^2 \ /(21000 \cdot 0.0284)) \ 0.23 \qquad = 2.99$ kNm/m

Because the minimum stiffness requirement is fulfilled no further check with regard to lateral torsional buckling is neccessary.

PARAMETER STUDIES

A continuous beam is investigated subjected to uniformly distribulted load, [1]. In order to calculate the moment capacity the design curve from DIN 18 800 is used [2]. For comparison the design curve of the British specification BS 5400 is taken into consideration, using the same elastic critical lateral torsional buckling Moment M_e. The results are shown in fig. 4.

The following conclusions may be drawn:

i)   compared to the beam without restraints (curve 1) the moment capacity increases significantly if the torsional restraint (curve 2) is taken into account,

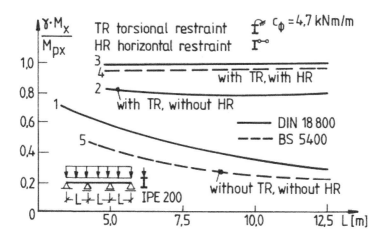

Fig. 4. Influence of restraints

ii)   if additionally the horizontal restraint of the top flange is taken into consideration (curve 3) the moment capacity reaches the full plastic moment,

iii) the dependency of the moment capacity from the beam length becomes nearly unimportant if restraints are taken into account (curves 2 and 3),

iv)   the differences between design curves of different specifications (curves 3 and 4, 1 and 5) decrease when restraints are taken into consideration

CONCLUSIONS

It was shown that beam design should not only take into account laterally unsupported beams but should make use of adjacent members. The adjacent members can lead to a horizontal restraint as well as to a torsional restraint.
The effect of these restraint are shown by a numerical emxample and by parameter studies on purlins. Specifications should allow for the favourable effects of adjacent members.

REFERENCES

1. Lindner, J.: Specifications and codes in western europe. 4th international colloquium on stability of metal structures, New York, 1989, proceedings

2. DIN 18 800 part 2. Stahlbauten; Stabilitätsfälle, Knicken von Stäben und Stabwerken (steel structures; stability, buckling of bars and skeletal structures), (11.1990)

3. Eurocode 3. Design of steel structures. General Rules and Rules for Buildings (draft nov. 1990)

4. Lindner, J., Scheer, J., and Schmidt, H. : Stahlbauten, Erläuterungen zu DIN 18 800 Teil 1 bis Teil 4 (steel structures, commentary to DIN 18 800 part 1 to part 4), Berlin, Beuth und Ernst & Sohn, 1992

5. Lindner, J., und Gregull, Th. : Torsional restraint coefficients of profiled sheeting. IABS colloquium Stockholm, 1986, Thin-walled metal structures in buildings, Proceedings pp 161-168, Zürich 1986

6. Lindner, J., und Gregull, Th.: Drehbettungswerte für Dachdeckungen mit untergelegter Wärmedämmung (Torsional restraint coefficients of roofing skin with termal insulation), Stahlbau 58(1989), S. 173-179

7. European Convention for Constructional Steelwork. ECCS: European Recommendations for the Stressed Skin Design of Steel Structures. ECCS-XVII-77-1E, Constrado 1977

8. Baehre, R., und Wolfram, R.: Zur Schubfeldberechnung von Trapezprofilen. Stahlbau 55 (1986), S. 175-179.

ELASTIC LOCAL BUCKLING OF I–SECTION BEAMS UNDER SHEAR AND BENDING

M.A. Bradford and M. Azhari
School of Civil Engineering
The University of New South Wales
Kensington, NSW 2033
AUSTRALIA

ABSTRACT

A complex finite strip method of analysis is described briefly which may be used to study the local buckling of plate structures subjected to bending and shear. The method is then used to investigate the interaction of bending and shear in precipitating the elastic local buckling of I–sections. Conclusions are drawn regarding the critical interaction of the shear and bending stresses.

INTRODUCTION

Local buckling of thin plate assemblies is characterised by localised distortions of the cross–section of the member, with the line junctions between intersecting plates remaining straight. This differs from lateral buckling [1], where buckling is of an overall mode and the cross–section does not distort, and from distortional buckling [2] where local and overall buckles interact. This paper is concerned with local buckling of I–section beams.

An extensive review of local buckling, starting from the work of Bryan in 1891, has been given in Ref. 3. For analysing general problems, the finite element method has been deployed over the last 20 years or so. A variation of this is the semi–analytical finite strip method, where polynomials have been retained for the transverse variation of displacements, but where the lengthwise variation of displacements is represented by harmonic functions. Shear may be included in this method by using spline functions instead of harmonics [4], or by prescribing the displacements using complex arithmetic [5]. The complex finite strip method, developed in Ref. 5, has received little attention since its development in 1974, and is used in this paper to treat the shearing actions. It is easily programmed, and results in solutions which require the extraction of the eigenvalue of a matrix with relatively few degrees of freedom.

This paper, then, deploys the complex finite strip method, as set out in Ref. 6, to study the elastic local buckling of I–beams under bending and shear. A brief description of the method is given, and some typical results are given for a range of I–section geometries.

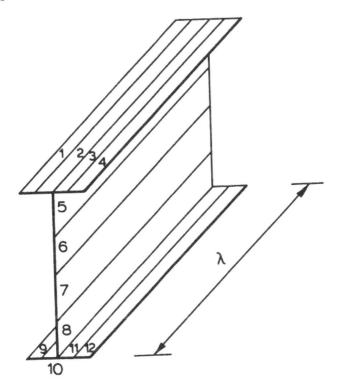

Fig. 1   Finite Strip Idealisation

FINITE STRIP METHOD

The I–sections considered in the finite strip buckling analysis are assumed to consist of a series of rectangular strips, whose buckling half–wavelength λ equals that of the member between simple supports. Each of the strips is connected to one or more of the other strips along one or both of its longitudinal edges, as shown in Fig. 1. In the finite element method, the domain would be further subdivided into a series of rectangular elements along the length of the strips. In that case, the displacement functions in the longitudinal direction of an element would normally be a polynomial.

The finite strip method used to study the local buckling of I–beams is based on that outlined in Ref. 6. The basic steps are similar to the finite element method, and involve:

(a) a definition of the (complex) displacement functions to describe the plate flexural and membrane deformations;

(b) a statement of the linear and nonlinear strain–displacement relation;

(c) selection of an appropriate elastic plate theory to describe the membrane and flexural behaviour of the plate strips;

(d) application of the principle of minimum total potential energy (or virtual work) to determine the stiffness and stability matrices; and

(e) solution of the eigenproblem by a suitable eigenvalue extraction routine; and determination of the buckled shape.

By using this procedure, the stiffness matrix [k] and stability matrix [g] for each strip may be determined, as set out in Ref. 6. These may then be assembled into global stiffness and stability matrices [K] and [G] respectively, which are generally complex. The variational form of the buckling equation is ultimately written as

$$\{ \delta R \}^T \left[[K] - \lambda^* [G] \right] \{R\} = 0 \tag{1}$$

where $\{R\}$ is the vector of global buckling displacements and λ^* is the buckling load factor. By noting the variations $\{\delta R\}$ are arbitrary,

$$\left[[K] - \lambda^* [G] \right] \{R\} = \{0\} \tag{2}$$

which represents the standard eigenvalue equation. For nontrivial $\{R\}$, the determinant of $[K] - \lambda^* [G]$ must vanish. This eigenproblem was solved using a complex eigensolver [6]. The value of λ^* represents the eigenvalue while the corresponding eigenvector {R} represents the buckled shape. The solution of Eq. 2 was performed on a desktop workstation, and was very rapid in terms of computer time.

BUCKLING SOLUTIONS

In order to determine the buckling interaction between bending and shear, it is firstly necessary to compute the critical shear stress in the absence of bending, and the critical bending stress in the absence of shear. The former is given by

$$\tau_{cro} = k_s \frac{\pi^2 E}{12(1-\nu^2)} \frac{1}{(b_w/t_w)^2} \tag{3}$$

and the latter by

$$\sigma_{cro} = k_b \frac{\pi^2 E}{12(1-\nu^2)} \frac{1}{(b_w/t_w)^2} \tag{4}$$

where k_s is the shear buckling coefficient for the web, determined from Ref. 7, and k_b is the bending buckling coefficient for the web, determined from Ref. 8. In the previous two equations, E is Young's modulus and ν is Poisson's ratio. Figure 2 shows the I–section subject to bending and shear stresses. The shear stress distribution in the flanges was determined from Ref. 1.

The variation of the dimensionless critical shear stress τ_{cr} / τ_{cro} is shown in Fig. 3 as a function of the dimensionless buckling half–wavelength λ/b_w and bending stress σ/σ_{cro} for a web plate with simply supported edges. Under pure shear, the minimum buckling half–wavelength is 1.252 b_w, which is in agreement with Bulson [9]. As the bending stress σ increases, the minimum shear buckling stress decreases, as does the buckling half–wavelength corresponding to the local nadir. Similar graphs are shown for the webs of I–sections in Fig. 4 ($b_f/t_f = 30$) and in Fig. 5 ($b_f/t_f = 20$). It can be seen that as the degree of web fixity increases (b_f/t_f decreases), the position of the local minimum λ_{min} decreases. In fact, for a built–in web plate under pure shear, the nadir was found to occur at $\lambda/b_w = 0.83$.

The finite strip method was used to calculate the relationship between σ_{cr} and τ_{cr} in an I–section, using values at the local buckling minima as determined above. Figure 6 was obtained for various values of the web slenderness b_w/t_w, with b_f/t_f set at 20,

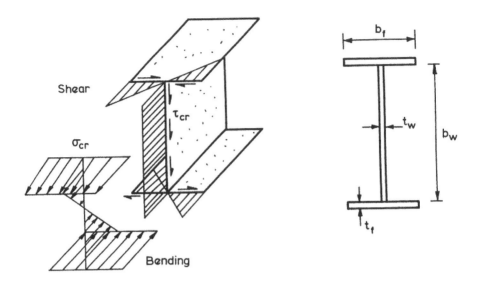

Fig. 2 Stress Distribution and Geometry

while Fig. 7 was obtained for various values of the flange slenderness b_f/t_f (including the simply supported case with b_f/t_f approaching infinity) with b_w/t_w fixed at 200. It is evident from these figures that the interaction at elastic buckling is very close to the circle

$$\left[\frac{\tau_{cr}}{\tau_{cro}}\right]^2 + \left[\frac{\sigma_{cr}}{\sigma_{cro}}\right]^2 = 1 \qquad (5)$$

Hence once k_b (and σ_{cro}) and k_s (and τ_{cro}) have been determined, it is easy to closely approximate the combinations of σ_{cr} and τ_{cr} which will cause elastic local buckling of an I–section beam.

CONCLUSIONS

A complex finite strip method of analysis has been described briefly. The method may be used to investigate local buckling in shear as well as bending, and for a beam with simply supported edges results in an eigenproblem whose solution on a desktop

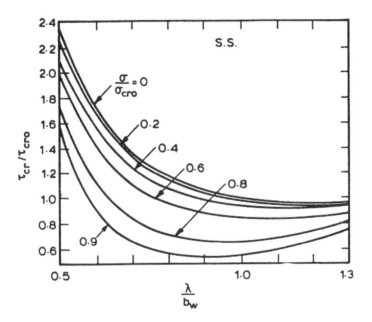

Fig. 3 Buckling Shear Stresses for Simply Supported Web

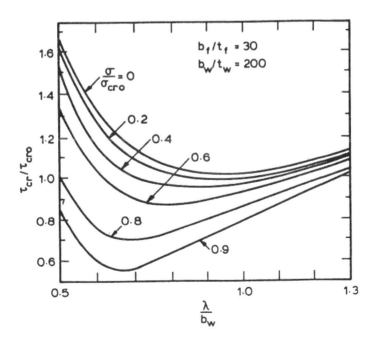

Fig. 4 Buckling Shear Stresses for $b_f/t_f = 30$

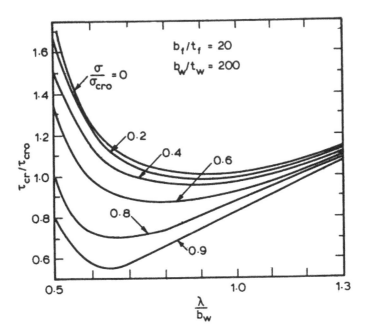

Fig. 5   Buckling Shear Stresses for $b_f/t_f = 20$

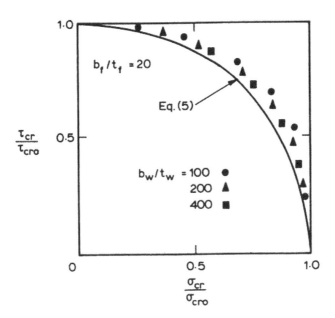

Fig. 6   Buckling Interaction for Fixed Flange Slenderness

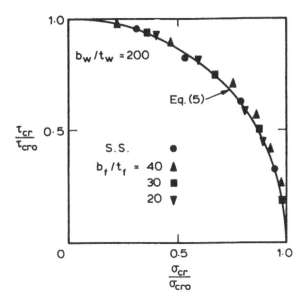

Fig. 7 Buckling Interaction for Fixed Web Slenderness

workstation is rapid.

The method was used to demonstrate the relationship between the buckling half–wavelength λ and the critical shear stress τ_{cr} in a web plate. The localised minimum value of λ was found to decrease as the flange slenderness decreased and the web became more built–in. The finite strip method was then used to determine the relationship between τ_{cr}/τ_{cro} and σ_{cr}/σ_{cro} at elastic local buckling. The relationship was found to be close to circular, making the prediction of the combinations of σ_{cr} and τ_{cr} at elastic local buckling a simple procedure.

REFERENCES

1. Trahair, N.S., and Bradford, M.A., "The Behaviour and Design of Steel Structures", Chapman and Hall, London, 1988.

2. Bradford, M.A., and Trahair, N.S., "Distortional buckling of I–beams", J.Struct.Div., ASCE, Vol. 107, No.ST2, 1981, pp.355–370.

3. Bradford, M.A., "Elastic local buckling of trough girders", J.Struct.Eng., ASCE, Vol. 116, No. 6, 1990, pp.1594–1610.

4. Lau, S.C.W., and Hancock, G.J., "Buckling of thin, flat–walled structures by a spline finite strip method", Thin–Walled Structs., Vol. 4, No. 4, 1986, pp.269–294.

5. Plank, R.J., and Wittrick, W.H., "Buckling under combined loading of thin, flat–walled structures by a complex finite strip method", Int.J.Num.Meth.Engg., Vol.8, 1974, pp.323–339.

6. Azhari, M., and Bradford, M.A., "Buckling modes in I–sections by a complex finite strip method", submitted for publication.

7. Azhari, M., and Bradford, M.A., "Local buckling of I–beams under pure shear", submitted for publication.

8. Hancock, G.J., Bradford, M.A., and Trahair, N.S., "Web distortion and flexural–torsional buckling", J.Struct.Div., ASCE, Vol. 106, No. ST7, 1980, pp.1557–1571.

9. Bulson, P.S., "The Stability of Flat Plates", Chatto and Windus, London, 1970.

NOTATION

b_f	width of flange
b_w	depth of web
E	Young's modulus of elasticity
[G]	global stability matrix
[g]	strip stability matrix
[K]	global stiffness matrix
k_b	bending local buckling coefficient
k_s	shear local buckling coefficient
[k]	strip stiffness matrix
{R}	vector of global degrees of freedom
t_f	flange thickness
t_w	web thickness

λ	buckling half–wavelength
λ^*	buckling load factor
ν	Poisson's ratio
σ	bending stress
σ_{cr}	elastic critical bending stress
σ_{cro}	elastic critical bending stress in the absence of shear
τ_{cr}	elastic critical shear stress
τ_{cro}	elastic critical shear stress in the absence of bending.

STRESSES IN THE WEB OF AN I-SHAPED BEAM DUE TO AN ATTACHED MEMBER

BRIAN W BERSCH
Bechtel Corporation
9801 Washingtonian Boulevard
Gaithersburg, MD 20878-5356
U.S.A.

ABSTRACT

When a member frames into the web of an I-shaped beam, stresses
are induced by the reactions, both globally throughout the beam,
and locally to the connection. The magnitude and distribution of
the global stresses are well known. However, the local stresses
are not as well understood. Within this paper, a study of the
local stresses induced by a moment reaction is made, using the
finite element method. It was found that the beam web behaves
similarly to a simply supported plate, of some effective width.
Using this analogy, and the results of the finite element study,
an expression calculating the maximum local stress was derived.

INTRODUCTION

When a member is connected to the web of an I-shaped beam,
using a moment connection, a shear force, V, and a moment, M_0,
developes at the connection to the beam, see Figure 1. These
reactions induce stresses in the beam, both locally to the
connection, and globally throughout the beam. The treatment of
the global stresses is well documented, however the local
stresses are not so well understood. The shear force will
induce stresses within the beam web, which may be evaluated

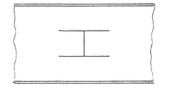
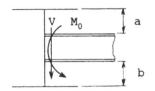

Figure 1. Elevation of a typical connection to a beam web.

using solutions developed by S. Timoshenko [1]. The moment reaction will also produce stresses. However, the magnitude and distribution of these stresses are not as well known.

Because of the uncertainity associated with these stresses, a conservative analysis is typically performed, with the result that stiffeners are frequently used to strengthen the web. Since these stiffeners are an expensive item, their elimination can result in a significant cost savings. Within this paper, a study will be made of these local stresses, using the finite element method, and an accurate technique to find the magnitude of the local stresses will be developed.

FINITE ELEMENT ANALYSIS

Analysis
The local web stresses due to the moment reaction, for the problem shown in Figure 1, will be investigated by performing a parametic study, using the finite element method. The key parameters to be studied are:

    Beam size - depth, width of flange, thickness of
                web, and thickness of flange
    Attachment shape - wide flange and tube steel
    Attachment size - width, depth, and thicknesses
    Attachment location - relative to beam flange

The combinations of attachment, beam, and attachment location that were analyzed, are summarized within Table 1. The finite element models were developed using thin plate elements. The attachment was made at mid-span of an "infinitely" long beam. This use of Saint-Venant's Principle does not remove any generality from the analysis, but it does eliminate any concern over the effects of the beam length and its supports. It also introduces a plane of symmetry, which simplifies the modeling. The flanges of the beam are unrestained. A typical model is shown in Figure 2.

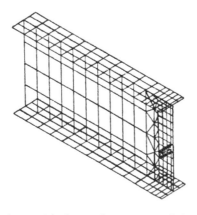

Figure 2. Typical finite element model.

TABLE 1
Summary of finite element analyses

BEAM	ATTACHMENT	a (in)	σ_{max} (ksi)
W36X194	TS 3x2x1/4	16.5	0.911
W24x100	TS10x10x1/4	7.00	0.801
	TS10x 8x1/4	7.00	0.887
	TS10x 6x1/4	7.00	0.990
	TS10x 4x1/4	7.00	1.167
	TS10x 2x1/4	7.00	1.391
	PL 10x1/4	7.00	1.603
	TS 8x 8x1/4	8.00	1.043
	TS 6x 6x1/4	17.00	1.674
	TS 6x 6x1/4	15.00	1.649
	TS 6x 6x1/4	13.00	1.576
	TS 6x 6x1/4	11.00	1.467
	TS 6x 6x1/4	9.00	1.346
	TS 6x6x1/16	9.00	1.333
	TS 4x 4x1/4	10.00	1.988
	TS 2x 4x1/4	11.00	2.694
	TS 2x 2x1/4	11.00	3.393
	W18x40	3.875	0.612
	W10x49	7.00	0.796
	W10x29	7.00	1.021
	W10x15	7.00	1.155
	W6x25	9.00	1.383
	W6x12	9.00	1.629
	WF w/d=b_f=6" t_w=t_f=1/16"	9.00	1.348
	W4x13	10.00	2.028
W24x68	TS 6x 6x1/4	17.00	2.259
	TS 6x 6x1/4	15.00	2.153
	TS 6x 6x1/4	13.00	2.034
	TS 6x 6x1/4	11.00	1.887
	TS 6x 6x1/4	9.00	1.728
W21x68	TS 6x 6x1/4	10.00	1.867
	W6x15.5	10.00	1.878
W18x105	TS 6x 6x1/4	10.50	1.179
	TS 6x 6x1/4	9.00	1.125
	TS 6x 6x1/4	7.50	1.034
	TS 6x 6x1/4	6.00	0.942
	TS 6x 4x1/4	10.50	1.394
	TS 6x 4x1/4	9.00	1.343
	TS 6x 4x1/4	7.50	1.213
	TS 6x 4x1/4	6.00	1.114
	TS 6x 3x1/4	10.50	1.507
	TS 6x 3x1/4	9.00	1.458
	TS 6x 3x1/4	7.50	1.322
	TS 6x 3x1/4	6.00	1.223
	W6x12	10.50	1.374
W12x65	TS 4x 4x1/4	7.00	3.707
	TS 4x 4x1/4	6.00	3.540
	TS 4x 4x1/4	5.00	3.202
	TS 4x 4x1/4	4.00	2.893
	W4x13	7.00	3.707

Taking advantage of symmetry, only one-half of the problem was modeled. Symmetric boundary conditions were used for all nodes in the beam, on the plane of symmetry. At the end of the beam, in order to model a clip angle connection, the nodal translations were restrained at the last row of nodes in the web. The boundary conditions used for the nodes defining the attachment, consisted of symmetric boundary conditions for all nodes lying in the plane of symmetry.

In order to apply the loading, the nodes in the outer face of the attachment were constrained to a common node at the centroid, by rigid links, and a concentrated moment of 1.0 kip-inch was applied to the common node. This approach eliminated the necessity of distributing the load in some approximate way to the nodes.

To represent mild steel, the material properties used for all elements, consisted of a modulus of elasticity of 29000 ksi, and a Poisson's ratio equal to 0.3.

A total of 50 analyses were made. The results of these analyses are summarized within Table 1.

Results
The models discussed in the previous section were analyzed using a linear, elastic finite element program [2]. The maximum and minimum principal stresses within the plate elements were obtained.

The wide flange and tube steel attachments showed the same general trends in their resulting stress distributions, and are shown in Figure 3. The maximum principal stress occured at point A, and decreased along the length of the beam. The stresses also varied from a maximum at the face of the attachment, to essentially zero at the flange.

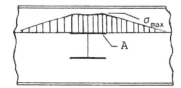 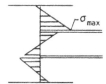

Figure 3. Distribution of maximum principal stresses.

Review of the resulting maximum principal stresses, and their distribution for these finite element analyses, produced the following observations:

Attachment Shape: Comparison of the maximum principal stresses for similar sized TS and wide flange (d and b_f) attachments on the same size beams, and at the same location, shows that the stresses are essentially equal, and are distributed in a similar manner.

This similarity in stress is concluded to be caused by the stiffening effect of the attachment's overall dimensions, and the location of the member's "web" has little effect on the resulting maximum local stress.

Attachment Size: It can be seen that the smaller the attachment, the higher the stress. This fact can be readily explained by applying a simply supported plate analogy. The problem can then be idealized as shown in Figure 4.

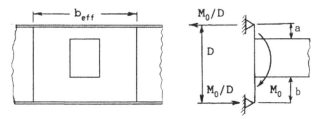

Figure 4. Simply Supported Plate Analogy.

Since the reactions M_0/D do not change for varying sizes of attachments, reducing the attachment's depth will cause the moments $(M_0a)/D$ and $(M_0b)/D$, and hence the stresses in the web, to increase. If the width of the attachment b_{att} decreases, the width of the beam web effective in resisting M_0 decreases, causing the stress to increase.

Attachment Location: The effect of the attachment's location, relative to the beam flanges, may be studied by considering the following cases:

$$\begin{array}{l}
\text{TS } 6\times6\times1/4 \text{ on } W24\times100 \\
\text{TS } 6\times6\times1/4 \text{ on } W24\times68 \\
\text{TS } 6\times6\times1/4 \text{ on } W18\times105 \\
\text{TS } 6\times4\times1/4 \text{ on } W18\times105 \\
\text{TS } 6\times3\times1/4 \text{ on } W18\times105 \\
\text{TS } 4\times4\times1/4 \text{ on } W12\times65
\end{array}$$

Considering the analogy shown in Figure 4, if the attachment is made at the mid-depth of the beam, $a = b$, and the maximum principal stress will be called σ_{cl}. If the attachment is now moved away from mid-depth, such that $a > b$, the maximum principal stress will be called σ.

Using the simply supported plate analogy discussed earlier, σ_{cl} and σ are related as follows:

$$\sigma = \sigma_{cl} \ \frac{a}{a_{cl}}$$

However, to account for the approximations inherent in the simply supported plate analogy, the relationship will be assumed to be of the following form:

$$\sigma = \sigma_{cl} \left[\frac{a}{a_{cl}} \right]^{\alpha}$$

The results of the finite analyses for the cases mentioned above were used to find α. The average α was found to be 0.42, and varied from a maximum is 0.498 to a minimum of 0.342. However, if α is taken as 0.5, the maximum difference for σ for the cases shown above is only 10.5%, and always differs on the conservative side. Thus, the stress for an attachment that is made at a point away from the mid-depth of the beam may be related to the stress for the attachment at the mid-depth by the square root of the ratio of the a dimensions, using the plate analogy.

GENERAL METHOD OF STRESS CALCULATION

The results of the finite element analyses, and the general trends observed in the previous section, will now be used to formulate a general relationship to be used for calculating the maximum local stress, due to the moment reaction, for attachments made to the webs of I-shaped sections.

Using the results from the previous section, the maximum local stress may be expressed as:

$$\sigma = \sigma_{cl} \left[\frac{a}{a_{cl}} \right]^{0.5} \tag{1}$$

Therefore, the problem reduces to one of finding an expression for σ_{cl}.

Consider the effective simply supported plate approximation shown in Figure 4, and let a be greater than b. Then, the maximum moment is $(M_{o}a)/D$. Using an effective section modulus, S_{eff}, the maximum stress in the web is:

$$\sigma = \frac{(M_{o}a)/D}{S_{eff}}$$

Solving for S_{eff}, gives,

$$S_{eff} = \frac{M_{o}a}{D\sigma} \tag{2}$$

For this problem, the effective section modulus may also be expressed in the following form, by introducing b_{eff} as the effective breadth of the web resisting M_{o}:

$$S_{eff} = \frac{b_{eff} t^2_{web}}{6} \qquad (3)$$

Equating equations (2) and (3), and solving for b_{eff} yields:

$$b_{eff} = \frac{6M_o a}{D\sigma t^2_{web}} \qquad (4)$$

This expression was used to find b_{eff} from the finite element analysis results, for all cases with the attachment located at mid-depth. The resulting b_{eff} is presented within Table 2, wherein the effective breadth minus the breadth of the attachment (b_f if WF) is also shown.

TABLE 2
Summary of calculated b_{eff} and $b_{eff} - b_{att}$

BEAM	ATTACHMENT	b_{eff} (in)	$b_{eff} - b_{att}$ (in)
W36x194	TS 3x 2x1/4	10.18	8.18
W24x100	TS10x10x1/4	19.95	9.95
	TS10x 8x1/4	17.45	9.45
	TS10x 6x1/4	15.44	9.44
	TS10x 4x1/4	13.44	9.44
	TS10x 2x1/4	11.44	9.44
	Plate 10x1/4	9.69	9.44
	TS 8x 8x1/4	17.51	9.51
	TS 6x 6x1/4	15.26	9.26
	TS 6x6x1/16	15.41	9.41
	TS 4x 4x1/4	11.48	7.48
	TS 2x 4x1/4	9.32	5.32
	TS 2x 2x1/4	7.40	5.40
	W10x49	20.08	10.08
	W10x29	15.65	9.65
	W10x15	13.84	9.84
	W6x25	14.86	8.86
	W6x12	12.61	8.61
	WF w/d=b_f=6" t_w=t_f=1/16"	15.24	9.24
	W4x13	11.26	7.26
W24x68	TS 6x 6x1/4	15.05	9.05
W18x105	TS 6x 6x1/4	13.84	7.84
	TS 6x 4x1/4	11.70	7.70
	TS 6x 3x1/4	10.66	7.66
W12x65	TS 4x 4x1/4	9.09	5.09

One trend that can be observed from the data contained within Table 2, is that $b_{eff} - b_{att}$ is essentially constant for constant combinations of d and D, with varying b_{att}. It can therefore be concluded that $b_{eff} - b_{att}$ is some function of d and D, and once this function is derived, S_{eff} may be found.

A function of the following form is proposed:

$$b_{eff} - b_{att} = \frac{C(D-d)}{(D/d)^{\alpha}}$$

where, C and α are constants.

Using the function shown above, along with the results of the finite element analyses, the following expression was derived:

$$b_{eff} = \frac{1.075(D-d)}{(D/d)^{0.6}} + b_{att} \tag{5}$$

Substituting (5) into (3), the effective section modulus can then be expressed as:

$$S_{eff} = \frac{\left[\dfrac{1.075(D-d)}{(D/d)^{0.6}} + b_{att} \right] t^2_{web}}{6} \tag{6}$$

As was shown previously,

$$M_{cl} = \frac{M_o a_{cl}}{D}$$

But, since

$$a_{cl} = \frac{D-d}{2}$$

M_{cl} is;

$$M_{cl} = \frac{M_o[(D-d)/2]}{D}$$

or,

$$M_{cl} = \frac{M_o}{2}(1-d/D) \tag{7}$$

Then, using (6) and (7), σ_{cl} is:

$$\sigma_{cl} = \frac{\dfrac{M_o}{2}(1-d/D)}{\left[\dfrac{1.075(D-d)}{(D/d)^{0.6}} + b_{att}\right] t^2_{web}}{6}} \qquad (8)$$

Substitution of equations (7) and (8) into (1), gives the following expression for the stress:

$$\sigma = \frac{\dfrac{M_o}{2}(1 - d/D)}{\dfrac{\left[\dfrac{1.075(D - d)}{(D/d)^{0.6}} + b_{att}\right] t^2_{web}}{6}} \left[\dfrac{2a}{D - d}\right]^{0.5}$$

Letting $\delta = d/D$, the expression for σ becomes:

$$\sigma = \frac{\dfrac{M_o}{2}(1 - \delta)}{\dfrac{\left[\dfrac{1.075(D - \delta D)}{(1/\delta)^{0.6}} + b_{att}\right] t^2_{web}}{6}} \left[\dfrac{2a}{D - \delta D}\right]^{0.5}$$

Rearranging this expression gives:

$$\sigma = \frac{3(2)^{0.5}M_o}{t^2_{web}}(a/D)^{0.5}\left[\frac{(1-\delta)^{0.5}}{1.075D(1-\delta)(\delta)^{0.6} + b_{att}}\right]$$

This expression was used to find the stress for all the cases analyzed within the finite element analysis. It was found that the stresses compared well, with a maximum percent error of only 15% on the conservative side, and less than 1% on the unconservative side. Thus it can be concluded that the expression derived above for the maximum stress in the web, will give a good estimate of the maximum local stress in the web of an I-shaped beam, loaded by the moment reaction of a wide flange or tube steel attachment, with the beam's top flange free to rotate. If the beam's top flange is restrained, by a floor slab for example, the expression will produce a conservative estimate for the maximum stress, since the boundary condition for the analogous plate will approach a fixed connection, thus reducing the moment, and hence the stress, at the face of the attachment.

SUMMARY AND CONCLUSION

Within this paper, the local stresses in the web of an I-shaped beam, resulting from the moment reaction of an attached wide flange or tube steel member, have been investigated. A parametic study of the problem has been performed, using the finite element method. The results of this study were then used to derive an expression that would give an accurate and conservative estimate of the maximum local stress. The maximum percent errors for the estimated stresses as compared to the finite element analysis results, were only 15% on the conservative side, and less than 1% on the unconservative side.

REFERENCES

1. Timoshenko, S.P., and Goodier, J.N., Theory of Elasticity, 3rd Edition, McGraw-Hill Book Company, New York, 1970.

2. Bechtel Structural Analysis Program (CE800), Versions E19-53 and F1-54.

3. Bechtel Corporation Calculation Number 6511 C-121.0, Revision 0.

ON SOME PARADOXICAL ASPECTS OF
ULTIMATE STRENGTH DESIGN MODELS

René MAQUOI
Professor
Université de Liège,Institut du Génie Civil
Quai Banning, 6, B-4000 Liège (Belgium)

ABSTRACT

The change from the "allowable stress" concept to the "limit states" design philosophy means more than warrant safety by means of load factors instead of by safety (penalty) coefficients affecting the material yield stress. The reference to ultimate design models can be source of ambiguities and apparent paradoxes. Present paper is mainly aimed at reviewing some of such aspects.

INTRODUCTION

Most of the recent codes and standards are based on the philosophy of limit states. Accordingly two kinds of limit states are considered: the serviceability limit states (SLS) and the ultimate ones (ULS). So much emphasis is put on the latter ones that many designers are prone to conclude - wrongly - that ULS are always governing the design.

The "allowable stress" design philosophy proceeds through elastic structural analysis and elastic verification of members and cross-sections; as far as strength is concerned, the determinative design requirement is to not exceed a specified proportion of the material yield stress. Of course ULS are paving the way for a less restrictive concept for strength; they allow for exploiting the possible ability to stress redistribution which is a consequence of material yielding. Such a redistribution may take place first within cross-sections, till plastic cross-sectional resistance be reached, and possibly amongst critical cross-sections, till the ultimate carrying capacity of the structure (onset of a plastic hinge mechanism) be obtained.

Of course to which extent stress redistribution is allowed for depends on section and material properties. Therefore the need for a classification of cross-sections according as : i) a plastic or an elastic structural analysis is permitted, and ii) the cross-sections can or cannot exhibit their plastic resistance (table 1).

TABLE 1

Classes of cross-sections - Method of analysis and verification of cross-sections

CLASS	ANALYSIS	CROSS-SECTION
1	PLASTIC	PLASTIC
2	ELASTIC	PLASTIC
3	ELASTIC	ELASTIC
4	ELASTIC	ELAST. RED.

In codes and standards, ULS refer in several circumstances to simplified design models, that are reputed to provide the designer with a satisfactory assessment of section resistance and/or structural carrying capacity. Most of these models are derived from physical modelling and are calibrated against test results.

Of course the design philosophy based on the concept of limit states reflects a better knowledge in both material and structural behaviour. It testifies undoubtfully to a larger accomplishment in the engineering profession. This evolution towards largely improved design concepts is not going without raising some ambiguities or anomalies. Present paper is aimed at pinpointing some of these paradoxical aspects.

ABOUT PLASTIC DESIGN ...

Requirement for plastic analysis is twice; it is related to the ability to stress redistribution within cross-sections, on the one hand, and amongst critical cross-sections, on the other hand. For sake of simplicity, reference is made in the following to frameworks where the members are subject to predominant bending.

Plastic adaptation within a specified cross-section allows for reaching the plastic cross-sectional bending resistance. The latter is the ultimate cross-sectional capacity because all the fibres have yielded. Such a fully yielded cross-section may then give rise to a plastic hinge that is nothing but an idealization of the real plastic zone. This hinge is likely to rotate only when experiencing the plastic cross-sectional resistance.

The onset of one plastic hinge in a structure results in a decrease - by one - of the degree of indeterminacy. Provided instability does not occur prematurously, a statically indeterminate structure of degree h generally needs the formation of (h + 1) plastic hinges before the structure transforms into a plastic collapse mechanism. The loading level at which this mechanism forms is the ultimate limit load; indeed the structure is no more able to sustain any increase in loading. Usually these hinges do not develop simultaneously but appear successively as a result of the evolving plastic adaptation amongst critical cross-sections. That implicates that any hinge, once formed, shall be able to rotate sufficiently to allow for the formation of the subsequent plastic hinges till the plastic collapse mechanism be fully developed.

Plastic adaptation requires material ductility and takes place only in stable conditions, i.e. before local buckling occurs. Therefore it is governed by material and section properties.

According to Eurocode 3 [1], there are four classes of cross-sections. Only Class 1 and Class 2 sections are concerned with plastic adaptation (table 1).

Class 1 sections - termed "plastic sections" - are those for which - in accordance with Eurocode 3 [1] - a plastic analysis may be conducted. They shall be able to develop their full plastic resistance; in addition, their available rotation capacity shall not be less than the one required to allow for the formation of a plastic collapse mechanism.

Class 2 sections - termed "compact sections" - are able to develop their full plastic resistance; however their rotation capacity is not sufficient to allow for the formation of a collapse mechanism. Thus the sections can be checked plastically but shall resist internal forces deduced from an elastic analysis of the structure subject to factored loads.

Approach used for Class 2 sections is thus based on an elastic internal force distribution. It consists in checking that the combination of the internal forces is plastically admissible (table 1) in any cross-section. This approach prevails as far as ULS are dealing with strength and not with displacements; it is safe because based on the static theorem of the limit analysis. The structural system is thus known *a priori* and it does not change whatever the loading level; its validity is subordinated to a satisfactory check of cross-sectional resistance.

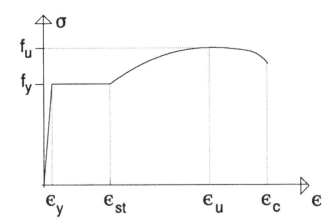

Figure 1 . Characteristic σ - ε diagram

Approach used for Class 1 sections is based on a plastic analysis and on a plastic check of cross-sections (table 1). Because the structural system evolves with the load factor magnitude, as a result of the onset of successive plastic hinges, a step-by-step procedure shall necessarily be used, even when a first order analysis is permitted . In addition plastic analysis shall require, as said above, a check of the rotation capacity in cross-sections where plastic hinges are likely to develop. Such a rather complex and lenghty detailed check has never been required till now. Indeed codes specified usually that plastic analysis may be utilised provided that the steel material properties (fig. 1) comply with some specific requirements, as those given in table 2.

TABLE 2

Some of the ductility requirements

EC3 [1]	ECCS [2]	ECCS [3]
$f_u/f_y \geq 1.2$ $\varepsilon_c \geq 15\%$ $\varepsilon_u/\varepsilon_y \geq 20$	$f_u/f_y \geq 1.2$ $\varepsilon_c \geq 15\%$ $(\varepsilon_{st} - \varepsilon_y)/\varepsilon_y \geq 6$	$f_u/f_y \geq 1.1$ - $(\varepsilon_{st} - \varepsilon_y)/\varepsilon_y \geq 6$

Regular mild steel grades, such as Fe 360 ($f_y \geq 235$ MPa), Fe 430 ($f_y \geq 275$ MPa) and Fe 510 ($f_y \geq 355$ MPa), may be accepted as satisfying these requirements *a priori*. Indeed their suitability for plastic analysis has been demonstrated in the past by means of appropriate experimental investigations.

Any set of the above requirements regarding material properties can therefore look like sufficient but probably not necessary. The consideration for improved requirements is especially needed when use is made of high strength steels, because there is a lack of experimental results.

In this respect, one must stress the fact that the limit b/t ratios for the plate components of a specified section subject to specified internal forces are highly dependent on the material yield stress; they are indeed proportional to a factor equal to $(235/f_y)^{0.5}$ (f_y in MPa). The higher the yield stress, the more restrictive the limit b/t values. The class to which a specified section belongs is not an intrinsic property of the section; it depends indeed not only on the type of loading but also on the yield stress magnitude. Hot-rolled sections drawn from existing catalogues have generally been accepted for plastic analysis, having in mind that they were, till recently, made of regular mild steel grades. Such sections may now be produced with much higher steel grades with the result that their suitability for plastic analysis and, possibly, for plastic cross-sectional resistance can become much more questionable. Should the use of high strength steels be promoted, then probably some changes in fabrication programmes could be expected accordingly in order to preserve the possible exploitation of plastic stress redistribution.

When plastic analysis is concerned, the available rotation capacity (ARC) cannot be less than the required rotation capacity (RRC).

RRC depends on the type of both structure and loading and on the material yield stress. For a specified structure subject to a specified loading, RRC increases slightly when the material yield stress increases (table 3) ; the additional demand amounts roughly 10 % when using high strength steel (case b) instead of regular mild steel (case a).

TABLE 3
Required rotation capacity in some typical structures

STRUCTURE		R	STRUCTURE		R
	a	3.0		a	1.3
	b	3.3		b	1.4
	a	3.0			
	b	3.3			
	a	2.5		a	2.5
	b	2.8		b	2.8

ARC is deduced from the moment-rotation curve (fig. 2) that is recorded in a simply supported beam subject to a point load at midspan. Research work was recently carried out [4, 5, 6] as well experimentally as theoretically. It is henceforth agreed that ARC of a specified section depends mainly on the following parameters :
- the stress ratio f_u / f_y between ultimate strength and yield stress of the material, which reflects the effect of strain-hardening;
- the size ratio L/b between the simply supported length of the beam and the width of the compression flange, which is a dimensionless measure of the moment gradient;
- the slenderness b/t of the compression flange, which is the governing parameter for local plate buckling of the flange;
- the thickness t of the compression flange, or preferably the boundary restraint k_θ provided by the web when the flange is prone to buckling ;
- the material yield stress f_y.

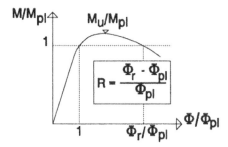

Figure 2 . Normalized moment - rotation curve

As a result, a provisional formula has been suggested [4] that gives the normalized available rotation capacity in H structural shapes:

$$R = R_o \left(\frac{f_u}{f_y}\right) + \Delta R(t) + \Delta R(b/t) + \Delta R(L/b) - \Delta R(k_\theta)$$

The first term describes the main influence of the material properties through (f_u/f_y) for a reference beam section. The other terms account for the influence of changes in geometric parameters - with respect to those of the reference beam - as well as for the parasitic role of (f_u/f_y) ratio. The reference beam is a doubly symmetrical I section of 3000 mm length subject to a point load at mid-span. The flanges are 15 mm thick and have a b/t ratio of 20. The web is 10 mm thick and has a b/t ratio of 35. This girder is made with steel grade FeE460, having a measured yield stress f_y of 530 MPa and a characteristic ratio $f_u/f_y \approx 1.26$. The ARC of a specified section decreases when the material yield stress increases (fig. 3)

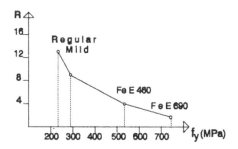

Figure 3 . Available rotation capacity vs. measured material yield stress [4]

Any increase in yield stress results thus in twice perverse effect: it generates a drop in ARC and is more demanding for RRC. Therefore the ability of existing structural shapes to exhibit their plastic resistance and to develop a plastic collapse mechanism in statically indeterminate structures, though nearly accepted without other reservation for usual steel grades,shall be examined carefully when use is made of high strength steels. There are thus possible conditions where the required rotation capacity can be larger than the available rotation capacity; in such circumstances plastic analysis shall be prohibited (fig. 4).

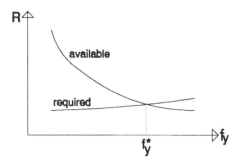

Figure 4 . Evolution of available - and required rotation capacity vs.material yield stress

The trends to the use of higher steel grades does not necessarily result in a proportional increase in carrying capacity.

Specifications given by existing codes and standards should be improved in order to give clear, sufficient and realistic information on :
a) the way to assess the available rotation capacity (ARC) of rolled H and I sections ;
b) the amount of required rotation capacity (RRC) in typical structural and loading conditions;
c) the value of the partial safety factor γ_M to be used when controlling the rotation capacity: RRC $\leq$ ARC/γ_M.

ABOUT GOVERNING LIMIT STATES ...

Plastic adaptation within the cross-sections, on the one hand, and possibly amongst critical sections, on the other hand, is likely to increase the carrying capacity of the structure compared to a full elastic design. When accounted for in the design, it allows for some economy and can result in smaller sizes of the structural shapes. Using higher strength steel preferably to regular mild steel has a similar effect.

Smaller sizes of structural shapes or sections result in smaller stiffness. As a consequence, the structure becomes more deformable with regard to as well vertical as horizontal deplacements. Serviceability of structures is especially concerned with the magnitude of such displacements in service conditions i.e. when the structure is subject to service (non factored loads).

Furthermore, the larger the sway displacements the more often second order elastoplastic analysis shall be required. Generally it is agreed implicitely that displacements in service conditions are computed based on a linear elastic analysis of the structure subject to non factored loads. Should this assumption - though not properly objective - be adopted, that would require a strength check in service conditions in order to verify whether the behaviour is elastic indeed.

Sake for economy leads to simplify the detailing of presumably rigid connections with the consequence that connections will sometimes exhibit a pronounced semi-rigid behaviour. The latter not only is between a real hinge and a rigid joint but also becomes highly nonlinear when bending at beam ends increases. Semi-rigid connections provide additional structural flexibility and can contribute a substantial increase of displacements in sway frames.

As a consequence the use of high strength steels and of semi-rigid connections, as well as plastic analysis are likely to render the serviceability limit states more determinative than the ultimate limit states.

ABOUT ULTIMATE SHEAR MODEL ...

Ultimate shear strength of transversely stiffened plate girders with slender webs is currently assessed by refering to tension field models. In Europe, the Cardiff model [7] has become the usual design model. Compared to the well-known BASLER model, it allows for an

to the original BASLER model, it allows for an additional shear strength due to the anchor of the tension band onto the flanges, provided the stress in the latter - due to axial force or/and bending moment - does not exceed the material yield stress. Therefore it is said to account for the significative influence of the flange flexural stiffness on the shear capacity.

According to the original model [7], ultimate shear resistance is composed of three components, respectively (fig. 5) :

a) Critical shear strength, as deduced from the linear theory of plate buckling ;
b) Tension field action, that develops till the web be fully yielded under the superimposed action of critical shear stress and tension field action ;
c) Frame action represented by the shear strength that is still available, once the web cannot be mobilized furthermore. This frame is composed of both flanges and transverse stiffeners and resists according as VIERENDEEL action till it collapses by formation of a plastic hinge mechanism .

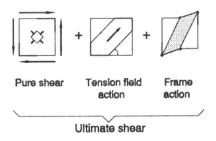

Figure 5 . Basic components of ultimate shear according to Cardiff model

For sake of simplicity, it is assumed that the resistance components can simply be added. That means that the respective actions take place successively; thus, tension field action occurs once the critical shear strength is reached and frame action takes up when the tension field action is exhausted.

Values of ultimate shear got from such a model - or from similarly based ones - have been found in a relatively good agreement with experimental results.

It is of paramount importance to keep in mind that only the amount of total shear force is concerned with this comparison. In other words, the reference to three resistance components only helps in the physical understanding of the basic phenomena and is a convenient tool because allowing for rather simple design. As a result, it would be unappropriate - that is at least the author's opinion - to draw conclusions from consideration of the individual strength components.

More especially tension field action is usually assessed by assuming a tension band, the width of which is governed by the anchor lengths onto the flanges. These lengths are derived from the assumption that plastic hinges develop at their ends in the flanges.

Though convenient, such an assumption is just a thrick. It leads to concentrate the tension field action within a limited band width where the tensile stress magnitude is presumed constant, while being zero outside the tension band. Tension field action exists actually in the whole web area with a varying stress magnitude.

On the other hand, experiments as well as numerical simulations demonstrate that the maximum shear capacity is reached well before the flanges are fully yielded [8]. That means that flange yielding develops mainly under increasing displacement rather than under increasing load. In addition it occurs with plastic zones of appreciable length exhibiting large strains and curvatures; therefore the concept of plastic hinges is not realistic and is nothing but a rough - though convenient - idealization.

As a conclusion, the actual behaviour does not evolve in full compliance with the design assumptions. That does not matter as far as ultimate shear is concerned (indeed calibration of design formulae enables a very satisfactory agreement between theoretical and experimental shear values). On the contrary, it does when some design requirements and detailing considerations are deduced from the aforementioned idealization.

More especially some codes [1,9,10] specify that the forces used to check the web-to-flange or web-to-stiffener welds shall be compatible with the stress fields in the web panels according to the method used to determine the shear buckling resistance. As a result, a larger fillet weld shall usually be required along the anchor length of the tension field at the web-to-flange or web-to-stiffener junction . Accordingly, in composite plate girders, additional shear studs should be necessary in this length portion, because experiencing additional shear force and pull-out force given respectively by the components, parallel to and perpendicular to the composite flange, of the tensile force in the tension band.

To the author's opinion, such rules governing the design of welds and stud connections are not fully justified because derived from a highly idealized combination of individual behaviour components. They are of course conservative for the length portion concerned but they are likely to increase appreciably the fabrication costs by preventing from automatic welding or by requiring locally more welding runs.

Recently, a novel approach has been developed [11] to assess the ultimate shear. In contrast to the Cardiff model, which gives satisfactory results in a limited range ($1<\alpha<3$) of web panel aspect ratios, UNGERMANN's model is shown applicable with a great accuracy whatever the α value. It is not especially subordinated to the assumption of plastic hinges in the flanges; therefore it looks like more to a tension field approach than to a properly tension band one. Should such a model have been the base for design specifications , it would not have led to aforementioned wanderings.

CONCLUDING REMARKS

No doubt that the limit states design philosophy is a more valuable concept than the former allowable stress method. Accordingly, ULS exploit much better the material capacities and are therefore expected to result in better economy. However increase in steel strength can render the plastic analysis questionable and require a more refined check of the rotation capacity of the critical sections. In addition, it can result in the fact that SLS are likely to be determinative for the design. Last, ultimate strength models, though very useful, are sometimes misused with regard to some of the design specifications that are drawn from. The author is of the opinion that there is still a need for theoretical and even experimental evidence with a view to implement codes and prevent these latter from being unduly coercible.

REFERENCES

[1]   Commission of the European Communities: <u>Eurocode 3 - Design of Steel Structures. Part 1 : General Rules and Rules for buildings</u>. Brussels, 1990.

[2]   E.C.C.S.: <u>European Recommendations for Steel Construction</u>. Brussels, 1978.

[3]   E.C.C.S.: <u>Ultimate Limit State Calculation of Sway Frames with Rigid Joints</u>. Publication n° 33, Brussels, 1984.

[4]   SPANGEMACHER, R.: <u>Zum Rotationsnachweis von Stahlkonstruktionen die nach dem Traglastverfahren berechnet werden</u>. Dissertation, R.W.T.H.Aachen, 1991.

[5]   E.C.S.C.: <u>Elastoplastic Behaviour of Steel Structures</u>. Contracts 7210-SA-508 and 204. Final Reports (to be published,1992).

[6]   KUHLMANN, U.: <u>Rotation kapazität biegebeanspruchter I-Profile unter Berücksichtigung des plastischen Beulens</u>. Institut für Konstruktiven Ingenieurbau, Rühr-Universität Bochum, Technical Report n°86-5, July 1986.

[7]   PORTER, D.M., ROCKEY, K.C. and EVANS, H.R.: The collapse behaviour of plate girders loaded in shear. In <u>The Structural Engineer</u>, August 1975, Vol. 53, n° 8, pp. 313-325.

[8]   CESCOTTO, S., MAQUOI, R. and MASSONNET, Ch.: Simulation sur ordinateur du comportement à la ruine des poutres à âme pleine cisaillées ou fléchies. In <u>Construction Métallique</u>, n°2, 1981, pp 27-40.

[9]   DASt Richtlinie: <u>Träger mit schlanken Stegen</u>. Draft 1988.

[10]   Commission of the European Communities: <u>Eurocode 4 - Design of Composite Steel and Concrete Structures. Part 1: General Rules and Rules for Buldings</u>. Brussels, 1990.

[11]   UNGERMANN, D.: <u>Bemessungsverfahren für Vollwand- und Kastenträger unter besonderer Berücksichtigung des Stegverhaltens</u>. Dissertation, Lehrstuhl für Stahlbau, R.W.T.H.Aachen, 1990.

GIRDERS COMPOSED OF CORRUGATED METAL SHEETS

BOGDAN O. KUZMANOVIĆ and MANUEL R. SANCHEZ
Bridge Department
Beiswenger, Hoch and Associates, Inc.
1190 N.E. 163 Street, N. Miami Beach, FL 33160, USA

ABSTRACT

A new girder composed only of light gage, flat-sided, corrugated metal sheets, bolted together using angles fabricated from the flat sheets of the same gage is presented. The behavior in bending of such girders was experimentally and theoretically investigated.

INTRODUCTION

In plate girders very thin webs should be used to be cost effective. The economy of such webs is in this case reduced by the prohibitive fabrication costs involved in attaching a system of stiffeners required for stability. As an alternative, to improve buckling strength of thin plates for conventional web plate and stiffener construction, corrugated sheet webs were tried, as early as 1924(1). For the web a corrugated ½ in. (13 mm.) thick steel plate was used with the troughs running perpendicular to the longitudinal beam axis. Unfortunately, the difficulties encountered in plate fabrication terminated their use. The next improvement came when light gage corrugated sheet metal was used for webs (2 to 7). The flanges remained as conventional, rigid flat plates.

An attempt to use light gage flat-sided corrugated metal sheets for both, flanges and webs is described here. The experimental part of the research was conducted in 1970 in the

Department of Civil Engineering, University of Kansas, Lawrence KS. The fabrication, testing and analytical approach are given in detail below.

CORRUGATED GIRDER

The flat-sided corrugated sheet metal used in the girder fabrication was 26 gage steel, t=0.021 in. (0,53 mm.) with the pitch of corrugations p=1 7/8 in. (48 mm) and rather shallow corrugations of 15/16 in. (23.8 mm.) as shown in Fig.1.

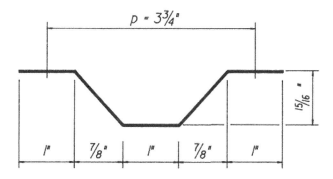

Figure 1. Typical cross-section of the sheet metal.

The cross section of the fabrication girder and its main dimensions with the location of electrical strain gages (at midspan) is shown in Fig.2. The 2"x2"x0.021", (50x50x0.53) angles, connecting the web and flanges were also fabricated from flat sheets of the same gage.

The method of attaching web and flanges to the angles can be either continuous, by fillet welding, or discontinuous, by spot welding or by bolting. The sheet gage 26, used for the girder, did not allow welding (7). Therefore, the discrete attachment with 3/8 in. (9.5 mm) bolts in every second corrugation at a spacing p=2x1 7/8 in.=3 3/4 in. (95 mm) was chosen. This pitch is quite large (10 diameters) and therefore the shear distortions were also expected to be large. If fillet welding could have been used instead of bolting, there would

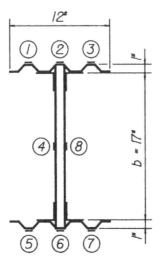

Figure 2. Girder cross-section with strain gages.

have been relatively small loss in shear stiffness, when compared with a flat web of the same weight, before the onset of buckling (5). Sherman and Fisher (7) have shown that in the case of a corrugated web and conventioned flat plate flanges, connected by welding, that girder deflections and shear stresses could be adequately predicted by using the thickness of the web, ignoring the corrugations and by applying the simplified technical bending theory. However, in this case, the shear distortional component of the deflections must be included in the calculations since it can account for more than half of the total deflection.

In our case of bolts, shear flexibility of web panels is composed of two deformation components, one due to the web shear and the other due to slip at the bolts. The sheet deformation is caused by bending of the corrugations and by the shear strain of the web. Flexibility due to the torsion of the corrugation faces and due to the membrane stresses in them normally can be ignored (5). The second component, the slip, was experimentally measured (6) and these data analytically treated (9) to obtain a general expression for slip flexibility. In general, the slip is a function of the corrugation geometry and bolt spacing and size.

When flanges are also corrugated some additional problems like their effective width and the integrity of the composed girder section take place.

To prevent premature lateral-torsional girder buckling a pair of girders was fabricated and connected together with L2x2x1/4 (L50x50x6mm) angles used as ties at supports and loading points. In Fig. 3 loading scheme is shown.

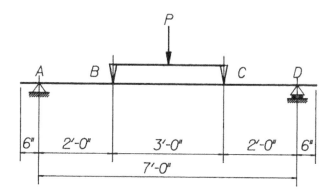

Figure 3. Girder loading scheme

The view of the coupled girders and loading set-up is given in Fig. 4. The total weight of one girder 8 ft. (2,438 m.) long with stiffeners and bolts was 60 lbs. (306.9 N)

Figure 4. View of coupled girders.

The loading force P was produced by a small manually operated hydraulic jack equipped with a pressure cell to double check the applied load. At the girder central cross section 8 foil strain gages of 120 ohms were installed. Girder deflections at mid-span were measured to 1 thousandth of inch (0,025 mm.). The test results of deflection magnitudes are given in Table 1 and Fig. 5. Maximum bending moment due to girder weight was 52.8 ft-lb (71,55 N.m).

METHOD OF ANALYSIS

The main object of the analytical approach was to determine beam deflections and not the stresses. The main reason is well known fact that the deflections demonstrate the integral of all the beam properties which include the material used, girder dimensions and fabrication technic.

As already said, in our case of an all-corrugated girder, the deflections can be analyzed by considering three major components in the deformational process, i.e., deflection components due to girder bending strain, shear strain of web panels, and shear slip of fasteners.

Bending strain component

In semi-corrugated girders with flat narrow flanges, the usual procedure to obtain girder moment of inertia were adequate. The corrugation of web could be ignored and the total web depth could be used as fully effective. When the flanges are no more rigid in transverse direction, as in our case, the effective cross section has to be obtained using procedures established for the design of cold-formed structured members in bending (8). Using corresponding formulas and table values from the specification, the effective girder moment of inertia was obtained as 21.64 in^4. (900,72 cm^4.) and the section modulus 2.29 in.^{3} (37,53 cm.^{3}). Without the application of this specification the corresponding values would be 57.89 in.^{4} (2409,56 cm.^{4}) and 6.13 in.^{3}(100,45 cm.^{3}), respectively.

Using known equations for bending deflections and corresponding to our loading case, this deflection component can be easily expressed as a function of the applied force P by the following expression:

$$\delta_1 = 0.01503\ P + 0.00074 \tag{1}$$

where the second term represents dead load deflection. In this expression P is in kips (4.444 kN) and deflections are obtained in inches (25.4 mm each). For several values of P, which were used in the testing, the corresponding deflection components are calculated and shown in Table 1.

Shear Strain

The development of a complete tension field action could not be expected in our case of discrete connection with bolts installed at every second corrugation with a pitch of p=3 3/4 in. (95 mm.) instead, the case of pure shear field action, i.e., without buckling was considered. This deflection component is obtained by applying well known expression,

$$\delta_2 = \frac{VA}{Ebt}\ .2\ (1 + \mu) \tag{2}$$

were V=shear force, E=Young's modulus, a=panel length, b=panel width, t=metal thickness and μ=Poisson's ratio (0.25 for sheet metal). When our values are substituted, the following expression for the second component is obtained

$$\delta_2 = 0.00290\ P \tag{3}$$

Shear Slip

In our case, the bending moment diagram has three distinct parts÷two parts where the moments are linearly variable on the length a=2ft (60 cm.), i.e., between the supports and loading points. In the middle part moments are constant and numerically equal to applied force P when expressed in ft-kips. Due to this moment change, the contribution to deflection due to shippage is considered separately for the constant and variable moment areas.

Constant Moments: Where the moments are constant, the force F acting on bolts is also constant and equal to M/d, where d=distance between angle gage lines. In our case, F is equal to

$$F = \frac{12P}{14} = 0.85714\ P \tag{4}$$

The slip flexibility, slip per unit force, as explained earlier, is equal to

$$S = 0.045 \frac{(0.028)}{t} \tag{5}$$

where t=sheet thickness=0.021 in. (0,53 mm). In our case S=0.060 in/kip (5.71 mm/kN). Finally, the first component of moment contribution to the slip, δ_{s1} when moments are constant, is

$$\delta_{si} = 2 \frac{FSp}{d} = 0.02755 \, P \tag{6}$$

Variable Moments: Where moments are variable the force on each bolt is different and also linearly by variable. In Fig.5 the defaults of bolt positions are shown.

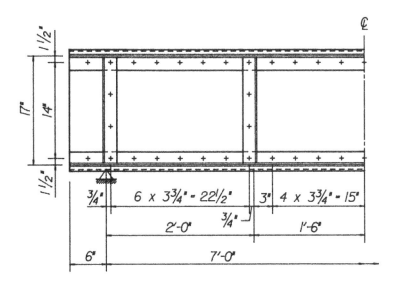

Figure 5. Details of Bolt Locations

If the first bolt, only 3/4 in. (19mm.) from the support, is disregarded, there are 6 bolts contributing to the slip and therefore the deflection. The bending moment at any abscissa x_i is in our case

$$M_i = 0.5X_iP \tag{7}$$

and the corresponding force acting on bolt $\underline{i}$ is

$$F_i = \underline{\underline{Mi}}{d} = 0.03571 \ X_i P \tag{8}$$

Therefore, the total second component of moment contribution to the slip, δ_{s2} is obtained by summation of all six bolt contributions as

$$\delta_{S2} = 2 \ \underline{\underline{SP}}{d} \sum_{i=1}^{6} F_i = 0.00115 \ P \sum_{i=1}^{6} X_i \tag{9}$$

Where x_i are in inches. As this sum in our case is equal to 83.25 in. (2,115 m.) the final expression giving second component as a function of the applied force P is

$$\delta_{S2} = 0.09556 \ P \tag{10}$$

The results = For the loads P, as applied in the girder test, all three major components (with two subcomponents for slip) are calculated and shown in the following Table 1. The same results are graphically reproduced in Fig. 6.

TABLE 1. Measured and Calculated Deflections due to girder bending in inches

Force P Kip.	0.250	0.650	0.900	1.150	1.400	1.650	1.900	2.150
Bending	.00450	.01051	.01427	.01803	.02179	.03774	.04459	.05031
Pure Shear	.00072	.00188	.00261	.00333	.00406	.00478	.00551	.00623
Shear Slip M=Const	.00689	.01791	.02480	.03168	.03857	.04546	.05235	.05923
Shear Slip M=variable	.02389	.06212	.08601	.10990	.13379	.15769	.18157	.20547
Deflection	.03600	.09242	.12769	.16294	.19821	.24567	.28402	.32124
Measured Deflections in Inches								
Test Values	0.040	0.125	0.145	0.190	0.240	0.285	0.295	0.546
% Error	10.0	26.1	11.9	14.2	17.4	13.8	3.7	41.2

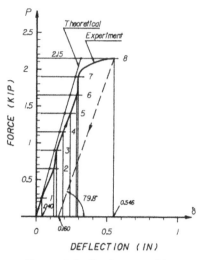

Figure 6. Deflections at midspan.

From the table it is seen that the theoretical deflections are underestimating the true deflections. The percentage error is quite large. It seems that there are two main reasons for errors. The fact is that deflections are rather small and small discrepancies produce large percentages. The other reason is probably due to Eq. (5) used for the slip flexibility, S. This expression as said earlier was developed by Heins and Blank (9) from the observed values given by Bryan and El-Dakhakhni (6). Obviously, these experimental values cannot be generalized and used for any type of bolted flange-web connection and sheet gage configuration irrespective of the magnitude of the corrugation pitch, bolt size and their spacing. This is especially true when the bolt spacing doubles the pitch with relatively deep corrugations. The use of Eq.(5) was necessary and obviously not the best choice. The slip per unit force was not determined during the performed test.

Comparing the relative magnitudes of four deflection components shown in Table 1 it is realized that the bending and pure shear components are about 6 times less than the two slip components. Besides this slip component dominance, it was noticed during the test, a slow degradation of the girder performance as an integral unit. This definitely contributed to the growth of the deflections. The main reason for this deterioration of the girder performance is due to the thin angles which could not prevent girder distortion. In

Fig.7 the side view of the girder is given when the testing was stopped. This photograph clearly demonstrates the lack of angle rigidity.

The rigidity of the girder configuration can be greatly improved by using double corrugated sheets connected back to back with their troughs running either parallels or perpendicular to each other. Utilization of resistance spot welding to connect together such sheets and their measured stiffness undoubtedly will eliminate many problems during the girder fabrication and later on when the girders are in service.

In building construction, with predominantly static loads and structural elements hidden from the sight by cladding material, the use of such corrugated girders could be cost effective. If the use of resistance spot welding and double sheets is made a better performance of such girders will be undoubtedly obtained and the magnitude of deflections reduced.

Figure 7. Condition of girder at the end of testing.

CONCLUSIONS

The following conclusions are suggested.

1) All corrugated sheet metal girders have a possibility to be used in building industry cost effectively.

2) The main problem of such girders is shear distortion.

3) Fabrication technique must be of an adequate standard to produce reliable girders.

4) For sheet metal gages 20 and less fillet welding technique can be used and only along parallel web flats (not diagonals) to achieve better girder integrity.

5) Certain advantages can be achieved by having double or triple sheets, spot welded along their contact planes, when used for the flanges.

6) If bolting techniques for web to flange connections are used, bolts spacing should correspond to the corrugation pitch.

7) Angles from standard mill production should be used as connecting elements to contribute to the girder integrity.

REFERENCES

1. Mensch, G., Das Bauingenieurwesen and der Britischen Reichsaustellungen in London-Wembly, Bautechnik, 1924, Vol.2, No.51, pp. 42-7.

2. Nilson, A.H., Shear diaphragms of light gage steel, Proceedings, ASCE, 1960, Vol. 86, No. ST11 (Nov.) pp. 111-39.

3. Peterson, J.P. and Card, M.F., Investigation of the buckling strength of corrugated webs in shear, NASA TN D-424, 1960.

4. Shanley, F.R., Investigation of shear-transmitting members=Design of a test beam, Report No. 68-20, Department of Engineering, University of California, Los Angeles, April 1968.

5. Rothwell, A., "The shear stiffness of flat-sided corrugated webs, "Aeronautical Quarterly, Vol. 19, Pt. 3, August 1968, pp. 224-34.

6. Bryan, E.R., and El-Dakhakhni, W.M., Shear of corrugated decks÷calculated and observed behavior, Proc. Instn. Civ. Engrs., Vol. 41, Nov. 1968., pp.523-40.

7. Sherman, D., and Fisher, James, beams with corrugated webs, <u>Proc.</u> of the 1st specialty conference on cold-formed steal structures, University of Missouri, Rola, 1971, pp.198-204.

8. American Iron and Steel Institute÷<u>Specification for the design of cold-formed steel structural members</u>, 1983, Washington, DC.

9. Heins, C.P. and Blank, D., Box beam stiffening using cold formed decks, <u>Proc.</u> 2nd specialty conference on cold formed steel structures, ASCE, Octob. 22-24, 1973, St. Louis, MO., pp. 537-71.

DESIGN OF COLUMNS IN NON-SWAY SEMI-RIGIDLY CONNECTED FRAMES

PATRICK A. KIRBY
Senior Lecturer
Department of Civil and Structural Engineering
The University of Sheffield
P O Box 600, Mappin Street, Sheffield, S1 4DU, U.K.

SHAKER S. BITAR
Lecturer
Department of Civil Engineering
The University of Manchester
Manchester, U.K.

CRAIG GIBBONS
Engineer
Ove Arup and Partners
13 Fitzroy Street, London, W1P 6BQ, U.K.

ABSTRACT

This paper traces a structured programme of work into semi-rigid action in steel frame structures which has been in progress at the University of Sheffield for over a decade. It has involved the testing of joints, subframes and complete frames for two and three dimensional response. Complementary analytical capabilities have been developed simultaneously and validated against the experimental results. These are now being used to investigate structural actions to develop design approaches suitable for everyday design office use. One such study involves the use of a finite element program which traces the full three-dimensional response of a column, with up to four beams semi-rigidly connected at its upper end, under the action of beam loads followed by concentric column loading to collapse. Use of the program has revealed that it may be possible to design columns in a very simple manner, without the need for a detailed knowledge of the beam to column connection moment-rotation response characteristics, thus giving significant economy not only in the material used (or enhanced load carrying capacity) but also in design effort.

DEVELOPMENT OF OVERALL SHEFFIELD RESEARCH PROGRAMME

Currently columns are almost always designed using interaction formulae which require, as input, values of end moments and axial loads. Although such formulae are normally based on a good representation of the maximum strength of a member under a fixed set of end actions, their determination is normally obtained by a scaling-up of the distribution of internal forces obtained at a lower load level; the overall frame analysis is usually nothing more than a simple linear calculation. Early work by Gent and Milner [1], on frames with rigid joints, first illustrated the concept of 'moment shedding'. To investigate this phenomenon in frames with semi rigid joints a structured program of research was initiated at the University of Sheffield in 1977 and has produced the following:-

i)   Sophisticated finite element programs which will predict the response of column subassemblages in both two and three-dimensional models.
ii)   Two full frame programs which are restricted to two-dimensional response (one developed at Sheffield and one obtained from Italy as a result of international collaboration).
iii) The conduct of some fifty tests on joints.
iv)   The assembly of a computerised data base with over 500 datasets from a world-wide literature search. This has led to collaboration with the University of Aachen on the production of a PC mounted version of the database.
v)   The conduct of 12 full-scale, two-dimensional and 10 full-scale, three-dimensional subassemblage tests to validate the programs listed under (i) above.
vi)   The conduct of 2 two-dimensional, full-scale frame tests to validate the programs tested under (ii) above.
vii) The conduct of 2 three-dimensional, full-scale frame tests.
viii) The development of hand methods for certain aspects of frame response.
ix)   The development of a finite element program to predict the response of a bounding frame with semi-rigid beam to column connections containing infill materials of brickwork or blockwork with gaps and openings.

Thus a large body of work has been carried out and a large number of analytical tools now exist.

ANALYTICAL STUDY

This study is particularly concerned with one element of the work, namely the use of the sophisticated finite program, as reported by Wang and Nethercot [2], to investigate the performance of internal, edge and corner columns in frames with a wide spectrum of joint characteristics. The early phases are a summary of work published elsewhere [3] in greater detail.

The program uses a finite element stiffness approach assuming an elastic, perfectly plastic stress-strain relationship; incorporates spread of yield through the cross-section and along the member; a multilinear joint moment-rotation relationship; lack of straightness in both planes and a parabolic residual stress pattern. The ability of the program to accurately trace the load-deflection response and to predict the ultimate load of semi-rigid frames has been demonstrated by checking against the results of a series of semi-rigidly connected column subassemblages.

The parametric study involved the analysis of a series of subassemblages incorporating two, three and four beams, representing corner, edge and interior columns respectively, of the form shown in Fig 1. The base of the column is assumed to be pinned. The beam ends remote from the column head represent beams which span approximately twice the length shown. The subassemblage models included an initial sinusoidal column deformation with a maximum amplitude of L/1000 about both the major and minor axes.

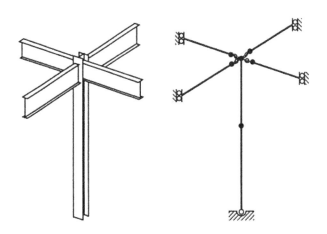

Figure 1. Subassemblage used

First Phase: The Influence of Moment Transfer on the Performance of the Subassemblages

This was investigated by considering various load arrangements for both one-way and two-way spanning floor systems. Figure 2 shows the assumed distribution of floor loading on each of the supporting beams for both types of floor system. The same characteristic dead weight of 3.8 kN/m^2 was assumed for each case. A live load value of 4.0 kN/m^2 was adopted as being typical of UK office loading. The design loads were assessed as 11.7 kN/m^2 maximum (1.6 x live plus 1.4 x dead) and 3.8 kN/m^2 minimum. Imbalanced loading was accentuated by considering extreme conditions of zero loading on some of the beams.

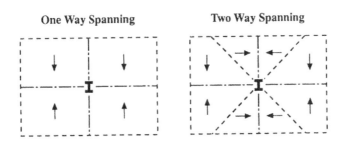

**Figure 2. Plan view showing one and two way spanning
arrangements**

The primary beams were 305 x 127 x 48 UBs spanning 5.5m
and the secondary beams 254 x 102 x 22 UBs spanning 4.0m. The
column was a 152 x 152 x 23 UC and was taken as 3.8m from its
base to the centre line of the beam members.

Two column orientations were considered; in type A the
primary beams framed into the column webs whilst in type B
they framed into the column flanges. The beneficial effect of
column continuity was conservatively ignored as one of the
aims of this study was to demonstrate the extent of the
adverse effects of moment transfer associated with semi-rigid
design.

For each load case, the performance of the subassemblage
was initially investigated for beam to column connections at
either end of the stiffness range. The flexible connection
was taken as a web cleat whilst the stiff connection was taken
as a rigid, full-strength connection. Figure 3 shows the
assumed moment rotation characteristic for the web-cleat
connection.

The failure loads are presented in Table 1 in a non-
dimensional form by dividing the ultimate loads computed by
the program, by the capacity determined using the program but
assuming the column to be pinned at its ends and subject to
axial load only. The ratio is designated the α_{pin} factor.

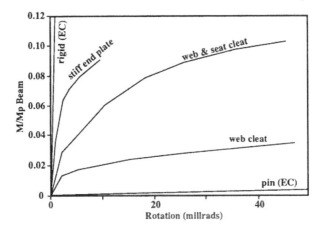

Figure 3. Assumed moment-rotation response

TABLE 1
Column failure load ratios, α_{pin}, from first phase
of the study

Connection	WEB CLEAT				RIGID			
Column Orientation	A		B		A		B	
$1/2$ way Spanning Slab Span	1	2	1	2	1	2	1	2
1	1.28	1.29	1.38	1.27	1.44	1.44	1.42	1.42
2	1.17	1.20	1.37	1.30	1.26	1.29	1.34	1.37
INTERIOR 3	1.22	1.24	1.38	1.30	1.31	1.34	1.38	1.40
4	1.24	1.25	1.38	1.25	1.34	1.35	1.40	1.33
5	1.26	1.27	1.38	1.35	1.37	1.38	1.41	1.47
- - -	- -	- -	- -	- -	- -	- -	- -	- -
6	1.03	1.06	1.35	1.29	1.06	1.18	1.23	1.28
EDGE 7	1.14	1.16	1.37	1.24	1.25	1.27	1.35	1.31
8	1.08	1.12	1.37	1.25	1.21	1.24	1.31	1.30
- - -	- -	- -	- -	- -	- -	- -	- -	- -
CORNER 9	1.12	1.14	1.28	1.14	1.25	1.25	1.33	1.22

In the first phase of the study the following conclusions
were drawn:

i)   In all of the 36 cases studied for the subassemblage
     employing web cleat connections, the failure load was
     greater than the failure load of the pin-ended column.
ii) All of the subassembalges in the study failed as a result
     of excessive minor axis distortion
iii) Higher failure loads were observed when the load applied
     to beams connected to the to the column flange was
     maximised whilst the load on the beams connected to the
     column web was minimised.
iv) By adopting a preferred column orientation, the minimum
     α_{pin} factor was increased from 1.03 to 1.14. If, in
     addition, a preferred distribution of beam loads was
     adopted, the α_{pin} factor was increased to 1.28.

**Second Phase: Influence of Different Connection Stiffnesses
and Beam/Column Stiffness Ratios on Ultimate Load Capacities**

The second phase of the study concentrated solely on the most
onerous arrangement found in the first study, A1 (see Table 1)
and, to optimise the use of computer processing time, only
load cases 1, 2 (interior column), 6 (edge column) and 9
(corner column) were considered.
     A total of three different primary beam sections were
investigated in conjunction with three different column

TABLE 2
Sections used in the study

Type of Member	Section Size
Beam B1	305 x 127 x 48 UB
Beam B2	305 x 165 x 54 UB
Beam B3	406 x 178 x 74 UB
Beam B4	610 x 305 x 179 UB
Beam B5	686 x 254 x 140 UB
Column C1	152 x 152 x 23 UC
Column C2	203 x 203 x 46 UC
Column C3	305 x 305 x 97 UC

sections, giving a total of nine different beam to column stiffness ratios. The sections used are included in Table 2. All other parameters are the same as those which were used in the initial study.

In the first instance, the web cleat connection characteristics remained the same as those used in the initial study so that a direct appraisal could be made of the influence of beam stiffness. A series of supplementary models were also studied incorporating enhanced web cleat-connection characteristics in which the stiffness of the connections were directly related to the size of the beam, giving a more realistic arrangement.

Analyses were also carried out using rigid connections. The non-dimensionalised failure loads, α_{pin}, for the various combinations are presented in Table 3. The corresponding results from the models employing enhanced web cleat connection characteristics are shown in parentheses.

As in the initial study, the α_{pin} factors for the web cleat connected subassemblages show a similar pattern to those which are rigidly connected. Comparison of the failure loads of the subassemblages using enhanced web cleat connections, with the corresponding unbracketed values in Table 3, shows an improvement in the failure load. This is primarily due to the increased connection restraint offered to the column. It is interesting to note that, when the enhanced web cleat characteristics are used in conjunction with beam B5, the improved failure load is similar to that achieved with fully rigid connections.

Unlike the initial study, there are several instances where the α_{pin} factor is slightly less than unity - load case 6, B1 C2 and B1 C3 of Table 3. This implies that the onerous effect of the disturbing moments from the connection can be marginally greater than the enhancing restraint effect.

TABLE 3
Values of α_{pin} from second phase of the study

| Column | Load case | Web cleat connection | | | Rigid connection | | |
		B1	B3	B5	B1	B3	B5
C1	1	1.28	1.35 (1.39)	1.40 (1.43)	1.44	1.44	1.45
	2	1.17	1.32 (1.34)	1.39 (1.42)	1.26	1.37	1.43
	6	1.03	1.25 (1.27)	1.37 (1.40)	1.06	1.32	1.41
	9	1.12	1.26 (1.32)	1.34 (1.41)	1.25	1.37	1.43
C2	1	1.09	1.13 (1.16)	1.16 (1.20)	1.20	1.21	1.21
	2	1.04	1.11 (1.14)	1.16 (1.19)	1.10	1.18	1.20
	6	0.98	1.08 (1.09)	1.14 (1.18)	0.97	1.14	1.20
	9	1.01	1.07 (1.12)	1.11 (1.18)	1.09	1.18	1.20
C3	1	1.02	1.03 (1.04)	1.04 (1.06)	1.05	1.07	1.07
	2	1.00	1.02 (1.03)	1.04 (1.05)	1.00	1.05	1.06
	6	0.99	1.01 (1.01)	1.03 (1.04)	0.94	1.03	1.06
	9	1.00	1.01 (1.02)	1.02 (1.05)	1.00	1.05	1.07

This is particularly apparent in the case of the rigidly connected subassemblage where the disturbing moment from the beams is significantly larger than for the equivalent subassemblage employing nominal web cleats. It is interesting to note, however, that this behaviour has been observed in instances where the column size has been increased and the beam size held constant.

From this phase it was concluded that:

i)   The reduced support rotations of stiffer beams subjected to the same load increases the axial load capacity of the column.

ii) It would appear that stocky columns (those with a high P_{ult}/P_{squash} ratio) receive less benefit than do slender columns, which have low P_{ult}/P_{squash} ratios.

Third Phase: Variation of Connection Stiffness

The values of the α_{pin} factors determined from the previous phases suggest that the development of a simple approach to column design is possible i.e. one in which the detrimental effect of transferred moment is ignored. In order to verify that the findings are valid over the complete spectrum of connection stiffness, the third phase was carried out in which the main variable was connection stiffness. The basic reference subassemblage used was as described earlier with some of the beam and column sizes shown in table 2.

The range of connection stiffness used were taken as

i)   Actually pinned - a lower bound reference datum (PIN)
ii)   Flexible - taken as web cleats (WC)
iii) Intermediate - taken as web and seat cleats (WSC)
iv)   Stiff - a stiff end plate (SEP)
v)   Rigid - an upper bound reference datum (RIG)

The characteristics used are shown as Fig.3. A further development examined columns increasing from the basic height of 3.8m to 5.7m and 7.6m. Beam spans were 4m. In all cases the subassemblages were subject to column loading only.
The results are presented as Table 4. It can be seen that a significant proportion of the potential benefit is obtained from even the most flexible connection used leading to the conclusion that it is the presence of a semi rigid connection that is the most important feature; the stiffness of that connection being of secondary importance.

TABLE 4
Values of α_{pin} for third phase of the study

H	CONN	C1			C2			C3		
(m)	TYPE	B2	B3	B4	B2	B3	B4	B2	B3	B4
	WC	1.33	1.35	1.42	1.11	1.12	1.20	1.02	1.03	1.06
3.8	WSC	1.39	1.40	1.42	1.16	1.17	1.20	1.04	1.04	1.06
	SEP	1.44	1.44	1.43	1.20	1.21	1.21	1.06	1.07	1.07
	RIG	1.44	1.44	1.44	1.21	1.22	1.22	1.07	1.07	1.07
	WC	1.56	1.60	1.76	1.22	1.25	1.44	1.04	1.05	1.16
5.7	WSC	1.68	1.70	1.78	1.34	1.36	1.45	1.09	1.09	1.17
	SEP	1.77	1.78	1.80	1.47	1.48	1.49	1.18	1.19	1.20
	RIG	1.78	1.80	1.80	1.49	1.50	1.50	1.20	1.21	1.21
	WC	1.69	1.73	1.90	1.30	1.37	1.66	1.07	1.09	1.26
7.6	WSC	1.81	1.83	1.91	1.50	1.52	1.68	1.14	1.14	1.29
	SEP	1.91	1.92	1.94	1.70	1.72	1.74	1.31	1.33	1.36
	RIG	1.92	1.93	1.94	1.73	1.75	1.77	1.36	1.38	1.40

Fourth Phase: Influence of Beam Span

The results given in Table 4 had been obtained using a very short beam span of 4m, to maximise the potential effect of the semi-rigid action. To investigate the influence of beam-span this was increased variously to 5m, 6m and 12m for flexible and stiff connections.

Table 5 shows the resulting α_{pin} factors which demonstrate a remarkable insensitivity of the results to beam stiffness.

TABLE 5
Values of α_{pin} for fourth phase of the study

H	CONN TYPE	SPAN			
		4m	5m	6m	12m
3.8	WC	1.35	1.33	1.32	1.31
	SEP	1.44	1.44	1.44	1.43
5.7	WC	1.60	1.57	1.56	1.54
	SEP	1.78	1.78	1.78	1.77
7.6	WC	1.73	1.70	1.69	1.65
	SEP	1.92	1.92	1.92	1.91

Fifth Phase: Influence of Load Pattern

As the cases considered in phases three and four had been concerned with axial column loading only, a further eight loading cases were considered with loading on one, two and all four beams. The columns used were C1 and the beams were type B3. Column loads were applied in increments up to 50kN/m before the columns were loaded axially to failure.

The column capacities were not sensitive to the pattern of loading indicating the relative insignificance of the applied moment for interior columns.

Conclusions

From the parametric study, a number of observations regarding three-dimensional subassemblage response can be made.

1) In all cases column failure appears to have been dominated by minor axis flexure caused by buckling, irrespective of the pattern of applied loads on the beams framing into the column head. No three-dimensional lateral torsional failures have been detected. (This finding has been confirmed by a number of experimental tests conducted by the Sheffield team). However, the use of beam sections as columns may render such action important.

2) Examination of the values of α_{pin} demonstrates that for almost every one of about 500 cases considered, the axial load in the column at failure exceeded the axial capacity of the column with pinned ends. Where this was not the case, the shortfall was only 2% for a semi-rigid connection and 6% for a rigid connection (both cases

related to edge rather than interior columns). This indicates that the beneficial effect of restraint provided by the beams through the semi-rigid connection outweights the disadvantageous effect of the moment transmitted through the connections.

3) It is the presence of connections which is a primary factor influencing column capacity. The precise magnitude of connection and beam stiffness appears to be of lesser importance.

4) The potential benefits for enhanced column capacity, above that of the corresponding pin-ended column, are greater for slender columns than for stocky columns. This is due to the relative importance of overall buckling which leads to greater column end rotations with the corresponding mobilisation of connection restraint when compared to a stocky column which has a tendency to squash rather than deform laterally.

5) In developing any design method suitable for a manual approach which properly reflects the end restraint, the connection stiffness must relate connection properties not only to the beam characteristics but also to the properties of the column.

6) The conclusion itemised under 2) above is potentially the most significant as it implies that columns could be designed as simple props with pinned ends carrying axial load only. Economies would thus arise from two sources. There would be
   a) an enhanced load carrying capacity accompanied by
   b) a reduction in design effort.

Acknowledgement: As was indicated in the early part of this paper, the Sheffield study has taken place over a significant period of time with the benefit of input from many investigators and several sponsors. The authors wish to acknowledge their contributions.

REFERENCES

1. Gent, A.R. and Milner, H.R., 'The Ultimate Load Capacity of Elastically Restrained H-Columns under Biaxial Bending'. Proceedings of the Institution of Civil Engineers, Vol.41, December 1968, pp 685-704.

2. Wang, Y.C and Nethercot, D.A., 'Ultimate Strength Analysis of Three-dimensional Column Subassemblages with Flexible Connections'. Journal of Constructional Steel Research, No.9, 1988, pp 235-264.

3. Gibbons, C., Nethercot, D.A., Kirby, P.A. and Wang, Y.C., "The Analytical Response of Partially Restrained Beam Column Connections in Non-Sway Steel Frames". Structures and Buildings Journal, The Institution of Civil Engineers (in press).

SUGGESTED MODIFICATIONS TO THE EC3 APPROACH FOR BEAM-COLUMN SIMPLIFIED DESIGN FORMULAE

ANTONELLO DE LUCA
Istituto di Ingegneria Civile ed Energetica, Università di Reggio Calabria
Via Cuzzocrea 48, 89100 Reggio Calabria, Italy.

CIRO FAELLA
Istituto di Ingegneria Civile, Università di Salerno
84080 Fisciano, Salerno, Italy.

ELENA MELE
Istituto di Tecnica delle Costruzioni, Università "Federico II"
P.le Tecchio, 80125 Napoli, Italy.

ABSTRACT

In this paper the approach of the recently drafted Eurocode 3 to analysis and design of semirigid sway frames is examined. In particular the two approximate methods (MAM and SMBLM), provided by the code to account for second-order effects, are compared and verified against results of advanced inelastic analyses carried out over an extensive range of imperfect sway frames. In the light of this comparison, a modification to one of the methods (the SMBLM) is suggested in order to improve the degree of accuracy.

INTRODUCTION

A large number of recent investigations in the field of steel structures is devoted to develop methods of second order inelastic analysis for the design of steel frames [1, 2]; a common feature of several of these studies is the attempt to provide analysis and design methods for steel frames involving global strength and stability considerations, such that the need of individual member strength and stability checks is avoided. In other words, these procedures, which are commonly called "advanced inelastic analysis and design methods", as recently pointed out by L.Argiris in [3], "... *incorporate directly within analysis techniques, the geometric, material, construction, loading effects which have until now been incorporated by specification provisions within the member capacity equations*."

Obviously, advanced analysis can model more accurately the behaviour of steel structures, and consequently provides more refined prediction concerning the ultimate capacity and the collapse mechanism of the structure as a whole. Furthermore, the design process becomes more rational and simple: in fact, including member mechanical and geometrical imperfections within the global analysis of the frame, the checks on member stability, and therefore the use of buckling curves and interaction formulae, become redundant.

Neverthless, advanced analysis requires considerably higher computational efforts than traditional approach of structural design based on the application of the capacity check relations for the design of the single member. Moreover, only in-plane analysis of two-dimensional frames can be performed in this way, remaining buckling curves and interaction formulae still necessary to verify out-of-plane behaviour. Finally, it is worthed pointing out that only two codes, the Australian AS4100 Specification [4] and the recently drafted Eurocode 3 [5] make explicit allowance for advanced inelastic analysis.

Despite of all these efforts concentrated into advanced analysis, it has to be pointed out that still a need does exist for simplified formulae to be used either for a pre-design of the structure or as an efficacious and simple tool in the design office applications. In this respect, most of the existing provisions provide several interaction formulae conceived in order to reconduct the frame check to a beam-column check (in [6] a review of the different approaches used around the world has been provided). Even the most advanced provisions (AS4100, EC3), while allowing advanced analysis, provide, at the same time, simplified methods to be used as an alternative.

From the research point of view it has to be said that still a considerable effort [7, 8] is spent in striving for interaction formulae capable of taking into account the different phenomena responsible for frame collapse. It is obvious in fact, that, while in the case of non-sway frames the problem is effectively reconductable to beam-column buckling, in the case of sway frames the interaction between column and frame is more significant than in non-sway structures, thereby is more difficult to take meaningfully it into account.

In this paper the simplified formulae suggested by EC3 for sway frames are examined. In particular, having briefly underlined the general approach suggested by EC3 in the following section of this paper, the beam-column problem in sway frames is analyzed and it is evidenced how, for a particular frame, the use of frame axial-bending interaction curves, introduced in [6, 9], allows to evidence the range of loading conditions in which strength or stability are the governing parameters.

Because of the large number of parameters involved in the problem, such as type and level of vertical loads, type and level of horizontal loads, frame slenderness, semirigidity of connections, value and distribution of imperfections, an extensive parametric analysis was carried out by means of inelastic approach, in order to achieve a better understanding of the problem and to check the pitfalls of the two methods given in EC3. At the same time, this has allowed to establish the range of validity and the degree of approximation of the methods proposed by EC3, and to suggest improvements to their simplified formulae.

THE EUROCODE 3

In the chapter of Eurocode 3 dealing with the design of frames, a classification of structures according to sway-resistance is given: a frame is defined "braced" if the resistance is supplied by a bracing system which is sufficiently stiff to carry all the horizontal loads, and therefore to provide overall stability, and as "unbraced" in the remaining cases.

A frame is classified as "non-sway" if its response to in-plane horizontal forces is sufficiently stiff such that deformations will not alter significantly the equations of equilibrium, thereby second-order effects on the overall stability may be neglected. To avoid calculating the deformations to justify this assumption, EC3 states that a frame is considered as non-sway frame if the critical vertical load is at least ten times the total design vertical load; in the opposite case the frame is classified as "sway", and the global structural analysis has to make allowance, exactly or approximately, for second-order effects associated with geometrical non-linearity. It is important to underline that the classification of a frame as sway or non-sway depends on the applied vertical load, while the classification as braced or unbraced is a characteristic of the stiffness of the structure itself.

In EC3 the global frame stability check must be carried out including the effects due to initial sway imperfections and considering the arrangement of variable loads which is critical for failure in a sway mode, by means of either exact (second order) or approximate (first order) analysis methods. In fact EC3 allows to perform first order elastic analysis even in the case of sway frames, provided that second order effects are indirectly accounted for by means of approximate methods. The two approximate methods suggested by EC3 are the Moment

Amplification Method (**MAM**) and the Sway Mode Buckling Length Method (**SMBLM**). The MAM includes approximately second order effects by amplifying the first order sway moments on girders and columns through a factor which is a function of the ratio between the total vertical load and the elastic critical load. Therefore this method can be associated with the LRFD approach where global P-Δ effects are introduced in the member check by means of the B_2 coefficient; the local P-δ effect, allowed for by means of B_1 coefficient in LRFD, is also present in the MAM. The buckling length of the column in the MAM is assumed equal to interstory height.

The second method, the SMBLM, does not need any evaluation of elastic critical load of the structure, since it takes into account second order effects by adopting a sway mode buckling length to be used in the interaction formula. Strength check are still needed and girder moments are increased of 20%, to take into account that with this approach girder moments are not affected by the amplification due to second order effects.

BEAM-COLUMN SIMPLIFIED FORMULAE SUGGESTED IN EC3

In the EC3, as in all other codes, the problem of sway frames subjected to vertical and horizontal loads, is addressed in a simplified manner by using interaction formulae, which have essentially the meaning of strength check for the member cross-section, and of stability check for the column. Therefore, the problem is tackled by using the following procedures:

Strength interaction formula
The member strength is checked by means of the ultimate interaction equation, which, in the common case of double-T section, can be written as:

$$\frac{N}{N_p} + \frac{M}{1.11 M_p} = 1 \tag{1}$$

being $N_p = A f_d$ and $M_p = Z f_d$, where A, f_d and Z are the the cross-section area, the design stress and the plastic modulus respectively. The bending moment to be used in this axial-bending interaction formula must include the second order effects.

Stability interaction formula
The single member buckling check is performed by means of the following interaction formula:

$$\frac{N}{N_c} + k \frac{M}{M_p} = 1 \tag{2}$$

where N_c is the squash load, and k is a factor which accounts both for the bending moment distribution law and for the amplification due to local P-δ effect. The bending moments to be introduced in this buckling interaction formula must include the second order effects.

If a second-order analysis has been performed to evaluate the internal forces and moments distribution, the values to be adopted in the equations (1) and (2) are the ones directly obtained by the analysis. On the contrary, if the first-order theory is used to analyze the frame, two alternatives, as before anticipated, are suggested by EC3 for taking approximately into account second order effects; their main features are synthesised hereafter.

1. In the MAM the buckling length is the interstorey height and sway moments are amplified; these moments are those associated with horizontal displacements of the top of a storey relative to the bottom, and therefore arise from horizontal loading, and also from vertical loading if either the structure or the loading is unsymmetrical. The moment amplification factor is provided by:

$$C = \frac{1}{1 - V/V_{cr}} \tag{3}$$

in which V is the design vertical load on the frame and V_{cr} is the elastic critical load. The amplified bending moments, have to be used in equations (1) and (2). The code allows to use this approximate method only when V/V_{cr} is less than 0.25.

2. In the SMBLM the second order effects are approximately taken into account adopting the sway mode length of columns in the stability check (2), and amplifying by 20% beams and connections sway moments in the capacity checks (1).

Both the approximate methods require an additional computational effort with respect to the case of non-sway frames; in fact in the first method (MAM) the evaluation of elastic critical load of the frame is required, while in the second one (SMBLM) the effective length must be defined, by means, for example, of alignment chart, even though "...*it is becoming more accepted that this approach is extremely inaccurate.*", as reported in [10]. The MAM, by introducing the distance between the design and the critical vertical load seems to be a more rational procedure, even though it suffers the disadvantage of a slight larger amount of calculations than the SMBLM. In fact, in addition to the critical load, sway moments need to be evaluated separately.

Frame interaction curves
In the case of sway frames subjected to vertical and horizontal loads, the strong interaction existing between plasticity and buckling phenomena makes the problem particularly complex. In [6, 9] it has been introduced the concept of frame axial-bending interaction curves, which allows to evidence the cases where either plasticity, and thereby the strength check formula of type (1), or buckling, and thereby the stability check formula of type (2), is governing. These curves are obtained by representing the vertical load multiplier, V, versus the horizontal load multiplier, α, such that the corresponding internal forces and moments distribution, satisfies the condition (1) or (2). Therefore, for a given frame, it is possible to have a complete view of the structural behaviour, as vertical and horizontal loads vary. Moreover, by introducing the non-dimensional ratios V/V_u and α/α_y, where V_u is the inelastic buckling load and α_y is the horizontal forces multiplier leading to the first column yielding in absence of vertical loads, it is possible to extend this representation to different frames in a single chart.

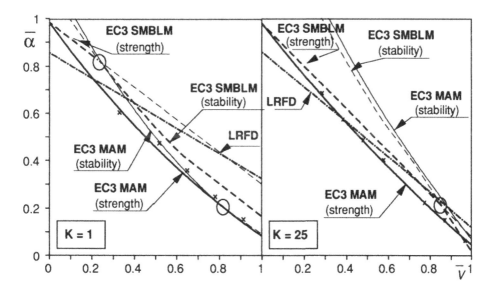

Figure 1. Frame interaction curves for 6-storey frames characterized by quasi-hinged (K=1) and fixed (K=25) connections.

Figure 1 shows the collapse domain obtained as the lower bound between the strength and stability checks both for the MAM and SMBLM. In the same diagram also the points resulting from advanced inelastic analyses provided together with the application of the well-known LRFD formula containing the B_1 and B_2 coefficients. More details regarding the data depicted in this figure and the way of obtaining this representation can be found in the next sections.

Two observations can be drawn from this figure: the first one is that the LRFD, being a method derived by interpolation of numerical results obtained for different frames, does not distinct strength and stability. The second, and more important consideration, is that the region of loading conditions in which stability is governing is very often limited to the case of large axial loads, while for most of the other loading conditions, the frame will collapse by plasticity, and only for very slender frames (case provided in the part of the figure 1 representing quasi-hinged conditions) the stability check becomes more important, as found in [6,9] by means of an extensive parametric analysis.

NUMERICAL ANALYSIS

In order to evaluate the range of validity and the degree of approximation of the methods suggested by EC3, a complete second order inelastic analysis was carried out on an extensive number of multistorey semirigid sway frames, affected by column and frame imperfections in accordance to the provisions of EC3.

The parametric analysis was devoted to investigate the following aspects:
• number of storeys,
• semirigidity of beam-to-column connections,
• loading conditions.

Regarding to this first point, three-storey and six-storey frames have been analyzed. The geometry of the schemes and the member cross section characteristics are represented in figure 2.

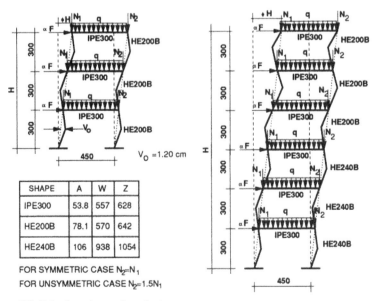

SHAPE	A	W	Z
IPE300	53.8	557	628
HE200B	78.1	570	642
HE240B	106	938	1054

FOR SYMMETRIC CASE $N_2=N_1$
FOR UNSYMMETRIC CASE $N_2=1.5N_1$

N.B. All the dimensions are in centimeters.

Figure 2. Analyzed frames.

Regarding the second point, different degree of partial restraint of the beam-to-column connection, ranging from the fixed to the hinged restraint, were considered for the analyzed frames. In fact, different values of the nondimensional connection stiffness K, defined as:

$$K = \frac{EI}{K_\varphi L} \qquad (4)$$

where K_φ is the connection stiffness and EI/L the beam stiffness, have been adopted for characterizing the semirigidity of joints (K=25, 10, 5, 2, 1, being K=25 and K=1 the nondimensional stiffnesses respectively of a fixed connection and of a quasi-hinged connection).

Finally, for what concerns loading conditions, the structure was loaded by distributed load on girders, equal to $0.5q^*$, being $q^* = 8Wf_d/L^2$ and W, f_d and L the elastic cross section modulus, the design stress, and the span of the girders respectively; for all frames also the case of distributed load equal to 0 was examined. Furthermore, concentrated vertical loads on all the column joints and concentrated horizontal loads at interstories were applied; different combinations of external loads, i.e. different relative amount of vertical and horizontal loads were considered, passing from pure horizontal forces to pure vertical load.

All the frames were affected by a global geometrical imperfection, simulating erection vertical out-of-alignment, equal to the one suggested by the EC3. All the columns presented an equivalent geometrical imperfection, which corresponds to the buckling curve b of EC3, in order to take into account the residual stress distribution and the bar out of camber.

The results of each numerical simulation, which makes allowance for geometrical and mechanical nonlinearities, can be represented by plotting the multiplier of horizontal forces α versus the top horizontal displacement δ. In figure 3 the curves obtained with the numerical analysis for two values of the nondimensional connection stiffness K, corresponding to the limit cases of fixed (K=25) and quasi-hinged (K=1) connections, and for different levels of vertical load, are depicted. It can be seen that the increase of both vertical load and connections stiffness produce a reduction in the ultimate multiplier of horizontal forces.

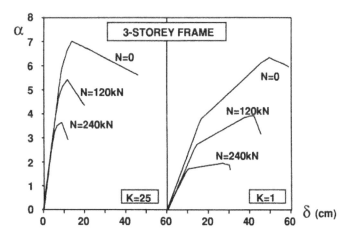

Figure 3. Influence of vertical loads and joints semirigidity on α-δ curve.

In order to compare the results of the numerical simulations with the values provided by EC3, nondimensional interaction curves, firstly introduced in [6], which represent the loading conditions producing the attainment of the design stress f_d, evaluated by stability and strength check equations, in the more stressed member of the frame, were plotted.

More into detail, the procedure adopted to obtain these curves, can be summarized as follows.
- Evaluation of internal forces and moments due to vertical load and, separately, of those due to horizontal forces, by means of an elastic first order analysis.
- Computation, by applying the effects superposition, of the bending moment and axial force to be adopted in the equations (1) and (2), through:

$$M = M_q + C(\alpha_o M_o + \alpha_v M_i^v + M_i^q) \qquad (5)$$

$$N = N_q + \alpha_o N_o + \alpha_v N_v + \alpha_v N_i^v + N_i^q \qquad (6)$$

where α_o and α_v are the horizontal and the vertical load multiplier respectively, and C is the amplification factor of sway moments, which is equal to 1 for columns and to 1.2 for beams in the SMBLM, while it varies with the total vertical load V, according to (3), in the MAM; M_q and M_o denote the bending moments due to distributed load and horizontal forces respectively, M_i^v and M_i^q (N_i^v and N_i^q) are the bending moments (the axial forces) due to equivalent horizontal forces which are introduced in order to account for the presence of imperfections as concentrated and distributed vertical loads respectively are applied; N_q, N_v and N_o are the axial forces due to distributed vertical loads, concentrated vertical loads and horizontal loads respectively.
- Then, fixed one value of the total vertical load V, and computed the corresponding value of bending moment and axial force by means of the relations (5) and (6), the values of the horizontal load multiplier α_o, satisfying the (1) and (2) conditions, are obtained.
- Thereby, for each value of vertical load V, the minimum between the two horizontal load multipliers α_o, respectively obtained by the (1) and (2) equations, represents a point on the interaction curve.
- Finally, the obtained MAM and SMBLM curves, have been nondimensionalized introducing the following parameters:

$$\bar{\alpha} = \frac{\alpha_o}{\alpha_{o\,max}} \qquad\qquad \bar{V} = \frac{V}{V_u} \qquad (7)$$

where α_{omax} is the maximum horizontal load multiplier in absence of vertical load, and V_u is the estimated value of the inelastic buckling load of the frame.

In figure 4 the previously explained procedure has been adopted for representing, together with the simulated points derived by the parametric analysis, the results of applying the MAM and the SMBLM procedure. In this figure only some of the results of the parametric analysis are provided due to conciseness. In particular, only loading case symmetrical and with $q=0.5q^*$ has been represented for quasi-hinged and fixed three and six storey frames.

The range of examined vertical-horizontal load combinations has been divided in three fields, depending on the ratio of vertical design load and the elastic buckling load V/V_{cr}. In particular in the first zone, where $V/V_{cr}<0.1$, the frame is not very sensitive to second order effects, thereby the member check is governed by strength considerations; in the second one, where $0.1<V/V_{cr}<0.25$, second order effects must be accounted for in the analysis, and collapse may occur due both to elastic buckling and to plasticity; finally, the structural cases included in the third field ($V/V_{cr}>0.25$) are generally not recommended, because of their excessive slenderness, or excessive level of design vertical load.

From these diagrams, it is possible to deduce the following considerations:
- The range of cases for which the strength check is governing appears generally very wide, and for the MAM it covers also the region in which vertical loads are prevalent ($V/V_{cr}>0.25$); for the SMBLM, an intersection between strength and stability check, which in the graphical representation is evidenced by a discontinuity point, generally occurs in the intermediate zone, where $0.1<V/V_{cr}<0.25$.

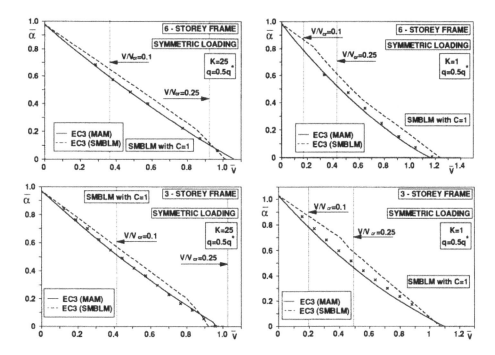

Figure 4. EC3 MAM and SMBLM interaction curves for 6- and 3- storey frames versus simulation.

- The MAM provides good results, very close to the ones obtained by numerical analyses, and enough conservative with respect to them. On the contrary, the SMBLM does not provide acceptable results, being the computed value of the horizontal load multiplier, for a given level of vertical load always non-conservative. This result can be explained in the light of what said above, as regard to strength check: in fact, in the SMBLM the correction suggested for including second order effects only consists of assuming the sway mode slenderness in the stability check, and of amplifying by 20% the bending moments to be adopted in beams and connections check. Therefore, in the column strength check, no allowance is made for second order effects, and, moreover, in the stability check the adoption of the sway mode length does not seem enough effective.

CONCLUDING REMARKS AND SUGGESTED MODIFICATIONS TO EC3

On the basis of the results presented in the previous section, where the non-conservativeness of the SMBLM has been proved, a modification of this method is proposed in this paper. Since the pitfall of the SMBLM is represented by the inadequateness of taking into account the second order effects by only amplifying the buckling length of the column, it is therefore suggested to adopt, also for column sway moments, an amplification factor equal to 1.2 (figure 5). It has to be emphasized that the application of the method in such a way, corresponds to assume, both in the strength and in the stability check equation, a moment amplification factor C constantly equal to 1.2, which, according to equation (3), corresponds to a ratio V/V_{cr} equal to 0.17. This means that the amplification factors for sway moments are equivalent between MAM and SMBLM at V/V_{cr}=0.17, while for larger V/V_{cr} ratios this factor will be larger in the MAM. In the field where

V/V_{cr} is greater than 0.17, the modified SMBLM still adopts the amplifying coefficient equal to 1.2, and thereby is less conservative than the MAM, as long as the governing equation is the one provided by strength check equation.

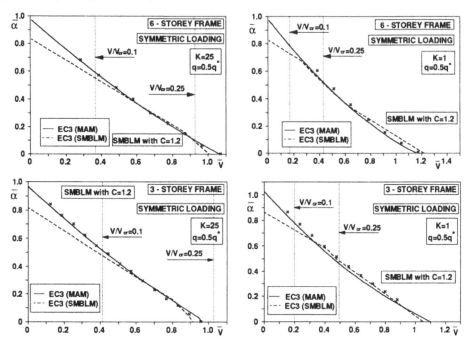

Figure 5. EC3 MAM and SMBLM (with the suggested modification) interaction curves for 6- and 3- storey frames versus simulation.

Thus, by applying the method in such a way, the results seems to be sufficiently in agreement with the numerical ones, and most times more conservative than the ones provided by MAM, which, being a more precise method, consequently allows for smaller safety margin.

In order to verify the suggested modification to the EC3 SMBLM method, all the data obtained by the parametric analysis have been plotted in figure 6. In particular, in this figure, the horizontal load multipliers obtained by means of MAM and SMBLM are directly compared with the values obtained by means of the numerical analysis. The points depicted in the chart represent, for a given level of the vertical loads, the values of the horizontal load multipliers $\overline{\alpha}_{MAM}$ and $\overline{\alpha}_{SMBLM}$ obtained by MAM and SMBLM versus $\overline{\alpha}_{sim}$ obtained by numerical simulation. It is obvious that if the value provided by the MAM or the SMBLM is coincident with the one obtained by the numerical analysis, then the point will be placed on the 45° sloped line, otherwise if the method provides either a more or a less conservative value, then the point will be placed either below or above the line respectively. Therefore the chart represents the scatters (positive or negative) between the numerical data derived by the parametric analysis, and the values computed as suggested by EC3.

It can be immediately emphasized that, the cases of both 6-storey and 3-storey frame, the SMBLM applied in the way recommended by the code leads to unconservative results; on the contrary, introducing the above proposed modification, the SMBLM appears to be more satisfactory, while preserving its semplicity.

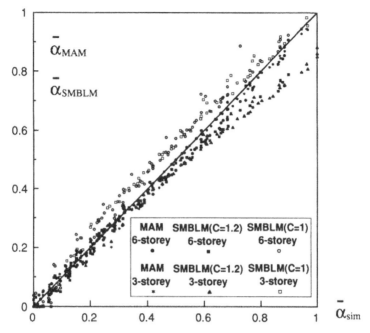

Figure 6. Comparison between non-dimensional horizontal load multipliers: numerical simulation versus EC3 approximate methods.

REFERENCES

[1]   **D. Anderson**: «Plastic hinge based methods for advanced analysis and design of steel frames european specifications and guides», <u>Proceedings</u>, SSRC Task Group 29 Workshop and Monograph on Plastic Hinge Based Methods for Advanced Analysis and Design of Steel Frames, Pittsburgh, April 5, 1992.

[2]   **M.J. Clarke, R.Q.Bridge, G.J.Hancock, N.S. Trahair**: «Australian trends in the plastic analysis and design of steel building frames», <u>Proceedings</u>, SSRC Task Group 29 Workshop and Monograph on Plastic Hinge Based Methods for Advanced Analysis and Design of Steel Frames, Pittsburgh, April 5, 1992.

[3]   **L. Argiris**: «Implementing advanced inelastic analysis and design methods in the design office», <u>Proceedings</u>, SSRC Task Group 29 Workshop and Monograph on Plastic Hinge Based Methods for Advanced Analysis and Design of Steel Frames, Pittsburgh, April 5, 1992.

[4]   **Standards Australia**: «AS4100-1990, Steel Structure», Standards Australia, Sydney, 1990.

[5]   **Commission of the European Communities**: «Eurocode 3 - Design of Steel Structures», Edited Draft, Issue 5, November 1990.

[6]   **A. De Luca, C. Faella, V. Piluso**: «Stability of Sway Frames: Different Approaches around the World», <u>Proceedings</u>, International Conference on Steel and Aluminium Structures, ICSAS 91, Singapore, 22-24 May 1991.

[7]   **Chen, W.F.**, «Design of Beam-Columns in Frames», <u>Proceedings</u>, 4th International Colloquium on Stability, SSRC, New York, 1989.

[8]   **Duan, L. Chen, W.F.**, «Design Interaction Equation for Steel Beam-Columns», <u>ASCE Journal of Structural Engineering</u>, Vol.115, No.5, May 1989.

[9]   **A. De Luca, C. Faella, V. Piluso**: «Second Order Inelastic Analysis and Design of Sway Frames», <u>Proceedings</u>, SSRC Annual Technical Session, Chicago, April 1991.

[10]   **P.J. Dowling, P. Knowles, G.W. Owens:** <u>Structural Steel Design</u>, The   Steel Construction Institute, Butterworths, 1988, page 192.

LATERAL-TORSIONAL BUCKLING STRENGTH OF TAPERED BEAM-COLUMNS UNDER COMPRESSION AND UNEQUAL END MOMENTS

MASAHIRO DOGAKI[1], TSUTOMU IZUTANI[2] and HIROSHI YONEZAWA[1]

[1] *Department of Civil Engineering, Kansai University*
3-3-35 Ymate, Suita, Osaka 564, Japan
[2] *Chuo Fukken Consultants Co., Ltd.*
3-5-26 Higashimikuni, Yodogawa-ku, Osaka 532, Japan

ABSTRACT

This paper is to aim at describing the load-carrying capacity of tapered beam-columns with I cross-section under a combination of compression and unequal end-moments. The finite difference algorithm is developed for analysing the geometrical and material non-linear behaviour of tapered beam-columns, whose web depth linearly varies in the longitudinal direction.

INTRODUCTION

Almost members in steel structures are beam-columns subjected to compression, biaxial moments and torsion simultaneously. They may respond in a variety of fashions to the imposed loads on the frame, depending upon the geometry of the cross-section, end-restraints, residual stress, initial out-of-straightnesses and material properties. A lot of theoretical and experimental investigations have extensively been carried out to understand the non-linear behaviour and load-carrying capacity of beam-columns.

In this paper, the load-carrying capacity for the lateral-torsional buckling of tapered beam-columns with I shaped cross-section, whose web depth varies in the longitudinal direction linearly, is predicted under a combination of compression and unequal end-moments about the strong axis. The finite difference approach is developed for predicting the ultimate strength of tapered beam-columns under combined loading, taking account of the geometrical and material non-linearities. A system of algebraic equations is precisely solved by the modified Newton-Raphson procedure for an incremental loading. Extensive parametric analysis is carried out to investigate the influences of the slenderness of members and geometrical and material imperfections on the non-linear behaviour and load-carrying capacity of tapered beam-columns.

METHOD OF ANALYSIS

Consider a lateral-torsional buckling phenomenon of tapered beam-columns subjected to a combination of compression N_0 and unequal end-moments of M_{z0} and βM_{z0} about the strong axis at both the ends, in which $\beta(\leq 1)$ is the end-moment ratio. A simply supported symmetrical beam-column is analysed of a length ℓ, a web of thickness t_w and depth linearly varying from b_{w1} to b_{w2} ($\alpha = b_{w2}/b_{w1}$) and flanges of width b_f and thickness t_f.

The following assumptions in this paper are introduced as: 1) material is elastic-perfectly-plastic without strain-hardening; 2) local plate buckling is not considered; 3) the length of beam-columns is enough long in comparison with the dimensions of cross-section; and 4) the movement of shear center in progress of yielding is ignored.

Equations of Equilibrium for Lateral-Torsional Buckling of Members

According to the Lin's and Massonnet's elasto-plastic method [1, 2], a set of the equations of equilibrium on the lateral-torsional buckling of tapered beam-columns are given as follows:

$$N_{x,x} = 0,$$
$$M_{z,xx} + N_x \bar{v}_{,xx} - M_y \bar{\phi}_{,xx} - M_{y,x} \bar{\phi}_{,x} = 0,$$
$$M_{y,xx} + N_x \bar{w}_{,xx} + M_z \bar{\phi}_{,xx} + M_{z,x} \phi_{,x} = 0,$$
$$M_{x,x} + (I_{cp}/A) N_x \bar{\phi}_{,xx} + M_z \bar{w}_{,xx} + M_{z,x} \bar{w}_{,x} - M_y \bar{v}_{,xx} - M_{y,x} \bar{v}_{,x} = 0 \quad \ldots \ldots (1)_{1-4}$$

in which $\bar{v}$ and $\bar{w}$ are the total displacements in the vertical and lateral directions given by a sum of initial out-of-straightnesses v_0 and w_0 and additional deflections v and w due to applied load respectively; $\bar{\phi}$ is the total angle of twist given by a sum of initial imperfection ϕ_0 and additional value ϕ due to applied load; N_x is the axial force; M_z and M_y are the bending moments about the z- and y-axes respectively; M_x is the torsional moment; I_{cp} is the sectional moment of inertia; A is the area at any cross-section; and the subscript preceding the comma describes the ordinary differentiation with respect to it.

The above-mentioned stress resultants at an elasto-plastic state are given by

$$N_x = EA \{ u_{,x} + (\bar{v}_{,x}^2 - v_{0,x}^2 + \bar{w}_{,x}^2 - w_{0,x}^2)/2 \} + EI_{cp} (\bar{\phi}_{,x}^2 - \phi_{0,x}^2)/2 - N_x^p,$$
$$M_z = -EI_z (\bar{v}_{,xx} - v_{0,xx} - \bar{w}_{,x} \bar{\phi}_{,x} + w_0 \phi_{0,x}) - M_z^p,$$
$$M_y = -EI_y (\bar{w}_{,xx} - w_{0,xx} - \bar{v}_{,x} \phi_{,x} + v_{0,x} \phi_{0,x}) - M_y^p,$$
$$M_x = EI_w (\bar{\phi}_{,xx} - \phi_{0,xx}) - M_x^p,$$
$$M_w = GJ(\bar{\phi}_{,x} - \phi_{0,x}) - EI_{w,x} (\bar{\phi}_{,xx} - \phi_{0,xx}) - EI_{w,xx} (\bar{\phi}_{,xxx} - \phi_{0,xxx}) - M_w^p \quad (2)_{1-5}$$

in which u is the displacement in the longitudinal direction; EI_z and EI_y are the flexural rigidities about the z- and y-axes respectively; GJ and EI_w are the St. Venant torsional constant and the warping constant respectively; and the stress resultants with the superscript p indicate the fictitious values introduced to compensate on actual elasto-plastic stress field and are calculated by summing up a surplus stress given as the difference between the stress obtained by the assumption of the elastic behaviour of beam-columns and the stress at yielding over the cross-section as follows:

$$N_x^p = \sum \{ \sigma_b + (\sigma^e + \sigma_r - \sigma_b)(\sigma_{eq} - \sigma_y)/\sigma_{eq}\} \Delta A,$$
$$M_z^p = \sum \{ \sigma_b + (\sigma^e + \sigma_r - \sigma_b)(\sigma_{eq} - \sigma_y)/\sigma_{eq}\} z \Delta A,$$

$$M_y^p = \sum \{ \sigma_b + (\sigma^e + \sigma_r - \sigma_b)(\sigma_{eq} - \sigma_y)/ \sigma_{eq} \} \, y \, \Delta A,$$
$$M_x^p = \sum \{ \tau_{xzb} + (\tau_{xz}^e - \tau_{xzb})(\sigma_{eq} - \sigma_y)/ \sigma_{eq} \} \, y \, \Delta A$$
$$- \sum \{ \tau_{xyb} + (\tau_{xy}^e - \tau_{xyb})(\sigma_{eq} - \sigma_y)/ \sigma_{eq} \} \, z \, \Delta A - M_w^p{}_{,x},$$
$$M_w^p = \sum \{ \sigma_b + (\sigma^e + \sigma_r - \sigma_b)(\sigma_{eq} - \sigma_y)/ \sigma_{eq} \} \, \omega_n \, \Delta A \qquad (3)_{1-5}$$

in which the stresses with the superscript e indicate those obtained by assuming that the beam-column is purely elastic even if it behaves within elasto-plastic region; the stresses with the subscript b are the fictitious values obtained at the previous loading step; σ_y is the yielding stress; σ_r is the residual stress due to welding; ΔA is a small area of elements on any cross-section of beam-columns; ω_n is the normalized unit warping with respect to the centroid; and the stress with the subscript eq is von Mises' equivalent stress given by

$$\sigma_{eq} = \{ (\sigma^e + \sigma_r - \sigma_b)^2 + 3 (\tau_{ij}^e - \tau_{ijb})^2 \}^{1/2} \qquad (4)$$

in which the shearing stress τ_{xy} is used as τ_{ij} when the yielding of material is checked in the web and τ_{xz} in the flanges.

Boundary Conditions

The beam-column is assumed to be simply supported at both the ends. It is subjected to a combination of compression and unequal bending moments at the ends. The corresponding boundary conditions at the end of x=0 are given as follows:

$$u = 0, \ \bar{v} = v_0, \ \bar{v}_{,x} - v_{0,x} = \theta_{z0}, \ \bar{w} = w_0, \ M_y = 0, \ \bar{\phi} = \phi_0, \ M_w = 0 \qquad (5)_{1-7}$$

and at the end of $x = \ell$,

$$u = -u_1, \ \bar{v} = v_0, \ M_z = \beta M_{z0}, \ \bar{w} = w_0, \ M_y = 0, \ \bar{\phi} = \phi_0, \ M_w = 0 \qquad (6)_{1-7}$$

in which u_1 is the compulsorily applied displacement in the longitudinal direction; and θ_{z0} is the rotation compulsorily applied about the strong axis.

NUMERICAL RESULTS AND CONSIDERATIONS

Fundamental equations (1) are solved together with the boundary conditions (5) and (6) to determine the non-linear behaviour of tapered beam-columns under combined loading. The numerical analysis is carried out by the finite difference method. A set of non-linear algebraic equations is solved by the modified Newton-Raphson procedure for an incremental load.

The geometrical imperfections in the lateral direction and with respect to the twist are assumed to be one half wave of sinusoidal series. The distribution of residual stress due to welding is assumed to be in Fig.1. Two cases for the magnitude of initial imperfections are considered, i.e., a)mean values of w_{0max} =$3\times10^{-4} \ell$, ϕ_{0max}=$-5\times10^{-4} \ell/b_{w0}$ and σ_{rc}/σ_y =-0.12, and b)higher values of w_{0max} =$8\times10^{-4} \ell$, ϕ_{0max}=$-8\times10^{-4} \ell/b_{w0}$ and σ_{rc}/σ_y =-0.24.

The predicted ultimate strengths are compared with the test results by Shiomi-Kurata [3] as shown in Fig.2, in which the tapered beam-columns are tested under a combination of compression and end-moment at one side. In this figure, the mark ● is the calculated results, and the ordinate and abscissa denote the non-dimensional theoretical ultimate strength M_{ult}^{th}/M_{pz0} and non-dimensional collapse value M_{ult}^{ex}

$/M_{pz0}$, in which M_{pz0} is the full plastic moment about the strong axis at x=0. It is found that the theoretical values agree closely with the test results and the finite difference procedure is available to predict the load-carrying capacity of tapered beam-columns under combined loading.

Fig. 3 shows the interaction curves of load-carrying capacity for the tapered beam-columns(α=0.75) under compression and unequal end-moments of β=0.75 with intermidiate modified slenderness ratio λ, $(= \ell \sqrt{\sigma_y A_0/EI_{y0}}/\pi$; A_0 is the area of cross-section at x=0; and EI_{y0} is the flexural rigidity about the weak axis at x=0), in which the figures (a) and (b) denote the interaction curves for the tapered beam-column with initial imperfection of case (a) and of case (b) respectively; and N_{p0} is the squash load at x=0.

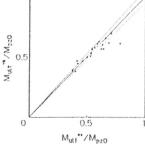

Figure 1. Distribution of residual stress. Figure 2. Comparison between theoretical and experimental ultimate strengths.

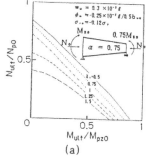

(a) (b)

Figure 3. Interaction curves of load-carrying capacity.

CONCLUSIONS

The elasto-plastic behaviour and load-carrying capacity of tapered beam-columns are determined under a combination of compression and unequal end-moments. The finite difference method is adopted to analyse the non-linear equilibrium equations for the lateral-torsional buckling of tapered beam-columns together with the boundary conditions.

REFERENCES

1. Lin, T. H., Theory of Inelastic Structures. John Wiley & Sons, 1968.
2. Massonnet, Ch., General theory of elastic-plastic membrane-plates. Engineering Plasticity, eds. by Heyman, J. and Leckie, F. A., Cambridge University Press, 1968, pp.443-471.
3. Siomi, H. and Kurata, M., Strength formula tapered beam-columns. J. Structural Engineering, ASCE, 1984, **110**, 7, 1630-43.

NONPROPORTIONALLY LOADED SWAY FRAMES WITH FLEXIBLE CONNECTIONS

SALEH ALI EIDAN
King Abdulaziz City for Science and Technology (KACST)
P.O. Box 9338, Riyadh-11413, SAUDI ARABIA

ZIA RAZZAQ
Old Dominion University, Norfolk, VA 23529, USA

ABSTRACT

Inelastic sway portal frames with semi-rigid beam-to-column and foundation connections are studied under the influence of nonproportional loads. Hollow rectangular sections are adopted. The strength of portal frames with flexible connections is found to be load path dependent.

INTRODUCTION

The current methods of inelastic frame analysis are based on the contention that the applied loads are proportional in nature. The actual structures, on the contrary, are subjected to nonproportional loads. This paper presents the results of a study of the influence of nonproportional loads on the behavior of sway portal frames. A rigorous elasto-plastic analysis is conducted based on an equilibrium procedure and includes initial member crookedness and residual stresses. The frame columns are hollow square and the beam is hollow rectangular, with $E = 29,000$ ksi and a yield stress of 46 ksi.

THE PROBLEM

Figure 1 shows schematically an imperfect unbraced portal frame with flexible joints at B and C, and flexible base connections at A and D. The beam is bent about its major axis. The initial crookedness of each member is taken as a half sine wave with a midspan amplitude of $L/1,000$. The residual stress distribution is shown in Figure 2. The frame flexible connections possess a moment rotation relationship as shown in Figure 3. The column and the beam section dimensions are , respectively, 7x7x0.375 in., and 6x8x0.375 in. As shown in Figure 1, the frame is subjected to a horizontal load H at B and a vertical load pair (P,P) at B and C. Three load paths, LP1 through LP3, are used and are defined as follow:

LP1: H and (P,P) applied proportionally while keeping H/P ratio equal to 0.10 until the frame collapses;

LP2: H applied first to the level obtained in LP1, followed by (P,P) until the frame collapses;

LP3: The load pair (P,P) applied first to the level obtained in LP2, followed by H until the frame collapses.

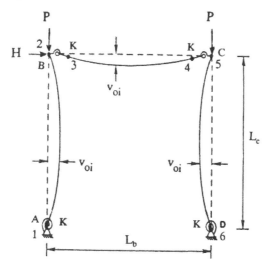

Figure 1. Imperfect unbraced portal frame with loading

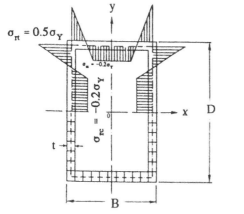

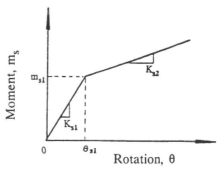

Figure 2. Cross section and residual stress distribution (Ref. 2)

Figure 3. Moment-rotations relationship

ANALYSIS OUTLINE

The analysis is conducted by using inelastic slope-deflection expressions derived from the inelastic differential equations of member equilibrium coupled with a finite-difference scheme. Elastic unloading of plastic material is incorporated. The global inelastic frame equations take the following form:

$$[K]\{ \delta \} = \{ M_c \} + \{ M_p \} + \{ M_a \}$$

in which [K] is the global tangent stiffness matrix, $\{ \delta \}$ is the deflection vector; $\{M_c\}$ is associated with the member crookedness; $\{M_a\}$ contains the applied moments; $\{M_p\}$ is the plastic force vector generated beyond the elastic range. To determine $\{ \delta \}$, the above equation is solved iteratively for each external load level.

RESULTS

With reference to Figure 3, the value of K_{s1} is kept constant at 24000 kip-in/rad., while K_{s2} is taken as 24000, 20000, and 16000 kip-in/rad., respectively, for Frames 1, 2, and 3 referred to in this paper. The stiffness of the frame connections is reduced when the spring plastic moment m_{s1} is reached. Here, m_{s1} is taken as 180 kip-in. Figure 4 shows the dimensionless load-deflection curves for Frame 3 with various load paths. The loads H and P are each nondimensionlized by the column squash load P_Y. The frame horizontal deflection, $\bar{\Delta}$, at the beam-level is given by the actual deflection divided by L_c. Each load path gives a different frame deflection configuration.

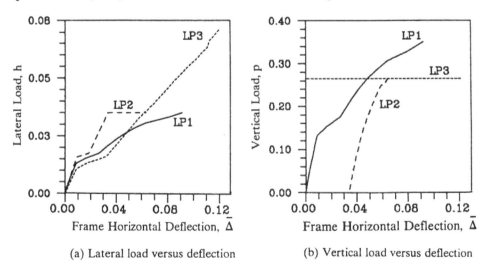

(a) Lateral load versus deflection   (b) Vertical load versus deflection

Figure 4. Load-deflection curves for Frame 3

Table 1 presents the maximum dimensionless external loads for Frames 1, 2, and 3. The effect of nonproportional loading on the frame strength is clear. For example, the dimensionless vertical load p for Frame 3 is decreased from 0.3500 with LP1 (proportional) to 0.2648 with LP2, a reduction of 24% in strength. Similarly, h = 0.0350 and 0.0741, respectively, with LP2 and LP3 for Frame 3, an increase of 2.11 times with LP3 as compared to LP2. Frame 1, however, does not show an appreciable effect of nonproportional loading. The results of Frame 1 compared to those for Frames 2 and 3 show the importance of the effect of connection flexibility on the nonproportionally arrived at peak loads.

Table 1: Numerical Results

Frame No.	Restraint, K_{s2} kip-in/rad	Load	Dimensionless External Loads for Load Path		
			LP1	LP2	LP3
1	24,000	p	0.4287	0.4054	0.4054
		h	0.0429	0.0428	0.0432
2	20,000	p	0.3730	0.3610	0.3610
		h	0.0373	0.0373	0.0434
3	16,000	p	0.3500	0.2648	0.2648
		h	0.0350	0.0350	0.0741

CONCLUSION

The load-carrying capacity of flexibly connected portal frames can be load-path dependent. For the results presented, the nonproportional loading effect on the peak loads is found to be significant when the connections have a bilinear moment rotation relation.

REFERENCES

1. Darhamulla, S. P., and Razzaq, Z., "Threshold Nonproportional Load-Moment Relations for Nonsway Frames," *Proceedings,* Fourth International Colloquium on Structural Stability-Mediterranean Session, Istanbul, TURKIYE, September 16-20, 1991.

2. Balio, G., and Companini, G., "Equivalent Bending Moments for Beam-Columns," *Journal of Constructional Steel Research,* Vol. 1, No. 3, May, 1981.

USE OF SIMPLIFIED TEMPERATURE PROFILES WITHIN STEEL BEAMS IN FIRE

by I W Burgess [a], J A El Rimawi [a] and R J Plank [b]

Department of Civil and Structural Engineering [a], School of Architecture [b], University of Sheffield, Mappin Street, Sheffield S1 3JD, U.K.

ABSTRACT

Predicted structural deflections during the growth of a fire of steel beams carrying concrete slabs on their top flanges are compared for different simplified representations of the internal temperature profiles. It is shown that very simple forms give extremely good accuracy, both in terms of deflection and failure temperature. Failure temperatures for a given load ratio are shown over a very large range of structural sections to be almost uniquely controlled by the span:depth ratios of the beams.

INTRODUCTION

Design attitudes to ensuring satisfactory performance of steelwork in the event of a building fire are gradually changing. Applied fire protection, despite its expense, is still commonplace, but quantitative approaches have been introduced recently [1], and these offer the prospect for designers to determine reduced levels of applied protection, or in some cases to demonstrate that the steel can remain unprotected. For instance, the implicit protection offered by concrete floor slabs to supporting beams results in a non-uniform temperature over the cross-section and the greater residual strength in the cooler parts of the section means longer survival times. A number of analytical methods have been developed to simulate the behaviour of steel structures under fire conditions. Those developed by the authors [2, 3] are based on a secant stiffness approach and provide a full description of the structural behaviour at increasing temperature up to collapse. These methods are not intended for routine design application, but can provide much data very efficiently for incorporation within simple design methods. One design approach for steel beams supporting a concrete slab requires the temperature distribution within the beam to be represented as a series of uniform temperature steps. The real temperature profile, whether determined by test or thermal analysis, is curvilinear and little guidance is available as to how this should be modelled for design purposes. This paper presents a resumé of results from an extensive analytical study on a wide range of beams using different temperature representations, spans and section sizes.

TEMPERATURE REPRESENTATION

A wide range of steel beam sections, covering the full product range manufactured by British Steel, has been considered. A finite element thermal analysis was conducted for each to determine the detailed temperature distribution within the steel as the furnace temperature is increased. These were then represented in various simplified forms as shown in fig. 1, using different step functions.

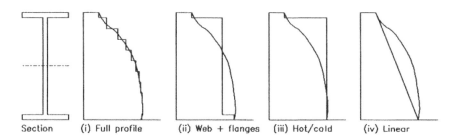

Section   (i) Full profile   (ii) Web + flanges   (iii) Hot/cold   (iv) Linear

Fig.1   Different temperature representations for fire analysis of floor beam carrying a concrete slab.

The structural analysis program was then used to determine the complete temperature-deformation history for each assumed temperature distribution and the results compared as illustrated for one section in fig. 2.

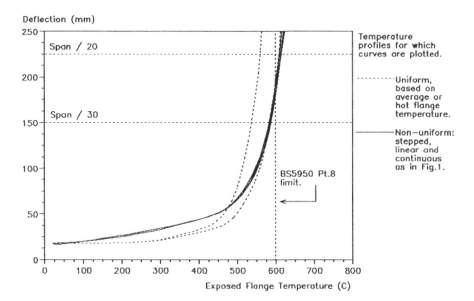

Fig. 2   Temperature-deflection plots for a 254x146UB43 section under different representations of temperature distribution.

In all cases the beams were loaded with their full nominal design load, and material properties based on BS5950 part 8 assuming grade 43 steel were used.

At lower temperatures the effect of thermal bowing can be clearly seen. The results based on uniform temperature profiles (for which there is no thermal bowing) indicate very little change in deflection, whilst the deformations calculated for the stepped profiles show a consistent growth. However it is the predicted failure temperature (based on a limiting deflection of span/20) which is of most interest to the designer, and here there is little perceptible difference between the results for the different representations. As would be expected the uniform temperature profile based on the maximum bottom flange temperature predicts the lowest failure temperature of about 560°C. The non-uniform representations give failure at 610-613°C, compared with the BS5950 load ratio method design estimate of 599°C. These results are typical of those determined for the full range of sections studied, and suggest that very simple temperature representations are adequate for design purposes.

THE INFLUENCE OF SPAN, SECTION, LOAD RATIO AND END CONDITIONS

In earlier work [3] it was found that span, section size, load ratio and support conditions all had an influence on failure temperatures. A comprehensive study was therefore undertaken to investigate these parameters, using a two-step temperature profile derived from thermal analysis results. The full range of steel cross-sections including both universal column and universal beam sections has been encompassed, with load ratios (applied bending moment as a proportion of ultimate moment capacity) varying from 0.2 to 0.67 (which is nominally fully loaded). In all cases the spans were varied so that the span:depth ratio ranged from 8 to 23, and both simply supported and fixed ended beams were considered. The results indicated that longer spans and shallower sections led to lower failure temperatures, and this suggested that span:depth ratio might be an important parameter. The results were therefore presented in the form of failure temperatures against span:depth ratio for different load ratios and these are shown in Fig. 3 for simply supported and fixed-ended beams.

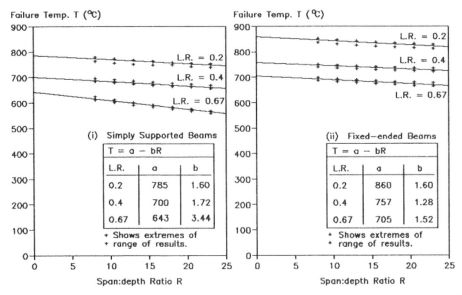

Fig.3 Failure temperatures plotted against span-depth ratio for a very large range of structural I- and H-sections for simple and fixed-ended beams.

Each 'point' on these curves corresponds to the results for some 27 universal beam and 42 universal column sections. It is clear that for a given span:depth ratio, regardless of the actual section size and span length, the results are remarkably consistent. There is also a reasonably linear relationship between failure temperature and span:depth ratio suggesting a simple way of extrapolating results of experimental studies for fixed length beams to more realistic conditions. The reduction in failure temperature corresponding to an increase in span:depth ratio from 8 to 23 is of the order 40-60°C depending on load ratio and end conditions.

Failure temperatures for fixed-end conditions are seen to be 70-100°C greater than those for simply supported beams. This is due to the redistribution of moments which can take place when there is a degree of redundancy within the structure. Clearly where design loads are based on simply supported conditions but real connections provide some continuity, there is an effective reduction in load ratio, and further increases in failure temperature can be realised. A considerable amount of work has been done recently at Sheffield [4] on the benefits to simply designed frames in fire accruing from the stiffness of real connections.

CONCLUSIONS

This study has demonstrated that for simple floor beams design procedures can be based on very simplified temperature patterns thoroughout the beam cross-section. The effect of span and beam depth is relatively small, but where necessary can be accounted for using a linear relationship between failure temperature and span:depth ratio. As expected reduction of the load ratio and provision of increased end restraint both lead to improvements in survival times. Current work is investigating other beam arrangements including shelf angle floors and slimfloors, as well as developing analytical methods to include real connection characteristics within the fire analysis of structural frameworks.

Acknowledgement

The work of Karl Zand, Andrew Hill and Nick Pierpoint, students in the Department of Civil and Structural Engineering at the University of Sheffield, is gratefully acknowledged.

REFERENCES

1. British Standard BS 5950, *The structural use of steelwork in building: Part 1: Code of practice for design in simple and continuous construction,* British Standards Institution, London, 1985.

2. Burgess I W, El-Rimawi J A & Plank R J, Analysis of beams with non-uniform temperature profile due to fire exposure, *J. Construct. Steel Research,* **16** (1990) 169-92.

3. Burgess I W, El-Rimawi J A & Plank R J, Studies of the behaviour of steel beams in fire, *J. Construct. Steel Research,* **19** (1991) 285-312.

4. El-Rimawi J A, Burgess I W & Plank R J, The analysis of steel beams in fire, *Proc. International Conference on Steel and Aluminium Structures, Singapore, 1991,* ed. S L Lee and N E Shanmugam, Elsevier Applied Science Publishers, Barking, 1991.

THE PRESENTATION AND COMPARISON OF SEVERAL CONSTITUTIVE LAWS OF STEEL IN FUNCTION OF TEMPERATURE

JOAQUIM C. VALENTE

Departamento de Engenharia Civil, Instituto Superior Técnico, Av. Rovisco Pais, 1096 Lisboa Codex, Portugal

ABSTRACT

The simulation of the behaviour of the steel structures submited to the fire action needs a good definition of the stress-strain relationships in function of the temperature and as near of the reality as possible.

In this paper we present stress-strain relationships of steel at elevated temperatures which try to model the behaviour of the steel from the elastic phase until colapse with the strain-hardening included. These relationships are the result of steady state tensile tests of Fe 360 at several levels of temperature.

TENSILE TEST

Mechanical and Chemical Characteristic

Table 1
Properties of the steel specimens

Chemical	Mechanical
C - 0.18 % Mn - 0.44 % P - 0.009 % S - 0.026 %	$\sigma_{y,20}$ = 278.6 N.mm^{-2} $\sigma_{r,20}$ = 410.05 N.mm^{-2} ΔL^{20} = 34.4 %

Specimen and Tests

The specimens were cilindrical and the dimension of the used specimens followed the rules suggested by the manufactor of the traction machine "Instron TTCL". These rules are in accordance with the BS 3633:part 1:1963. The specimens were heated by an electric cilindrical furnace. The control of the temperature was based on value the measured by two thermocouple, one on each end of the specimen. The deformation speed of the specimens was 1 mm/min. For each level of temperature three or five tests were done.

RESULTS

The stress-strain relationships of the fig. 1 are the result of the tests from the room temperature until 700 °C. These curves represent the behaviour of the steel until the maximum strength.

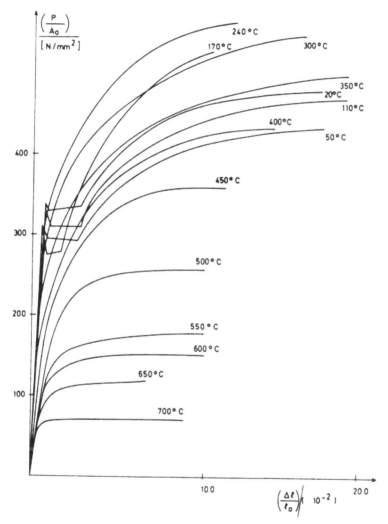

Figure 1. Stress-strain relationships of the steel Fe 360 at elevated temperatures.

Proposal for a Mathematical Model

Based on the curves shown in figure 1 we tried to develop an analitical model which represents the full development of the steel tensile tests, from the elastic phase until the colapse. As usually, we assume that the behaviour at elevated temperatures of all steels is of the same type but the values, for each level of temperature, of the proportional limit stress and the maximum stress are proportional to the value of the yield stress at room temperature.

Table 2.
Definition of the parameters of the mathematical model

Range	$\epsilon_{\epsilon,T}$	$E_{\epsilon,T}$
I linear	$E_T \cdot e$	$E_{,T}$
II elliptical	$\dfrac{a}{b}\sqrt{a^2-(e_{y1}-e)} + \sigma_p - C$ $a^2 = \dfrac{E_T(e_{y1}-e_p)^2 + C(e_{y1}-e_p)}{E_T}$ $b^2 = E_T(e_{y1}-e_p)\,C + C^2$ $C = \dfrac{(\sigma_{max}-\sigma_p)^2}{2(\sigma_p-\sigma_{max}) + E_T(e_{y1}-e_p)}$	$\dfrac{b(e_{y1}-e)}{a\sqrt{a^2-(e_{y1}-e)^2}}$
III horiz. plateau	σ_{max}	0
IV decrea. branch	$\sigma_{max} + E_T(e-e_{y1})$	$-\dfrac{\sigma_{max}}{e_R-e_{y1}}$

We chose a mathematical model similar to the one of Robert and Schaumann, (2), and the one proposed by the Eurocodes 3 and 4, (7,8). Our model is define in figure 2 and table 2 and it is mainly based on the parameters presented in table 3.

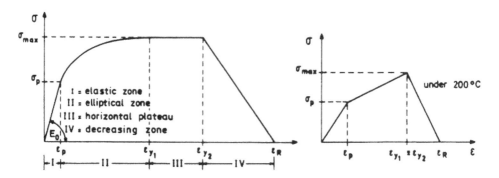

Figure 2. Mathematical model for stress-strain curves of steel at elevated temperatures.

Table 3
Values of the parameters in function of the temperature to the
mathematical model

Temp.	$\dfrac{E_{,T}}{E_{,20}}$	$\dfrac{\sigma_P}{\sigma_{Y,20}}$	$\dfrac{\sigma_{max}}{\sigma_{Y,20}}$	e_{Y1}	e_{Y2}	e_R
20.	1	1	1.35	0.15	0.15	0.20
100.	1	1	1.35	0.15	0.15	0.20
200.	0.9	0.9		0.15	0.15	0.20
250.		0.6	1.7	0.15	0.15	0.20
300.			1.61	0.15	0.15	0.20
350.			1.45	0.15	0.15	0.20
400.	0.7			0.15	0.15	0.20
450.			1.07	0.10	0.15	0.20
500.			0.76		0.20	0.25
550.			0.50	0.05	0.20	0.25
600.			0.44		0.20	0.25
650.		0.125	0.33	0.034	0.20	0.25
700.	0.15		0.20	0.02	0.20	0.25
1000.	0.0	0.0	0.00	0.02	0.20	0.25

These parameters have a linear change in function of the
temperature between each two consecutive values shown in the
table 3. The main difference between (2,7,8) and the proposal of
the author is that the elliptical zone also progresses in the
hardening phase. Also in the author's proposal the maximum stress
is for strain of 0.15 and for a temperature levels under 400 °C,
(fig. 1). This value of the strain is higher than the strain of
0.02 taken by other authors, (2,3,8). Up to 400 °C the curves do
not have an horizontal plateau of maximum stress level, (fig. 4).
The cut of the stress-strain relationships for a strain of 0.02
and the introduction of an horizontal plateau, after this strain,
gives very conservative maximum stress values, (2,3,7,8).
In the present proposal the value of the strain corresponding to
the maximum stress decreases from 0.15 at 400 °C to 0.02 at
700 °C. The strain 0.02 in the begining of the horizontal zone
seems good for temperatures higher than 700 °C.
For strains between 15% and 20%, and for temperatures less than
500 °C a decreasing branch between ϵ_{y2} to ϵ_R was introduced.
These strain values change to 20% and 25% for temperatures higher
than 550 °C.
The stress-strain curves obtained from the tensile tests and due
to the mathematical model are drawn together in figure 3 and the
development of these two curves is almost the same.

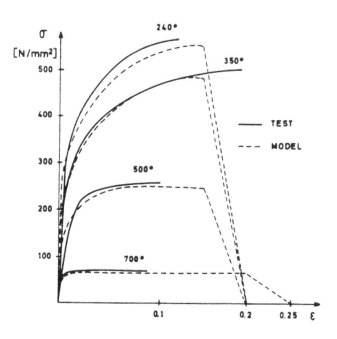

Figure 3. Comparison between the mathematical model and the results of the tensile tests

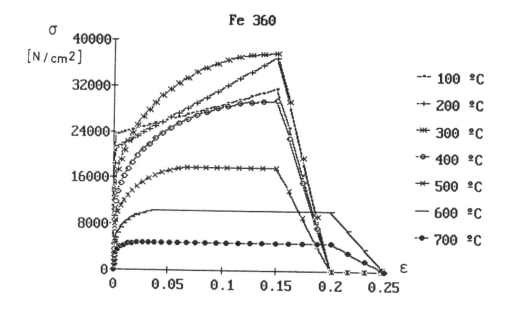

Figure 4. The stress-strain relationships due to the proposed model for the steel Fe 360

**Comparison of various parameters in the different proposals
for stress-strain relationships of steel**

The Elastic Modulus
The laws of variation of the elastic modulus of the steel in
function of the temperature proposed in (2,3,5,6,8,10) do not
have great differences for temperatures under 500 °C, (fig. 5).
For higher values of the temperature the elastic modulus proposed
in (6,8,10) and the author's assumes almost the same values and
it is zero for 1000 °C. The functions proposed in (2,3,5) show
higher values for these temperatures.

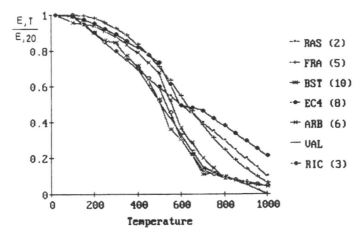

Figure 5. Comparison of several proposals for variation in
function of the temperature of the elastic modulus of the steel

The proportional limit stress
The comparison of the proportional limit stress proposed in
(2,5,6,8) and by the author shows the same shape in the range of
the studied temperatures, (figure 6), but the gap between the
highest values proposed, (5), and the smallest, (author), is
about 60%, at 400 °C.

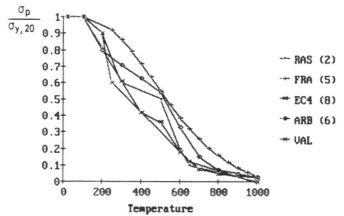

Figure 6.   Comparison of the proportional limit stress of
steel in function of temperature

The stress for a strain of 0.02

The study of the stress for this strain is very interesting because many authors propose this stress for the begining of the horizontal plateau corresponding to the maximum stress.

Figure 7 shows that for temperatures higher than 400 °C the stress has almost the same value for all the proposals. Up to 400 °C the stress-strain relationships with hardening strain, (4,5,6,author), that stress takes greater values than the $\sigma_{y,20}$, (fig. 7).

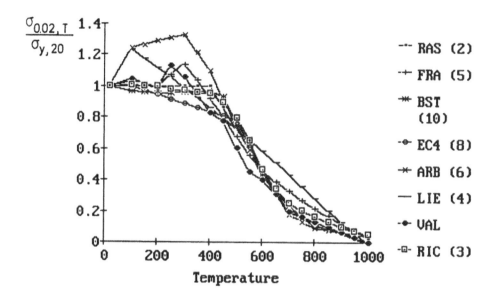

Figure 7. Comparison of the stress for a strain of 0.02 in function of the temperature

Conclusions

The proposed of stress-strain relationships have a non-linear evolution between the proportional limit stress and the maximum stress level for temperatures above 200 °C . This non-linear evolution has the strain-hardening phase included.

The evolution, in function of the temperature, of the value of the strain corresponding to the maximum stress, (begining of the horziontal plateau), seems near to the reality, (figures 1,4). Moreover, the begining of the decreasing branch moves from the strain values 0.15 - 0.20 to 0.20 - 0.25, (table 2). This change represents the increasing of the elongation of the specimen for higher temperatures.

In the future, it would be usefull to compare the numerical simulation of the behaviour of steel structures under fire conditions using these different stress-strain relationships and the results of structural fire tests. This comparasion would give more information about the real values of the strain-hardening phase.

REFERENCES

1. Valente, J. C.; Margarido, F.; Pádua Loureiro, A.; Tovar de Lemos, A. F., Estudo do Comportamento Mecânico à Tracção do Aço A235 desde a Temperatura Ambiente até 700 º C, Materiais87, 3º Encontro Nacional da Sociedade Portuguesa de Materiais, Braga, Abril 1987.

2. Rubert, A.; Schaumann, P., Temperaturabhängige Werkstoffeigenschaften von Baustahl bei Brandbeanspruchung, Stahlbau 54(1985), H. 3, PP 81-86, Verlag Wilh Ernt & Sohn, Hohenzollerndamm 170, 1000 Berlin B1.

3. Richter, E., Stress-strain Relationships for Calculation of the Load Bearing and Deformation Behaviour of Fire Exposed Structural Elements, Institut Für Baustoffe, Massivbau und Brandschutz, Technische Universität Braunschweig, August 1988.

4. Lie, T. T., A Procedure to Calculate Fire Resistence of Structural Members. Fire and Materials, vol 8 nº. 1, 1984, p. 40-48.

5. Franssen, J-M., Etude du Comportement au feu des Structures Mixtes Acier-Beton, Thèse de doctorat, Faculté des Sciences Apliquées, Université de Liège, 1986.

6. Ianizzi, R.; Schleich, J. B., Mechanical Properties of Structural Steel at Elevated Temperature - Comparisons by Numerical Simulations, Final Report, Report n. 5/90 Arbed Recherchers, Luxemburg, 1991.

7. Eurocode n.3, Design of Steel Structures Part 10: Structural fire Design, April 1990.

8. Eurocode n.4, Design of Composite Structures Part 10: Structural fire Design, April 1990.

9. Kirby, B. R.; Preston R. R., High Temperatures Properties of Hot-rolled, Structural Steel for Use in Fire Engineering Design Studies, Fire Safety Journal, 13(1988), pp 27-37.

10. BS4360: Grade 43A, 1979 and BS4360: Grade 50B, 1979.

Plate and Box Girders

BEHAVIOUR AND DESIGN OF TRANSVERSE STIFFENERS OF STIFFENED COMPRESSION FLANGES

W HINDI
W S Atkins (Services) Ltd
Regent Centre, Regent Road, Aberdeen, AB9 8UQ, UK
formerly Research Fellow University of Surrey

J E HARDING
Professor of Structural Engineering
University of Surrey, Guildford, Surrey, GU2 5XH, UK

ABSTRACT

The behaviour of and design requirements for the transverse stiffeners of stiffened compression flanges has been studied using a fully non-linear finite element package. Using nodal line representation of the transverse stiffeners, the form and magnitude of the forces applied to the stiffener during the collapse of the stiffened flange were investigated. Further studies concentrated on the behaviour of the flange with varying sizes of stiffener cross-section. This paper looks at the behaviour defined by these results and compares the findings with the requirements of the relevant British Code, BS5400 Part 3, related to the design of steel bridges. Based on this comparison suggestions have been made for modified design recommendations.

INTRODUCTION

The compression flanges of steel box girder bridges are usually stiffened both by longitudinal and transverse stiffeners. The former enhance the basic buckling capability and hence the strength of the flange, while the primary role of the latter is to provide support for the longitudinally stiffened flange, enabling the higher buckling capability to be obtained. This latter requirement means it is very important to ensure that the deflection of the transverse stiffeners is limited so that they maintain the support assumed for the longitudinally stiffened panels. In the British Code [1]. this is accommodated by the use of a stiffness requirement based on a factor of safety against critical buckling, combined with a strength requirement which incorporates a proportion of the in-plane flange load, applied as a lateral load to the transverse stiffeners representing the destabilising action of the flange, combined with any direct load applied. The latter requirement is consistent with the approach often taken for definition of the load applied to a bracing system acting in a secondary plane to the main loading of a structural member.

The British Code adopts a lateral load equal to 0.5% of the flange axial force multiplied by a factor of 1.3 which is applied to an effective section of transverse stiffener plus a width of flange, acting as a beam between the webs.

ANALYTICAL PREDICTIONS OF THE BEHAVIOUR OF THE TRANSVERSE STIFFENERS

The finite element package LUSAS was used to carry out an extensive parametric study on the non-linear behaviour of orthogonally stiffened compression flanges incorporating discretely represented longitudinal and transverse stiffeners. The size of the transverse stiffener was treated as one of the key variables. Four equally spaced longitudinal stiffeners were used with their local cross-section slenderness limited to avoid any local outstand instability. The material properties adopted were typical of grade 50 steel.

After a preliminary study carried out with nodal line representation of the transverse stiffeners to identify the main parameters, a main study using two longitudinally stiffened flange panels with a single transverse stiffener, with in some cases nodal line representation of the latter was undertaken. Plate panel initial imperfections were incorporated with a maximum magnitude of the sub-panel width over 200 combined with a single half wave longitudinal stiffener imperfection of the cross-frame spacing over 500 with a 3mm imperfection at the transverse stiffener position taken for the preliminary study, changed to the transverse stiffener spacing divided by 250 for the main study.

In the main study analyses were carried out on 120 orthogonally stiffened compression flanges divided into 3 categories with local plate slenderness values of 20, 40 and 60. Stiffener area to panel area ratios were varied and column slendernesses, length over radius of gyration of 40, 80 and 120, were incorporated. Nodal line representation of the transverse stiffener as well as 3 different stiffener sizes were separately studied. The dimensions of the stiffeners were once, twice and three times the dimensions of the longitudinal stiffeners both in depth and thickness. Flat stiffeners were used in all cases to simplify the analytical work.

The finite element model used was based around a rectangular 8-noded semi-loof shell element for both plate and the stiffener sections. A typical finite element mesh is shown in fig 1. For the two panel cases examined in the main study advantage was taken of symmetry by only analysing one half of the panel width. While this does make assumptions about the buckling mode in the transverse direction it is not thought that it would have any significant influence on the results. The flanges were assumed to be simply supported around the three boundaries shown in the figure. An incrementing uniform stress was applied to both the plate and the stiffener outstands at the ends of the model which was felt to be better than applying a displacement loading configuration because the latter would restrict rotation of the stiffened panel at the end boundaries which would reduce the panel deformation and hence affect the load applied to the stiffener. The number of elements used was as shown in the figure, with the exception that the refinement of the mesh in the longitudinal direction was dependent on the panel aspect ratio, ranging between five and fifteen between the end and the transverse stiffener. The figure shows the intermediate value of ten.

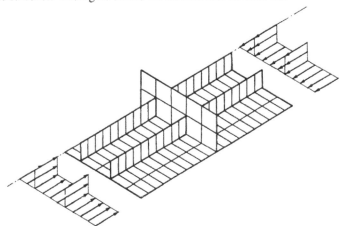

Figure 1   A typical finite element mesh of a two panel flange.

RESULTS OF THE STUDY

Table 1 shows the geometries considered in the main study with variations in values of stiffening ratio, panel slenderness and stiffener column slenderness. The stiffener column slenderness was based on the stiffener section combined with a plate width equal to the stiffener spacing. The three pairs of dimensions given for each parameter combination are the sizes of the transverse stiffeners studied as well as the nodal line analysis. These dimensions are the depth and thickness of the stiffener in each case and have been referred to as stiffener 1, 2 and 3 in describing the results in the paper with stiffener 1 being the smallest in each case.

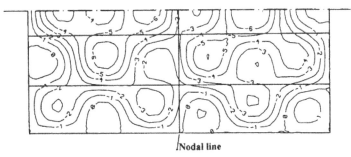

Nodal line

Figure 2   Deformation at collapse
$$(b'/t = 20, a/i = 40, A_s/b' \, t = 0.2).$$

A range of collapse modes was obtained from the analyses. For stocky configurations, a panel slenderness of 20 and a column slenderness of 40, the buckling mode for the nodal line analysis and for the two large transverse stiffeners corresponded to a plate failure mode similar to that shown in fig 2.

TABLE 1
Stiffened Plate Parameters

b'/t	a/i	As b't	Transverse Stiffener	As b't	Transverse Stiffener	As b't	Transverse Stiffener	As b't	Transverse Stiffener
20	40	0.1	90*9	0.2	127*12.7				
			180*18		254*25.4				
			270*27		381*38.1				
	80	0.1	90*9	0.2	127*12.7				
			180*18		254*25.4				
			270*27		381*38.1				
	120		90*9		127*12.7				
			180*18		254*25.4				
			270*27		381*38.1				
40	40		60*6		90*9		130*13		155*15.5
			120*12		180*18		260*26		310*31
			180*18		270*27		390*39		465*46.5
	80	0.1	60*6	0.2	90*9	0.4	130*13	0.6	155*15.5
			120*12		180*18		260*26		310*31
			180*18		270*27		390*39		465*46.5
	120		60*6		90*9		130*13		155*15.5
			120*12		180*18		260*26		310*31
			180*18		270*27		390*39		465*46.5
60	40		50*5		75*7.5		105*10.5		130*13
			100*10		150*15		210*21		260*26
			150*15		225*22.5		315*51.5		390*39
	80	0.1	50*5	0.2	75*7.5	0.4	105*10.5	0.6	130*13
			100*10		150*15		210*21		260*26
			150*15		225*22.5		315*51.5		390*39
	120		50*5		75*7.5		105*10.5		130*13
			100*10		150*15		210*21		260*26
			150*15		225*22.5		315*51.5		390*39

All stiffener dimensions in mm, b=2000 mm, b'=400mm, b=flange width
b'=sub-panel width, t=plate thickness, a=transverse stiffener spacing, As=stiffener area

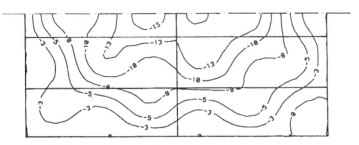

Figure 3   **Deformation at collapse**
(b'/t = 20, a/i = 40, A_s/b' t = 0.2, stiffener 1).

With the smallest transverse stiffener this changed to an overall buckling mode of the whole flange with the transverse stiffener deforming significantly as shown in fig. 3. In the case of the plate panel mode, a mode change occurred after peak load to a column buckling mode with deformations away from the stiffener outstand on either side of the transverse stiffener position. Neither the plate panel mode nor the orthotropic buckling mode was apparent when the column slenderness was increased substantially and a longitudinally stiffened panel failure was obtained for all stiffener sizes at the column slenderness of 120 as illustrated in fig 4. This figure corresponds to the smallest of the stiffeners but even in this case the deformations shown, which include the initial shape, showed little amplification from the imperfection value, and the strengths obtained were virtually constant for all four analyses showing little influence from the size of the stiffener itself.

Figure 4   Deformation at collapse
(b'/t = 20, a/i = 120, A$_s$/b' t = 0.2, stiffener 1).

An increase in the panel slenderness value to 40 produced little change in the behaviour for panels with column slenderness of 40, compared with the previous analyses. However, there was generally more evidence of a mixed mode buckling form and for stiffeners 2 and 3 plate panel action was more in evidence prior to peak load. After peak load the same types of failure mode were obtained for all the stiffener sizes as for the stocky panel analyses. With an increase in column slenderness to 120 the results followed closely the trend in behaviour with the stockier panel geometry, again with little increase in deformation at the line of the transverse stiffener for any of the stiffener cross-sections and hence little sensitivity in collapse strength to stiffener size. For this intermediate panel slenderness, figs 5 and 6 demonstrate the dependence of strength on stiffener size as the column slenderness changes from 40 to 120. Fig 7 shows the forces applied at the location of the transverse stiffener from the nodal line analysis relating to the geometry of fig 6, and it can be seen that the forces concentrate substantially at the locations of the longitudinal stiffeners where the column action applies the highest loads. The plate panels being flexible out-of-plane do not have the ability to apply significant forces to the transverse stiffener. The common assumption therefore of a uniformly distributed load is a relatively poor model of the actual forces applied.

In moving to panels with a local plate slenderness of 60 it was expected that plate action would dominate the failure behaviour of flanges for all but the highest column slenderness. Indeed all flanges with a column slenderness of 40 failed by pure plate panel buckling, although for the smallest stiffener there was still some increase in the deflection at the transverse stiffener location prior to peak load, and after peak load there was a sudden change in the mode shape to overall flange buckling away from the stiffener outstands. For this case deformation at peak load is shown in fig 8.

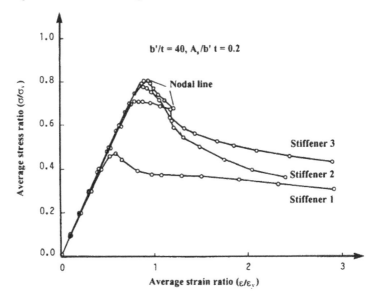

Figure 5   Dependence of response on stiffener size (a/i = 40).

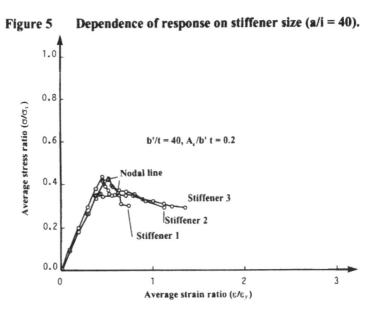

Figure 6   Dependence of response on stiffener size (a/i = 120).

With flanges having a column slenderness of 120, for all but the flange with the smallest transverse stiffener, a mixed mode occurred at collapse, with column buckling tending to dominate. This can be clearly seen in the example of fig. 9. The smallest transverse stiffener with this column slenderness, however, produced a very clear failure mode corresponding to orthotropic panel buckling as shown in fig. 10.

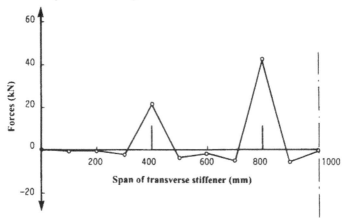

Figure 7   **Forces at nodal line.**

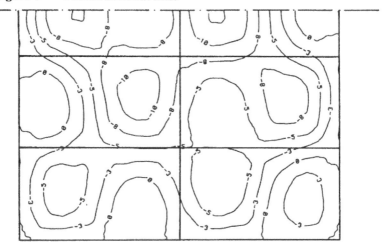

Figure 8   **Deformation at collapse**
(b'/t = 60, a/i = 40, A$_s$ /b' t = 0.2, stiffener 1).

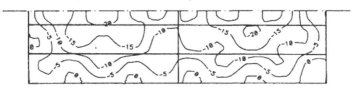

Figure 9   **Deformation at collapse**
(b'/t = 60, a/i = 120, A$_s$ /b' t = 0.2, stiffener 2).

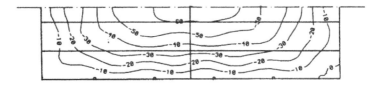

Figure 10   Deformation at collapse
(b'/t = 60, a/i = 120, A_s/b' t = 0.2, stiffener 1).

It is impossible within the confines of this paper to look in detail at the variation of performance with the various parameters but it can be noted that the variation of strength with transverse stiffener size diminished substantially as the stiffening ratio increased between 0.1 and 0.6 with the latter figure showing no variation in strength between any of the analyses with actual transverse stiffeners and the nodal line result. The general variation can be appreciated from fig 11 which shows the peak stress ratios corresponding to each of the geometries and transverse stiffener sizes for the intermediate panel slenderness of 40. The other points marked on the graphs will be discussed in the next section.

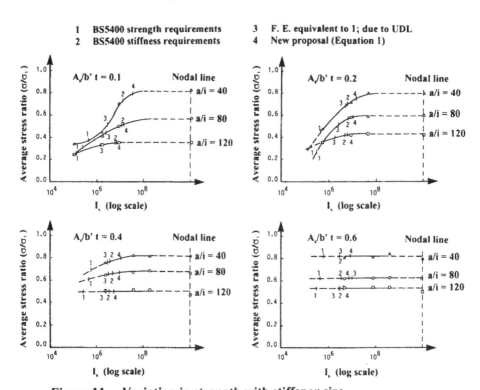

Figure 11   Variation in strength with stiffener size.

DESIGN OF TRANSVERSE STIFFENERS

The nodal line analyses produced force distributions applied to the transverse stiffener due to failure of the longitudinally stiffened panel in the flanges. Following the type of design framework present in the strength requirements of BS5400, the magnitude of these forces was translated to an equivalent uniformly distributed load applied to the transverse stiffener acting as a beam, in order to produce the same maximum bending moment as that produced by the actual reactions. In performing this calculation a simply supported beam section comprising the stiffener and an effective width of plate equal to the lesser of the transverse stiffener spacing or one-quarter of the spacing between the main beam webs, as given in BS5400, was used. The uniformly distributed load varied between 0.7 and 4.2% of the average longitudinal collapse load of the flange per unit width, compared with the 0.5% in the code. Low figures were generally obtained with higher plate panel slendernesses and low column slendernesses confirming it was really the stiffener behaviour acting as a column which produces the high transverse forces. The range of figures encountered agrees generally with the conclusions of Leonhardt and Hommel [2] who indicate that the load on the transverse stiffener can be as high as 2 to 6% of the flange axial force.

Fig 11 shows a comparison between the stiffener inertia required to satisfy the strength and stiffness requirements of the code and the stiffener inertia of the transverse stiffener which just reaches first extreme fibre yield at the collapse load of the flange. These are respectively points 1, 2 and 3 on each of the curves. It can be seen that the code stiffness requirement is always significantly more onerous than the code strength requirement, with the finite element prediction being generally in between the two with the exception of the case with very high stiffening ratio where requirement 3 is the largest for the case of a column slenderness of 80. The other important point is that this finite element requirement does not achieve the nodal line collapse strengths of the flange, with resulting strength in the extreme cases being only about half the actual capacity. Clearly therefore a more onerous inertia requirement is needed. This difference seems to be caused by the fact that the loading on the transverse stiffener must be significantly affected by the deflections of the stiffener itself through an amplification effect. The nodal line forces therefore are generally a lower bound for relatively small stiffeners. An empirical model for the lateral load was therefore produced that gave a stiffener that did not significantly weaken the failure capacity of the flange.

The equation produced gives a lateral load as a function of the average collapse load of the flange, the flange width, the plate slenderness and the stiffener ratio and is given by equation 1 below.

$$w = \frac{0.387\bar{N}_u}{b}\sqrt{\frac{b'/t}{A_s/b't}} \tag{1}$$

where b is the flange width, $\bar{N}_u$ is the flange collapse load, b' is the panel width and A_S is the area of each stiffener.

This is then applied as a UDL to the effective beam section with a simple elastic check on extreme fibre yield. The resulting inertias, without effective width of plate, are also shown in figure 11 as number 4 on each line and it can be seen that these generally compare well with the stiffness requirement from the code. Despite the simplicity of the lateral load equation the overall effect of this requirement in terms of optimising stiffener size versus flange strength is excellent, especially in the cases of panels with low stiffening ratio and low column slenderness.

The stiffness of the resulting transverse stiffeners was also investigated by looking at the deflections at peak load of these stiffeners through reference to the finite element results. By interpolation or extrapolation from the deflections of the transverse stiffener corresponding to the three sizes of stiffener adopted in each finite element analysis, the predicted deflections of the stiffener corresponding to the design proposal could be approximately established. In the same way the deflections corresponding to the strength and stiffness requirements from the code could also be obtained. The values of the central deflections resulting from the design proposal are generally small ranging between 0.5 and 6.8mm which can be compared with a serviceability limit for beam deflection of b/200 multiplied by 1.5 to notionally allow for conversion to ultimate limit state giving a total of 15mm. The predicted deflections are therefore clearly very reasonable.

The results for the two code criteria are in some cases substantially greater, particularly for the stiffness requirement where in some cases deflections significantly exceed this notional serviceability limit. Deflections are actually very sensitive to the inertia of the stiffener.

DISCUSSION AND CONCLUSIONS

The authors believe this to be a unique study of the non-linear behaviour of the orthotropically stiffened flanges, giving for the first time an overall view of the performance of the transverse stiffeners in the non-linear regime up to ultimate collapse. The results clearly show the non-uniform distribution of force applied to the transverse stiffeners due primarily to the action of the longitudinal stiffeners. While the nodal line analyses apparently give significantly lower force magnitudes than are obtained with deflecting stiffeners, the former give a clear picture of the type of force distribution that will occur. The analyses with actual stiffener cross-sections have given sufficient information to enable a design model to be formulated, giving a lateral load in line with the general philosophy of the strength requirement in BS5400 that should be applied to the transverse stiffener to obtain the required stiffener inertia. This lateral load is a simple function of the geometric parameters of the flange as well as the collapse loading of the longitudinally stiffened panel.

Comparison between the resulting inertia requirement and those required by the stiffness and strength criteria in the code, shows that the stiffness criteria is really very close in general terms but still produces strengths in some cases that are slightly below the longitudinally stiffened panel capacity and hence could result in some degradation of strength in the flange. In comparison, the strength requirement of the code produces inertias that are well below those needed to maintain strength, but of course it should not be forgotten that the strength requirement will be affected in a very major way by direct loads acting on the stiffener and therefore the destabilising influence will in many cases be a minor component of the resulting lateral load. However, it is clearly more desirable to have this base level correct so that the increase in load beyond this level truly reflects the response of the system. It should also not be forgotten that in the hogging region of a continuous beam there may be no directly applied additional loading. The authors would therefore recommend the use of a single requirement in future design codes, based around the strength criterion with additional loading being superimposed. This would then give the desired representation for cases where no additional loading is present and an increasing requirement in cases where additional loading is substantial. As the results produced are compatible with the stiffness requirement and produce acceptably low deflections in the transverse stiffener there seems no logical reason to have a separate additional requirement limiting deflection. It is reassuring, however, in the context of the code that the stiffness requirement in itself satisfies the base requirement for stiffener inertia.

REFERENCES

[1]   British Standards Institution, 1982, Code of Practice for Design of Steel Bridges, BS5400, Part 3, London, BSI.

[2]   Leonhardt, F. and Hommel, D., 1973,"The Necessity of Quantifying Imperfections of all Structural Members for Stability of Box Girders", Steel Box Girder Bridges, Proc. of the International Conference organised by the Institution of Civil Engineers in London, London, pp. 11-19.

ACKNOWLEDGEMENTS

The authors gratefully acknowledge the sponsorship of the SERC for the investigations conducted during this study. The work was carried out using the finite element program LUSAS which is a general purpose program commercially available from Finite Element Analysis Ltd in Kingston upon Thames.

ON ULTIMATE STRENGTH OF HORIZONTALLY CURVED BOX GIRDER BRIDGES

Hiroshi NAKAI, Toshiyuki KITADA
Department of Civil Engineering, Osaka City University,
Sugimoto 3-3-138, Sumiyoshi-ku, Osaka, 558 JAPAN

and Yasuo MURAYAMA
Design Part of Steel Bridge, Kurimoto Iron Works Co. Ltd.,
Sibatani 2-8-45, Suminoe-ku, Osaka, 559, JAPAN

ABSTRACT

Presented in this paper is an experimental and analytical study on the ultimate strength of the thin-walled box girder with the longitudinal stiffeners subjected to bending and torsion. An experimental apparatus for applying both the bending and torsional moments to box girder specimens was newly developed. Through this experimental apparatus, the behaviors up to the collapse and the ultimate strength of these thin-walled box girder specimens with the longitudinal stiffeners were investigated. An interaction curves for the combined actions of bending and torsion at the ultimate state of curved box girders were also proposed.

INTRODUCTION

At present, the Japanese Specification for Highway Bridges (thereafter referred to as JSHB) [1] is trying to be revised from the allowable stress design method to the limit state design method. According to the limit state design method, it is an important problem to know the ultimate strength of structures. Then, the ultimate strength of steel members under the combined actions of various stress-resultants are investigated and discussed by many researchers [2]-[3]. Particularly, the ultimate strength of box girder subjected to bending and torsion becomes one of an essential problem in designing the horizontally curved girder bridges [4].

The authors have experimentally studied the ultimate strength of thin-walled box girder without longitudinal stiffeners under the combined actions of bending and torsion and have proposed the interaction curve of them [5]. In addition to these studies, a series of experimental studies were carried out to inquire the characteristics of the ultimate strength of box girder with longitudinal stiffeners for applying to the design of horizontally curved girder bridges as the following manners [6].

Firstly, the experimental apparatus, which enables to apply arbitrary bending and torsional moments to the test specimens, were newly developed. For the cross-section of the test specimens, a trapezoidal box-section having the super-elevation at the compression flange was selected based on the survey of the typical horizontally curved girder bridges in Japan [4].

Secondly, four test specimens were fabricated and two of the specimens were tested under pure bending and pure torsion, respectively, while the remaining two test specimens being tested under combinations of bending and torsion.

Finally, a formula to predict the ultimate strength for bending, torsion and their combinations was proposed by examining the test data, assuming the failure modes of test specimens and comparing with test results.

OUTLINE OF TESTS

In this study, a loading equipment, shown in Fig. 1, was built to apply bending and torsional moments to the test specimens, simultaneously.

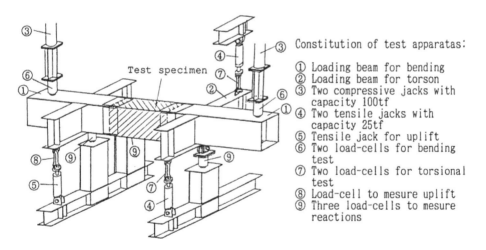

Constitution of test apparatas:

① Loading beam for bending
② Loading beam for torson
③ Two compressive jacks with capacity 100tf
④ Two tensile jacks with capacity 25tf
⑤ Tensile jack for uplift
⑥ Two load-cells for bending test
⑦ Two load-cells for torsional test
⑧ Load-cell to mesure uplift
⑨ Three load-cells to mesure reactions

Figure 1. Test apparatus.

Four test specimens were fabricated by the same steel plate having the identical cross-section in order to compare the ultimate strength under four different loading conditions. In the compression flange plates of test specimens, the super-elevation was provided to realize the normal stress distribution, as illustrated in Figure 2, according to the survey of the horizontally curved girder bridges [4]. The thickness of flange and web plates of test specimens were determined by the provisions of JSHB [1].

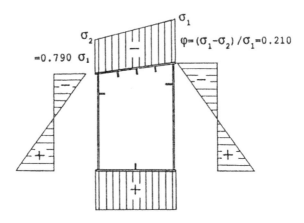

Figure 2. Normal stress distribution in test specimen.

Table 1 lists the name of test specimens corresponding to loading conditions and the calculated ultimate strength without considering the buckling of flange and web plates.

Table 1. Name of specimens, limit stress-resultants and loading conditions

Loading Condigions	Pure bending (M)	Pure torsion (T)	T/M≒0.86	T/M≒0.29
Test specimen	M1	T2	MT3	MT4
Critical stress-resultant: (1tf·m=9.8kN·m)	Yield bending moment		M_y=19.4 tf·m,	
	Fully plastic bending moment		M_p=24.6 tf·m,	
	Fully plastic torsional moment		T_p=15.1 tf·m	

Table 2. Results of ultimate strength for each specimens

Test specimens	M_u(tf·m)	T_u(tf·m)	M_u/M_p	T_u/T_p	ℓ_u
M1	23.53	0	0.958	0	0.958
T2	0	12.03	0	0.796	0.796
MT3	13.60	11.36	0.553	0.752	0.933
MT4	21.06	6.19	0.856	0.409	0.949

$\ell_u=\sqrt{(M_u/M_p)2+(T_u/T_p)^2}$: Distance from origin to interaction curve

(1tf·m=9.8kN·m)

While, the ultimate strength obtained from tests are summarized in Table 2, where the non-dimensionalized ultimate bending and torsional moments, M_u/M_p and T_u/T_p are also indicated in this Table.

From the test result, the following facts can be pointed-out concerning the ultimate strength of stiffened box girders:

i) For the case of pure bending, the influence of normal stress gradient in the compression flange plate can be ignored. However, in evaluating the ultimate strength of overall cross-section of stiffened box girder, the interaction of buckling between flange and web plates must be considered.

ii) For the case of pure torsion, the ultimate strength of stiffened box girder is governed by the shear buckling strength of a single panel of web plates.

iii) For the case of combined actions of bending and torsion, the ultimate strength of stiffened box girders are affected by normal as well as shearing stresses, and their ultimate strength becomes small in comparisons with pure bending or pure torsional ultimate strength. Thus, the interaction curve can be approximated as the elliptic curve.

PREDICTION OF ULTIMATE STRENGTH

ULTIMATE BENDING STRENGTH

To simplify the prediction of the ultimate bending strength of stiffened box girder, an asymmetric cross-section is converted into a symmetrical and rectangular box cross-section as shown in Figure 3.

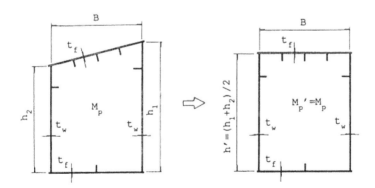

(a) Original cross-section   (b) Converted cross-section

Figure 3. Conversion of box cross-section.

For this cross-section, the ultimate bending strength, $\alpha=M_u/M_p$ can be estimated by using width-thickness parameter, R_b taking into accounts of the interaction of buckling between flange and web plates as follows [7]:

$$\alpha=M_u/M_p=\{1+0.5(R_b-0.5)\}^{-2} \tag{1}$$

, in which R_b is the width-thickness parameter and put as:

$$R_b = \frac{0.8}{(R_f/R_w)^{h/B}} (R_b^e - 0.5) + 0.5 \qquad (2)$$

, where the width-thickness parameter, R_f of compression flange is given as:

$$R_f = \frac{B}{nt_f} \sqrt{\frac{12(1-\mu^2)}{k_f \pi^2}} \sqrt{\frac{\sigma_Y}{E}} \qquad (3)$$

by denoting μ: Poisson's ratio, E: Young's modulus, B: width of flange plate, n: number of panel in stiffened flange plates, t_f: thickness of flange plate, and K_f: buckling coefficient (=4). The relative stiffness of longitudinal stiffener is designed to have larger than that of JSHB [1].

In addition, the width-thickness parameter, R_w of web plate for pure bending is determined by:

$$R_w = \frac{B}{t_w} \sqrt{\frac{12(1-\mu^2)}{k_w \pi^2}} \sqrt{\frac{\sigma_Y}{E}} \qquad (4)$$

, where h: depth of web plate, t_w: thickness of web plate, and k_w: buckling coefficient of web plate. For example, this value is taken as $k_w=23.9$ for unstiffened web plate, and $k_w=110.8$ for stiffened web plate with a longitudinal stiffener located at 0.2h.

Therefore, the width-thickness parameter, R_b^e of overall box girder can be set as:

$$R_b^e = \frac{B}{nt_f} \sqrt{\frac{12(1-\mu^2)}{k_b \pi^2}} \sqrt{\frac{\sigma_Y}{E}} \qquad (5)$$

, in which the coefficient, k_b is defined as the buckling coefficient taking into considerations of the interaction between flange and web plates, and it is clarified that the following approximate formula is available:

$$k_b = 0.036 \times \{(\frac{t_f}{t_w}) - 3.5\}^4 + 4.0 \qquad (6)$$

within the ranges where

$$k_b \geq 4, \quad 0.5 < t_f/t_w < 3.5 \qquad (7)$$

according to the parametric survey of the actual curved bridges as shown in Figure 4.

The numerical results of a test specimen are plotted in Figure 5 together with those of the unstiffened box girders in Ref.[5].

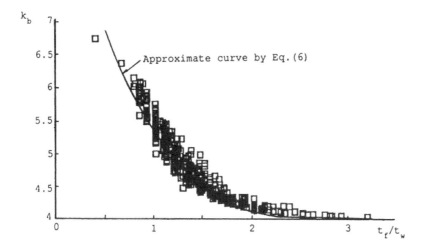

Figure 4. Relationship between k_b and t_f/t_w in actual curved bridges.

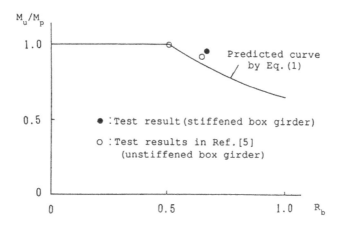

Figure 5. Ultimate bending moment.

ULTIMATE TORSIONAL STRENGTH

The ultimate torsional strength, T_u/T_p of stiffened box girder is decided from the shear strength of a panel having the most large width-thickness ratio, in shown in Figure 6. The ultimate torsional strength T_u/T_p is determined from an approximate formula as [6]:

$$\beta = T_u/T_p = \tau_u/\tau_Y \qquad (8)$$

where τ_u is the shear strength of the panel, and this strength is consisted of shear buckling strength, τ_{cr} and post buckling

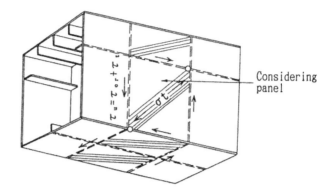

Figure 6. Limit state for torsion.

strength, τ_t according to the theory of diagonal tension field, i.e.:

$$\tau_u = \tau_{cr} + \tau_t \tag{9}$$

In this equation, the shear buckling strength, τ_{cr} is given by:

$$\tau_{cr} \quad = \tau_Y \qquad\qquad\qquad\qquad\quad : R_\tau \leq 0.6$$
$$= \{1-0.614(R_\tau-0.6)\}\tau_Y \quad : 0.6 < R_\tau \leq 2 \tag{10}$$
$$= (1/R_\tau^2)\tau_Y \qquad\qquad\qquad : 2 < R_\tau$$

according to the proposal of FHWA [8], where R_τ is the width-thickness parameter of a single panel, and put as [9]

$$R_\tau = \frac{b}{t_w} \sqrt{\frac{12(1-\mu^2)}{k_\tau \pi^2}} \sqrt{\frac{\tau_Y}{E}} \tag{11}$$

, and k_τ is the buckling coefficient for pure shear given by [9]:

$$k_\tau = 5.34 + 4.00/(a/b)^2 \quad : a/b \geq 1$$
$$= 4.00 + 5.34/(a/b)^2 \quad : a/b < 1 \tag{12}$$

, where   b: width of a single panel, a: spacing of the transverse stiffener.

The post buckling strength, τ_t is estimated by assuming the plastic hinges are occurred in the vicinity of the corner part of panel, which leads to:

$$\tau_t = \sigma_t\{\sin\theta\cdot\cos\theta - (a/b)\sin^2\theta\} \tag{13}$$

, where

$$\sigma_t = \{1 - (\tau_{cr}/\tau_Y)^{1.2}\}\sigma_Y \qquad (14)$$

: tensile strength in diagonal tension field

$$\theta = (1/2)\tan^{-1}(b/a) \qquad (15)$$

: angle of diagonal tension field

In Figure 7, the numerical results of the test specimen by using Eqs.(8) through (15) together with those of the unstiffened box girder as reported in Ref.[5].

From this figure, both the test results of stiffened and unstiffened box girders seem to coincide with the calculated results.

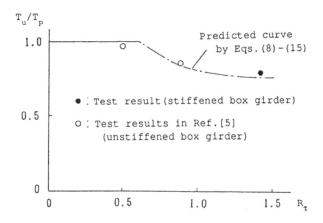

Figure 7. Ultimate torsional moment.

INTERACTION CURVE OF ULTIMATE STRENGTH

As mentioned above, the ultimate strength of stiffened box girder, in the case where bending and torsional moments coexist, is represented by the following elliptic interaction curve as proposed in Ref.[5]:

$$(\frac{M_u}{\alpha M_p})^2 + (\frac{T_u}{\beta T_p})^2 = 1 \qquad (16)$$

, where α and β are, respectively, defined as the parameter corresponding to ultimate bending and torsional moments, i.e., parameter α can be determined by Eqs.(1) through (6), while parameter β can be decided by Eqs.(8) through (15). For instance, the parameters α and β result in:

$$\left.\begin{array}{l} \alpha = 0.850 \\ \beta = 0.757 \end{array}\right\} \qquad (17)$$

for the test specimens.

Figure 8 illustrates the interaction curve of Eqs.(16) and (17) with the non-dimensionalized by ordinate as M/M_p and abscissa as T/T_p together with the test results. Although there find a few errors within 10%-15% between tested and calculated results, the proposed interaction curve of in Eq.(16) has the tendency to coincide with the test results, and gives the safety prediction of ultimate strength.

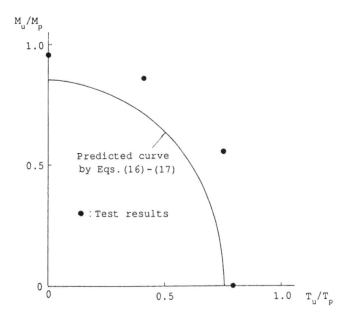

Figure 8. Interaction curve for combined actions of bending and torsion.

CONCLUSION

In this paper, the characteristics of ultimate strength of thin-walled box girder subjected to bending, torsion and combined actions of them are clarified through the experimental study. By referring the test results, an approximate formula for predicting the ultimate bending and torsional moments and their interaction curve for the use of horizontally curved box girder bridges are proposed. Further, the validity of the above equations shall be checked by FEM analysis based on finite displacement and elasto-plastic theory.

REFERENCES

1. Japanese Road Association, Japanese Specification for Highway Bridges, Part II, Steel Bridges, Maruzen, Tokyo, Feb. 1990.
2. Yoo, C.H. and Heins, C.P., Plastic Collapse of Horizontally Curved Bridge Girders, Proc. of ASCE, Vol.98, No-ST4,

April 1972, pp.899-914.

3. Ohta, T. and Hino, S., Elasto-Plastic Analysis of Uniform Rectangular Section Bar Subjected to Bending and Torsional Moments, Proc. of JSCE, No.285, May 1979, pp.37-44.

4. Nakai, H. and Yoo, C.H., Analysis and Design of Curved Steel Bridges, McGraw-Hill, New York, 1988.

5. Nakai,H., Murayama,Y., Kitada,T., and Takada,Y., An Experimental Study on Ultimate Strength of Thin-Walled Box Beam subjected to Bending and Torsion, Journal of Structural Engineering, JSCE, Vol.36A, March 1990, pp.63-70.

6. Nakai, H., Murayama, Y. and Kitada, T., An Experimental Study on Ultimate Strength of Thin-Walled Box Beams with Longitudinal Stiffener subjected to Bending and Torsion, Journal of Structural Engineering, JSCE, Vol.38A, March 1992, pp.155-165

7. Nishimura,N.,Yoshida,N. and Takeuchi,S., Strength Formula for Steel Beam-Columns with Box Section Considering Local Buckling, Proc. of JSCE, No.416/I-13, April 1990, pp.385-393.

8. Wolchuk, R. and Mayrbaurl, R.M., Proposed Design Specification for Steel Box Girder Bridges, Report No.FHWA-TS-80-205, U.S. Dept. of Transporta tion, Federal Highway Administration, Office of Research and Development, Washington, D.C., Jan. 1980.

9. Subcommittee on Stability Design Committee on Steel Structures of JSCE, Edited by Fukumoto,Y., Guidelines for Stability Design of Steel Structures, JSCE Steel Structural Series 2, Oct. 1987.

ULTIMATE LOAD BEHAVIOUR OF STEEL PLATE GIRDERS SUBJECT TO (i) CONSTANT AND (ii) REPEATED PATCH LOADING

KAREL JANUŠ, IRENA KUTMANOVÁ
formerly Klokner Institute, Czech Technical University
Šolínova 7, 166 08 Praha 6, Czechoslovakia

and

MIROSLAV ŠKALOUD
Institute of Theoretical and Applied Mechanics,
Czechoslovak Academy of Sciences
Vyšehradská 49, 128 49 Praha 2, Czechoslovakia

ABSTRACT

The objective of the paper is to summarize the main results of the recent stages of the Prague extensive research on the ultimate limit state of steel webs under the action of various kinds of partial edge loading. The research programme comprised tests on (i) 50 plate girders without longitudinal stiffeners subject to constant patch loading, (ii) 184 plate girders with longitudinal stiffeners subject to constant patch loading, and (iii) 80 plate girders, with and without longitudinal ribs, subject to repeated patch loading, the size of the loaded flange, the depth-to-thickness and aspect ratios of the web, and - in the case of longitudinally stiffened webs - the position and size of the longitudinal stiffener being varied. During the experiments, the impact of the partial edge load on the deformation and stress states of the test girders and on their failure mechanisms was studied. In the case of girders under repeated loading, also the impact on the initiation and propagation of cracks, and thus on the "erosion" of the constant loading collapse mechanism as a result of web "breathing", was analysed.

INTRODUCTORY REMARKS

One of the buckling problems that attracted plenty of attention over the last years was that of the ultimate load behaviour of thin steel webs subject to partial edge loading. Some time ago, the third author, jointly with P.Novák, carried out an extensive investigation into the problem. These studies, representing the first stage of our research on the performance of plate girders under the action of a patch load, were concerned mainly with the ultimate limit state of webs without longitudinal stiffeners. The limited scope of this contribution does not allow us to present here the results of this first (and oldest) stage (comprising 50 tests) of the Prague research on the ultimate load behaviour of steel plate girders subject to patch loading; therefore, in this respect the reader is referred to [1].

However, the webs of deep plate girders are often stiffened by longitudinal ribs. At least, this is frequent practice with webs loaded by combined shear and bending, and sometimes the use of longitudinal stiffeners is necessary also for other reasons, for instance, to reduce web deflection or even initial "dishing". Then, of course, the question arises whether the presence of longitudinal stiffening can also favourably influence the behaviour of a web if this is loaded by a patch load, or by such a combination of loading in which the effect of patch loading predominates.

Little information was available in regard to this problem; therefore, the authors decided to contribute to its solution.

CONSTANT PATCH LOADING

Test Girders

The first part of the authors' experimental investigation, focused on constat loading, comprised 184 tests, sub-divided into six stages, the following quantities being varied in them:
- the distance of the longitudinal stiffener from the loaded flange,
- the size of the longitudinal stiffener,
- the character (single or double-sided) of the longitudinal stiffener,
- the depth to thickness ratio of the web,
- the aspect ratio of the web,
- the size of the loaded flange and,
- for some test girders, even the load length, c (but in most cases and the load length was constant c=a/10).

The general details of the test girders can be seen in Fig. 1, their main characteristics in Table 1.

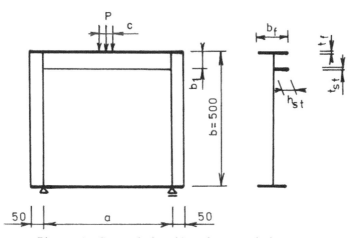

Figure 1. General details of test girders

TABLE 1
General details of test girders (constant loading test)

Test series	Number of test	Web thickness (mm)	Constant quantities	Variable quantities
1	26	2	Load length c=a/10 Flange size Single-sided longi- tudinal stiffener	Aspect ratio α of the web: $\alpha=a/b$ ($\alpha=1$, $\alpha=2$) Distance b_1 of the longi- tudinal stiffener from the loaded flange ($b_1 = 0.1 - 0.5b$) Longitudinal stiffener size
	24	6	$c = a/10$ $\alpha = 1.245$ Flange size Double-sided longi- tudinal stiffener	Distance b_1 of the longi- tudinal stiffener from the loaded flange ($b_1 = 0.15 - 0.4b$) Longitudinal stiffener size
2	42	2,4,6	$c = a/10$ $\alpha = 1$ $b_1 = 0.2b$ Single-sided longi- tudinal stiffener	Longitudinal stiffener size Flange size
3	36	2,4,6	$c = a/10$ $\alpha = 1$ $b_1 = 0.2b$ Single-sided longi- tudinal stiffener	Longitudinal stiffener size Flange size
4	24	4	$c = a/10$ $\alpha = 1$ $b_1 = 0.1b$ Single-sided longi- tudinal stiffener	Longitudinal stiffener size Flange size
5	16	4	$\alpha = 1$ No longitudinal stiffener	$c = 0.1a, 0.2a, 0.3a$ Flange size
6	16	4	$c = a/10$ $\alpha = 1$	$b_1 = 0 - 0.2b$ Longitudinal stiffener size Flange size

Test Set Up and Experimental Apparatus

The measurements were concentrated mainly on
(i)   strains at a number of characteristic places of the web and
     longitudinal stiffener,

(ii) web buckling, and
(iii) stiffener deflection.

Strains were measured by means of electric resistance strain gauges C 120 positioned at various places of interest for our study of the progression of plasticization in the test girder, and mounted on both surfaces of the web sheet and of the flat stiffening element. The measurements were facilitated by using a measurement unit Peekel controlled by a computer PDP 11/40.

The buckled pattern of the web was measured by means of a special piece of apparatus, which transforms the measured quantity, i.e. deflection, into an electric signal and then plots it via a plotter. Also part of web initial curvatures were detected with the aid of this device.

The other initial curvatures and all post failure plastic residues in the webs were measured using the stereophotogrammetric method, which already proved to be of advantage in previous experiments conducted by the authors.

Ultimate Loads of Test Girders

The relationship between (i) the average values of the experimental ultimate loads, P_{ult}^{exp}, of the test girders and (ii) the distance, b_1, of the stiffener from the loaded flange is shown in Fig.2, P_{ult} denoting the ultimate loads of the related longitudinally unstiffened girders.

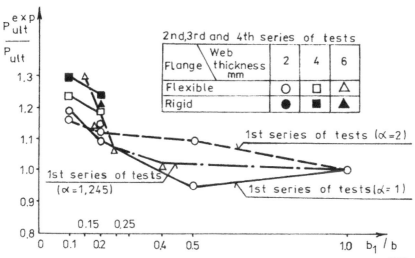

Figure 2. Relationship between the experimental ultimate loads, P_{ult}^{exp}, and the distance, b_1, of the longitudinal stiffener from the loaded flange

Failure Mechanisms of Test Girders

All test girders collapsed by the formation of a failure mechanism which consisted of (i) a segmental line plastic hinge in the web sheet and (ii) three point plastic hinges in the loaded flange (as the patch load was in practically all cases very narrow, only one hinge developed in the middle of the flange under this load).

The segmental line plastic hinge in the web formed in the vicinity of the loaded flange, the size of this flange considerably influencing the configuration of the line hinge. The larger the size of the flange, the longer and deeper the segment. The effect of the longitudinal rib was then reflected in shortening the length and depth of the arc of the line hinge, if compared with the situation in a web having the same dimensions but not being reinforced by a longitudinal rib.

The plastic hinges in the loaded flange occurred (i) at mid-span, i.e. under the partial edge load, and (ii) at those flange sections from which the segmental line plastic hinge in the web sheet emanated. The distance of the outer hinges from the central one was influenced by the flange size; i.e. the larger the size of the flange, the greater was the distance of the outer hinges.

Effect of the Size of the Loaded Flange

An examination of the obtained results indicates that the effect of the size of the loaded flange upon the load-bearing capacities of the webs (and of the whole plate girders) is very significant. It considerably influences the dispersion of the load into the web, its buckling and even the collapse mechanism of the plate girder.

Effect of the Position and Size of the Longitudinal Stiffener

An analysis of the results reveals that a longitudinal rib can substantially improve the behaviour of a plate girder web subject to partial edge loading only when the rib is located in the vicinity of the loaded flange, i.e. if $b_1 < b/4$. The conclusion follows from the observation that, in the case of a web under the action of a discrete edge load, the buckling is pronouced merely in the region immediately adjacent to the applied patch load. If the longitudinal stiffener is positioned off this region, it is not able to influence the web buckling and, consequently, the ultimate load behaviour of the girder.

As for the effect of the size of the rib, this was pronounced only in the range of flexible ribs, where the curve plotting the relationship between ultimate load P_{ult}^{exp} and stiffener rigidity γ exhibited a conspicuous rising tendency. In the domain of larger stiffeners, the curve was flat, so that the growth of ultimate strength with a further increase in stiffener size was very slow. The average increase in the ultimate loads of the girders tested with a longitudinal stiffener was of 24.2% for $b_1 = 0.1b$ and 15.7% for $b_1 = 0.2b$, if compared with the ultimate loads of girders without longitudinal stiffeners.

Effect of the Aspect Ratio of the Web and of the Character of the Longitudinal Stiffener

Neither the aspect ratio of the web, nor the character (single or double-sided) of the stiffener influenced in a significant way the ultimate load behaviour of the webs of the test girders.

Effect of the Longitudinal Stiffener on Web Deflection

One of the objectives of the tests was to study how the presence of a longitudinal stiffener can help to reduce the deflection of webs under partial edge loading if compared with the buckled patterns of unstiffened webs. An analysis of the obtained data indicates that, in the case of the writers' experiments, this reduction was pronounced and amounted on an average to 42%.

REPEATED PATCH LOADING

Test Girders

The second part of the authors' experimental research was focused on the behaviour of webs under the action of repeated patch loading, with the view to look into the effect of their "breathing". The test girders had the same dimensions, and were fabricated from the same material, as those used in the constant patch loading tests (see above), this being so as to enable the authors to compare the results of both respective experimental series.

80 test girders subject to repeated patch loading were tested, their webs being fitted with a longitudinal stiffener positioned at (i) one tenth, (ii) one fifth of the web depth or (iii) not being stiffened longitudinally at all (Table 2).

TABLE 2a

Geometrical and material characteristics of the test girders
of the first stage of "breathing" experiments

Web		Flange		Longitudinal stiffener			Notation of test girder
t_w (mm)	$R_{y,w}$ (MPa)	b_f/t_f (mm/mm)	$R_{y,f}$ (MPa)	b_1/b (-)	h_{st}/t_{st} (mm/mm)	$R_{y,st}$ (MPa)	
					16/5	-	TG 341-1, TG 341-1'
					20/8	-	TG 341-2, TG 341-2'
					25/8	-	TG 341-3, TG 341-3'
		120/8	277.5	0.1	30/8	290.3	TG 341-4, TG 341-4'
					35/8	271.6	TG 341-5, TG 341-5'
					40/8	275.3	TG 341-6
				0			TG 341-6'
4	303.9				16/5	-	TG 342-1, TG 342-1'
					20/8	-	TG 342-2, TG 342-2'
					25/8	-	TG 342-3, TG 342-3'
		120/20	244.4	0.1	30/8	290.3	TG 342-4, TG 342-4'
					35/8	271.6	TG 342-5,
					40/8	275.3	TG 341-6
				0			TG 342-5', TG 341-6'

TABLE 2b

Geometrical and material characteristics of the test girders
of the second stage of "breathing" experiments

Web		Flange		Longitudinal stiffener			Notation of test girder
t_w (mm)	$R_{y,w}$ (MPa)	b_f/t_f (mm/mm)	$R_{y,f}$ (MPa)	b_1/b (-)	h_{st}/t_{st} (mm/mm)	$R_{y,st}$ (MPa)	
4	303.9	120/8	277.5	0			TG 441-0/1, TG 441-0/1' TG 441-0/2, TG 441-0/2'
				0.2	32/8	268.4	TG 441-1/1, TG 441-1/1' TG 441-1/2, TG 441-1/2'
					45/8	268.4	TG 441-2/1, TG 441-2/1' TG 441-2/2, TG 441-2/2'
		120/20	244.4	0			TG 442-0/1, TG 442-0/1' TG 442-0/2, TG 442-0/2'
				0.2	32/8	268.4	TG 442-1/1, TG 442-1/1' TG 442-1/2, TG 442-1/2'
					45/8	268.4	TG 442-2/1, TG 442-2/1' TG 442-2/2, TG 442-2/2'

Test Set Up and Experimental Apparatus

The repeated loading was materialized by means of a 1000 kN AMSLER pulsator, the frequency of loading cycles being of 3.75 Hz.
    The following quantities were registered during the tests:
(i)   The initial imperfections of the experimental girders,
(ii)   the values of deflections and strains at a number of selected places,
(iii) the initiation and propagation of cracks,
(iv)   the acoustic emission in the web in the neighbourhoud of the applied load, and
(v)   the temperature changes in the same portion of the web.
    The load P cycled between (i) zero and (ii) a value P_{max}, which in turn varied between (α) the constant loading ultimate load P_{ult}^{exp} and (β) the onset-of-surface-yielding load, detected also in the related constant loading test. Thus, under the above loading, the webs of the girders tested operated in the elasto-plastic range and, consequently, their performance was expected to be governed by low-cycle fatigue. Therefore, the basic number of loading cycles was chosen so as to be equal to 5×10^4.
    In the case of girders where after 5×10^4 cycles no failure (whether through initiation of cracks or through excessive plastic buckling of the girder web) occurred, the experiment was continued under a higher load level. If, however, a crack appeared in a certain loading cycle, the experiment went on, under the same load level, as long as the load-carrying capacity of the test girder was not exhausted.

TABLE 2c
Geometrical and material characteristics of the test girders
of the third stage of "breathing" experiments

Web		Flange		Longitudinal stiffener		b_1/b (-)	Notation of test girder
t_w (mm)	$R_{y,w}$ (MPa)	b_f/t_f (mm/mm)	$R_{y,f}$ (MPa)	h_{st}/t_{st} (mm/mm)	$R_{y,st}$ (MPa)		
4	328.2	120/8	277.5	35/8	268.4	0	TG 541-0/1, TG 541-0/1' TG 541-0/2, TG 541-0/2'
						0.1	TG 541-1/1, TG 541-1/1' TG 541-1/2, TG 541-1/2' TG 541-1/3, TG 541-1/3'
						0.2	TG 541-2/1, TG 541-2/1' TG 541-2/2, TG 541-2/2' TG 541-2/3, TG 541-2/3'
		120/20	244.4	35/8	268.4	0	TG 542-0/1, TG 542-0/1' TG 542-0/2, TG 542-0/2'
						0.1	TG 542-1/1, TG 542-1/1' TG 542-1/2, TG 542-1/2' TG 542-1/3, TG 542-1/3'
						0.2	TG 542-2/1, TG 542-2/1' TG 542-2/2, TG 542-2/2' TG 542-2/3, TG 542-2/3'

Failure Mechanisms of the Girders

The failure mechanism of each test girder was very attentively studied during the experiments. In doing so, the authors concluded that this mechanism was in principle very similar to that they had previously observed in their constant loading tests (see above); viz. it consisted of set of three plastic hinges in the loaded flange and a segmental line plastic hinge in the adjacent zone of the web sheet.

However, in addition to that, in the case of most cyclic loading tests, the initialion of a crack in the zone immediately adjacent to the applied load was detected. This initiation and propagation of cracks was very carefully studied. In all experiments, this was done visually with the aid of a magnifying glass, but in the third series of tests, in the case of 12 test girders, the acoustic signal emission method was employed (via a dosimetr SEDO 2.1) to verify visual observations. It was concluded that, in comparison with the visual observations, the acoustic emission approach gave more accurate results. For example, this method was able to detect the initiation of cracks by 4500-9000 cycles (i.e. by 20-40 minutes) earlier. This means that, within the regime of 5×10^4 repeated cycles, the difference was of 9-18%.

Two general conclusions may be drawn from the evidence obtained: (i) the propagation of cracks was not uniform and frequently proceeded with more or less long interruptions, (ii) the appearance of a crack by far did not herald the end of the useful life of the girder concerned. The girder was thereafter able to sustain a good many more loading cycles before its load-bearing capacity was exhausted.

The Magnitude of Load versus the Number of Loading Cycles

On the basis of the tests conducted, low-cycle fatigue curves were plotted, and a limit load, P_{fat}, was determined so that it was the highest load under which no failure of the girder yet occurred in the course of the 5×10^4 loading cycles applied.

The main data obtained are listed in Fig. 3, where - in dependance on the placing of the longitudinal rib - the following quantities are given for all girders tested:
a) The onset-of-surface-yielding loads, $P_{y,w}^{su}$ (see the black portions of the bars), and
b) the low-cycle fatigue limit loads, P_{fat}, determined as the maximum load values under which no appearance of cracks (or of other kind of failure) was detected (black and white portions of the bars) after 5×10^4 load cycles.

Both aforesaid quantities are related to P_{ult}^{exp}, i.e. to the ultimate loads obtained in the foregoing constant loading tests.

An examination of the results obtained shows that (i) the "breathing" of the web led to a 10 - 35% reduction of the ultimate load of the girder if compared to the associated constant loading experiment, (ii) even in the case of webs "breathing" under the action of a repeated partial edge load a certain (and quite significant) degree of plastification in the webs did not jeopardize the safety of the girders.

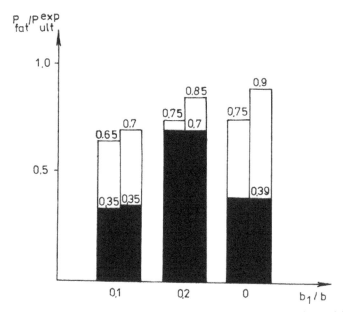

Figure 3. Summary of the test results of the authors' "breathing" experiments

REFERENCES

1. Škaloud, M. and Novák, P., Post-buckled behaviour of webs under partial edge loading. Transaction of the Czech. Acad. of Sci., Series of Techn. Sci., Academia, Prague, 1975, 3, pp. 1-94.

2. Škaloud, M. and Kárníková, I., Experimental research on the limit state of the plate elements of steel bridges. Transactions of the Czech. Acad. of Sci., Series of Techn. Sci., Academia, Prague, 1985, 1, pp. 1-141.

3. Januš, K., Kárníková, I. and Škaloud, M., Experimental investigation into the ultimate load behaviour of longitudinally stiffened steel webs under partial edge loading. Acta technica ČSAV, 1988, 2, pp. 158-195.

4. Januš, K., Kutmanová, I. and Škaloud, M., "Breathing" of slender steel webs. Acta technica ČSAV, 1991, 5, pp. 637-649.

DESIGN OF TRANSVERSE WEB STIFFENERS IN
PLATE AND BOX GIRDER BRIDGES

K RAHAL
Brown & Root Civil
Dorking, Surrey RH4 2J2

J E HARDING
Professor of Structural Engineering
University of Surrey, Guildford, Surrey, GU2 5XH

ABSTRACT

The behaviour of and design requirements for the transverse stiffeners of transversely stiffened web plates in plate and box girders has been studied using a fully non-linear finite element package. Using nodal line representation of the transverse stiffeners, the form and magnitude of the forces applied to the stiffeners during the collapse of the web plates were investigated. Analyses were also carried out with discrete stiffeners so that the variation in strength with stiffener size could be determined and the effect of the deflection of the stiffener assessed. A limited study was also carried out with a complete girder arrangement including flanges, so that the contribution of the transverse rigidity of the flanges to girder shear strength could be established. Analyses were undertaken for webs in shear, bending and direct compression.

Using the results of the finite element study for validation purposes, a simple beam model was formulated with the stiffener and associated width of web plate acting as a beam spanning between the girder flanges, subjected to a lateral load resulting from the destabilising action of the buckling web plates. This simple design model can be used to optimise the stiffener size and is compatible with the basic approach adopted in the British Code BS5400, Part 3, which can therefore be modified to allow for the findings of this study.

INTRODUCTION

It has been known for some time that the approach in BS5400, Part 3 [1] for the design of transverse web stiffeners leads to very large stiffener cross-sections for slender webs because the panel destabilising action produces very high loading. In this code formulation the destabilising forces from web buckling which basically apply transverse loading to the stiffening elements are converted, via a critical buckling equivalence, to an axial column load. The way in which this method produces a P-Δ magnifier on the deflection of the transverse stiffener is doubtful and it is this action which can lead to excessively large stiffeners for slender web geometries.

In the early 1980's work by Rockey et al [2] and Thorne and Grayson [3] demonstrated that transverse stiffeners were being over-designed by possibly substantial factors but these studies stopped short of providing simple well founded design formulations. In order to provide such a basis a wide ranging parametric study of the behaviour of web panels with transverse stiffeners subjected to shear loading, in-plane bending and axial compression was carried out using a non-linear finite element package LUSAS, the results of which were validated against previous work by the second author and that of Grayson [4]. These correlations were considered in reference 5.

ANALYTICAL STUDY

A rectangular 8 noded isoparametric semi-loof shell element was used to model both the plate and stiffener elements and the flanges, when present, in the study. The majority of cases used a single transverse stiffener and two adjacent plates with the surrounding boundary considered to be simply supported as shown in fig. 1. This model was deemed to be appropriate if stiffener deflections were modest, so that there was no significant carry over between the individual stiffener and the location of adjacent stiffeners, modelled in this case by simply supported boundaries. The model was sufficient to enable the optimum rigidity of the stiffener to be identified but in cases where the stiffener was very small the interaction between adjacent stiffeners could not be regarded as being exactly reproduced. Strengths predicted therefore for the zero stiffener case should not be used as accurately representing an unstiffened web configuration.

A flat stiffener was used for all cases with an outstand slenderness d_s/t_s equal to 10 to avoid any local stiffener buckling complications. In some cases the stiffener was replaced by a nodal line so that the transverse forces exerted by the panel at this location could be identified, although it must be acknowledged that this then ignores any deflections at this location and their effect on the force configuration. A number of models were also analysed with complete discrete flanges to look at the effect of the flange rigidity on behaviour.

Steel with a yield stress of 275N/mm^2 and an E value of 205000N/mm^2 was considered for most of the study but the effect of varying yield stress was investigated. Imperfections were adopted for both the panels and for the stiffener in line with the specifications of BS5400. Panel imperfections of G/165 $\sqrt{\sigma_y/355}$ were added to an overall half sine curve giving a stiffener imperfection of G/750. In all the cases considered G was the panel width. Different panel modes were incorporated in order to identify the wave-form leading to the maximum lateral load on the stiffener. The main parameters varied in the study were the panel slenderness b/t, the aspect ratio $\phi = a/b$ and the size of the transverse stiffener. Shear loading was applied as a tangential displacement to the vertical boundaries. Bending was applied as a varying linear displacement and compression as a uniform displacement. As well as flanges of varying sizes, two extreme simple boundary conditions were applied to the more approximate model, corresponding to the restrained and unrestrained situations found in the British Standard. The former represents an infinitely rigid flange situation while the latter represents no edge restraint to in-plane transverse forces. The restrained condition was found to impose substantial constraint on the stiffener behaviour and gave results totally unlike those obtained from the unrestrained condition and the conditions with actual flange modelling, and was therefore generally discounted.

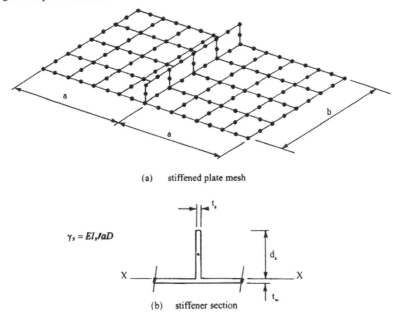

(a)   stiffened plate mesh

$\gamma_s = EI_s/aD$

(b)   stiffener section

Figure 1   Stiffened plate geometry.

FINITE ELEMENT BEHAVIOUR UNDER SHEAR LOADING

Fig 2 shows the stress-strain response of a relatively slender panel $\lambda = b/t \sqrt{\sigma_{yw}/355} = 180$ under shear loading with varying flange sizes and a stiffener which corresponds to the design size established later in this paper. The flange rigidity is expressed in terms of m_{fw} which is a relative measure of flange to web bending rigidity and is as defined in the British Code, $m_{fw} = 0.25 \, \sigma_{yf}b_f t_f^2/(b^2 t_w \sigma_{yw})$ where b_f and t_f are the flange width and thickness and the other dimensions are as shown in fig. 1. It is clear that increasing flange rigidity reduces the deflection of the stiffener for a given level of web shear and also causes a modest but not substantial increase in shear strength. It should be noted that the shear strength shown in this figure is that carried by the web panel and does not include the component carried by the flanges. It can be concluded that the additional deflections resulting from the axial forces on the stiffener due to tension field action are more than countered by reduction in transverse bending of the stiffener because of the flange support, remembering also that the flange in practice will normally have a significant level of direct stress reducing the effective restraint. It was therefore concluded that the use of the unrestrained lower bound would be quite adequate in terms of web modelling so that the optimum size of the stiffener could be determined.

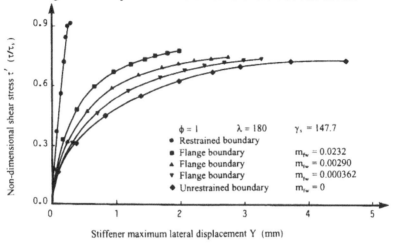

Figure 2   Effect of flange rigidity on stiffener lateral deformation.

One interesting feature of the results is that there is negligible evidence of significant tension field action producing any substantial net column force in the stiffener, although this conclusion may not apply if the sub-panel is significantly more slender and the decision to proceed with unrestrained boundary conditions might not be viable for very slender web panels. However, these would be outside the range of normal design.

Fig 3 shows the panel deflections at plate ultimate shear capacity for a girder of medium web slenderness with an imperfection mode which has two half waves both in the vertical and horizontal direction producing a line of zero deflection from the panel mode along the line of the transverse stiffener prior to loading. There was still imperfection in the stiffener, however, because of the single superimposed overall half wavelength mode. Because the deflection form mirrors the initial shape, the imperfection mode in this case was critical for the behaviour of the stiffener for this aspect ratio.

Figure 4 shows the variation of stiffened plate ultimate shear capacity with stiffener size and differing aspect ratio for a λ value of 180. It can be noted that the critical imperfection mode for $\phi = 0.5$ is different to that for higher values. The stiffener size parameter γ_s is a ratio of the stiffener inertia and that of the plate, $\gamma_s = EI_s/aD$. The figure demonstrates clearly the effect of reducing the stiffener size and it is not difficult to envisage a design requirement which specifies a minimum stiffener rigidity below which overall panel strength is significantly affected. A corresponding set of graphs for the case of a stocky panel, $\lambda = 60$, is shown in fig. 5, and in this case it is clear that the effect of stiffener size is far less important because the panel under shear essentially fails by in-plane action subjecting the stiffener to very small forces. However, there is still an indication of strength loss as the stiffener size becomes very small.

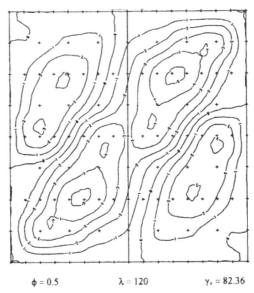

$\phi = 0.5$ $\lambda = 120$ $\gamma_s = 82.36$

Figure 3   **Lateral displacement at the plate ultimate shear capacity.**

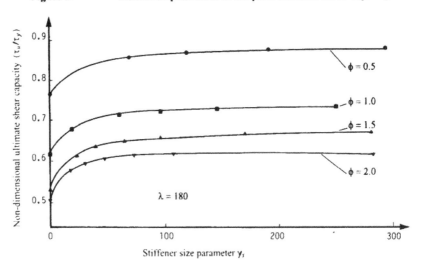

Figure 4   **Variation of stiffened plate ultimate shear capacity with stiffener size parameter γ_s.**

A study of the effect of variation in yield stress indicated very clearly that the true non-dimensionalising parameter in terms of evaluating the transverse stiffener force, and hence its behaviour was the yield stress to the power of 3/2. This is different to the non-dimensionalising parameter for web panel behaviour given in the code as the square root of the yield stress, and this different parameter has been incorporated in the final design model for determining the stiffener lateral loading.

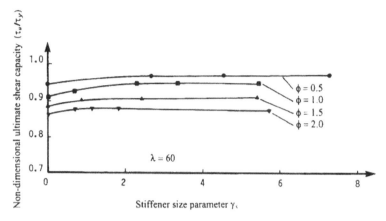

Figure 5   **Variation of stiffened plate ultimate shear capacity with stiffener size parameter γ_1.**

FINITE ELEMENT BEHAVIOUR UNDER COMBINED SHEAR AND IN-PLANE COMPRESSION

Because levels of compression are generally modest in transversely stiffened girder webs a limited study was carried out restricted to plates loaded by shear with low levels of combined in-plane compressive stress. As mentioned previously, loading was applied by uniform displacement. A notional compressive stress limit of around 20% of the yield stress at collapse was applied. The variation in response with the mode of the panel imperfection was again studied and critical mode forms adopted which produced loading in the stiffener in a direction compatible with the cases for shear alone. It can be noted that the presence of the compression has a significant effect on the form of the buckles for high aspect ratios due to the destabilising influence of the compressive stress. Because of shortage of space the detailed results of this stress combination will not be considered in this paper.

FINITE ELEMENT BEHAVIOUR UNDER IN-PLANE BENDING

Again, imperfection studies were carried out in order to identify the mode which produced the highest stiffener stresses but at the same time was compatible with the worst shear situation, namely the maximum tensile stresses in the stiffener outstand. Fig 6 shows the plate deformations at peak bending capacity for a bending only loading with the critical imperfection mode corresponding to single half wave panel forms with the same sign of imperfection on opposite sides of the transverse stiffener. This symmetry of form can be seen clearly in the resulting deflection profile, but it is also apparent that a significant buckle has gone through the actual stiffener location because the size of the stiffener in this example is relatively small. The studies of bending combined with shear confirmed that the non-dimensionalising yield stress parameter held good throughout the full interaction behaviour and also showed that the collapse stress of the beam in bending was very much less dependant on the size of the transverse stiffener. Because of this it was very much more difficult to identify a particular stiffener size appropriate for design and indeed, using the design philosophy established later, actually led to some potentially conservative stiffeners but the view was taken that as the majority of bridge web panels are loaded by coexistent shear and bending stresses and the critical design case occurs when the stiffener outstand stresses are in sympathy under this combination, it was reasonable to apply the same design approach to the combined situation.

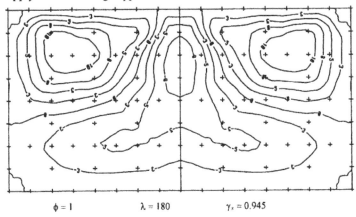

$$\phi = 1 \qquad \lambda = 180 \qquad \gamma_s = 0.945$$

Figure 6   Lateral displacement at peak bending capacity.

STIFFENER DESIGN MODEL

The background to the British Standard clauses for transverse stiffener design came from an original study carried out by Richmond [6]. The study of Richmond used the initial concept of a beam model with the stiffener and an effective width of web loaded by a line load of intensity w representing a resolved component of the in-plane shear, the in-plane compression and a smaller percentage of the in-plane bending stress. This load representation is shown in fig. 7.

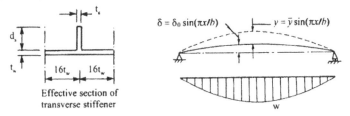

Figure 7   Simple beam model under Richmond load.

The effective width shown of $32t_w$ is adopted in the bridge code. The beam deflection can be calculated but the procedure is iterative because of the inter-dependence between the load and deflection. The lateral load is a function of the deflection and increases non-linearly with increasing web deformation. The background to these calculations can be found in reference 7. It was found by comparing the finite element results with the Richmond predictions that for slender panels the latter underestimates both lateral displacement and stiffener stresses. This is illustrated in fig. 8 which shows the shear stress versus displacement for a typical case.

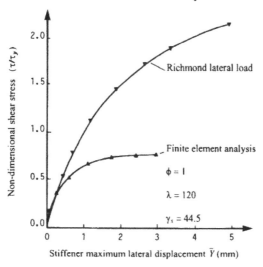

Figure 8   **Comparison of Richmond lateral load with finite element analysis.**

It was also of great interest to look at the finite element results in terms of the predictions of the Rockey tension field approach and it became very clear that there was no significant evidence of overall tension field forces being applied to the stiffener. This is evident in fig. 9.which shows the distribution of stress through the stiffener outstand and the proximity of the zero bending condition to the elastic neutral axis calculated for the effective section. A beam model was therefore produced with a lateral loading that approximated the forces found from the nodal line analyses. This process showed that the form of the lateral force was dependant on the panel aspect ratio, but this dependency was ignored and a single half wave sinusoidal force representation used, as in the Richmond approach. The lateral load expression which was developed for shear only is shown in equation 1 and it can be seen that it is a non-linear function of the shear stress and a relatively simple function of the dimensions of the web.

$$W(N/mm) = \left(\frac{\tau}{\tau_y}\right)^3 \frac{b\lambda^{1/3}\phi}{35.5} \left(\frac{\sigma_y}{355}\right)^{3/2} \sin\frac{\pi x}{b} \tag{1}$$

A typical comparison between the beam model and the finite element results is shown in fig. 10 where outstand yield of the stiffener can be shown to provide realistic modelling of the collapse. In using the beam model the shear stress used in the equation is the shear failure stress of the unstiffened sub-panel and can be taken as the code value for design purposes.

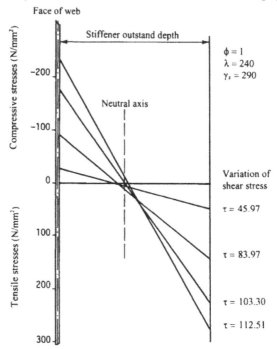

Figure 9   Variation of axial stresses across stiffener at its mid-height.

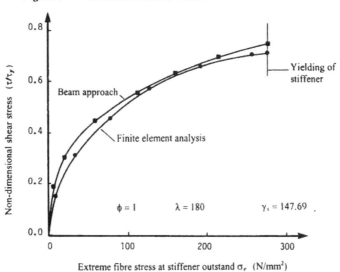

Figure 10   Comparison between finite element results and beam approach.

By using this equation, a stiffener size can be evaluated at which first outstand yield just occurs at the collapse stress of the sub-panel. This provides the optimum design rigidity for the stiffener. Figure 11 shows a comparison between the design rigidity produced by the beam model and that required by BS5400 and it can be seen that for slender panels the requirements from the beam approach are very substantially less than those of the code, although it must be recognised that the γ_s parameter is not a linear function of stiffener area and the difference with the code is exaggerated by this form of presentation.

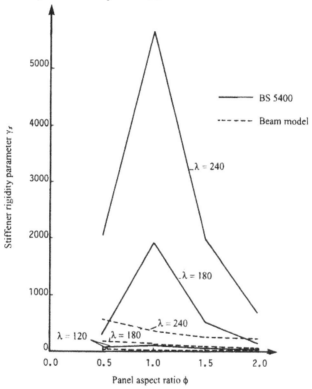

Figure 11   Comparison between BS5400 and beam model.

Using the results of the compression and bending cases the design equation for the lateral load was expanded to take account of the possible interactive combination, and this modified equation is shown in equation 2.

$$W(N/mm) = \frac{b\lambda^{1/3}\phi}{35.5}\left(\frac{\sigma_y}{355}\right)^{3/2}[(\tau' + \psi\,\sigma_c')^3 + (\psi'\,\sigma_b')^3]\sin\frac{\pi x}{b} \tag{2}$$

The parameter ψ representing the proportional influence of the compressive stress was evaluated as a simple function of the panel slenderness and aspect ratio, equation 3.

$$\psi = 0.12\sqrt{\lambda/\phi} \tag{3}$$

ψ', the corresponding multiplier on bending action, was found to be reasonably represented by a constant value of 0.415.

Figure 12 shows the excellent comparison obtained between the beam approach and the finite element analysis for a combination of shear and compression and the agreement is generally as good for the case of bending. A fuller account of the derivation of the design model and the resulting correlations can be obtained from reference 8.

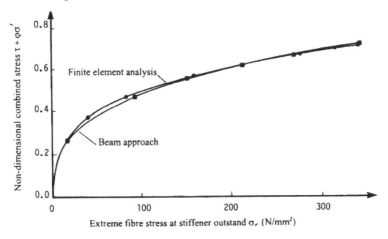

Figure 12   Comparison between finite element analysis and beam approach for plates under combined shear and compressive displacement ($\phi = 1$, $\lambda = 180$, $D_s = 100$, $T_s = 10$, $\sigma_y = 355$ N/mm^2).

THE DESIGN PROCESS

This paper has summarised the establishment of a process for the design of intermediate transverse stiffeners in webs subjected to combined shear and in-plane direct loading. The method is summarised below to clarify the various stages and as with any other design process is iterative.

A. Establish starting dimensions for the stiffened plate by reference to the load applied including provision of a first estimate of possible stiffener size.

B. Using the known ratio between the applied stresses and an available design method, e.g. that of BS5400, evaluate the collapse values of the applied stresses σ_c, σ_b and τ for the unstiffened sub-panels. If these collapse values are insufficient, the panel dimensions should be adjusted and step B repeated.

C. Use these collapse values of the three stresses to evaluate the lateral load distribution given by equation 2.

D. Calculate the effective inertia of the stiffener section with a flange of total width equal to 32 times the web thickness.

E. Using a simple linear elastic beam analysis evaluate the outstand bending stress at a level of applied lateral load specified in C above. As the load is sinusoidal and the beam determinate, a simple analytical expression can be derived.

F. Check this stress against the yield stress. If less than σ_y the stiffener size should be reduced and vice versa. Return to step C with new stiffener dimensions.

Steps A and B establish whether the web thickness and panel aspect ratios are appropriate for the applied loading. The remaining steps relate to optimisation of the stiffener size. However, it is possible that following step F there may be a requirement to change the panel aspect ratio, e.g. minimise overall cost and there would then be a need to return to step B. This iterative process is best implemented via a simple micro-computer program on a desk-top machine.

DISCUSSION AND CONCLUSIONS

This paper has described a wide ranging parametric finite element study of the behaviour of transverse stiffeners in girder webs and the forces to which they are subjected by the buckling of the web panels. A simple linear beam model has been formulated with a loading expression as a function of the applied web stresses and the panel parameters which is based on the results of the analyses. This beam model can be used in design to optimise the size of the transverse stiffeners so that stiffeners are provided which do not weaken the buckling performance of the webs and which essentially are just large enough to provide nodal lines acting as boundaries to the web panels.

While it has not been discussed in this paper, reference 8 also looks at the possibility of designing the stiffener only for shear stresses at maximum panel capacity and removes the need to consider the more complex interactive loading in the stiffener loading expression. The reader is directed to this reference for more information on this aspect of the work.

In practical design there are often direct loads acting on the stiffener, in addition to the destabilising forces from the webs, e.g. via cross girders. These can easily be incorporated in the design formulation using a beam-column type approach combining the lateral load from destabilising actions with any direct axial load on the stiffener acting as a column between the webs.

REFERENCES

[1]   British Standards Institution. Code of practice for design of steel bridges, BSI, London, 1982, BS5400, Part 3.

[2]   Rockey K.C. et al. The design of transverse stiffeners on webs loaded in shear - an ultimate approach. Proc. Instn Civ. Engrs, Part 2, 1981, 71, Dec., 1069-1099.

[3]   Horne M.R. and Grayson W.R. (Morris L.J. (ed.)) Parametric finite element study of transverse stiffeners for webs in shear. Instability and plastic collapse of steel structures. Granada Publishing, London, 1983, 329-341.

[4]   Grayson W.R. Behaviour and design of stiffened web panels. University of Manchester, 1981, PhD thesis.

[5]   Rahal K.N. and Harding J.E. Transversely stiffened girder webs subjected to shear loading. Part 1: behaviour. Proc. Instn Civ. Engrs, Part 2, 1990, 89, Mar., 47-65.

[6]   Richmond B. Report on parametric study on web panels. Maunsell and Partners, London, 1972, Report for Department of the Environment.

[7]   Rahal K.N. and Harding J.E. Transversely stiffened webs subjected to shear loading. Part 2: stiffener design. Proc. Instn Civ. Engrs, Part 2, 1990, 89, Mar., 67-87.

[8]   Rahal K.N. and Harding J.E. Transversely stiffened girder webs subjected to combined in-plane loading, Proc. Instn Civ. Engrs., Part 2, 1991, June, 237-258.

ACKNOWLEDGEMENTS

The authors gratefully acknowledge the sponsorship of the Hariri Foundation for the first author of this paper and also the assistance provided by Finite Element Analysis Ltd who developed the package used for the parametric study.

NONLINEAR BEHAVIOUR OF LONGITUDINALLY STIFFENED WEBS IN COMBINED PATCH LOADING AND BENDING

MASAHIRO DOGAKI[1], YOSHITAKA NISHIJIMA[2] and HIROSHI YONEZAWA[1]

[1] *Department of Civil Engineering, Kansai University, 3-3-35 Yamate, Suita, Osaka 564, Japan*

[2] *The Japan Research Institute Ltd., Osaka 555, Japan*

ABSTRACT

The ultimate strength of longitudinally stiffened plate girders under combined patch load and bending moment is predicted by the finite difference procedure which is developed for numerically analysing their non-linear behaviour. The geometrical and material non-linearities are taken into account. A set of algebraic equations is precisely solved by the modified Newton-Raphson method. The theoretically predicted ultimate strengths are compared with test results. The numerical solutions are presented to examine the influences of the length of patch load, width-to-thickness ratio of web, stiffness of flanges, location and stiffness of longitudinal stiffener, and geometrical imperfection at the web on the post-buckling behaviour and ultimate strength of plate girders under combined loading.

INTRODUCTION

Plate girders must often sustain loads on flange, which produce compressive membrane stress in their direction at web. Such loads are frequently encountered in practice, e.g., wheel loads on crane girders, purlins transmitting loads to main frame members of buildings, diaphragms on bearing in box-girders, and during the erection of large plate- and box-girder bridges by launching. The loads may be distributed over a wide range of distances varying from whole length of plate girders to extremely narrow area. As the loads act on partly, a plate girder receiving them may undergo local plate buckling called as crippling at its web. In general, transverse stiffeners may be located in positions at which large concentrated loads come to bear. In some cases, however, adoption of transverse stiffeners is not appropriate, e.g., during the erection of plate girders by launching.

Since Roberts and Rockey [1] proposed a simple and practical method for predicting the ultimate strength of plate girders under patch load using a collapse mechanism, a number of studies [2-11] have been performed on their ultimate strength. But there is a few research on the analytical prediction of the elasto-plastic behaviour of longitudinally stiffened plate girders under patch load [12,

13] . The present paper is to aim at developing the finite difference algorithm for predicting the post-buckling behaviour and ultimate strength of plate girders with longitudinal stiffener under combined patch load and bending moment.

METHOD OF ANALYSIS

A plate girder between transverse stiffeners is analysed, which consists of a web of length a, depth b and thickness t, flanges of width b_f and thickness t_f , and a longitudinal stiffener of depth b_s and thickness t_s being attached on a surface of the web at the distance b_1 from the bottom flange, as shown in Fig.1. The plate girder under patch load and bending moment is modelled as one rigidly supported in the transverse direction over a length c at the center of bottom flange and subjected to both shearing force and bending moment at the sides. Its web is assumed to be simply supported on the transverse stiffeners and elastically restrained by the flanges. The following assumptions are introduced in this paper: 1) Kirchhoff-Love's hypothesis is valid; 2) the material is elastic-perfectly-plastic with no strain-hardening; and 3) no local plate buckling of flanges is considered.

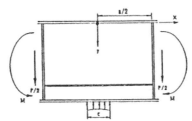

Figure 1. A longitudinally stiffened plate girder between transverse stiffeners under a combination of patch load and bending moment.

Equations of Equilibrium at Elasto-Plastic Web

Lin [14] and Massonnet [15] separately proposed a useful method for analysing an inelastic bending of thin plates with moderately large deflection. They extended the Kármán-Marguerre's plate theory to a large deflection theory of elasto-plastic plates. According to the above Lin-Massonnet's method, the equations of equilibrium in the x-, y- and z-directions for an elasto-plastic web with moderately large deflection can be obtained as follows:

$$u,_{xx} +(1-\nu)u,_{yy}/2 +(1+\nu)v,_{xy}/2+(\bar{w},_x\bar{w},_{xx}-w_0,_xw_0,_{xx})$$
$$+(1-\nu)(\bar{w},_x\bar{w},_{yy}-w_0,_xw_0,_{yy})/2+(1+\nu)(\bar{w},_y\bar{w},_{xy}-w_0,_yw_0,_{xy})/2-(N_x^P,_x+N_{xy}^P,_y)/B=0,$$

$$(1+\nu)u,_{xy}/2 +(1-\nu)v,_{xx}/2+v,_{yy}+(1+\nu)(\bar{w},_x\bar{w},_{xy}-w_0,_xw_0,_{xy})/2$$
$$+(1-\nu)(\bar{w},_y\bar{w},_{xx}-w_0,_yw_0,_{xx})/2+(\bar{w},_y\bar{w},_{yy}-w_0,_yw_0,_{yy})-(N_{xy}^P,_x+N_y^P,_y)/B=0,$$

$$\bar{w},_{xxxx} + 2\bar{w},_{xxyy} + \bar{w},_{yyyy} - (w_0,_{xxxx} + 2w_0,_{xxyy} + w_0,_{yyyy})$$
$$-(B/D)\,[\bar{w},_{xx}\,\{\quad u,_x+\bar{w},_x\bar{w},_x/2-w_0,_xw_0,_x/2 + \nu\,(v,_y+\bar{w},_y\bar{w},_y/2-w_0,_yw_0,_y/2)\,\}$$
$$+\bar{w},_{yy}\,\{\,\nu\,(u,_x+\bar{w},_x\bar{w},_x/2-w_0,_xw_0,_x/2)+\quad v,_y+\bar{w},_y\bar{w},_y/2-w_0,_yw_0,_y/2\,\}$$
$$+ (1-\nu)\bar{w},_{xy}(u,_y+v,_x+\bar{w},_x\bar{w},_y-w_0,_xw_0,_y)]$$
$$+ (N_x^P\bar{w},_{xx}+2N_{xy}^P\bar{w},_{xy}+N_y^P\bar{w},_{yy} + M_x^P,_{xx}+2M_{xy}^P,_{xy}+M_y^P,_{yy})/D = 0 \quad \cdots \cdots \quad (1)_{1-3}$$

in which u and v are the displacements in the x- and y-directions; $\bar{w}$ is the total deflection given by a sum of initial deflection w_0 and deflection w due to applied load; N_x and N_y are the membrane forces per unit length in the x- and y-directions; N_{xy} is the membrane shearing force per unit length; M_x and M_y are the bending moments per unit length about the y- and x-axes; M_{xy} is the twisting moment per unit length; the superscript p on the stress resultants means fictitious values being introduced to compensate on actual elasto-plastic stress field; B and D are the extensional and flexural rigidities of the web; ν is the Poisson's ratio; and the subscript preceding the comma indicates the partial differentiation with respect to it.

Boundary and Continuous Conditions

Boundary conditions: The plate girder is assumed to be simply supported on the transverse stiffeners and subjected to patch load over the length c on the central portion of bottom flange and bending moment at the sides. The displacements in the longitudinal and transverse directions is applied at the sides so that the plate girder is subjected to patch load and bending moment. The corresponding boundary conditions at the sides are given as follows:

$$u \pm u_a \pm (y - \tfrac{1}{2}b)\phi = 0, \quad v = v_a, \quad \bar{w} - w_0 = 0, \quad \bar{w}_{,xx} - w_{0,xx} = 0,$$

$$\int_0^b N_x dy + N_{f(y-0)} + N_{f(y-b)} + N_s = 0,$$

$$\int_0^b N_x(y - \tfrac{1}{2}b)dy + \tfrac{1}{2}b(-N_{f(y-0)} + N_{f(y-b)}) + (\tfrac{1}{2}b - b_1)N_s = M_0 \qquad \ldots\ldots (2)_{1-6}$$

in which ϕ is the unknown inclination with respect to the y-axis and determined so as to satisfy the condition $(2)_6$ that the resultant moment equals the applied bending moment M_0 ; the displacement u_a is the unknown for adjusting the displacement in the longitudinal direction so that the resultant axial force is not induced, i.e., condition $(2)_5$; v_a is the displacement applied in the transverse direction; N_x is the membrane force of web in the x-direction; and N_f and N_s are the axial forces of the flanges and longitudinal stiffener and can be calculated by

$$N_f = EA_f(u_{,x} + \bar{w}_{,x}\bar{w}_{,x}/2 - w_{0,x}w_{0,x}/2) - N_f{}^P \qquad \ldots\ldots (3)$$
$$N_s = EA_s(u_{,x} + \bar{w}_{,x}\bar{w}_{,x}/2 - w_{0,x}w_{0,x}/2) - N_s{}^P \qquad \ldots\ldots (4)$$

where A_f and A_s are the areas of the cross-section of the flanges and longitudinal stiffener; and E is the modulus of elasticity.

Consider the boundary conditions at the junctions of the web and flange. At the portion subjected to patch load on the bottom flange, the web is assumed to be rigidly supported in the transverse direction and clamped, i.e.,

$$N_{f,x} - N_{xy} = 0, \quad v = 0, \quad w - w_0 = 0, \quad w_{,y} - w_{0,y} = 0 \qquad \ldots\ldots (5)_{1-4}$$

in which N_{xy} is the membrane shearing force of the web at the bottom edge.

The flanges are analysed as beam-columns subjected to axial force, biaxial bending and torsion. The boundary conditions can be obtained at the top flange and bottom flange where patch load does not act on as follows:

$$N_{f,x} \pm N_{xy} = 0, \quad M_{fz,xx} \pm N_y + N_f v_{,xx} + N_{f,x}v_{,x} = 0,$$
$$M_{fy,xx} \pm (M_{y,y} + 2M_{xy,x}) + N_f \bar{w}_{,xx} + N_{f,x}w_{,x} = 0, \quad M_{ft,x} + M_y = 0 \ldots\ldots (6)_{1-4}$$

in which N_y, M_y and M_{xy} are the membrane force, bending moment about the x-axis and twisting moment of the web at the boundary of $y=0$ or $y=b$; and M_{fz}, M_{fy} and M_{ft} are the bending moments about the z- and y-axes and twisting moment of the flanges, and calculated by

$$M_{fz} = -EI_{fz}v_{,xx} - M_{fz}^P, \quad M_{fy} = -EI_{fy}(\bar{w}_{,xx} - w_{0,xx}) - M_{fy}^P,$$
$$M_{ft} = GJ_f(\bar{w}_{,xy} - w_{0,xy}) - M_{ft}^P \qquad \ldots (7)_{1-3}$$

where EI_{fz} and EI_{fy} are the flexural rigidities about the z- and y-axes; and GJ_f is St. Venant' torsional constant.

Continuous conditions: The longitudinal stiffener is also treated as an equivalent beam-column subjected to axial force, biaxial bending and torsion as well as the flanges. The continuous conditions at the junction of two web panels separated by the longitudinal stiffener can be obtained as follows:

$$N_{s,x} + (N_{xy2} - N_{xy1}) = 0, \quad M_{sz,xx} + (N_{y2}-N_{y1}) + N_s v_{,xx} + N_{s,x} v_{,x} = 0,$$
$$M_{sy,xx} + (M_{y2,y}-M_{y1,y}) + 2(M_{xy2,x}-M_{xy1,x}) + N_s \bar{w}_{,xx} + N_{s,x}\bar{w}_{,x} = 0,$$
$$M_{st,x} - (M_{y2}-M_{y1}) = 0 \qquad \ldots (8)_{1-4}$$

in which the subscripts 1 and 2 on the stress resultants of web indicate the number of web panel separated by the longitudinal stiffener; and M_{sz}, M_{sy} and M_{st} are the bending moments about the z- and y-axes and twisting moment of the longitudinal stiffener, and calculated by

$$M_{sz} = -EI_{sz}v_{,xx} - M_{sz}^P, \quad M_{sy} = -EI_{sy}(\bar{w}_{,xx} - w_{0,xx}) - M_{sy}^P,$$
$$M_{st} = GJ_s(\bar{w}_{,xy} - w_{0,xy}) - M_{st}^P \qquad \ldots (9)_{1-3}$$

where EI_{sz} and EI_{sy} are the flexural rigidities about the z- and y-axes; and GJ_s is St. Venant' torsional constant.

Fictitious Stress Resultants

As a plate girder is beyond elastic range, the fictitious stress resultants in the fundamental equations, and boundary and continuous conditions must be given. Such values can be obtained by numerically integrating surplus stresses over the thickness of web or over the cross-section of the flanges and longitudinal stiffener. The surplus stresses are given as the difference between the elastic stresses and stresses at the yielding. The fictitious stress resultants are given by

$$N_x^P = \sum \sigma_x^P \Delta t, \quad N_y^P = \sum \sigma_y^P \Delta t, \quad N_{xy}^P = \sum \tau_{xy}^P \Delta t,$$
$$M_x^P = \sum \sigma_x^P z\Delta t, \quad M_y^P = \sum \sigma_y^P z\Delta t, \quad M_{xy}^P = \sum \tau_{xy}^P z\Delta t \qquad \ldots (10)_{1-6}$$
$$N_f^P = \sum \sigma_{fx}^P \Delta A_f, \quad M_{fz}^P = \sum \sigma_{fx}^P y\Delta A_f, \quad M_{fy}^P = \sum \sigma_{fx}^P z\Delta A_f,$$
$$M_{ft}^P = \sum \tau_{fxz}^P z\Delta A_f \qquad \ldots (11)_{1-4}$$
$$N_s^P = \sum \sigma_{sx}^P \Delta A_s, \quad M_{sz}^P = \sum \sigma_{sx}^P y\Delta A_s, \quad M_{sy}^P = \sum \sigma_{sx}^P z\Delta A_s,$$
$$M_{st}^P = \sum \tau_{sxz}^P z\Delta A_s \qquad \ldots (12)_{1-4}$$

in which Δt is the length of small layers in the direction of web thickness; ΔA_f and ΔA_s are the areas of small elements in the flanges and longitudinal stiffener; and the stresses with superscript p are the surplus values of the web, flanges and longitudinal stiffener.

When elastic stresses, i.e., the stresses obtained by assuming the plate girder to be purely elastic even if it is within elasto-plastic state, gradually increase under incrementally applied loading, the surplus stresses can be calculated as follows:

$$\sigma_x^{\,P} = \sigma_{xb} + (\sigma_x^{\,e} - \sigma_{xb})(\sigma_{eq} - \sigma_Y)/\sigma_{eq},$$
$$\sigma_y^{\,P} = \sigma_{yb} + (\sigma_y^{\,e} - \sigma_{yb})(\sigma_{eq} - \sigma_Y)/\sigma_{eq},$$
$$\tau_{xy}^{\,P} = \tau_{xyb} + (\tau_{xy}^{\,e} - \tau_{xyb})(\sigma_{eq} - \sigma_Y)/\sigma_{eq} \qquad \cdots\cdots (13)_{1-3}$$
$$\sigma_{fx}^{\,P} = \sigma_{fxb} + (\sigma_{fx}^{\,e} - \sigma_{fxb})(\sigma_{feq} - \sigma_Y)/\sigma_{feq},$$
$$\tau_{fxz}^{\,P} = \tau_{fxzb} + (\tau_{fxz}^{\,e} - \tau_{fxzb})(\sigma_{feq} - \sigma_Y)/\sigma_{feq} \qquad \cdots\cdots (14)_{1,2}$$
$$\sigma_{sx}^{\,P} = \sigma_{sxb} + (\sigma_{sx}^{\,e} - \sigma_{sxb})(\sigma_{seq} - \sigma_Y)/\sigma_{seq},$$
$$\tau_{sxz}^{\,P} = \tau_{sxzb} + (\tau_{sxz}^{\,e} - \tau_{sxzb})(\sigma_{seq} - \sigma_Y)/\sigma_{seq} \qquad \cdots\cdots (15)_{1,2}$$

On the other hand, for the case of unloading or when the elastic stresses decrease under applied loading, the surplus stresses are given by

$$\sigma_x^{\,P} = \sigma_{xb}, \qquad \sigma_y^{\,P} = \sigma_{yb}, \qquad \tau_{xy}^{\,P} = \tau_{xyb} \qquad \cdots\cdots (16)_{1-3}$$
$$\sigma_{fx}^{\,P} = \sigma_{fxb}, \qquad \tau_{fxz}^{\,P} = \tau_{fxzb} \qquad \cdots\cdots (17)_{1,2}$$
$$\sigma_{sx}^{\,P} = \sigma_{sxb}, \qquad \tau_{sxz}^{\,P} = \tau_{sxzb} \qquad \cdots\cdots (18)_{1,2}$$

in which σ_Y is the yielding stress; the superscripts e and p indicate the elastic and fictitious stresses; the stresses with the subscript b are the fictitious values calculated at the previous loading step; and the stresses with the subscript eq are the von Mises' equivalent stresses given as follows:

$$\sigma_{eq} = \{ (\sigma_x^{\,e} - \sigma_{xb})^2 - (\sigma_x^{\,e} - \sigma_{xb})(\sigma_y^{\,e} - \sigma_{yb})$$
$$+ (\sigma_y^{\,e} - \sigma_{yb})^2 + 3(\tau_{xy}^{\,e} - \tau_{xyb})^2 \}^{1/2} \qquad \cdots\cdots (19)$$
$$\sigma_{feq} = \{ (\sigma_{fx}^{\,e} - \sigma_{fxb})^2 + 3(\tau_{fxz}^{\,e} - \tau_{fxzb})^2 \}^{1/2} \qquad \cdots\cdots (20)$$
$$\sigma_{seq} = \{ (\sigma_{sx}^{\,e} - \sigma_{sxb})^2 + 3(\tau_{sxz}^{\,e} - \tau_{sxzb})^2 \}^{1/2} \qquad \cdots\cdots (21)$$

Numerical Procedure

The elasto-plastic large deflection of longitudinally stiffened plate girders with initial deflection is clarified by precisely solving Eqs.(1) together with the boundary conditions (2), (5) and (6), and continuous conditions (8) for an incremental load. It is difficult to obtain the exact solution because of complex non-linear problem. A finite difference method is used in this paper. To express the fundamental equations, and boundary and continuous conditions into finite difference form, the web is divided by the number of subdivisions n_x and n_y in the x- and y-directions. Eqs.(1)$_1$ and (1)$_2$ for the in-plane displacements u and v, and Eq.(1)$_3$ for the deflection $\bar{w}$ are expressed in terms of the central finite difference at inner nodal points in the web. The boundary and continuous conditions are expressed in terms of an appropriate finite difference using forward or backward technique. As a result of rewriting the equations in terms of finite differences, a set of non-linear algebraic equations for the unknown incremental displacements is obtained as follows:

$$[K] \ \{\Delta d\} = \{R\} \qquad \cdots (22)$$

in which $[K]$ is a band matrix consisting of constant coefficients relating to the dimensions of plate girders; $\{\Delta d\}$ is a row vector of unknown incremental displacements; and $\{R\}$ is a row vector of residual quantities calculated by substituting displacements obtained at each iteration into the equilibrium equations, and boundary and continuous conditions.

To obtain the solution of Eq.(22), the modified Newton-Raphson method is adopted. Acceleration techniques such as Jacobi method and Aitken method are used to improve convergent characteristics. Incremental displacements can be obtained by solving Eq.(22) with a residual quantity by the method of triangular decomposition. It is checked whether a new iterative solution satisfies the equilibrium equations, and boundary and continuous conditions.

NUMERICAL RESULTS AND CONSIDERATIONS

The parametric analysis is carried out to obtain the ultimate strength of longitudinally stiffened plate girders under combined patch load and bending moment. The initial deflection at the web is assumed to be one half wave of sinusoidal functions in both its longitudinal and transverse directions being of a maximum value of $w_{0max} = b/2500$. The half of plate girder between transverse stiffeners is analysed, because of the symmetry of the geometry of plate girders and the applied load with respect to its transverse coordinate. An iterative solution is obtained with a specified relative allowance of 0.1% for all the displacements in the x-, y- and z-directions at each step of applied load. Yielding stress, modulus of elasticity, and Poisson's ratio are $\sigma_Y = 235$MPa, $E = 205$GPa and $\nu = 0.3$.

Accuracy of Finite Difference Solutions

To decide the appropriate number of subdivisions for the finite difference analysis, the initially deflected web with no longitudinal stiffener is analysed of an aspect ratio $a/b=1$, width-to-thickness ratio of web $b/t=150$, width-to-thickness ratio of flanges $b_f/t_f = 30$, and ratio of the area of flange to web $A_f/bt=0.5$ under only patch load. It is found that the accurate ultimate strength can be predicted if the web is divided into the number of subdivisions $n_x = 20$ and $n_y = 30$ in the x- and y-directions. To trace the propagation of yielding zone over the cross-section of plate girder, the web is divided into small layers of $n_z = 10$ in the z-direction, and the flanges and longitudinal stiffener are divided into small elements of the number of subdivisions $n_{fy} = n_{sy} = 6$ and $n_{fz} = n_{sz} = 10$ in the y- and z-directions.

The theoretically predicted ultimate strengths are compared with the collapse loads for the 37 specimens carried out by authors [16-18], Takimoto, Imagawa and Moriwaki [19] and Janus, Karnikova and Skaloud [20]. The tested plate girder models were longitudinally stiffened. Figure 2 shows the comparison between the theoretical and experimental ultimate strengths, in which the ordinate and abscissa describe the non-dimensional theoretical and experimental ultimate strengths respectively where $P_{ult}{}^{th}$, $P_{ult}{}^{ex}$ and V_y are the theoretical and experimental ultimate patch strengths and full plastic shearing force given by $\sigma_Y bt/\sqrt{3}$ of the web; and the marks $\bigcirc$, $\triangle$ and $\square$ mean the authors', Takimoto's and Janus's test results respectively. It can be seen from this figure that the analytical predictions $P_{ult}{}^{th}/2V_y$, scattering around $P_{ult}/2V_y = 0.15$ in this figure coincide with the experimental values $P_{ult}{}^{ex}/2V_y$. In these cases, the width-to-thickness ratio of the web was about 250. On the other hand, the analytical method overestimates the ultimate patch strength for the cases of the plate girder web of $b/t=150$ plotted at the range of $0.3 \leqq P_{ult}/2V_y \leqq 0.5$ in comparison with the experimental results.

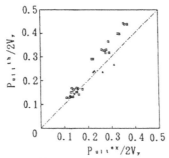

Figure 2. Comparison between theoretical and experimental ultimate strengths.

Load-Displacement Curves under Combined Patch Load and Bending

As a plate girder is erected by launching, it may be subjected to various combined patch load and bending moment. Figure 3 shows the relation between the non-dimensional patch load $P/2V_y$ and non-dimensional displacement $v_0 b/t^2$ in the transverse direction at the sides for the plate girders without and with longitudinal stiffener respectively, in which the mark ● indicates the first yielding. The combination of patch load and bending moment is considered of the range of $0 \leq (M_0/M_p)/(P/2V_y) \leq 10$, in which the value of $(M_0/M_p)/(P/2V_y)=0$ indicates the case of which the plate girder is subjected to only the patch load; and M_p is the full plastic moment of plate girder. The plate girder with initial deflection $w_{0max}=b/2500$ is analysed of $a/b=1$, $b_f/t_f=25$ and $A_f/bt=0.5$. In the case of the plate girder with longitudinal stiffener shown in Fig.3(a), the longitudinal stiffener of $\gamma/\gamma^*=1$ and $b_s/t_s=12.5$ is attached on the position of $b_1/b=0.2$ from the bottom flange in which γ is the ratio of the flexural rigidity of longitudinal stiffener to stiffness of web plate, i.e., EI_s/bD; and γ^* is its optimum value specified by the Japanese Specifications of Highway Bridges, i.e., $30a/b$. The load-displacement curves for the plate girders under both patch load and bending moment are almost similar to that for the plate girder under only patch load. The relation between the patch load and transver se displacement becomes linear before the first yielding.

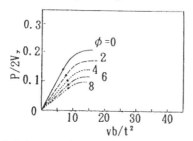

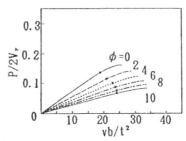

(a) Web without longitudinal stiffener (b) Web with longitudinal stiffener
Figure 3. Applied patch load versus transverse displacement curves at the sides.

Distribution of Deflection and Propagation of Yielding

The distribution of deflection and propagation of yielding zone in the web at the maxmum patch load are shown in Fig.4 for three different combinations of patch load and bending moment, i.e., $(M_0/M_p)/(P/2V_y)=0$, 4 and 8, in which figures (a) and (b) describe for the plate girder without and with longitudinal stiffener respectively; and the mark ● indicates the point of yielding. For the plate girder with no longitudinal stiffener, it deforms with a half wave in both the directions similar to geometrical imperfection. As the web is reinforced by a longitudinal stiffener of $\gamma/\gamma^*=1$, it deflects with the shape of two half waves in the transverse direction. The spread of yielding zone for the longitudinally stiffened plate girder is localised in comparison with that for the plate girder with no longitudinal stiffener.

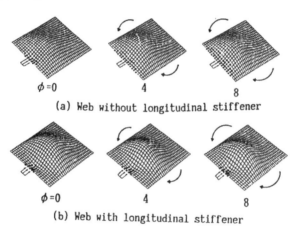

$\phi=0$ 4   8

(a) Web without longitudinal stiffener

$\phi=0$ 4   8

(b) Web with longitudinal stiffener

Figure 4. Distribution of deflection and propagation of yielding zone in web at failure for plate girders without/with longitudinal stiffener.

Interaction Curve of Ultimate Strength

The interaction curves for the ultimate strength of plate girders under combined patch load and bending moment are shown in Fig.5 for three different lengths of patch load, i.e., $c/a=0.1$, 0.2 and 0.3, in which the figures (a) and (b) show those for the web of $b/t=150$ without longitudinal stiffener and for the web of $b/t=250$ with longitudinal stiffener of $\gamma/\gamma^*=1$ and $b_s/t_s=12.5$ attached at the location of $b_1/b=0.2$ respectively. The plate girder is analysed of of $a/b=1$, $b_f/t_f=25$ and $A_f/bt=0.5$. It can be seen from this figure that the relation between the ultimate patch strength and ultimate bending strength becomes almost linear. The ultimate strength of unstiffened plate girders greatly decreases with the increase of in-plane bending , espectially for the case of which the length of patch load is long.

Flexural Rigidity of Longitudinal Stiffener

Figure 6 shows the variation of the ultimate patch strength $P_{ult}/2V_y$ with the flexural rigidity γ/γ^* of longitudinal stiffener for the plate girder in three different combinations of patch load and bending moment. The length of patch load is assumed to be of $c/a=0.1$. The dimensions of plate girder is $a/b=1$, $b/t=250$, $A_f/bt=0.5$, $b_f/t_f=25$, $b_1/b=0.2$ and $b_s/t_s=12.5$. It is obvious that the ultimate patch strength $P_{ult}/2V_y$ significantly increases with the increase of the flexural rigidity of longitudinal stiffener within the range of $0 \leq \gamma / \gamma^* \leq 0.5$. As the flexural rigidity γ / γ^* is larger than unit, the ultimate patch strength does not almost increase.

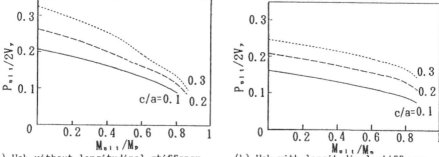

(a) Web without longitudinal stiffener   (b) Web with longitudinal stiffener

Figure 5. Interaction curves of plate girder under patch load and bending moment.

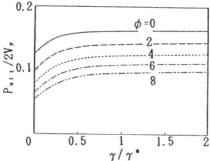

Figure 6. Relation between the ultimate patch strength
and flexural rigidity of longitudinal stiffener.

CONCLUSIONS

The finite difference approach was developed for predicting the ultimate strength of Longitudinally stiffened plate girders under combined patch load and bending moment, taking account of the geometrical and material non-linearities. The equations of equilibrium of stress resultants at the web within elasto-plastic region were introduced together with the boundary and continuous conditions relating to the equilibrium of stress resultants on the flanges and longitudinal stiffener modelled as elasto-plastic beam-columns. A system of non-linear algebraic equations was solved by the modified Newton-Raphson procedure with incremental algorithm, in which some acceleration techniques were adopted to improve the convergent characteristics. The influences of the length of patch load, width-to-thickness ratio of web, stiffness of flanges, and geometrical imperfection on the elasto-plastic large deflection behaviour and ultimate strength of plate girders under patch load and bending moment were clarified numerically.

The numerical analysis was performed by the digital computers FACOM M-780 and VP-50 in Kansai Univ..

REFERENCES

1) Roberts, T.M., and Rockey, K.C., A mechanism solution for predicting the collapse loads of slender plate when subjected to in-plane patch loading, Proc. of ICE, Part 2, Vol.67, 1979, 155-75.

2) Roberts, T.M., Patch loading on plate girders, Plated Structures - Stability and Strength, ed. R. Narayanan, Applied Science Publishers, London, 1983, pp.77-102.

3) Elgaaly, M., Web design under compressive edge loads, Engineering Journal, AISC, 1983, 153-71.

4) Dubas, P. and Gehri, E.(eds.), Behaviour and Design of Steel Plated Structures, ECCS, TCS, SSTWG8.3, Applied Statics and Steel Structures, ETH-Honggerberg, Zurich, 1986.

5) Fukumoto, Y.(ed.), Guidelines for Stability Design of Steel Structures, Japan Society of Civil Engineering, 1987(in Japanese).

6) Galambos, T.V.(ed.), Guide to Stability Design Criteria for Metal Structures, 4th ed., John Wiley & Sons, 1988, pp.223-230.

7) Stein, E., Lambertz, K.-H., und Plank, L., Traglastberechnung dunnwandiger Strukturen bei grossen elastoplastischen Deformationen, Der Stahlbau, Jg.54, Ht. 1, 1985, 9-13.

8) Ramm, E. und Weimar, K., Traglasten unversteifter Tragerstegbleche unter konzentrierten Lasten, Der Stahlbau, Jg.55, Ht.4, 1986, 113-18.

9) Weimar, K. and Ramm, E., Ultimate loads of unstiffened plate girders subjected to concentrated loads, ECCS Colloquium on Stability of Plate and Shell Structures, Ghent University, 6-8 April, 1987, pp.79-84.

10) Shimizu, S., Horii, S., and Yoshida, S., The collapse mechanisms of patch loaded web plates, Jour. Construct. Steel Research, Vol.14, 1989, 321-37.

11) Dogaki, M., Kishigami, N., and Yonezawa, H., Ultimate strength analysis of plate girder webs under patch loading, Steel Structures - Recent Research and Developments, eds. S. L. Lee and N. E. Shanmugam, Elsevier Applied Science, 1991, 192-201.

12) Shimizu, S., Horii, S., and Yoshida, S., Behaviour of horizontally stiffened web plates subjected to the patch load, Jour. of Structural Engineering, Vol.37A, 1991, 229-38(in Japanese).

13) Dogaki, M., Tanabe, T., and Yonezawa, H., Ultimate strength of plate girders with longitudinal stiffener under patch loading, The 3rd Pacific Structural Steel Conference, Oct., 1992.

14) Lin, T.H., Theory of Inelastic Structures, John Wiley & Sons, 1968.

15) Massonnet, Ch., General theory of elastic-plastic membrane-plates, Engineering Plasticity, eds. J. Heyman and F.A. Leckies, Cambridge University Press, 1968, pp.443-471.

16) Dogaki, M., Murata, M., Kishigami, N., Tanabe, T., and Yonezawa, H., Ultimate strength of plate girders with transverse stiffeners under patch loading, Technology Reports of Kansai University, No.32, 1990, 139-50.

17) Dogaki, M., Murata, M., Nishijima, Y., Okunura, T., and Yonezawa, H., Ultimate strength of plate girders with longitudinal stiffener under patch loading, Technology Reports of Kansai University, No.33, 1991, 121-32.

18) Dogaki, M., Tsuda, H., Matsuoka, Y., Aoki, T., and Yonezawa, H., Experiments of longitudinally stiffened plate girders under patch loading, Technology Reports of Kansai University, No.35(in press).

19) Takimoto, T., Imagawa, Y., and Moriwaki, Y., Experimental study of stiffened plate girders under patch loading, The 42nd Annual Meeting of JSCE, 1987(in Japanese).

20) Janus, K., Karnikova, I., and Skaloud, M., Experimental investigation into the ultimate behavior of longitudinally stiffened steel webs under partial edge loading, Acta Technica CSAV, No.2, 1988, 158-95.

INTERACTION CURVE OF ULTIMATE STRENGTH OF STEEL PLATES UNDER IN-PLANE COMBINED LOADING

SATOSHI NARA
Associate Professor, Dr.Eng./Department of Civil Engineering
Gifu University
1-1 Yanagido, Gifu 501-11, JAPAN

ABSTRACT

This paper aims to propose an interaction curve of ultimate strength of steel plates under in-plane combined loading. The ultimate strength of the plates is estimated statistically, because statistical studies are essential to the promotion of a limit-state design approach. The paper has two parts. Basic strength curves of the plates are explained in the first part. The second part deals with the interaction curve of the ultimate strength of the plates. Based on the precise analytical data for the ultimate strength and the three basic strength curves, the interaction equation is proposed in an exponential form on the basis of the mean values of the ultimate strength.

INTRODUCTION

There have been extensive investigations on the compressive ultimate strength of plates both for unstiffened and stiffened. A wide variety of load carrying capacity curves of compressive steel plates is used for different specifications in many countries. Making a comparison between the curves, there is a difference in ultimate strength in all range of plate slenderness. Therefore, evaluation in the ultimate strength is essential to promote the ultimate limit-state design method.

On the other hand, there have been comparatively few studies on the ultimate plate strength under in-plane combined loading. A book edited by Dubas and Gehri was published as an accomplishment of a technical working group in ECCS[1] six years ago. The purpose of the book is to present a state-of-the-art of present day knowledge of the behaviour up to collapse of steel plated structures. Results in this book give the background of Eurocode No.3. In Japan, guidelines for stability design of steel structures edited by Fukumoto were issued by Japan Society of Civil Engineers[2]. Moreover, a study group, the chair of which he took, were organized in 1987, financially supported by the Grand-in-Aid for Co-operative Scientific Research Sponsored by the Ministry of Education, Science and Culture of the Japanese Government. Final report of the group had been published[3].

This paper develops a chapter of the above mentioned final report which proposed mean value curves of ultimate strength of steel plates under individual in-plane loading[4]. Making use of SGST(Study Group of Steel Structures in Tokai) Format, these curves become useful to evaluate the ultimate strength for the limit-state design method. Concerning the plates subjected to in-plane bending and compression, the ultimate strength is estimated statistcally, based on the analytical results and sufficient data associated with the measured imperfections of steel bridges. Because, it is impossible to deal with unsufficient experimental data statistically.

MAIN FEATURES OF STUDY

The following informations are essential to promoting the limit-state design of steel structures,
  a) Properties of ultimate strength of individual structural element,
  b) Appropriate resistance factor, and
  c) Standards of fabrication tolerances.
Studies on the tolerances were carried out in Europe[5] and Japan[6].
  Fig.1 shows a general procedure of the statistical evaluation. This procedure is divided into the following three main parts.

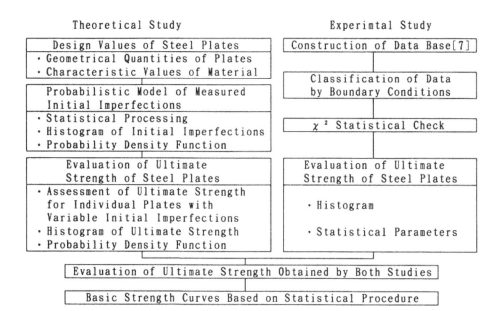

Figure 1. Statistical evaluation of ultimate strength

153

EXPERIMENTAL DATA
Experimental data base constructed by Itoh-Fukumoto[7,8] is available
concerning compressive plates. After classification of these data by
boundary conditions, statisitical check is used based on the χ^2 test.
As a result, statistical properties of ultimate strength are clarified.

THEORETICAL DATA
A statistical study on ultimate strength of compression plate members
has been presented by Komatsu-Nara[9]. In order to estimate statistical
properties of ultimate strength theoretically, needed are informations
of practical design variables and initial imperfections as well as an
appropriate analytical method.

Information of practical design variables: Ranges of design variables
which dominates the ultimate strength of steel plates are decided based
on actual steel bridges constructed in Japan.

Information of practical data of initial imperfections: Concerning
initial deflections[6,9] and residual stress[2,10], numerous data had
been collected through measurement of actual steel bridges as well as
pilot tests.

Appropriate analytical method: An elasto-plastic finite displacement
analysis has been developed[11]. Based on this analytical method,
evaluated are sensitivity curves which express relations between
ultimate strength and initial imperfections. Using these sensitivity
curves, histograms of ultimate strength are obtained easily.

BASIC STRENGTH CURVES
Compared theoretical data with experimental ones, mean value curve of
ultimate strength is proposed in the following form,

$$\frac{N u}{N y} \quad , \quad \frac{M u}{M y} \quad or \quad \frac{S u}{S y} \quad = \quad (\frac{\overline{\lambda} pcr}{\lambda p})^{\beta} \tag{1}$$

in which, $\overline{\lambda}$ pcr is the critical plate slenderness ratio when $N u/N y$,
$M u/M y$ or $S u/S y$ is equal to 1. By means of the least square method,
the formulas are obtained. Moreover, resistance factors are caluculated.
Interaction curves of ultimate strength are proposed in the exponential
equation based on these mean value curves of ultimate strength of the
plates under individual in-plane loading.

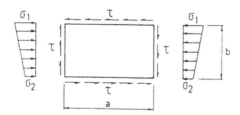

Figure 2. Steel plate under in-plane combined loading

ANALYTICAL MODEL

STEEL PLATE PANELS

Fig. 2 shows a steel plate under the in-plane combined loading. These complicated loading conditions are divided into 3 kinds of in-plane loading, that is, compression, in-plane bending and shear. The models of the plate are simply supported along the four edges. The in-plane boundary conditions are introduced along the edges with rigid bars, as shown in Fig. 3.

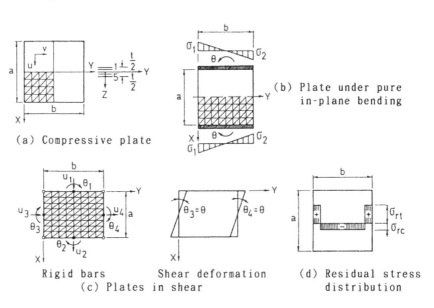

(a) Compressive plate

(b) Plate under pure in-plane bending

Rigid bars   Shear deformation   (d) Residual stress
(c) Plates in shear   distribution

Figure 3. Analytical model of steel plates

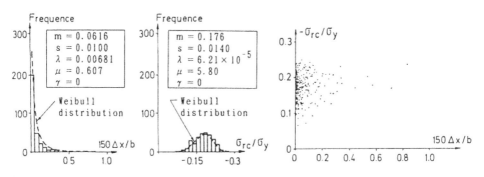

(a)Initial deflection (b)Residual stress

Figure 4. Histograms and probability
           density functions of initial
           imperfections

Figure 5. Scatter of Initial
           Imperfections

STATISTICAL MODEL OF INITIAL IMPERFECTIONS

Fig. 4 shows histograms and probability density functions of residual
stress and initial deflection which were obtained by several research
projects in Japan[2,6,9,10]. Initial deflections were measured over the
gauge length equal to width of plate panel b. After examining the
fitness of the Weibull distribution by means of the χ^2 test, the
Weibull distribution, whose probability density function is expressed by
eq. (2), has been found most comfortable to all the histograms of initial
imperfections. Making use of these probability density functions and
600 random numbers, 300 pairs of initial imperfections which consist of
initial deflection and residual stress are obtained, as shown in Fig. 5.

$$f(x) = \frac{\mu}{\lambda}(x - \gamma)^{\mu-1}\exp\left\{\frac{-(x-\gamma)^{\mu}}{\lambda}\right\} \qquad (2)$$

BASIC STRENGTH OF STEEL PLATES

COMPRESSIVE PLATES

Many investigations on the ultimate strength of steel compressive plates
have been performed in recent years. A wide variety of load carrying
capacity curves of the plates is used for different specifications in
many countries, as shown in Fig. 6.

Theoretical Study: A steel structure contains both residual stresses
and initial deflections, which considerably influence the ultimate
strength of the structure and its structural members. Therefore, the
effect of such initial imperfections should be introduced as a primary
factor in the estimation of the ultimate strength by means of an
appropriate analysis based on elasto-plastic finite displacement theory.
For this reason, a statistical study on the initial deflections of plate

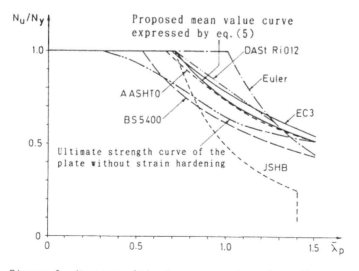

Figure 6. Average ultimate compressive strength curve

members in steel girder bridges fabricated in Japan had been carried out
[6].   On the basis of the studies on the initial deflections[6] and the
residual stresses[2,10] in Japan, statistical evaluation of the ultimate
strength of steel plates have been made possible[9].   Three ultimate
strength curves are given through the theoretical studies on the
ultimate strength curve of the plates with strain hardening[4,12].
As shown in Fig.6, eq.(3) shows the average ultimate strength curve.

$$\frac{N u}{N y} = (\frac{0.7}{\overline{\lambda} p})^{0.865} \leq 1 \quad , \quad \overline{\lambda} p \leq 1.3 \quad\quad (3)$$

Experimental Study: Based on the numerous experimental data in many
countries, numerical data-base approach to the ultimate strength of the
plates has been established[7,8].   After categorized by the cross-
sectional shapes and loading equipments, obtained are two ultimate
strength curves by means of the χ^2 statistical test, that is, the
average ultimate strength and (m-1.65 s) curve.

$$\frac{N u}{N y} = (\frac{0.7}{\overline{\lambda} p})^{0.862} \leq 1 \quad\quad (4)$$

Eq.(4) shows the average ultimate strength curve.   "m" and "s" denote
mean value and standard deviation of the ultimate strength, respectively.
If the normal distribution is suitable for the histogram of the ultimate
strengths, (m-1.65 s) curve is equal to 95% fractile one.

Basic Strength Curve: The average ultimate strength curve expressed by
eq.(4), which was obtained by the experimental data, is in good
agreement with the analytical data.   Therefore, the following mean value
curve of ultimate strength are proposed.

$$\frac{N u}{N y} = (\frac{0.7}{\overline{\lambda} p})^{0.86} \leq 1 \quad\quad (5)$$

Using SGST Format, the average strength curve becomes useful to evaluate
the ultimate strength for the limit-state design method.

Plates under In-Plane Bending

Ultimate Strength of Steel Plates with Tolerances: Based on the
numerical results[13] obtained for the ultimate strength, the curves are
expressed by the following equations, as shown in Fig.7.   These two
curves greatly differ with the residual stress.

$$\frac{M u}{M y} = (\frac{1.21}{\overline{\lambda} p})^{0.420} \leq 1.5 \quad (6) \qquad \frac{M u}{M y} = (\frac{0.790}{\overline{\lambda} p})^{0.594} \leq 1.5 \quad (7)$$
$$\text{for } \sigma rc/\sigma y = -0.4 \qquad\qquad\qquad \text{for } \sigma rc/\sigma y = 0$$

Histograms of Ultimate Strength: Relations between ultimate strength
and initial imperfections are calculated by an numerical analysis based
on the elasto-plastic finite displacement theory.   Sensitivity curves
and the 300 pairs of initial imperfections which consist of initial
deflection and residual stress produce histograms of ultimate strength
which are classified by plate slenderness $\overline{\lambda} p$.

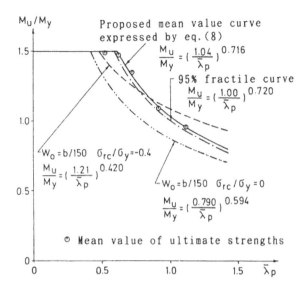

Figure 7. Proposed ultimate strength curves of steel plates
under pure in-plane bending

Basic Strength Curve: Mean value and 95% fractile value of the 300
ultimate strength data are calculated for $\overline{\lambda} p \leq 1.3$. These values are
approximated accurately by the following equations,

$$\frac{Mu}{My} = (\frac{1.04}{\overline{\lambda} p})^{0.716} \leq 1.5 \quad (8) \qquad \frac{Mu}{My} = (\frac{1.00}{\overline{\lambda} p})^{0.720} \leq 1.5 \quad (9)$$

Eq. (8) and (9) show the mean value and 95% fractile curve, respectively.
Both ultimate strength curves are dominated by initial deflection more
than by residual stress, as shown in Fig. 7.

PLATES IN SHEAR

Ultimate Strength of Steel Plates with Tolerances: In the case of
design for steel plates under the in-plane combined loading, it is
important to clarify the ultimate strength of steel plate panels in
shear. A study on characteristics of the ultimate strength of the
panels under uniform shearing stress on the basis of the elasto-plastic
finite displacement theory has been carried out[14]. Magnitude and
shape of the initial deflections affect the ultimate shear strength of
the panels. The following ultimate shear strength curve is proposed.
to represent the theoretically obtained ultimate strength, as shown in
Fig. 8.

$$\frac{Su}{Sy} = (\frac{0.486}{\overline{\lambda} p})^{0.333} \leq 1, \qquad 0.486 \leq \overline{\lambda} p \leq 2 \qquad (10)$$

Basic Strength Curve: After obtaining histograms of ultimate strength,
mean value and 95% fractile value of the 300 ultimate strength data are

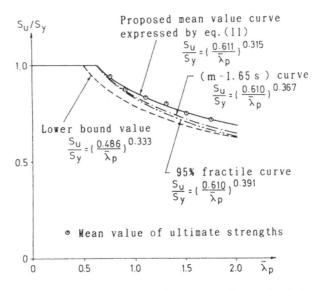

Figure 8. Proposed ultimate strength curves of steel plates in shear

calculated for $\bar{\lambda}p \leq 2$. These values are approximated accurately by the following equations,

$$\frac{S_u}{S_y} = \left(\frac{0.611}{\bar{\lambda}p}\right)^{0.315} \leq 1.5 \quad (11) \qquad \frac{S_u}{S_y} = \left(\frac{0.610}{\bar{\lambda}p}\right)^{0.391} \leq 1.0 \quad (12)$$

Eq.(11) and (12) show the mean value and 95% fractile curve, respectively. These two curves and $(m-1.65s)$ curve are shown in Fig.8. The difference between 95% fractile and $(m-1.65s)$ curve becomes large as $\bar{\lambda}p$ increases. Both ultimate strength curves are dominated by initial deflection more than residual stress.

RESISTANCE FACTOR

<u>SGST format</u>: SGST Format[15] defines the term of resistance as follows,

$$\phi R_n = \phi_1 (1 - k_R V_R) R_m \leq (\text{Term of Load}) \qquad (13)$$
where
$$V_R = \sqrt{V_F^2 + V_M^2 + V_P^2} \qquad (14)$$

Resistance factor ϕ is calculated using the following equation,

$$\phi = \phi_1 (1 - k_R V_R) \psi \qquad (15)$$
where
$$\psi = R_m / R_n \qquad (16)$$
$$R_m / R_n = F_m M_m P_m \qquad (17)$$

F_m, M_m, V_F and V_M are given by experimental data[7,8].

<u>Resistance factor</u>: P_m and V_P are unknown statistical parameters. The parameters are derived from individual ultimate strength of the plates and the basic strength curve. Because P_m and V_P are mean value and

coefficient of variance of P which is defined as a ratio of $\bar{R}$ to $\bar{R}n$. In the case of compressive plates, $\bar{R}$ and $\bar{R}n$ are individual ultimate strength obtained by experimental data and basic strength expressed by eq.(5), respectively. $\bar{R}$ of the plates under in-plane bending and shear is individual ultimate strength obtained by the statistical simulation using the sensitivity curves of ultimate strength and 300 pairs of initial imperfections. Moreover, $\bar{R}n$ is given by eqs.(8) and (11). After Pm and Vp are calculated, substituting eqs.(14), (16) and (17) into eq.(15), resistance factor is obtained as shown in TABLE 1. According to relations between mean value and 95% fractile curve of ultimate strength, resistance factor of compressive plates is smaller than that of plates under in-plane bending or shear. However, there are still few experimental data of steel plates under in-plane bending or shear. Therefore, it is appropriate to use $\phi = 0.85$ at present.

TABLE 1
Resistance factor of steel plates

Load	Compression	Bending	Shear
ϕ	0.85	0.92	0.91

PLATES UNDER IN-PLANE COMBINED LOADING

PLATES UNDER IN-PLANE BENDING AND COMPRESSION

Interaction curves of ultimate strength of steel plates with tolerances: Author had already proposed an evaluation method and interaction curves of ultimate strength of steel plates under in-plane bending and compression[13]. These interaction curves were introduced based on ultimate strength of steel plates with tolerances in the follwing equations,

$$(Nu*)^p + (Mu*)^q = 1 \tag{18}$$

where

$$p = 0.468\,\bar{\lambda}p^2 - 1.63\,\bar{\lambda}p + 2.00$$
$$q = 0.041\,\bar{\lambda}p^2 + 0.340\bar{\lambda}p + 0.974 \tag{19}$$
$$\text{for } \sigma rc/\sigma y = -0.4, \ 0.5 \leq \bar{\lambda}p \leq 1.3$$
$$p = 0.782\,\bar{\lambda}p^2 - 1.77\,\bar{\lambda}p + 1.88$$
$$q = -0.250\,\bar{\lambda}p^2 + 0.523\bar{\lambda}p + 0.968 \tag{20}$$
$$\text{for } \sigma rc/\sigma y = 0, \ 0.5 \leq \bar{\lambda}p \leq 1.3$$
$$Nu* = \frac{Nu}{(Nu)_{\varphi = 0}}, \quad Mu* = \frac{Mu}{(Mu)_{\varphi = 2}} \tag{21}$$

The shape of the interaction curve is dependent on the plate slenderness $\bar{\lambda}p$. p and q are obtained by means of the least square method.

Histograms of Ultimate Strength: Relations between ultimate strength and initial imperfections are calculated by an numerical analysis based on the elasto-plastic finite displacement theory. Sensitivity curves are obtained in the following form,

$$Ku = (e_1 + e_2 d + e_3 d^2)(f_1 + f_2 r + f_3 r^2) \tag{22}$$

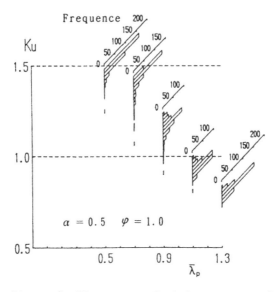

Figure 9. Histograms of ultimate strength

in which $d = 150 \Delta x / b$ and $r = -\sigma rc / \sigma y$. Substituting the 300 pairs of initial imperfections which consist of initial deflection and residual stress into eq. (22), obtained are histograms of ultimate strength which are classified by plate slenderness $\overline{\lambda} p$, as shown in Fig. 9 for $\varphi = 1$.

Basic interaction curve of ultimate strength: Mean value and 95% fractile value of the 300 ultimate strength data are calculated for φ and $\overline{\lambda} p \leq 1.3$. The mean values are plotted in Fig. 10. These values are approximated accurately by eq. (18) and the following equations,

$$p = 1.10 \, \overline{\lambda} p^2 - 2.24 \, \overline{\lambda} p + 1.91$$
$$q = -4.05 \, \overline{\lambda} p^2 + 7.32 \, \overline{\lambda} p - 1.00 \tag{23}$$

The solid lines in the figure show the interaction curves expressed by eq. (18) and (20). The broken line shows the elastic interaction curve as follows.

$$N u* + M u* = 1 \tag{24}$$

Comparison with other study: Fig. 11 for $\alpha = 0.5$ shows the comparison of the proposed interaction curves with those obtained by Horne-Dowling-Ogle[1]. Their interaction curves, which are denoted by the solid line in the figure, are as follows.

$$\frac{\sigma_{c,u}}{S c \sigma y} + \left(\frac{\sigma_{B,u}}{S b \sigma y} \right)^2 = 1 \tag{25}$$

in which,
$$\sigma_{c,u} = N u \sigma y / N y$$
$$\sigma_{B,u} = M u \sigma y / M y \tag{26}$$

and, S c and S b are the nondimensionalized ultimate strength of the

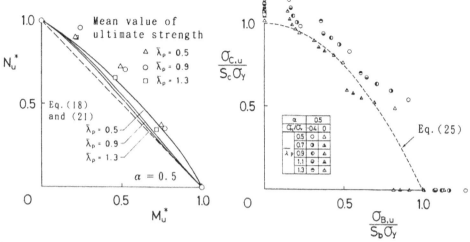

Figure 10. Proposed interaction
curves of ultimate strength
of the plates under in-plane
bending and compression

Figure 11. Obtained ultimate strength
compared with the interaction curve
proposed by Horne-Dowling-Ogle[1]

plates by σy under uniaxial compression and pure in-plane bending,
respectively. S c and S b are given by the function of (b / t)$\sqrt{}$ (σ y/
355). Many plotted points in the figure show the present numerical
results. Compared with Eq. (25), the numerical points for α =0.5 fall
inside of the interaction curve for high bending moment and σ rc=0.

FURTHER STUDY

Authors had already proposed an evaluation method and interaction
curves of ultimate strength of steel plates under in-plane combined
loading[16]. These interaction curves are introduced based on ultimate
strength of steel plates with tolerances. However, they will be
applicable to evaluation method of mean value of ultimate strength,
because they are expressed by generalized non-dimension parameters.

Further work is to be confirmed that the proposed method and
interaction curves are valid for evaluating mean values of ultimate
strength of steel plates under in-plane combined loading with shear.

CONCLUSION

This study aims at proposing the design formulas of the steel plates
under in-plane loading by means of the numerous experimental data as
well as the large number of numerical results based on the elasto-
plastic finite displacement theory.

The main conclusions obtained in this study are as follows.
(1) Proposed are the interaction curve of mean value of ultimate
strength of the plates under in-plane bending and compression,
based on mean value curves of ultimate strength of the plates

162

under uniaxial compression, pure in-plane bending and shear,
(2) By using the proposed ultimate strength curves and the original interaction curves which have been proposed, it is possible to obtain precise and actual design formulas,
(3) Further study is necessary to clarify validity of the proposed evaluation method and interaction curves of ultimate strength of steel plates under in-plane combined loading for evaluating mean values of ultimate strength.

REFERENCES

1. Technical Working Group TWG 8.3 of the ECCS, Behaviour and design of steel plated structures, ed. Dubas,P. and Gehri,E., Applied Statics and Steel Structures, Swiss Federal Institute of Technology, Zürich, 1986.
2. Subcommittee on Stability Design of the JSCE, Guidelines for stability design of steel structures, ed. Y.Fukumoto, Japan Society of Civil Engineers, Tokyo, Oct., 1987.
3. Fukumoto,Y.(ed.), Unified evaluation of ultimate strength of steel framed structures, Final Report of the Grand-in-Aid for Co-operative Scientific Research Sponsored by the Ministry of Education, Science and Culture of the Japanese Government, Mar., 1990.
4. Nara,S. and Fukumoto,Y., Étude statistique de la résistance ultime des plaques en acier sollicitées dans leur plan. Construction Métallique, 1991, 3, 15-24.
5. Massonnet, Ch., Tolerances in steel plated structures. IABSE Surveys, Aug., 1980, S-14/80, 49-76.
6. Committee on Initial Deformation Measurement of the JSSC, Statiscal study on the initial deformations and the ultimate strengths of steel bridge members. JSSC, Apr., 1980, 16, 10-43(in Japanese).
7. Itoh,Y. and Fukumoto,Y., Basic compressive strength of steel plates from test data. Proc. of JSCE, 1984, 340, 129-139.
8. Itoh,Y., Ultimate Strength Variations of Structural Steel Members, Doctoral Thesis, Presented to Nagoya University, Nov., 1984.
9. Komatsu,S. and Nara,S., Statistical study on steel plate members. Journal of Structural Engineering, ASCE, Apr., 1983, 109, 977-992.
10. Komatsu,S., Ushio,M. and Kitada,T., An experimental study of residual stresses in stiffened panels. Transactions of JSCE, 1977, 9, 70-72.
11. Nara,S. and Yamaguchi,H., Elasto-plastic analysis of steel plates under in-plane combined loading. to appear.
12. Nara,S., Deguchi,Y. and Komatsu,S., Ultimate strength of compressive plates with strain hardening. Journal of Structural Engineering, Mar., 1987, 33A, 141-150(in Japanese).
13. Nara,S. and Fukumoto,Y., Ultimate strength of steel plates under in-plane bending and compression. Der Stahlbau, 1989, 57, 179-185.
14. Nara,S., Deguchi,Y. and Fukumoto,Y., Ultimate strength of steel plate panels with initial imperfections under uniform shearing stress. Proc. of JSCE, Apr., 1988, 392/I-9, 265-271(in Japanese).
15. SGST, Evaluation of strength of steel structural members and its application to a reliability design. Bridges and Foundations, 1980, 14, 33-41(in Japanese).
16. Nara,S. and Fukumoto,Y., Evaluation of ultimate strength of steel

plates under in-plane combined loading. Proc. of Pacific Structural Steel Conference. AISC. Gold Coast. 28-31. May. 1989. 450-460.

NOTATION

NOTATION FOR ULTIMATE STRENGTH

a : length of plate. b : width of plate.

E : Young's modulus($=2.06 \times 105 N/mm^2$).

k_1, k_s : buckling coefficients based on the elastic buckling theory, expressed by the following equations.

$$k_1 = \begin{cases} 8.4/(2.1-\varphi), & 0 \le \varphi \le 1 \\ 10(1-\varphi)^2 - 6.27(1-\varphi) + 7.63, & 1 < \varphi \le 2 \end{cases} \quad k_s = \begin{cases} 5.34 + 4.0/\alpha^2, & \alpha \ge 1 \\ 4.0 + 5.34/\alpha^2, & \alpha < 1 \end{cases}$$

Ku : index of ultimate strength($=(Nu/Ny)+(Mu/My)$).

Mu : in-plane bending moment at the ultimate state.

$(Mu)_{\varphi=2}$: ultimate in-plane bending moment under pure bending.

My : yield in-plane bending moment.

Nu : axial force at the ultimate state. Ny : yield axial force.

$(Nu)_{\varphi=0}$: ultimate axial force under uniaxial compression.

Su : shearing force at the ultimate state. Sy : yield shearing force.

t : thickness of plates.

W_0 : maximum amplitude of initial deflection.

α : aspect ratio of plate($= a/b$).

Δx : initial deflection obtained according to the measuring method specified IDWR.

φ : in-plane normal stress gradient($=2(Mu/My)/\{(Nu/Ny)+(Mu/My)\}$)

ν : Poisson's ratio($=0.3$).

σcr: elastic plate buckling stress($= k_1 \sigma_E$).

σ_E : Euler reference stress($=(t/b)^2 \cdot (\pi^2 E)/\{12(1-\nu^2)\}$).

σrc: residual compressive stress. σrt: residual tensile stress.

τcr: elastic plate buckling stress($= k_s \sigma_E$).

σy : measured direct yield stress. τy : measured shear yield stress.

$\bar{\lambda} p$: plate slenderness parameter($= \sqrt{(\sigma y/\sigma cr)}$ or $\sqrt{(\tau y/\tau cr)}$).

NOTATION FOR RESISTANCE FACTOR

k_R : constant which expresses lower bound ($= 1.65$).

F : coefficient for fabrication ($= S/Sn$).

Fm : mean value of coefficient for fabrication ($= 1.00$).

Fy : nominal yield stress.

M : coefficient for materials ($= \sigma y/Fy$).

Mm : mean value of coefficient for materials ($= 1.15$).

Pn : coefficient for initial imperfections ($= \bar{R}/\bar{R}n$)

Pm : mean value of coefficient for initial imperfections.

R : resistance. Rm : mean value of resistance.

$\bar{R}$: resistance nondimensionalized by measured values.

$\bar{R}m$: mean value of resistance nondimensionalized by nominal values.

Rn : resistance specified by design code.

S : actual section modulus. Sn : nominal section modulus.

V_F : coefficient of variance for fabrication ($= 0.05$).

V_M : coefficient of variance for materials ($= 0.11$).

V_P : coefficient of variance for initial imperfections.

V_R : coefficient of variance for resistance.

ϕ : resistance factor.

ϕ_1 : corrective coefficient ($= 1.00$).

ψ : coefficient of reliability of design formulas.

DESIGN OF PLATE GIRDERS ACCORDING TO EUROCODE 3

T. ABECASIS; D. CAMOTIM; A. REIS
Invited Lecturer , Associate Professor and Professor
Department of Civil Engineering, Instituto Superior Técnico,
Technical University of Lisbon , Av. Rovisco Pais, 1096 Lisboa Codex, Portugal

ABSTRACT

In this paper a systematic approach for the use of the provisions of Eurocode 3 (Part 1) concerning the design of plate girders is proposed. A sequential diagram is presented which contains the main steps, options and decisions associated with the corresponding safety checking or determination of an ultimate load combination. Unstiffened and tranversally stiffened plate girders are dealt with and from the several aspects that must be considered in the establishment of such diagram attention is paid exclusively to the ones involved in ultimate limit states. Finally, some suggestions are made with the aim of clarifying the choices and/or lowering the amount of work required to design plate girders according to Eurocode 3.

INTRODUCTION

The coming into effect of Eurocode 3 (Part 1:1 General Rules and Rules for Buildings) makes it very important for designers to be familiar with the philosophy, principles and rules contained in this Design Code. Particularly in what concerns the design of plate girders the use of Eurocode 3 involves a process of sequential analyses, options and decisions which is not clearly identified.

In this paper a list of the main steps to be followed in the safety checking of unstiffened and transversally stiffened plate girders is presented (longitudinally stiffened plate girders are not covered by Part 1 of Eurocode 3). Some of these steps are associated with a number of possible options which are discussed and illustrated by means of a sequential diagram.

One final remark to mention that no reference is made either to serviceability limit states or to ultimate limit states induced by fatigue or involving joints and connections.

CLASSIFICATION OF THE CROSS-SECTION

The classification of a cross-section depends on the proportions and yield strength of their *compression elements* (elements which are either totally or partially in compression and that, depending on their position with respect to the axis of bending, may be designated as *webs*, *internal flange elements* or *outstand flanges*), the number and location of which depend on the combination of axial force and bending moment acting on the cross-section, i.e., on the load case. The most efficient girders have slender plate elements, very often Class 3 flanges and Class 4 webs. Here it is assumed that the plate girder cross-section belongs to Class 4 (a cross-section is classified by quoting the highest class of all its compression elements). As a result it is also implicitly assumed that the internal forces and moments used in checking the resistance of the plate girders are obtained by means of an *Elastic Global Analysis*, and that the resistance of the cross-section is verified *elastically*. In the case of a girder with a cross-section of Class 4 it is always necessary, in order to make allowances for the effects of local buckling, to replace it by an *effective cross-section*. The member buckling and cross-section resistances are then determined as for a girder with a Class 3 cross-section.

Effective Cross-Section Properties

The effective cross-section properties (A_{eff}, W_{eff}) must be based on the values of the *effective widths* b_{eff} of the compression elements. These values are obtained from the widths of the compression elements which are under compressive stresses through the multiplication by a reduction factor ρ which depends on a normalized plate slenderness $\overline{\lambda}_p$ which depends on the yield strength and on the critical plate-buckling stress. The effective cross-section depends on the load case.

In many cases the neutral axis of the effective cross-section shifts by a dimension e_N compared to the neutral axis of the gross cross-section. This fact must be taken into account (i) in the calculation of the properties of the effective cross-section and (ii) by including in the bending moment acting on the cross-section an additional term $\Delta M = N.e_N$, where N is the axial force.

RESISTANCE OF THE CROSS-SECTION (I - BENDING AND COMPRESSION)

It is assumed that the plate girder is subjected to a combination of compressive axial force, shear and uniaxial bending moment (about the strong axis of the cross-section). Moreover, it is also assumed that (i) *shear lag effects* may be neglected and that (ii) *flange induced buckling* does not occur.

Compression

The resistance of a cross-section is verified if $N_{Sd} \leq A_{eff} f_y/\gamma_{M1}$ ($\equiv N_{c.Rd}$) ,where N_{Sd} is the design value of the axial force, A_{eff} is the area of the effective cross-section f_y is the yield strength and γ_{M1} is the partial safety factor associated to the buckling resistence of members or to the resistance of Class 4 cross-sections (γ_{M1}= 1.1 in Eurocode 3).

Bending Moment

In the absence of shear force the resistance of a cross-section without holes for fasteners is verified if $M_{Sd} \leq W_{eff} f_y/\gamma_{M1}$ ($\equiv M_{c.Rd}$) ,where M_{Sd} is the design value of the bending moment and W_{eff} the elastic modulus of the effective cross-section.

Bending and Compression

In the absence of shear force the resistance of a cross-section without holes for fasteners is verified if
$$[N_{Sd}/(A_{eff} f_{yd})] + [(M_{Sd} + N_{Sd} e_N)/(W_{eff} f_{yd})] \leq 1 \tag{1},$$
where $f_{yd} = f_y/\gamma_{M1}$, and A_{eff} and W_{eff} are the area and elastic modulus of effective cross-sections associated to pure compression and pure bending respectively.

RESISTANCE OF THE CROSS-SECTION (II - SHEAR BUCKLING)

It is assumed that the shear resistance of the cross-section is limited by the shear buckling of the web. This assumption is valid for *thin* webs, i.e., webs that satisfy one of the following conditions: (i) $d/t_w > 69 \varepsilon$ (unstiffened web) and (ii) $d/t_w > 30 \varepsilon$ $(k_\tau)^{1/2}$ (transversally stiffened web) where d and t_w are the web depth and thickness, $\varepsilon = (235/f_y)^{1/2}$ and k_τ is the buckling factor for shear, which incorporates the influence of the spacing between stiffeners. Eurocode 3 requires that webs satisfying (i) must be provided with transverse stiffeners at the supports. As a result, in the context of the verification on the shear buckling resistance of a cross-section the designation "unstiffened web" actually means "web with transverse stiffeners only at the supports".

Eurocode 3 proposes two methods for the verification of the shear buckling resistance of a cross-section, namely the *simple post-critical method* (SPCM) and the *tension field method* (TFM). The TFM may only be used for cross-sections with transversally stiffened webs (TSW) and in one of the following situations: (i) internal panels such that $1 \leq (a/d) \leq 3$ and (ii) end panels such that $1 \leq (a/d) \leq 3$ provided that a suitable (stiff) end post is provided where a is the clear spacing between transverse stiffeners. The SPCM may be used in all cases, i.e., is an alternative in the situations mentioned above and is the *only* method applicable in the case of: (i) unstiffened webs (UW), (ii) internal or end panels of TSW such that $(a/d) < 1$ or > 3 and (iii) end panels of TSW such that $1 \leq (a/d) \leq 3$ if no suitable end post is available.

Simple Post-Critical Method
According to this method the resistance of a cross-section is verified if

$$V_{Sd} \leq dt_w \, \tau_{ba}/\gamma_{M1} \ (\equiv V_{ba.Rd}) \tag{2},$$

where V_{Sd} is the design value of the shear force an τ_{ba} is the simple post-critical shear strength, which depends on the web slenderness $\overline{\lambda}_w = [(f_{yw}/\sqrt{3})/\tau_{cr}]^{1/2}$ (f_{yw} is the web yield strength and τ_{cr} is the elastic critical shear strength).

Tension Field Method
According to this method the resistance of a cross-section is verified if

$$V_{Sd} \leq [(dt_w \, \tau_{bb}) + 0.9 \, (gt_w \, \sigma_{bb} \sin \emptyset)]/\gamma_{M1} \ (\equiv V_{bb.Rd}) \tag{3},$$

where σ_{bb}, g and $\emptyset$ are the strength, width and inclination of the tension field, and τ_{bb} is the initial shear buckling strength, which is a function of $\overline{\lambda}_w$ ($\tau_{bb} \leq \tau_{ba}$). Although, due to space limitations, most of the expressions involved in the calculation of $V_{bb.Rd}$ are not presented here, it should be pointed out that: (i) it is necessary to calculate the anchorage lengths of the tension field along the compression and tension flanges (s_c and s_t, respectively), (ii) the value of each anchorage length may not exceed a and depends on the values of $\emptyset$, σ_{bb} and M_{Nf}, which is the *reduced plastic resistance moment* of the corresponding flange, (iii) the appropriate value of the inclination $\emptyset$ must be comprised between $\theta/2$ and θ (θ is the slope of the panel diagonal) and must be such that it maximizes the value of $V_{bb.Rd}$, (iv) since the determination of the appropriate value of $\emptyset$ requires an iterative procedure, Eurocode 3 proposes an alternative to avoid this work, which consists of considering $\emptyset = \theta/1.5$ and usually leads to a slight underestimation of the shear resistance and (v) when a suitable end post is provided the determination of the design shear buckling resistance of an end panel follows the procedure outlined above, the difference with respect to internal panels consisting in the fact that a different expression must be used to calculate s_c and that the anchorage length along the end post s_s must be determined.

RESISTANCE OF THE CROSS-SECTION (III - INTERACTION BETWEEN SHEAR FORCE, BENDING MOMENT AND AXIAL FORCE)

The procedure for taking into account the interaction depends on the method used to calculate the shear buckling resistance.

Simple Post-Critical Method

The cross-section is able to resist any combination of shear force V_{Sd} and bending moment M_{Sd} corresponding to a point lying within the boundaries of the curve represented in Fig. 1 (a). Observe that, concerning the different regions of the interaction diagram,

(i) The resistance of a cross-section is verified if $M_{Sd} \leq M_{f.Rd}$ and $V_{Sd} \leq V_{ba.Rd}$, where $M_{f.Rd}$ is the design plastic moment resistance of a cross-section consisting only of the flanges and allowing for the presence of the axial force N_{Sd}.

(ii) If $V_{Sd}/V_{ba.Rd} \leq 0.5$ the design resistance of the cross-section for bending moment and compression need not be reduced to allow for the shear force.

(iii) If $V_{Sd}/V_{ba.Rd} > 0.5$ the following criterion must be verified

$$M_{Sd} \leq M_{f.Rd} + (M_{pl.Rd} - M_{f.Rd}) [1 - (2 V_{Sd}/V_{ba.Rd} -1)^2] \qquad (4),$$

where $M_{pl.Rd}$ is the reduced plastic resistance moment of the cross-section allowing for the presence of N_{Sd}.

(iv) In any case, M_{Sd} cannot exceed the design resistance for bending and compression, determinated previously by means of eq. (1) and designated by M_u in Fig. 1 (a).

Tension Field Method

The interaction diagram corresponding to the resistance to combinations of shear force and bending moment (with allowance for axial force) is represented in Fig.1 (b). The following observations are appropriate:

(i) The resistance of a cross-section is verified if $M_{Sd} \leq M_{f.Rd}$ and $V_{Sd} \leq V_{bw.Rd}$ where $V_{bw.Rd}$ is the "web only" shear buckling resistance of the cross-section.

(ii) If $V_{Sd}/V_{bw.Rd} \leq 0.5$ the design resistance of the cross-section for bending moment and compression need not be reduced to allow for the shear force.

(iii) If $V_{Sd}/V_{bw.Rd} > 0.5$ a criterion identical to the one expressed in eq. (4) (with $V_{ba.Rd}$ replaced by $V_{bw.Rd}$) must be verified. Again M_{Sd} cannot exceed the value M_u.

(iv) If $M_{Sd} < M_{f.Rd}$ allowance can be made for the contribution of the flanges to the tension field action (V_{b0} in Fig. 1 (b) is the value of $V_{bb.Rd}$ corresponding to $M_{Sd} = 0$).

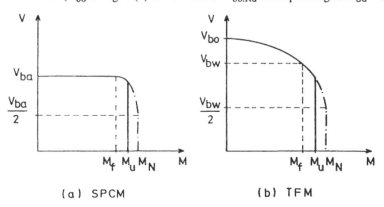

(a) SPCM   (b) TFM

Figure 1. Interaction of shear buckling and moment (allowing for axial force) resistances

TRANSVERSE STIFFENERS REQUIREMENTS

In plate girders with TSW the transverse stiffeners (TS) may be classified according to their *location* and according to their *function*. According to their location TS may be (i) *intermediate stiffeners* if they are adjacent to two web panels or (ii) *end stiffeners* if they are adjacent to a sole web panel. According to their function TS may be (i) *load bearing stiffeners* if external concentrated loads (applied forces or support reactions) act on them or (ii) *non load bearing stiffeners* if only forces transmitted by webs and flanges in the postcritical range are applied.

Intermediate Non Load Bearing Stiffeners

In order to check to safety of these stiffeners the following requirements have to be met:
(a) The second moment of area (I_s) must be such that (i) $I_s \geq 1.5\ d^3 t_w/a^2$ (a/d < $\sqrt{2}$) or
    (ii) $I_s \geq 0.75\ d^3 t_w/a^2$ (a/d $\geq \sqrt{2}$).
(b) The out-of-plane buckling resistance must be verified. In order to do this it is necessary to take into account that: (i) the end compressive force, acting in the middle plane of the web, is given, for both SPCM and TFM, by $N_s = V_{Sd} - dt_w \tau_{bb} \geq 0$, where the value of τ_{bb} used is the lower for the two adjacent panels, (ii) buckling curve c and a buckling length $l \geq 0.75$ d must be used, (iii) the effective cross-section includes a width of web plate equal to 30 ε t_w (half in each side), for the case of a symmetric stiffener and (iv) in asymmetric stiffners (one sided, for instance), the resulting load eccentricity must be allowed for (the stiffener acts as a beam-column).

Intermediate Load Bearing Stiffeners

These stiffeners must always be symmetric about the centre line of the web and are subjected to a compressive force, acting in the middle plane of the web, given by $N_s = R_{Sd}$, where R_{Sd} is the design value of the bearing reaction or external applied load. It is necessary to verify: (i) the out-of-plane buckling resistance using curve c, a buckling length $l \geq 0.75$ d and the effective cross-section mentioned previously and (ii) the cross-section resistance adjacent to the loaded flange. The width of web plate included in the effective cross-section must be limited to s_y, which is a length associated with the crushing resistance of UW subjected to tranverse forces applied through a flange.

End Load Bearing Stiffeners (Stiff End Posts)

Besides the requirements that must be satisfied by an intermediate load bearing stiffeners (with $N_s = R_{Sd}$) a stiff end post of a panel, the shear buckling resistance of which is determined using the TFM, is also required to resist the corresponding tension field anchorage force Fbb. In the case of a single plate end post this may be achieved by satisfying a criterion presented in Eurocode 3. No design expressions are given for the case of a twin-stiffener type of end post.

WEB CRIPPLING - TRANSVERSE FORCES

The resistance of an UW to concentrated (or distributed over short lengths) transverse forces applied, through a flange, in its plane is governed by (i) *crushing* of the web close to the flange, (ii) *crippling* of the web in the form of localised buckling and crushing of the web close to the flange, or (iii) *buckling* of the web over most of the depth of the member. The corresponding web resistances are designated by $R_{y.Rd}$, $R_{a.Rd}$ and $R_{b.Rd}$, respectively. They are influenced by, among other quantities, the length of stiff bearing (s_s). In its determination a load dispersion with slope 1:1 is admitted.

    Eurocode 3 makes a distinction between two types of load application, namely (i) forces applied through one flange and resisted by web shear forces and (ii) forces applied to one flange and transferred through the web directly to the other flange (at member intersections, for instance). In the first case, only the crushing and crippling resistances must be verified. In the second case it is sufficient to verify the crushing and buckling resistances. Finally, in the case of a TSW with the force applied between stiffeners, the web crippling resistance is still governed by the same phenomena but its value increases due to the presence of the stiffeners (no quantitative indication is given concerning this effect).

    Transverse forces in UW or between the stiffeners of TSW may also affect the resistance of the cross-section to bending and compression by lowering (i) the local buckling resistance of the web and (ii) the value of the longitudinal stresses that can be applied before yielding takes place in the web.

BUCKLING RESISTANCE OF THE PLATE GIRDER

Since the plate girder is subjected to a combination of compressive axial force, shear and uniaxial bending moment about the strong axis of the cross-section (axis y), the corresponding buckling resistance must be also verified. Two situations are considered, namely that the girder is (i) *laterally unrestrained* (i.e., such that lateral-torsional buckling is a potential failure mode) or (ii) *laterally restrained*. In the first case the following criteria must be satisfied:

$$[N_{Sd}/(\chi_{min} A_{eff} f_{yd})] + [(k M_{Sd} + N_{Sd} e_N) / W_{eff} f_{yd})] \leq 1 \qquad (5)$$

$$[N_{Sd}/(\chi_z A_{eff} f_{yd})] + [(k_{LT} M_{Sd} + N_{Sd} e_N) / (\chi_{LT} W_{eff} f_{yd})] \leq 1 \qquad (6),$$

where χ_{min} is the lesser of the reduction factors associated with flexural buckling (χ_y and χ_z), χ_{LT} is the reduction factor for lateral-torsional buckling, and k and k_{LT} are quantities which take into account the influence of, among others, the shape of the bending moment diagram. In the second case, only the first criterion need to be satisfied. The calculation of the values of χ_y, χ_z, χ_{LT}, k and k_{LT}, which is not dealt with here, may be performed using expressions given is Annex F of Eurocode 3.

CONCLUDING REMARKS

1) A sequential diagram was presented which corresponds to the use of Eurocode 3 in the safety checking or determination of an ultimate load combination of unstiffened or transversely stiffened plate girders acted by a combination of axial compression, shear force and uniaxial bending moment about the strong axis.
2) It was initially assumed that the cross-section was proportioned in such a way that (i) it belonged to Class 4, (ii) shear lag effects could be neglected and (iii) no flange induced buckling would occur. As a result it was necessary to define and characterize an effective cross-section.
3) Three main aspects were involved in the verification of the cross-section resistance (Bending and Compression, Shear Buckling and Interaction between Shear force, Bending moment and Axial force). Particular attention was paid to the application of the two methods proposed for the verification of the shear buckling resistance (simple post-critical method and tension field method).
4) Tranverse stiffeners were classified and the requirements that the different types must fulfill were briefly referred.
5) The effects of the application of concentrated tranverse forces to unstiffened plate girders were described. Particular emphasis was given to web crippling.
6) The criteria that must be satisfied in the verification of the buckling resistance of the plate girder were just presented.
7) The practical application of the rules listed in Eurocode 3 will be made significantly easier by the elaboration of design aids. Examples of design aids that it would be convenient to prepare are:
   - Design aids for the *determination of the shear buckling resistance of internal and end web panels* according to the two methods proposed in Eurocode 3. The corresponding stiffener requirements could also be provided.
   - Design aids for the *determination of the flexural and lateral-torsional buckling resistances of compression members and beams*, respectively, particularly when non-standard features are present.

REFERENCES

1. Eurocode nº 3 - Design of Steel Structures, Part 1.1 General Rules and Rules for Buildings, 1992.

2. Dubas, P. and Gehri, R. (Eds.), "Behaviour and Design of Steel Plated Structures", ECCS Publication nº 44, Zürich, 1986.

The Collapse Behaviour of Web Plates on the Launching Shoe

SHIGERU SHIMIZU
Department of Civil Engineering
Shinshu University
500 Wakasato, Nagano 380 JAPAN

SATOSHI SAKATA
Osaka Prefectural Goverment
Osaka, JAPAN

ABSTRACT

An elasto-plastic large deflection analysis is made on plates subjected to a patch load and in-plane bending on purpose to study the collapse behaviours of web plates on a launching shoe. An FEM procedure with the arc-length increment technique is used as a numerical method. Magnitude of in-plane bending moment and the existence and the location of a horizontal stiffener are adopted as parameters. Through the analysis, influence of the in-plane moment and contributions of stiffeners to the strength of plates are found. Effects of the parameters to the collapse modes are also found from the study.

INTRODUCTION

In designing a web panel on a launching shoe, attention should be paid on strength or collapse behaviours because such a panel generally has no vertical stiffener on the shoe. Such a panel is often modelized as a patch loaded plate. Some studies have been made on the patch loaded plate, for example, by Roberts et al.[1]. Authours also made a sereis of studies on the patch loaded plates; in some of the authors' previous studies[2,3], numerical analyses by DRM (Dynamic Relaxation Method) are made and behaviour of such plates are being cleared. However, with the DRM procedure, it is not easy to take the complex conditions into account because this method adopts the Finite Difference technique to discretize a plate.

Following to authors' previous studies, a sereis of elasto-plastic large deflection FEM analyses on the patch loaded plates are made. This paper presents results of the analysis.

ANALYTICAL PROCEDURE AND MODELS

In the analysis, triangle planner-shell elements for the elasto-plastic
large deflection analysis presented by Komatsu, Kitada et al.[4] are used
and the arc-length increment technique is adopted.

A typical analized models is shown in Fig.1. This model is corresponding to a web panel subtended by two vertical stiffeners, and has top
and bottom flanges and a horizontal stiffener with a launching shoe of
length c. This model also has the initial out-of-plane deflection in its
larger panel. Dimensions of this model are ; b=100 cm, t_f=2.0 cm, t_w=0.6
cm, t_s=0.6 cm and w_{omax}=0.06 cm where notations are indicated in the figure. Magnitude of in-plane bending moment is described with ratio ϕ of
normal stress σ due to the in-plane bending to shear stress τ, that is
$\phi = \sigma / \tau$. Parameters adopted in this analysis are shown in Table 1.

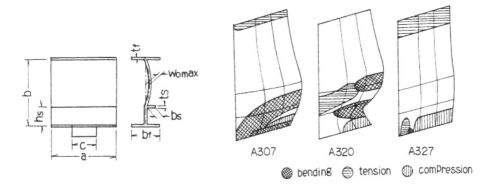

Fig.1 A typical model   Fig.2 Yielded zones with deformations

Table 1. Parameters

parameters	definition	values & notations		
aspect ratios	$\alpha = a/b$	1.0[A]	0.6[B]	
patch length	$\beta = c/a$	0.1[1]	0.3[3]	0.5[5]
location of stiffener	$\gamma = h_s/b$	non[0]	0.1[1]	0.2[2]
bending moment	$\phi = \sigma / \tau$	non[0]	4.1[4]	7.3[7]

In Table 1, notations indicated in [] are used as model names, for
example, Model A304 has its aspect ratio α=1.0, dimensionless patch
length β=0.3, no stiffener and bending moment ϕ=4.1.

Material properties used in this study are Young's modulus E=206 GPa,
yield stress σ_y=235 MPa and Poisson's ratio ν=0.3 which are corresponding to the Grade SS400 steel.

NUMERICAL RESULTS

Fig.2 shows yielded zones of web plates of models A307, A320 and A327 at
their maximum loads. Out-of-plane deformation shapes are also shown in
this figure. It is found from this figure that the model A307, which has
no horizontal stiffener and is subjected to larger in-plane bending moment,
has two zones yielded by bending. That is this model has two yield lines
in its web plate; one of them appears along the lower flange, and the
other in the lower part of the web plate. This layout of yielded zones is
similar to models A300 and A304.

On model A320 which has a horizontal stiffener, its deformation mode
has a nodal line at the horizontal stiffener and the web plate deforms for
alternate directions with two half waves. This model also has two yield
lines; one is located in the lower panel subtended by the horizontal
stiffener and the lower flange, and the other in the upper panel.

The figure for model A327 indicates that this model has no yield line
and is collapsed due to the compression of the lower region of the web
plate. This type of collapse mode is similar to one of model A324. However, on model A324, a yield line begins to develope in the upper panel.

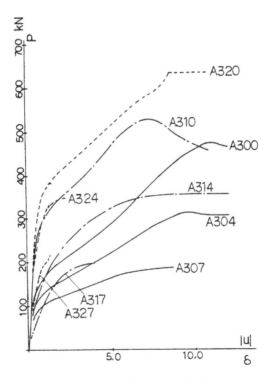

Fig.3 Load-deflection curves

Model A310 is collapsed with two yield lines, and Model A314 has a yieded zone due to compression and also has tow yield lines. Model A317 collapsed by yielded due to compression. That is, models A310-A317 have their yielded zone layouts with middle states of those of corresponding models of A300-A307 and A320-A327.

Fig.3 shows P-δ curves of analyzed models. The horizontal axis of this figure denotes $\sqrt{u \cdot u}$ where u is the displacement vector of FEM analysis. It is clear from these figures that the existence of in-plane moment reduces strength of plate. It is also found that existence of a horizontal stiffener generally increases the maximum loads. However, on models with in-plane bending, stiffeners have small contributions to increase the Pmax. This figure also indicates that models collapsed due to compession of the lower region of web plates, such as models A324, A327 and A317, have small "deformation capacity".

On the model A320, the P-δ curve seems to be breaked at its maximum load. On this model, almost all stiffener elements fully yielded at this stage, and after this stage the web plate begins to deform rapidly for theout-of-plane direction.

CONCLUSION

An elasto-plastic large deflection analysis is made on the web plates subjected to a patch load and an in-plane bending. Through the analysis, it is found that the existence of in-plane bending reduces strength of plates considerably. It is also found that, on models subjected to the in-plane bending, stiffener has small contribution to increase the maximum load.

The numerical analysis of this study is made with use of Computer Center, University of Tokyo, and Shinshu University Computer Center.

REFERENCES

1) Roberts,T.M. & Rockey,K.C., A mechanism solution for predicting the collapse load of slender plate girders when subjected to in-plane patch loading, Proc.ICE,2(2)(1977) 155-75
2) Shimizu,S., Horii,S. & Yoshida,S., The collapse mechanisms of patch loaded web plates, JCSR, 14(1989) 321-37
3) Shimizu,S., Horii,S. & Yoshida,S., Behaviour of stiffened web plates subjected to the patch load, Preliminaly Report of IUTAM Symposium, Prague, 1990
4) Komatsu,S.,Kitada,T. & Miyazaki,S., Elastic-plastic analysis of compressed plate with residual stress and initial deflection, Proc.JSCE, 244(1975) 1-14

NUMERICAL SIMULATION OF PRE-CRITICAL AND POST-CRITICAL BUCKLING BEHAVIOUR OF STEEL PLATED STRUCTURES

PERE ROCA AND ENRIQUE MIRAMBELL

Universitat Politècnica de Catalunya
Department of Constructional Engineering
c/ Gran Capitán s/n. 80834 Barcelona, Spain.

INTRODUCTION

Design of steel plated structures requires to study the possible bidimensional buckling in order to ensure their safety as well as their correct performance in serviciability conditions. However, an accurate treatment of such problems remains partly unavailable due to the difficulties that the nonlinear geometric nature of the phenomena introduces. Furthermore, eventual yielding of steel may strongly couple with the geometric inestabilities, thus giving a complex behaviour which keeps so far uncompletely known.

In this paper, a numerical model for the nonlinear geometric and material analysis of steel shell and plate structures, based in the finite element method, is presented together with its application to the analysis of an example of plate buckling behaviour. The complete formulation of the model, jointly with a set of studied examples may be found in [1].

TREATMENT OF NONLINEAR GEOMETRIC EFFECTS

Geometric nonlinear effects caused by large displacements are considered, while strains are assumed to keep infinitesimal. To account for that, an Updated Lagrangian description is adopted which, once the major order terms are neglected when formulating the Virtual Work Equation, a simple process results which can be summarized in the following items: First, the geometry of the structure must be actualized after each iteration in the analysis, so that the equilibrium condition is always introduced on the deformed shape of the structure. Second, the compatibility equations that relate the displacement and strain components are considered including their quadratic terms. Third, a Geometric Stiffness is introduced and formulated in matrix form to be added to the global Material Stiffness of the structure in order to enhance the convergence of the process used for the solution of the global system of nonlinear equations.

SHELL ELEMENT OF STEEL FOR NONLINEAR MATERIAL ANALYSIS

Due its proven efficiency, the Ahmad isoparametric Lagrangian 9 node shell element has been adopted for the present analysis. Consistently with the use of a Ahmad element, the stress component normal to the mid-surface of the shell is assumed to be approximatelly null at any point into the element. This way, the element may be regarded as a multi-layered system where each layer paralel to the mid-surface is subjected to an state of planar stresses. The material nonlinearity is included by allowing each layer to reach variable values for the mechanical properties as the state of the material is modified throughout the analysis due to the loading process. Besides the in-plane set of stresses, transverse shear stresses are also included to permit the whole equilibrium of the element under bending effects, as well to account for the global deformation due to shear behaviour.

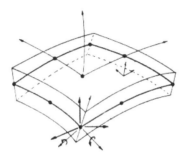

Fig. 1— Adopted Lagrangian 9 node shell element.

CONSTITUTIVE EQUATION FOR STEEL

The adopted constitutive equation for steel combines an incremental elastic biaxial model with a kinematic yield surface and an ultimate strain envelope.

A modified Von Misses criterium is adopted as yielding surface, which has the form:

$$\sigma_{co} - \sigma_e = 0$$
$$\sigma_{co}^2 = (\sigma_x - \sigma_{xp})^2 + (\sigma_y - \sigma_{yp})^2 - (\sigma_x - \sigma_{xp})(\sigma_y - \sigma_{yp}) + \\ + 3[(\tau_{xy} - \tau_{xyp})^2 + (\tau_{xz} - \tau_{xzp})^2 + (\tau_{yz} - \tau_{yzp})^2] \tag{1}$$

where σ_e is the uniaxial yielding stress for steel, and $\boldsymbol{\sigma}_p = \{\sigma_{xp}, \sigma_{yp}, \tau_{xyp}, \tau_{xzp}, \tau_{yzp}\}$ are the first yielding set of stresses (Fig.2).

The kinematic nature of this yield surface follows from the continuous actualization of the components of $\boldsymbol{\sigma}_p$ that define its new position, so that after each iteration, vector $\boldsymbol{\sigma}_p$ is made equal to the last total stresses $\boldsymbol{\sigma}$. This yield surface, as defined, may move as a solid, but not have shape or size changes.

The stress-strain behaviour for biaxial states is defined through an isotropic incremental relationship having as only parameters the tangent uniaxial modulus E and the Poisson's coefficient ν. For the tangent modulus two values are possible: the uniaxial initial slope E_0, and, once the yield criterium is reached, a hardening modulus

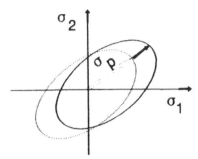

Fig. 2 —Movement of the yield surface in the σ_1, σ_2 space.

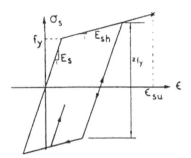

Fig. 3 —Bilinear stress-strain relation for steel.

E_{sh} (Fig. 3). Unloading and reloading are assumed to occur with the initial modulus E_0. Failure is assumed to be controlled by a fixed ultimate strain envelope given by

$$\varepsilon = \sqrt{\varepsilon_x^2 + \varepsilon_y^2 - \varepsilon_x \varepsilon_y + 3(\gamma_{xy}^2 + \gamma_{xz}^2 + \gamma_{yz}^2)} \leq \varepsilon_u \tag{2}$$

where ε_u is the steel uniaxial ultimate strain.

NUMERICAL EXAMPLE

A rectangular steel plate simply supported on its four sides, with a ratio between the maximum to the minimum sides of $a/b=2$, and a thickness $t = 0.01b$, is studied under the effect of a lateral in-plane compression as shown in Fig. 4.

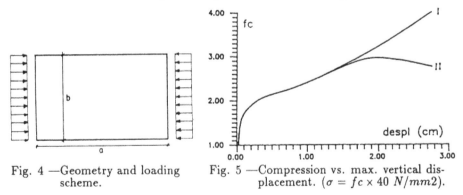

Fig. 4 —Geometry and loading scheme.

Fig. 5 —Compression vs. max. vertical displacement. ($\sigma = fc \times 40\ N/mm2$).

The following material properties have been adopted: $\sigma_e = 260 N/mm^2$, $E_0 = 210\ 000\ N/mm^2$, $E_{sh} = 10\ 000\ N/mm^2$, and $\nu = 0.3$

To study this pannel a mesh made up of 4×4 shell elements has been defined. Despite its relative coarseness, this mesh shows to yield accurate results when compared with more refined meshes for a similar type of study [1].

Two kinds of analysis are carried out and then compared: First, a Nonlinear Geometric Analysis (type I) from which a value for the pre-crictical buckling is achieved. Second, a Nonlinear Geometric and Material Analysis (type II) which allows to also characterize the post-critical failure taking into account the effect of the eventual yielding of steel.

To excite the buckling modes, a small geometric imperfection with a magnitude comparable to that of the thickness is introduced in the mid-surface of the studied pannel. Different positions for the initial imperfection may call to different buckling modes. Thus, the two modes shown in Fig. 6 are obtained for the present example when the imperfection is placed, first, in the center of the pannel and, second, close to a loaded border. The two-peaks buckling mode yields the lowest values for both pre-critical buckling stress (76 N/mm^2) and post-critical buckling stress (124N/mm^2), thus being the one more likely to occur. For this last, the relationship between the applied compression and the maximum vertical displacement is shown in Fig. 5.

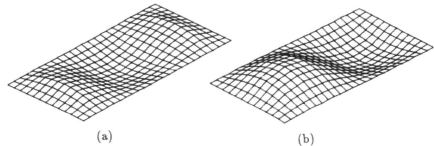

(a)   (b)

Fig.6—Deformed shape for a imperfection located at (a) the center
of the pannel, and (b) at $b/8$ from the loaded border

The accuracy of the obtained results may be appraised when compared with the available analytical solution by Bleich [2]. The analytical value for the stress σ_c causing the pre-critical buckling follows from the well known expression

$$\sigma_c = \frac{\pi E}{12(1-\nu^2)} \frac{t^2}{b} K \qquad \text{where} \qquad K = (\frac{n}{\alpha} + \frac{\alpha}{n}) \qquad \alpha = \frac{b}{a} \qquad (3)$$

and n is the number of peaks (or semi-waves) shown by the deformed shape. The different pre-critical buckling stresses numerically obtained by relocating the geometric imperfection show to match very satisfactorily with the values which are obtained for $n = 1$ and $n = 2$ using equation (3).

REFERENCES

1. Mirambell, E., Roca, P., Casado, F. "Análisis del comportamiento de paneles metálicos frente al fenómeno de abolladura" Departament d'Enginyeria de la Construcció, Universitat Politècnica de Catalunya, Barcelona 1992.
2. Bleich, F. "Buckling Strength of Metal Structures". McGraw Hill, New York, 1952.

COLD-FORMED STEEL SECTIONS:
THE STATE OF THE ART IN ITALY

DE MARTINO A.[1], LANDOLFO R.[2], MAZZOLANI F.M.[2]

(1) Architecture Faculty, University of Naples, Italy
(2) Engineering Faculty, University of Naples, Italy

ABSTRACT

The usage of cold-formed steel sections in the building industry in Italy is briefly discussed with particular reference to the application in civil and industrial buildings, long span roofs, seismic systems and refurbishment operations.

Side by side, the research development supported these applications by means of theoretical and experimental activities. They are mainly devoted to analyse the cyclic behaviour of beam-to-column connections and the bending capacity of cold-formed members and long span sheetings, in order to emphasize the influence of the b/t ratio on the maximum strength and the correlation between the design of the effective section and the ductility requirement, for the use in dissipative zone of seismic resistant structures.

INTRODUCTION

The use of cold-formed steel sections is more and more spreading in the Italian constructional market both for civil and industrial buildings, but also in the recent new field of retrofitting which has been growing up in the last ten years.

On the other side the new European codes (EC3, EC8), with the introduction of different behavioural classes interpreting the plastic performance of the cross section of members, strongly penalize the use of thin walled sections in seismic areas. Since seismic areas involve a large part of Italian territory, the penalization is very strong and the use of these profiles is at this time confined to traditional fields.

For this reason in the last years many researchers have devoted a big effort to theoretical and experimental investigations in order to deepen the main aspects arising from the use of this kind of profiles in dissipative zone of seismic-resistant structures with the aim to find economic solutions.

The more outstanding applications of various typologies of structures, made of cold-formed sections with thin and medium thickness, are briefly presented in this paper together with a synthetic review of theoretical and experimental studies.

SOME RECENT DEVELOPMENT

Building system

Presented in 1982, the constructional metasystem named BASIS (Building Activities Steel Integraded System) represented a significant proposal in the field of industrialized buildings based on the use of steel elements. The main components of this system are:

- one type of column, made of HEB 140 rolled section, single for buildings of up to 4 stories and double for those up to 8;
- cold-formed steel sections for the main and secondary beams made of back-to-back channel in ranges of thickness from 4 to 8 mm and depth from 120 to 280 mm;
- corrugated steel sheets integrated with reinforced concrete slab for deck.

This constructional system can be used to resist horizontal loads both with steel bracing structures and with concrete core. The BASIS has been used in Italy and abroad for apartment buildings [1].

Long-span steel decking and roof

The development of steel sheetings over the past years has been characterized by three generations of profiles.

The third generation profiles are trapezoidal unit with both longitudinal and transversal stiffeners, which provide suitable solution for spans up to 12 m without purlins. Among the profiles belonging to the third generation, the so-called TRP 200, manufactured by Plannja of Lulea, Sweden, is one of the most representative. This trapezoidal profile, with depth 199 mm and width 800 mm, is made from hot-dipped galvanized steel sheet with thickness range from 0.75 to 1.50 mm. TRP 200 is also produced in perforated form, with small holes for approximately 17% of the surface of the profile, for improving acoustic performance.

The mentioned profile has been used in Italy for roofing in the field of industrial buildings or in floor structures of multistory buildings by completing its cross section with casted concrete. Some applications have been also proposed for the rehabilitation of old roofs and floor structures [2].

In the field of long span roofs a new constructional system is based on the use of a particular corrugated sheet [3]. This system is particularly used for cylindrical shells spanning over 40 mt between the supports to cover spaces mainly devoted to sport activities. The shell section is made by two external corrugated but flexible sheets connected by a spatial truss built with circular hollow members. The mechanical properties of this system have been derived by an experimental procedure.

Refurbishment

The main part of the storical heritage in Italy is given by masonry buildings. Several earthquakes in the far and in the recent past have caused a continuous deterioration of such constructions and a new activity of reinforcing and upgrading is now in progress. With this purpose some researches have been sponsored by the Ministry of University in order to find out the suitable way of application of steelworks in the refurbishment operations. Cold-formed shapes and trapezoidal sheets have been considered as new materials which can be conveniently used instead of the traditional technologies to give integrated structural system [4].

RESEARCH BACKGROUND IN ITALY

Profiles

A general research project devoted to study the bending behaviour of the cold-formed thin walled sections, with the aim to provide useful data on the use of these profiles in seismic resistant structures, has been carried out at University of Naples. One of the main goal of this project is to check the provisions given in the European codes which greatly penalize thin walled sections because of their thickness. Cold-formed profiles are mainly included in the class of slender sections and their dissipative capacity is consequently not considered, so their use is practically not allowed in seismic resistant structures.

From the theoretical point of view the study consisted in simulating the flexural elasto-plastic behaviour of a section, by developing numerical procedures able to follow its deformation history, taking into account local buckling phenomena and various other parameters such as residual stresses, material law etc.. Starting from the ideal moment-curvature relationship for given sections, the program allows the introduction of the different numerical model and the related parameters chosen to interpret the main phenomena under consideration. The influence of each parameter has been examined on sections specially selected in order to cover different values of b/t ratios both for flanges and webs.

The experimental tests were finalized to calibrate the proposed model in order to make it available for extrapolation in simulating the behaviour of other sections of the same type with different geometrical and mechanical characteristics.

From the global analysis of experimental results and from the comparison with design values provided by Italian and European codes it is possible to draw some final remarks:
- the design values of bending moment provided by above codes appears to be in general slightly safe and in good agreement with the experimental ones;
- the classification proposed by EC3 is only partially confirmed by the observed Mexp/My ratios;
- the load-deflection and moment-curvature curves put in great evidence the different behaviour of stiffened versus unstiffened sections.

For all these reasons, the choice of design moment significantly lower than the observed one (and also lower than the one allowed by the present codes) may be proposed in order to achieve a significant ductility level. A calculation method based on appropriate reduction factors can allow the use of cold-formed thin gauge sections also in dissipative zones of seismic resistant structures, provided that the design moment should be correlated to the requested ductility [5].

Sheets
In order to investigate on the main aspects of bending behaviour of long span sheetings several series of tests were carried out at University of Naples, by considering different conditions, such as type of load (concentrated or distributed), overall dimensions and influence of the concrete slab as a composite section.

These test results allowed to develop mechanical models for the evaluation of the ultimate bending moment. Further on, a simulation program has been implemented and calibrated on the same testing data both for sheets and for composite sections [6].

Connections
Several studies have been devoted to the cyclic behaviour of beam-to-column moment resisting joints connecting cold-formed members. Experimental tests have been also performed to evaluate the stiffness, the strength and the ductility of this system [7], following the ECCS test procedure. The joints were particularly designed to avoid local buckling phenomena which should reduce energy dissipation capacity.

The tests have been done on different typologies to be used for residential buildings. The results have shown a very interesting behaviour of these joints with ductility values varying from 18 to 40, stating the possibility to use these semi-rigid connection in seismic areas. Similar results have been also underlined by different authors [8].

Composite structures
The encased cold-formed beams are used sometimes in order to contained within the floor depth, according to modern architectural and constructional requirements.

For such a building type, with reduced cross-sectional height, bond slip between steel and concrete is to be expected at relatively small loads. An analysis to determine elastic stiffness and ultimate loads has been carried out in [9]. The proposed method follows the behaviour of the beam from the beginning of cracking through the phase of bond slip and up to yielding. The results from the first set of experimental tests on beams, reinforced with pairs of C-shaped profiles of different thickness values, turned out to be in agreement with the theoretical predictions.

Code
The Italian code for cold-formed steel section (CNR 10022/1984) is based on the previous AISI specification (1980). This code has been a mixed approach for design of cold-formed steel members for compression element, being stiffened elements treated by an effective width approach, while unstiffened elements by an allowable stress approach. The first recent European (EC3 - Annex A/1990) and American (AISI/1986) codes, on the contrary, are based on a unified approach for all compressed members.
The consequence of the different approaches are emphasize in a comparison among the calculation methods which are used in the above codes by considering several cases, which are representative of the main typologies coming from the use of cold-formed profiles [10].

REFERENCES

1. BALDACCI F., Metasistema per l'edilizia industrializzata residenziale in acciaio. Costruzioni Metalliche, 6/1984.

2. LANDOLFO R., MAZZOLANI, F.M., Behaviour of third generation trapezoidal steel sheetings. In Testing of Metals for Structures , E & FN Spon, London, 1992, pp. 444-52.

3. BENUSSI F., Coperture cilindriche realizzate con sistema sandwich di profili piegati a freddo. Proceedings of the " Giornate Italiane sulla Costruzione in Acciaio C.T.A." Abano Terme, october 1991.

4. MAZZOLANI, F.M., Refurbishment and Extensions: The case for steel. Proceedings of the International Symposium of I.C.S.C., Luxembourg, may 1990.

5. DE MARTINO A., GHERSI A., LANDOLFO R., MAZZOLANI, F.M., Calibration of a bending model for cold-formed sections. Proceedings of the 11th International Specialty Conference on Cold-Formed Steel Structures, University of Missouri-Rolla, october 1992.

6. DE MARTINO A., GHERSI A., LANDOLFO R., MAZZOLANI, F.M., Bending behaviour of long-span steel sheeting: test and simulation. Proceedings of International Conference on Steel and Aluminium Structures, Singapore, may 1991.

7. BERNUZI F., DE MARTINO F.P., ZANDONINI R., Comportamento ciclico di giunti in strutture realizzate con profili in acciaio sagomati a freddo da coils. Proceedings of the " Giornate Italiane sulla Costruzione in Acciaio C.T.A." Abano Terme, october 1991.

8. CREMONINI P., Sistema strutturale di giunti rigidi per trave-colonna mediante staffe bullonate. Acciaio, 3/1990.

9. BOZZO E., CAPURRO M., Analysis of static behaviour of encased cold-formed beams. Costruzioni Metalliche, 6/1988.

10. GHERSI A., LANDOLFO R., Aspetti innovativi dell'EC3 nella verifica delle sezioni in parete sottile. Proceedings of the " Giornate Italiane sulla Costruzione in Acciaio C.T.A." Abano Terme, october 1991.

Connection Design

FIELD STUDIES OF BOLT PRETENSION

GEOFFREY L. KULAK
Department of Civil Engineering
University of Alberta, Edmonton, Canada T6G 2G7

PETER C. BIRKEMOE
Department of Civil Engineering
University of Toronto, Toronto, Canada M5S 1A4

ABSTRACT

Pretensioned bolts must be used when it is required that a connection not slip when load is applied. However, studies that report on the attainment of bolt pretension are generally available only for laboratory conditions. This paper reports on the results of two independent studies that used an ultrasonic bolt length measurement device to establish the pretension in bolts installed in the field. Six different sites were used by one researcher to obtain the pretensions in 232 bolts of different sizes, grades, and which used different installation methods. A total of 104 bolts at three different sites were monitored by the other team. In all cases, installation was done by the erector of the structure; the research teams were not involved in the installation process in any way.

INTRODUCTION

Bolted connections that must resist movement under applied load, termed slip-critical in North American practice, must use pretensioned bolts and have faying surfaces that meet specified frictional criteria. The level of pretension is set at a minimum value by the governing standard. For example, both North American practice [1] and the rules provided by Eurocode 3 [2] require that the bolt pretension be at least 0.70 times the ultimate tensile strength of the bolt material. It is of obvious importance, then, to know whether the specified pretensions are being attained in practice. Indeed, in order to establish the reliability of the slip-critical joint, that is, the probability of slip, the mean value of the pretension in the bolts and its standard deviation must also be known [3].

Studies that report on the actual tension of bolts in joints made up in the laboratory are available, especially for bolts installed by the turn-of-nut or calibrated wrench methods [3]. However there are few studies of bolt pretension that reflect field practice. One recent study did report on pretension in bolts installed in field joints [4], but in that case the installation of the bolts in which the pretension was measured was done by the researchers. One reason that studies of bolt tension in field joints are lacking is because until recently there has not been a method of measuring these forces that is both convenient and reliable. Only in about the last decade have devices been developed that enable bolt tension to be determined in the field in a relatively simple and reliable way [5]. These devices use ultrasound to read the bolt length before and after installation or removal. The change in length determined in this way can be related to the bolt tension. Although the ultrasonic measurement device is relatively expensive, with careful use and calibration it gives reliable values of bolt tension. Two recent studies [6, 7] have used the electronic bolt gage to measure the pretension in bolts installed under a variety of field conditions, including different types and grades of fasteners, methods of installation, different sites, and so on. The results of those studies are reported in this paper.

SCOPE

The work was conducted by two groups of researchers who worked independently but along similar lines. Team 1, located at the University of Toronto, measured bolt pretensions during the period June to November 1988. The measurements were made at six different sites in eastern Canada or eastern U.S.A., and a variety of structures were included. The specified methods of bolt installation included turn-of-nut [1, 2] and use of load indicating washers, otherwise known as direct tension indicators [1]. Both ASTM A325 and A490 bolts (ISO grades 8.8 and 10.9, respectively) were included. Team 2, located at the University of Alberta, measured bolt pretension during the period March to April 1991 at three different bridge sites in western Canada. Only A325 bolts installed using the turn-of-nut method were involved. Full details of the studies must be obtained from the original source material [6, 7], but a brief summary follows. Sites 1 to 6 were those done by Team 1, and Sites 7 to 9 were done by Team 2.

Site 1 - Multi-storey building, A490 bolts, installation by turn-of-nut method, 19 bolts examined.

Site 2 - Bathtub girder bridge, A325 bolts, installation by turn-of-nut method, 32 bolts examined.

Site 3 - Nuclear generating plant, A325 bolts, installation by turn-of-nut method, 22 bolts examined.

Site 4 - Bathtub girder bridge, A325 bolts, installation by turn-of-nut method, 24 bolts examined.

Site 5 - Long span sports stadium, galvanized A325 bolts, installation by turn-of-nut method, 75 bolts examined.

Site 6 - Stadium, A325 bolts, installed using load indicating washers, 60 bolts examined.

Site 7 - Short span girder bridge, A325 bolts, installation by turn-of-nut method, 22 bolts examined.

Site 8 - Short span girder bridge, A325 bolts, installation by turn-of-nut method, 44 bolts examined.

Site 9 - Medium span girder bridge, A325 bolts, installation by turn-of-nut method, 38 bolts examined.

BOLT TENSILE STRENGTHS

It was important to determine the tensile strength of the bolts in which the pretensions were measured so as to establish a benchmark for evaluation of the pretension load levels. Accordingly, both investigation teams measured the tensile strength of a sample of the bolts in the study. All tests were conducted on full-size bolts, not coupons. In some cases, the bolts used to determine the direct tensile strengths were the same bolts for which the pretension had been measured. In other cases, the bolts used for the direct tension tests were a representative sample of the bolts for which the pretension was measured.

The results of the tensile strength tests are shown in Table 1. The *Guide to Design Criteria for Bolted and Riveted Connections* [3] quotes mean values of this ratio of measured tensile strength to specified minimum tensile strength of 1.18 for A325 bolts and 1.10 for A490 bolts. (The standard deviations of the means are given as 0.035 and 0.045 for A325 and A490 bolts, respectively). Thus, the results reported herein by the two research groups are consistent with one another and consistent with the values reported by the *Guide*.

TABLE 1

Measured ultimate tensile strength / specified ultimate tensile strength

	Team 1 ASTM A325 Bolts	Team 1 ASTM A490 Bolts	Team 2 ASTM A325 Bolts
No. of Bolts Tested	43	8	9
Mean	1.21	1.13	1.21
Standard Dev.	0.05	0.01	0.03

MEASURED BOLT PRETENSIONS

The ultrasonic measurement device, or "bolt gage", is simply an electronic instrument that delivers a voltage pulse to an acoustic transducer. The transducer emits a very brief burst of ultrasound that passes through the bolt, echoes off the far end, and returns to the transducer. The electronic instrument very precisely measures the time required for this round trip. The measured "transit time" can then be converted into length or into load if the transit time is measured before and after tightening. Achieving a suitable level of accuracy requires proper settings for electronic compensations and careful calibration of the device; however, it is considered [5] that bolt pretensions can be established with an error less than 5%.

Both teams of researchers in this program used the bolt gage to first measure the fastener length in bolts that had already been installed in the field. The fastener was then removed by loosening the nut and another reading of length taken. There are practical advantages to this sequence, but, in addition, it eliminates a certain bias that might result if the bolting crews know that the bolts they install will be monitored for pretension. Moreover, in some cases in this study, the fasteners had been installed several years before the pretensions were measured.

Once the change in length of the bolt had been established as the fastener was unloaded, laboratory studies on samples of bolts from the same lots (or, the same bolts in some cases) enabled calculation of the pretension that had existed. These laboratory studies established the stiffness of the fasteners in the unloading mode. With this stiffness established, the change in length taken in the field was used to calculate the pretension.

For those cases in which the bolting up preceded the field study by only a short period (the majority of cases), observations were taken on how the installation actually proceeded. Of course, the specified methods of installation and inspection were also identified from the requirements that were set out by the owners of each particular structure.

Team 1 had the largest number of variables in their test program. However, because the field sites had to be accepted as they became available, it was not possible to control the parameters in a systematic way. Nevertheless, it is worthwhile to examine both the overall results and to make some comparisons between certain parameters. Tables 2 through 5 summarize the results and allow the comparisons.

TABLE 2

Measured tension / specified minimum tension: **Team 1** − overall results

Grouping	Number of Bolts	Measured Tension / Spec. Minimum Tension
All Bolts	232	1.15 (0.16)
All A325 Bolts	213	1.16 (0.16)
All A490 Bolts	19	1.08 (0.08)

Table 2 gives the overall results obtained by Team 1, tabulated as the ratio of the measured tension to the specified minimum tension. The value in parentheses is the standard deviation of the mean value in each case. The results show that the mean value of the ratio for all 232 bolts exceeded the specified minimum by about 15%. For A325 bolts only, 213 in number, the ratio was 1.16. The sample of A490 bolts was relatively small; here the ratio was 1.08. These values are consistent with those reported in the *Guide* [3], that is, they confirm the data that was based entirely on measurement of pretensions in joints made up in the laboratory.

In Table 3, the data obtained by Team 1 are broken down by bolt diameter. This puts 138 bolts (all A325) into the category 7/8 inch diameter (22 mm) and 94 bolts (both A325 and A490) into the 1 inch diameter (25 mm) category. It shows that the ratio of measured tension to specified minimum tension was greater for the 7/8 inch diameter bolts (1.18 versus 1.11). It is likely that bolt length plays a stronger role than does diameter, per se, but these data have not yet been processed.

TABLE 3

Measured Tension / Specified Minimum Tension: **Team 1** − Effect of Bolt Diameter

Grouping	Number of Bolts	Measured Tension / Spec. Minimum Tension
A325 7/8 in. dia.	138	1.18 (0.15)
A325 & A490 bolts 1 in. dia.	94	1.11 (0.16)

The effect of the method of installation is summarized in Table 4 for the Team 1 results. There is about an equal number of bolts in each of the three categories of turn-of-nut installation, direct tension indicator (load indicating washer), and a category designated "sound." In this latter case, bolting crews installed the bolts to a level associated with the feel or the sound of the impact wrench; this level may have been established by first performing the snug plus turn procedure in a hydraulic bolt calibrator. This is, of course, not a method acknowledged by any specification, but it does seem to be used frequently; turn-of-nut had been specified in the cases cited. It was also observed by Team 2 at some of their sites. It gave the largest value of the pretension ratio of the three methods.

TABLE 4.

Measured Tension / Specified Minimum Tension: **Team 1** – Effect of Method of Installation

	Number of Bolts	Measured Tension / Spec. Minimum Tension
A325 Bolts, Galvanized, Turn-of-Nut Installation	75	1.12 (0.18)
A325 Bolts, Direct Tension Indicator	60	1.12 (0.13)
A325 Bolts, "Sound"	78	1.23 (0.15)

Finally, the results obtained by Team 1 are broken down as a function of the type of structure. Table 5 indicates that the ratio of measured tension to specified minimum tension is significantly higher at bridge sites than at all non-bridge locations. There is no objective way to establish why this is so, but it is likely that more care in both installation and in inspection of high-strength bolts is usually taken at bridges sites than at other locations. Perhaps this is because the need for slip-critical connections is clearly established for bridge structures. As well, access to joints is often better in bridges than in buildings.

TABLE 5.

Measured Tension / Specified Minimum Tension: **Team 1**– Influence of Type of Structure

Grouping	Number of Bolts	Measured Tension / Spec. Minimum Tension
Bridges (all A325 bolts)	56	1.24 (0.13)
Non-Bridges	176	1.11 (0.15)

It has already been noted that Team 2 investigated only bridges and only A325 bolts installed by the turn-of-nut were encountered. Three different bridges were examined, although the erection crew and inspector were the same on two of them. Since the parameters were so limited, the most helpful comparisons can be made with Team 1 results. Table 6 shows the Team 2 results.

TABLE 6

Measured tension / specified minimum tension: **Team 2** – overall results

	Number of Bolts	Measured Tension / Spec. Minimum Tension
All Bolts (all A325)	104	1.30 (0.21)

Comparison of the Team 2 results (all bridges, all A325 bolts) with the corresponding results from Team 1 shows that the ratio of measured pretension to specified minimum tension is similar for these cases. For Team 1, the result was a mean value of 1.24, standard deviation 0.13 (Table 5) and for Team 2, the mean value was 1.30, standard deviation 0.21 (Table 6). If all the test results are taken together for this particular case of A325 bolts installed in bridges, the mean value of the ratio of measured pretension to specified minimum tension is 1.27, standard deviation 0.20. The bridge results for both teams are plotted in the frequency histogram of Figure 1.

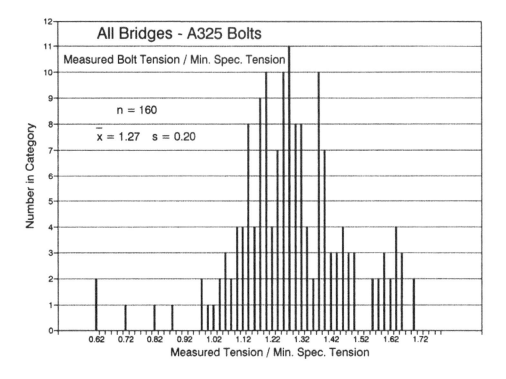

Figure 1. Measured bolt pretension / minimum specified bolt pretension for A325 bolts in bridges

APPLICATION OF THE RESULTS

An example of how the data can be used will be made with reference mainly to one case, that of A325 bolts installed in bridges using the turn-of-nut method of installation. North American design rules [1] allow the designer to input a slip coefficient that is the expected mean value and a clamping force (bolt pretension) that is the specified minimum value. Neither of these is deterministic, however, and the approach taken is to recognize this and to accommodate the variation of each quantity by means of a multiplier. In the case of the clamping force, the multiplier allows for the increase between the bolt pretension actually attained as compared with the specified minimum value of bolt pretension. The approach allows the identification of a slip probability level. Up to this time, the evaluation has been made entirely on the basis of bolt pretensions measured in the laboratory. With the data now available, it is possible to comment on the suitability of the use of laboratory test data only.

In the *Guide* [3], the mean value and standard deviation of the ratio of measured bolt pretension to specified minimum bolt pretension have been taken as 1.35 and 0.12, respectively. Using the results obtained in this study, and combining the data obtained by the two investigators, the same ratio has a mean value of 1.27. This is reasonably close to the value used by the *Guide*, but it is unconservative. The difference is only about 6%, however. Since the bolt pretensions measured at bridge sites were significantly larger than those at non-bridge sites, the degree of unconservatism increases as the sample is widened, however.

European practice [2] is similar in formulation to that just described; the designer selects a slip coefficient and assumes that the bolt pretension will be at least 70% of the ultimate tensile strength of the bolt. The design equation includes a partial safety factor rather than specifying a probability of slip. It is not known by the authors if the partial safety factor was selected on the basis of any knowledge of pretension levels and variability that might be expected to occur in practice.

SUMMARY AND CONCLUSIONS

The pretensions in bolts installed in the field under normal erection conditions have been measured in two separate investigations. Nine different sites were included in the study, which encompassed bridges, a high-rise building, long span stadium structures, and a heavy industrial plant. Bolt grades included ASTM A325 and A490 (ISO Grades 8.8 and 10.9), both galvanized and black bolts, and bolts of 7/8 and 1 inch (22 and 25 mm) diameter. The methods of installation specified at the various sites included turn-of-nut and use of direct tension indicators (load indicating washers). Pretensions were established using an ultrasonic bolt length measurement device that was used after, and independent of, the bolting-up procedure. The results are therefore considered representative of bolt pretensions in current construction.

The A325 bolts examined in this study were about 21% stronger than the minimum specified strength for that grade. A490 bolts were about 13% above their minimum specified strength. These results are consistent with previously reported values.

The measured pretension of all 317 A325 bolts (both studies) was 1.21 times the specified minimum pretension. For A490 bolts (19 in number), the corresponding figure was 1.08. In the case of A325 bolts, installation by turn-of-nut or by use of direct tension indictors gave very similar results. However, many installations were observed to be controlled mostly by the sound of the impact wrench, even though turn-of-nut had been prescribed. The pretension in these cases was significantly greater than for the other methods of installation.

Bolts installed in bridge structures had significantly higher pretensions than bolts in all other types of structures. Both research teams found a similar level of pretension in the bridge bolts.

The pretension levels for field-installed bolts can be compared with the values used in the past to calibrate design rules for slip-critical joints. For the case of A325 bolts installed in bridges (turn-of-nut installation), which is the largest group reported herein, the field condition shows a slightly lower pretension level than had been determined previously on the basis of laboratory studies. The differences are larger for the other cases described in this report. The turn-of-nut method of installation controls the amount of extension of the bolt and generally brings the bolt shank beyond its proportional limit. This inherently produces pretensions greater than the minimum specified value, in proportion to the actual strength of each fastener.

It should also be recognized that the level and variability of installed pretensions likely are influenced by such factors as the fit-up procedure and the flatness and thickness of the steel in the connection. These aspects are contained within the study, but it has not been possible to consider them quantitatively.

ACKNOWLEDGMENTS

The work described in this report was supported financially by the Research Council on Structural Connections. That support is acknowledged with thanks. The field work and processing of data were carried out largely by Nick Grgas (Team 1) and Khalid Obaia (Team 2), and their very significant contributions are also acknowledged with thanks. Grgas was also the recipient of a student scholarship provided by the Natural Sciences and Engineering Research Council of Canada.

REFERENCES

1. Research Council on Structural Connections, Load and Resistance Factor Design Specification for Structural Joints Using ASTM A325 or A490 Bolts, (distributed through the American Institute of Steel Construction, Chicago, Illinois), 1988.

2. Commission of the European Communities, Eurocode No. 3, Design of Steel Structures, draft November 1990.

3. Kulak, G.L., Fisher, J.W., and Struik, J.H.A., Guide to Design Criteria for Bolted and Riveted Joints, Second Edition, Prentice-Hall, 1987.

4. Piraprez, E., Bolt Preloads in Laboratory and in Field Conditions of Acceptance, Proceedings, Second International Workshop on Connections in Steel Structures, Pittsburgh, Pennsylvania, April 1991.

5. Bickford, J.H., An Introduction to the Design and Behavior of Bolted Joints, Second Edition, Marcel Dekker Inc., 1990.

6. Grgas, N., Field Investigation and Evaluation of the Pretension of High Strength Bolts, draft thesis, Department of Civil Engineering, University of Toronto, 1991.

7. Kulak, G.L. and Obaia, K.H., A Field Study of Fastener Tension in High-Strength Bolts, Structural Engineering Report No. 177, Department of Civil Engineering, University of Alberta, Edmonton, Canada, April 1992.

MODELLING OF BEAM-TO-COLUMN JOINTS FOR THE DESIGN OF STEEL BUILDING FRAMES

R. MAQUOI and J.P. JASPART
Department M.S.M.
University of Liège
Quai Banning, 6, B-4000 Liège, Belgium

ABSTRACT

Present paper is aimed at presenting and classifying well known second-order inelastic computer programs for the analysis of building frames with semi-rigid beam-to-column joints according to their type of numerical joint modelling and at higlighting and discussing the influence of this modelling on the structural frame response.

INTRODUCTION

For sake of economy, beam-to-column bolted joints without any column web stiffener become a common practice (joints between H or I sections). Such a joint has a non-linear behaviour: when the beam is subject to bending, the axes of the connected members do not rotate a same angle, what results in a relative rotation that is not proportional to the beam bending moment.

In a strong axis beam-to-column joint, two main sources of deformability are identified (fig. 1) :
a) The deformation of the connection associated to the deformation of the connection elements (end plate, angles, bolts,....), to that of the column flange and to the load-introduction deformability of the column web ;
b) The shear deformation of the column web associated mostly to the pair of forces F_b carried over by the beam(s) and acting on the column web at the level of the joint; these forces are statically equivalent to the beam moment M_b.

These components are illustrated in figure 2 for the particular case of a joint between a column and a one-sided beam. The flexural deformability of the connection elements is concentrated at the end of the beam (fig. 2.a). The associated behaviour is expressed in the format of a M_b - ϕ curve.

The deformation of the ABCD column web panel is divided into :
- The load-introduction deformability which consists in the local deformation of the column web in both tension and compression zones of the joint (respectively a lengthening and a

shortening) and which results in a relative rotation ϕ between the beam and column axes; this rotation concentrates mainly along edge BC (fig. 2.b) and provides also a deformability curve M_b - ϕ.

The shear effect - due to shear force V_n - which results in a relative rotation γ between the beam and column axes (fig. 2.c); this rotation makes it possible to establish a second deformability curve V_n - γ.

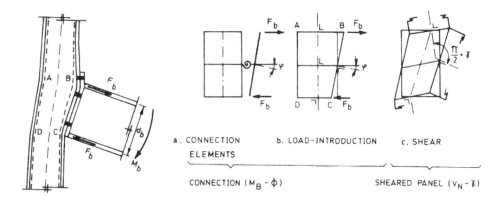

a. CONNECTION   b. LOAD-INTRODUCTION   c. SHEAR
ELEMENTS

CONNECTION $(M_B - \phi)$ SHEARED PANEL $(V_N - \gamma)$

Figure 1 -   Deformation of   Figure 2 -   Joint deformability
          a strong axis joint   components

It is important to stress that the deformability of the connection (connection elements + load-introduction) is only due to the forces carried over by the flanges of the beam (equivalent to the beam moment M_b), while the shear in a column web panel is the result of the combined action of these equal but opposite forces <u>and</u> of the shear forces in the column at the level of the beam flanges. In fact, the actual value of the shear force V_n may be obtained from the equilibrium equations of the web panel [1]; it is given by the following formula (figure 3) :

$$V_n = \frac{M_{b1} + M_{b2}}{d_b} - \frac{V_{c1} + V_{c2}}{2} \qquad (1)$$

As highlighted herein under, most of the researchers refer to another formula:

$$V_n = \frac{M_{b1} + M_{b2}}{d_b} \qquad (2)$$

which is nothing but a rough approximation of the actual one (formula 1). The validity of formula (1) has been clearly demonstrated in [1].

The difference between the loading of the connection and that of the column web in a specified joint requires, at a theoretical point of view, that account be taken separately of both deformability sources when designing a building frame (figure 4.a). However doing so is only practicable when the frame is analysed by means of a sophisticated computer program allowing for the separate modelling of both deformability sources. In all other cases, the

TABLE 1

Joint representations in second order inelastic programs

Modelled components of joint deformability	Joint representation			Computer program	
	Element(s) used	V_n definition	Figuration	From	Plastic hinge (H) or plastic zone (Z) theory
	Rotational springs and in-finitely rigid beam studs	formula 2	a : $M_b - \phi$ b : $V_n - \gamma$	Lausanne [4]	Z
Separate modellings of connection(s) and sheared web panel	Set of axial springs and rigid bars	formula 2	a : $M_b - \phi$ b : $V_n - \gamma$	Innsbruck [5]	H
	Set of trusses	formula 2		Aachen [6]	Z

Modelling of the connections (or of the joints after "concentration")				
MINDLIN beam element, linear constraints and spring block	formula 1		Lausanne* [7]	Z
Spring + internal constraints incorporated in a single beam finite element	formula 1		Liège [8] (see following section)	Z
Spring block	formula 2		Milan*[9] Liège [10] Warwick[11]	Z Z H
Macro-element	formula 2		Cachan/ CTICM**[12]	H

*: stiffnesses of the connection under normal and shear forces may also be simulated by the spring block

**: interaction between normal force and bending moment in the connection may be taken into account.

actual behaviour of the joints must be simplified by concentrating the whole deformability at the beam end (figure 4.b).Reference [2] gives guidelines on how to "concentrate" the joint deformability in an accurate and safe manner for design practice.

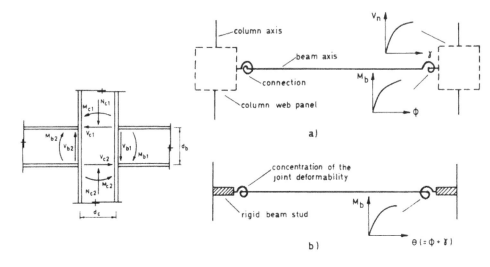

Figure 3 -   Loading of an   Figure 4 - Actual and simplified joint modelling
              interior joint   (a) Actual one
                                               (b) Simplified one (after concentration [2])

JOINT REPRESENTATIONS IN COMPUTER PROGRAMS

Several joint representations have been adopted in the last years by different authors. They have to be classified in two categories (table 1) :
- representations allowing for the separate modelling of the connection and column web panel behaviours ;
- representations allowing for the modelling of the connection behaviour only (or of the joint behaviour by means of the "concentration" concept).

Table 1 presents an outline of the main representations proposed as well as of the computer programs in which they have been implemented. For each of these programs, informations relative to :
- the definition of the shear force V_n to which it is referred when simulating the behaviour of the sheared column web panel or when concentrating the joint deformability (formula 1 or 2) ;
- the way in which the spread of plasticity in the connected beam and column members is taken into account (plastic zone or plastic hinge theory);
are also reported in table 1. People interested in is begged to consult the references given in table 1 as well as reference [3] where a detailed description of the programs and joint representations is given. The sake of consideration for the real dimensions of the joint in the representations presented in table 1 has to be pointed out.

FINELG FINITE ELEMENT PROGRAM

The non-linear finite element program FINELG developed jointly at the University of liège, Belgium, and at the Polytechnic Federal School of Lausanne, Switzerland, has been recently implemented in Liège to simulate accurately the non-linear behaviour of connections and sheared column web panels [8]. The flexural behaviour of the connection and of the adjacent beam as well as that of the column web panel are gathered into a <u>single</u> "plane beam + connection + sheared panel" finite element. Any type of non-linear response may be associated to the behaviour of the beam, of the connection and of the web panel respectively. Aforementioned element can be used according anyone of the three different manners sketched in figure 5.

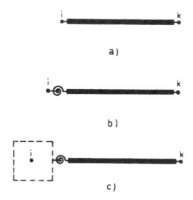

Figure 5 - Use of the finite element

      (a) "Beam" element
      (b) "Beam+connection" element
      (c) "Beam+connection+sheared web panel" element

It exhibits several superiorities :
- in contrast to approaches [4], [5] and [6], it fulfills the equilibrium equations of the web panel ;
- it makes possible an accurate and realistic picture of the actual macroscopic behaviour of the column web panel ;
- it does not need a more refined discretization than that just required when rigid joints (in contrast to [4], [5], [6] and [7]);
- the number of equations which have to be solved at each step of the structural analysis is not increased due to the non-linear semi-rigidity of the joints.

NUMERICAL APPLICATIONS

In a recent document issued by ECCS TWG 8.1/8.2 ([3]), most of the numerical procedures described in the previous section (except [4], [7] and [8]) have been applied to the study of two unbraced and sway plane building frames with semi-rigid joints (figure 6). The comparison diagrams, on which the curves resulting from the application of procedures [7] and [8] have been superimposed, are presented in figure 7.

It is worthwhile stressing that: i) the deformability curves which characterize the web panel, connection and joint semi-rigid response (figure 6.b) have been analytically obtained by means of a mechanical prediction model proposed by TSCHEMMERNEGG in Innsbruck and, ii) it is explicitly referred, in this model, to formula (2) for the determination of the shear force V_n. That means, in practice, that the V_n - γ curve which has to be introduced as a data in all the computer programs allowing for the separate modelling of connections and sheared column web panels has to be simply derived from the shear curves given in figure 6.b by dividing the values of the bending moment M_b by the beam height d_b (figure 1); this V_n - γ curve must thus be considered as the actual one.
At each step in the non-linear numerical computation, the loading of the column web panel (figure 3) is assessed and the shear force V_n is evaluated by means of formula 1 (for procedures [7] and [8]) or of approximate formula 2 (for procedures [5] and [6]); the relative rotation γ is then deduced from the V_n - γ data curve.

These informations are useful when interpreting the comparison diagrams (figure 7) which clearly show that (figure 6):
- The λ - v_1 curves obtained by means of procedures [9], [10], [11] and [12] (concentration of the joint deformability) are in close agreement;
- Procedures [5] and [6], which adopt more refined models but with a shear force V_n evaluated by means of the approximate formula 2, provide quite similar results;
- Procedures [7] and [8], which are based on a separate modelling of the joint deformability components and on an actual definition of V_n (formula 1), lead to higher values of the ultimate load factor λ and of the frame stiffness at each step of loading.

CONCLUSIONS

From this study, it may be concluded that :
- The concentration of the joint deformability into rotational elements at beam end provides generally a safe appraisal of the frame response (overestimation of the transverse displacement of the beams and of all the storeys under service loads/underestimation of the ultimate resistance under factored loads). This safe character results mainly from the overestimation of the shear force resulting from the evaluation of V_n by means of formula 2, when "concentrating" the joint deformability.
This conclusion is supported by an extensive parametric study recently performed at the University of Liège [2]. This study shows that the concentration provides an accurate prediction of the actual response of braced and unbraced frames, except when the beam-to-column connections have a stiffness and a strength similar to (or higher than) those of the web panel in shear, as for frames A and B (see figure 6.b). In the latter case, the actual frame collapse is somewhat underestimated. The concentration is however sufficiently accurate to be recommended for practical applications.
- The separate modelling of the joint deformability components provides only a better assessment of the actual frame response if reference is made to the exact definition of the shear force V_n (formula 1). The use of procedures [5] and [6], which, in contrast to procedures [9], [10], [11], and [12], require special "sheared panel" elements (what increases, often substantially [8], the size of the numerical system which has to be solved

at each loading step in the computation) becomes consequently somewhat questionable.

These conclusions have to be stressed just as databases for semi-rigid joints are developed in different countries with the view to validate existing analytical models for the prediction of the joint response. As a matter of fact, due to the dimensions of the connected members commonly used for joint laboratory tests, the influence of the V_n definition on the experimentally obtained V_n - γ curve may be sometimes especially significant (26 % and 37 % for instance for the two austrian tests on welded joints reported and discussed in [1] and [4]), what reduces considerably, when use is made of formula 2, the interest of such a validation process.

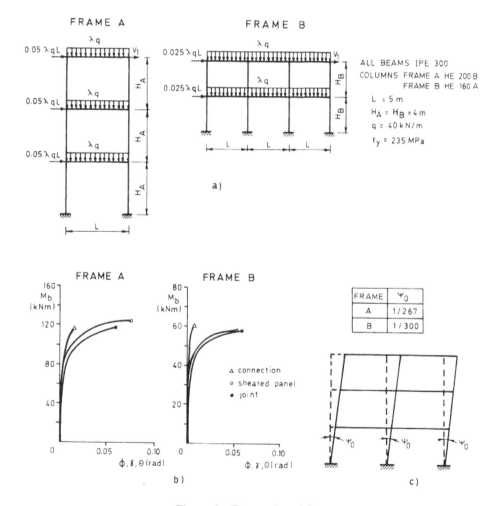

Figure 6 - Frames A and B

(a) Configurations and loadings
(b) Deformability curves
(c) Initial deflected shape

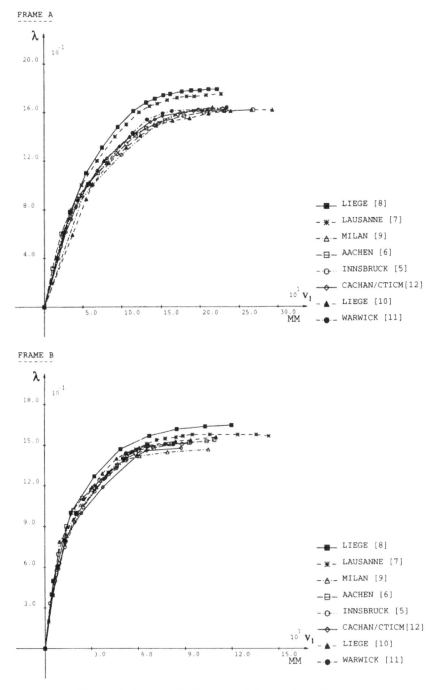

Figure 7 - Load-drift diagrams of frames A and B

REFERENCES

[1] Atamaz Sibai, W. and Jaspart, J.P., Etude du comportement jusqu'à la ruine des noeuds complètement soudés. Internal report IREM N° 89/7, Polytechnic Federal School of Lausanne, Switzerland, and MSM N° 194, University of Liège, Belgium.

[2] Maquoi, R. and Jaspart J.P., Influence of the non-linear and bi-linear modellings of beam-to-column joints on the structural response of braced and unbraced steel building frames. Fourth Int. Colloquium on Structural Stability, Istanbul, Turkey, Sept. 16-20, 1991, 133 - 151.

[3] ECCS TC8 on Structural Stability, TWG 8.1/8.2 Skeletal Structures, Analysis and Design of Steel Frames with Semi-Rigid Joints. ECCS Publication n° 67, 1992.

[4] Atamaz Sibai, W. and Frey, F., Numerical simulation of the behaviour up to collapse of two welded unstiffened one-side flange connections. Connections in Steel Structures, Elsevier Applied Science Publishers, London and New-York, 1988, 85 - 92.

[5] Tautschnig, A., Entwicklung eines neuen, makromechanischen Knotenmodels und Erstellung eines darauf aufbauenden EDV-programmes zur Berechnung von Stahlskelettragwerken unter Beruecksichtigung nichtlinear Nachgiebigkeiten des Verbindungselemente insbesondere bei "Steifenloser Bauweise", Ph.D. Thesis, University of Innsbruck, Austria, July 1983.

[6] Stutzki, C., Traglastberechnung raeumlicker Stabwerke unter Beruecksichtigung verformbaren Anschluesse. Ph. D. Thesis, University of Aachen, Germany, 1982.

[7] Atamaz Sibai, W. and Frey, F., Non-linear analysis of the flexibility connected frames using new joint elements. Fourth Int. Colloquium on Structural Stability, Istanbul, Turkey, Sept. 16-20, 1991, 168-175.

[8] Jaspart, J.P., Etude de la semi-rigidité des noeuds poutre-colonne et son influence sur la résistance et la stabilité des ossatures en acier. Ph. D. Thesis, University of Liège, Belgium, January 1991.

[9] Poggi, C., A finite element model for the analysis of flexibly connected steel frames. Int. Jnl. for Numerical Methods in Engineering, Vol. 26, N° 10,October 1988,2239-2254.

[10] Jaspart, J.P. and de Ville de Goyet, V., Etude expérimentale et numérique du comportement des structures composées de poutres à assemblages semi-rigides. Construction Métallique, N° 2, 1988, 31-49.

[11] Anderson, D. and Kavianpour, K., Analysis of steel frames with semi-rigid connections. Structural Engineering Review, Vol. 3, N° 2, June 1991, 79 - 87.

[12] Galea, Y., Colson, A. and Pilvin, P., Programme d'analyse de structures planes à barres avec liaisons semi-rigides à comportement non-linéaire. Construction Métallique, n° 2, 1986, 3 - 16.

Hollow Section Column to Open Section Beam Connections – Design Appraisal

M Saidani and D A Nethercot
Department of Civil Engineering,
University of Nottingham, University Park, Nottingham UK

ABSTRACT

An appraisal has been conducted of existing research into the structural behaviour of connections between I-section beams and RHS columns. Its purpose was to decide whether sufficient information is available to serve as the basis for an authoritative design guide.

INTRODUCTION

It is becoming increasingly accepted in the field of structural steelwork design that connections should receive as much attention as the individual members. Arguments based on economy, structural integrity, ease of construction etc., may all be cited in support of this. Whilst codes of practice have traditionally provided significant amounts of assistance with design of the various types of structural member, they have also tended to contain relatively little material relating directly to connection design. Within the area of the traditional I-section beam to H-section column type of connection a number of design documents have, however, been produced by the industry. Similarly, for connections between RHS and CHS tubular members a significant amount of design guidance has been provided by CIDECT. Forms of connection involving jointing of open to closed sections have, however, so far received comparatively little attention. This paper is concerned principally with an appraisal of existing research information on the behaviour and design of connections between I–section beams and RHS columns. It results from a comprehensive review of the subject [1], aimed specifically at identifying information directly suitable as the basis for the preparation of a design guide.

Literature Survey

In a recent, wide-ranging review of previous work on tubular connections of all types, the SCI [2] have identified several different forms of connection between I-section beams and SHS columns referred to in some form or other in the technical literature. Figure 1, which gives a few example types, shows how several of these mimic traditional forms of connection between I-section beams and H-section columns. Thus in preparing design guidance for connections between I-beams and SHS columns it is natural to start by examining the most up to date and reputable material dealing with the design of connections between open sections.

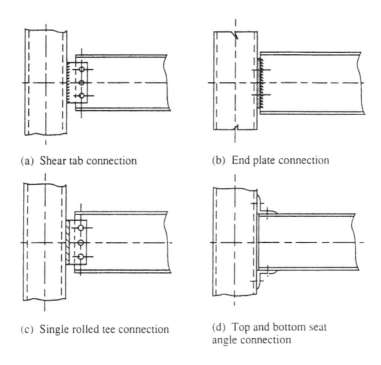

(a) Shear tab connection   (b) End plate connection

(c) Single rolled tee connection   (d) Top and bottom seat angle connection

Figure 1. Example types of connections to hollow section columns

Three such documents have been published in the last few years: by the BCSA/SCI [3], by the Australian Institute of Steel Construction [4] and by the American Insitute of Steel Construction [5]. In addition, the very recent EC3 [6] contains much more information on actual connection design than has traditionally been provided in national codes of practice.

The authors [1] have recently conducted an appraisal of all relevant available information as a way of assessing the feasibility of basing detailed design guidance for connections between I-section beams and RHS columns on existing information. The overall conclusion to this work was that coverage of the subject is extremely patchy. In certain cases – such as the shear tab connection that is considered in some detail later in this paper – failure modes are largely similar to those for the equivalent connection using open sections and thus similar design checks may be used. The principal difference consists of connecting to the face of the RHS column rather than to the flange of an H-section; in certain cases guidance on the special structural problems introduced by this may be obtained from established sources. For many other forms of connection, however, comparatively little detailed quantitative information could be located. In some instance this was not due simply to a lack of investigation of the particular detail but rather to the fact that previous testing had been of a somewhat ad-hoc nature, leading simply to illustrations of behaviour rather than serving as the basis for design rules. In certain cases e.g. where concrete filled tubular columns were employed, extensive investigations had been conducted – but with particular applications in mind, with the result that the design proposals were both complex and specialised.

A summary of the most significant previous investigations is provided in Table 1. This shows that something over a hundred full scale tests have been conducted – small numbers of results from the ad hoc test programmes have not been included in Table 1 – but that comparatively little use has been made of the findings in producing design procedures. For example the six tests on tees welded to the column face and bolted to the beam web reported in reference 9 were all considered to exhibit satisfactory behaviour. The design recommendations were, however, restricted to comments of the form:

- The width of the tee flange should be sufficient to permit welding along the corners of the tube wall.
- The ratio of tee flange width-to-thickness should not exceed 10 (9 was the maximum value tested).
- Tee web thickness, weld design and beam web bolts should be sized by conventional techniques.

Whilst these are helpful in establishing the general basis of design, a more complete understanding - almost certainly requiring additional testing and/or the application of theoretical techniques such as yield line analysis, finite elements etc.– is necessary before a full procedure can be established.

TABLE 1

Major investigations of behaviour of connections between open and closed sections

Connection type	Design procedure	Experimental studies	Numerical/ Theo.studies	Comments
shear-tab connection	yes	2 tests (7) 13 tests (8) 8 tests (9)	elastic finite element (8) finite difference (9)	for open to open section connections several design procedures exist
single rolled T-connection	no	6 tests (9)	none	–
angle cleat connection	suggested that for open to open	2 tests (9)	none	connection extremely flexible
moment connection with strap angles	no	23 tests (10)	theoretical analysis using simple theory of beams	some broad recommen- dations and conclusions
welded tee- end connection	for tube in compression design based on stability criteria, for tube in tension design basis uncertain	18 tests (11)	yield line analysis	–
semi-rigid connection between a W beam and an RHS column- using a rein- forcing plate	based on "assumed" stress distribution	10 tests (12)	theoretical analysis based on "assumed stress distribution" in the elements	design guide- lines with some observations
externally stiffened box column to I- beam connection	no	15 tests (13)	elastic finite element analysis	T-stiffeners more efficient than other forms of stiffeners
concrete filled square tubular column to steel composite beam connection	static design and dynamic design by assumed yield line mechanism	analysis of a 31 storey office building + testing of subassemblages (14)	yield line analysis	good resis- tance to seismic loading

The Shear Tab Connection

The shear tab connection is very popular in the USA and Australia and is increasing in popularity throughout Europe (particularly in the UK). It is simple to fabricate and very easy to erect on site. The connection is known by several different names: shear tab or single plate framing (American), fin plate (British), web side plate (Australian) etc. Figure 2 shows the general layout of the connection.

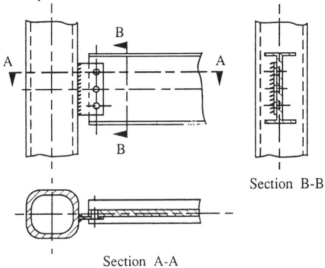

Section B-B

Section A-A

Figure 2. General layout of the shear tab connection

For connections between open sections complete design procedures are given in references 3-5. That presented in the BCSA/SCI Manual [3] is based on a conservative simplification of the approaches of references 4 and 5, supplemented by additional testing [15]. It identifies the four necessary design checks (corresponding to failure in each of the main components) of Table 2. In the case of check III this should include the possibilities that the actual form of failure could be lateral-torsional buckling of the fin plate if its length to thickness ratio is large (>15 in ref. 3).

TABLE 2

Design checks for Shear Tab Connections to Open Section Columns

Check	Design Condition/Possible Failure Mode
I	• bearing failure of the bolts
II	• shear and/or bending failure of the supported beam at net section
III	• shear and/or bending failure of the fin plate
IV	• failure of the weld

The checks of Table 2 do not include failure of the column wall as load transfer is assumed to occur directly into the column web, see Figure 2. However, for connections between open and closed sections the flexibility of the tube wall will introduce additional possible forms of failure.

It is therefore of interest to look closely at the available test data for shear tabs connected to RHS columns (7, 8, 9). From the 23 test results available, the failure modes listed in Table 3 were identified by the authors.

TABLE 3

Observed Failure Modes from Tests

Reference	Failure Modes Reported	Relationship to Table 2
7	–	–
8	• yielding of the gross area of the tab	III
	• bearing failure of the tab	I
	• fracture and yielding of the welds	IV
	• punching shear failure of the tube wall	V
	• surface tearing of the tube wall material beneath the weld	V
	• lateral buckling of the tab	III
	• shear yielding and fracture of the bolts	I
9	• local buckling of tube	V
	• web crippling of beam	II
	• bearing on bolt holes in beam web	I
	• weld tearing	IV

These may be consolidated into the four design checks of Table 2 plus an additional check V on the column face as indicated by the last column of Table 3.

The full design procedure for a shear tab to an RHS column therefore needs to address the five potentially critical areas of the joint shown in Figure 3, checks I, II and IV address basic strength reqirements; check III may need to include for buckling of the tab; check V must include all possible forms of tube wall failure.

In principle, it should be possible to utilise established methods for checks I-IV [3-5], supplemented by suitable procedures for design check V. In terms of the connection's ability to support the beam in a way that accords with the normal assumptions of "simple

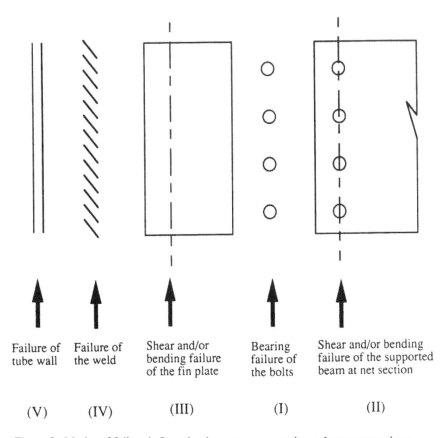

Failure of tube wall   Failure of the weld   Shear and/or bending failure of the fin plate   Bearing failure of the bolts   Shear and/or bending failure of the supported beam at net section

(V)   (IV)   (III)   (I)   (II)

Figure 3. Modes of failure in I-section beam to open-section column connections

construction" – limited rotational stiffness and therefore limited moment transfer to the column – test evidence [8,9] is encouraging. Both test series provide information on moment–rotation characteristics, indicating that column face deformation actually increases connection flexibility, thereby making actual behaviour closer to that of the assumed pin.

Three possible column wall failure modes have been observed:

- local buckling
- bending
- punching shear

The first of these has been observed in one of the tests of reference 9, whilst the other two modes are both amenable to theoretical analysis using respectively yield line analysis [16] or an approach that assumes that the tube face develops the yield strength of a unit length of the shear tab [8], leading to a requirement for a wall thickness t_w of:

$$t_w \geq t_{pl} \frac{\sigma_{ypl}}{1.2\,\sigma_{ut}} \qquad (1)$$

in which t_{pl} = thickness of shear tab plate

  σ_{ypl} = yield strength of shear tab

  σ_{ut} = ultimate strength of tube wall

Sherman [8] did, however, observe two tube wall failures of a material nature caused by the need to resist through thickness stresses induced by the tab. These resulted in separation of the weld from the tube wall and may be attributed either to lamellar tearing or lack of weld penetration. Greater care with the welding process eliminated this in subsequent tests.

Unfortunately only two of the tests of reference 8 identified punching shear as the cause of failure; in all cases the test description refers to "all of the connections exhibited possible multiple failure modes". Thus the number of results available for verifying design proposals is very limited.

CONCLUSIONS

The findings of a review of existing information on the behaviour of connections between I-beams and RHS columns have been summarised. Coverage of the subject in a way that is suitable for direct use in preparing a design guide has been found to be patchy. For the shear tab connection, for which the largest amount of useful information is available, the adaptations to established design procedures for the open section connection necessary to cover the additional failure modes possible when an RHS column is used have been discussed.

ACKNOWLEDGEMENTS

The investigation on which this paper is based was financed by CIDECT.

REFERENCES

1.   Saidani, M. and Nethercot, D.A.., Beam to Column Connections Between Tubular and Open Sections – Design Appraisal, CIDECT, Programme 5AY, January 1992.

2.   Steel Construction Institute, Survey of existing connection types and design methods using hollow sections, Document No. SCI-RT-224, July 1991.

3.   BCSA/SCI, Joints in simple construction, Vol. 1: Design methods, The Steel Construction Institute, 1991.

4.   Hogan, T.J. and Thomas, I.R., Design of structural connections, Australian Institute of Steel Construction, 4th edition 1991.

5.   AISC, Manual of steel construction, American Institute of Steel Construction, 1986.

6.   Commission of the European Communities, Eurocode 3: Design of steel structures, Part 1: General rules and rules for buildings, Volume 1, 1991.

7.   Herlache, S.M. and Sherman D.R., Beam connections to rectangular tubular columns", National Steel Construction Conference, AISC Miami Beach, Florida, June 1988, pp. 23.1–23.

8.   Sherman, D.R. and Ales, Jr. J.M., The design of shear tabs with tubular columns, National Steel Construction Conference AISC, Washington, D.C., June 1991, pp. 1.1–1.22.

9.   White, R.N. and Fang, P.J., Framing connections for square structural tubing, Journal of the Structural Division, ASCE, Vol. 92, No. ST2, April 1961, pp. 175–194.

10.   Picard, A . and Giroux, Y.M., Moment connections between wide flange beams and square tubular columns, Canadian Journal of Civil Engineering, Vol. 3, No. 2, 1976, pp. 174–185.

11.   Stevens, N.J. and Kitipornchai, S., Limit analysis of welded Tee end connections for Hollow tubes, Journal of the Structural Division, ASCE, Vol. 116, No. 9, September 1990, pp. 2309–2323.

12.   Dawe, J. L. and Grondin, G.Y., Semi-rigid connection between a W-beam and an HSS column, CIDECT Report No. 5AK-84/8-E, March 1985.

13.   Ting, L.C., Shanmugam, N.E. and Lee S.L., Box-Column to I-Beam connections with External Stiffeners, Journal of Constructional Steel Research, Vol. 18, No. 3, 1991, pp. 209-226.

14.   Teraoka, M. and Morita, K., Structural design of high rise building consisting of concrete filled square tubular columns and steel composite beams and its experimental verification. 4th International Symposium on Tubular Structures, The Netherlands, June 25-26, 1991.

15.   Moore, D.B., and Owens, G.W., Verification of design methods for finplate connections, Structural Engineer, Vol. 70, No. 3, February 1992, pp. 40–53.

16.   Stockwell, F.J., Yield line analysis of column webs with welded beam connections, Engineering Journal AISC, Vol. 12, No. 4, 1975 pp. 12–17.

REINFORCED BRANCH PLATE-TO-RHS CONNECTIONS IN TENSION AND COMPRESSION

J.L. DAWE
Professor, Department of Civil Engineering,
University of New Brunswick
P.O. Box 4400, Fredericton,N.B.,E3B 5A3, Canada

S.J. GURAVICH
Graduate student, Department of Civil Engineering,
University of New Brunswick

ABSTRACT

The tension and compression zones of moment plate connections were studied separately by testing branch plate-RHS connections under each type of loading. RHS flanges were reinforced by doubler plates fillet-welded all around. Results from thirteen specimens tested in tension indicate that branch plate to reinforcing plate width ratio is an important parameter in joint behaviour. The dominant failure mode was punching shear of the reinforcing plate. The importance of reinforcing plate and RHS wall thicknesses was apparent from results of thirteen specimens tested in the compression series.

NOTATION

b_1 width of branch plate
b_p width of reinforcing plate
b_o width of RHS
c end distance of reinforcing plate from junction with branch plate
t_1 thickness of branch plate
t_p thickness of reinforcing plate
t_o thickness of RHS wall

INTRODUCTION

The use of rectangular hollow sections (RHS) in steel frame construction is increasing globally because of a growing recognition of the cost savings compared with open-profile structural sections. RHS are lighter for equivalent capacity, cost less to paint, and have obvious aesthetic appeal for exposed frameworks. W shape sections perform well under strong axis loading such as occurs in beam applications while square hollow sections make excellent columns because of equal resistance with respect to both axes.

There are no guidelines for the design of moment connections between these two types of section. Typical methods of connecting W shape beams and columns such as welding beam flanges or moment plates directly to the column face are not suitable because of the flexible RHS flange [1]. Commonly used references for hollow structural steel design in Canada provide examples of moment connections between W shape beams and RHS columns that are expensive to fabricate and are difficult to use for multiplanar joints [2].

Earlier work at the University of New Brunswick evaluated the performance of a connection similar to typical W shape beam-column joints but modified to reflect the inherent weakness of the RHS face [3]. The present research programme has examined this connection in greater detail by modelling tension and compression zones separately to determine governing design parameters. Some limited testing of reinforced branch plate-RHS connections in tension has been performed in the past [4]. The connection has not been tested in compression previous to this research.

LITERATURE REVIEW

There has been extensive research carried out on connections between hollow section members. Wardenier compiled a large body of this work in text book form in 1982 [5]. Most of the research has been focussed on the behaviour of typical lattice framework connections. Test programmes undertaken at a number of European, Japanese, and North American universities under the auspices of CIDECT (the International Committee for the Development and Study of Tubular Construction) are summarized in CIDECT Monograph No 6 [6].

Relatively little attention has been paid to connections between W shape beams and RHS columns in past research. Five different shear connections were tested by White and Fang in 1966 and the authors noted the susceptibility of the column face to deformation because of its relative flexibility [1]. Picard and Giroux in 1976 designed and tested a W beam to RHS column moment connection which bypassed the column flange by employing coped strap angles welded to the beam flange and column web [7].

Korol and El-Zanaty found that chord flange stiffeners increased the strength and stiffness of RHS to RHS moment connections [8]. Dawe and Grondin reported on the performance of two different configurations of a moment plate connection between W beam and reinforced RHS column [3]. Test results indicated that reinforcment of the column flange with a flange doubler plate permitted attainment of the full capacity of the beam.

Unreinforced branch plate-RHS connections have been tested in tension by Wardenier, Davies, and Stolle [9]. Davies and Packer used yield line analysis and punching shear criteria to derive an expression for strength of these connections [10].

TEST PROGRAMME

General
Two series of specimens were tested: 13 in tension, denoted the HSST series and 13 in compression, denoted the HSSC series. All 26 specimens were constructed by a local steel fabricating firm. The RHS was reinforced by a doubler plate fillet-welded all around to the flange. A branch plate was welded perpendicular to the reinforcing plate by a complete penetration groove weld. Typical tension and compression series specimens are shown in Figure 1.

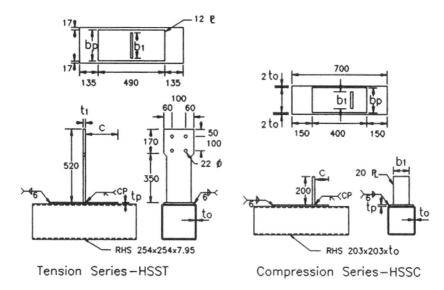

Tension Series—HSST　　**Compression Series—HSSC**

FIGURE 1. Typical test specimens

Parameters investigated in the tension series were width and thickness ratios of branch plate to reinforcing plate and end distance of reinforcing plate to the branch plate junction. In addition to these parameters, the thickness ratio of branch plate to RHS wall, was included in the compression series.

Specimen description

Tension series: RHS dimensions were held constant for all 13 HSST specimens in order to isolate the relationship between branch plate and reinforcing plate. A comprehensive range of width ratios was tested by selecting a wide RHS, 254 x 254 x 7.95 mm, for the series.

Branch plate width was varied from 220 mm to 100 mm in 20 mm increments for HSST1 to HSST7. HSST8 to HSST13 were all fabricated with a branch plate width of 140 mm. Branch plate thickness was 16 mm for HSST1 to HSST7 and HSST11 to HSST13. Thicknesses of 20, 12, and 10 mm were used for HSST8, HSST9, and HSST10, respectively.

Reinforcing plate thickness was held constant for all 13 specimens at 12 mm. Reinforcing plate width was 220 mm for HSST1 to HSST12. This dimension was reduced to 210 mm for HSST13. The branch plate was centered on the reinforcing plate length for HSST1 to HSST10 and HSST13. Edge distances of 75 mm and 150 mm were used for HSST11 and HSST12, respectively.

Compression series: The 13 specimens were configured similarly to those of the tension series. The basic components; branch plate, reinforcing plate, and RHS were joined by the same types of welds specified for tension specimens. The overall dimensions of the RHS were held constant for all 13 specimens, at 203.2x203.2 mm.

Connections were full width (b_1=b_p) for all but HSSC2 and HSSC11 with branch plate widths of 120 mm and 68 mm, respectively. Branch plate thickness was 16 mm for all 13 specimens.

Reinforcing plate thickness was 8 mm for HSSC4, HSSC5, HSSC7, and HSSC12, 12 mm for HSSC1 to HSSC3, HSSC8 to HSSC11, and HSSC13, and 16 mm for HSSC6. Reinforcing plate width was set at b_o-$4t_o$ [2] for all specimens. The branch plate was centered on the reinforcing plate length for HSSC2, HSSC3, HSSC6, HSSC9, and HSSC11 to HSSC13. Edge distance was 50 mm for HSSC4 and HSSC10, 100 mm for HSSC5 and HSSC8, and 150 mm for HSSC1 and HSSC7.

RHS wall thickness was 6.35 mm for HSSC1, HSSC4 to HSSC8, HSSC10, HSSC12, and HSSC13. Wall thickness of 7.95 mm was used for HSSC2, HSSC3, and HSSC11. HSSC9 had an RHS wall thickness of 9.53 mm.

Test Arrangement and Instrumentation

Tension series: All tests were performed in an 890 kN capacity universal testing machine. Each specimen was centered on top of the machine crosshead. It was secured by threading 4 tension rods through a reaction beam below the crosshead and up through 2 knife-edge beams positioned across the top of the RHS at the ends. Nuts at top and bottom of the tension rods were tightened. The branch plate was inserted in a grip and bolted in place.

Strain gauges were installed across the branch plate parallel to its longitudinal axis, on a line 40 mm up from the base. Linear Variable Displacement Transducers (LVDT) and dial gauges were used to monitor movement of the reinforcing plate relative to the RHS and to measure deformation of RHS webs. Experimental setup and instrumentation are illustrated in Figure 2. The testing machine scale and a load cell were used for recording load.

FIGURE 2. Tension series arrangement and instrumentation

Compression series: Specimens were whitewashed at potential yield zones before testing. The RHS was seated horizontally on the load table of the testing machine. A grooved compression plate was centered on top of the branch plate for transferring load to the specimen.

LVDTs were positioned to monitor movement of the reinforcing plate, and top flange and sidewalls of the RHS. Five of the 13 specimens were strain-gauged along the web of the RHS, perpendicular to the longitudinal axis and in close proximity to the branch plate connection. Test setup and typical configuration of LVDTs and strain gauges are shown in Figure 3.

Loading Procedure

Tension series: A tensile load was applied to the branch plate in 18 kN increments until the specimen began to yield and then in 9 kN increments to failure. The movement of specimen and corresponding load were acquired continuously by the LVDTs and load cell, and stored on diskette. Dial gauge and strain gauge readings were taken at each 18 kN increment after the load stabilized.

Compression series: The loading procedure and data acquisition were similar to those for the tension series tests.

FIGURE 3. Compression series arrangement and instrumentation

Mechanical Properties

Tensile coupons from plates and RHS members were prepared and tested in a 15000 kg capacity testing machine according to ASTM A370-91 [11]. The results for both the tension and compression series specimens (yield strength, ultimate strength, and Young's Modulus) are presented in Table 1 for a limited number of tests performed to date.

TABLE 1
Mechanical properties

Test Series	Component	Yield Strength (MPa)	Ultimate Strength (MPa)	Young's Modulus (MPa)
Tension	RHS-7.95mm	359	470	-
Series	Plate-20mm	368	501	218323
	Plate-16mm	360	530	224849
	Plate-12mm	363	521	230432
	Plate-10mm	359	540	220860
Compression	RHS-6.35mm	395	459	205459
Series	RHS-7.95mm	376	477	202366
	RHS-9.53mm	405	486	205459
	Plate-16mm	366	518	-
	Plate-12mm	337	514	-
	Plate-8mm	335	502	-

Test Results

Tension series: The ultimate load and punching shear load for each specimen are given in Table 2. HSST5 through HSST13 exhibited similar modes of failure. The reinforcing plate failed by punching shear. Then, after a variable increase in load (0.5% to 15% of ultimate), the specimens failed by tearing of the fillet weld. HSST1 and HSST2 also failed by tearing of the fillet weld but the reinforcing plate remained intact. HSST4 failed by punching shear of the reinforcing plate at ultimate load.

TABLE 2
Tension series test results

Spec. no.	HSST1	HSST2	HSST3	HSST4	HSST5	HSST6	HSST7
Ult. Ld.(kN)	587	718	547	489	449	455	439
Punching Ld.(kN)	-	-	543	489	*	413	429

Spec. no.	HSST8	HSST9	HSST10	HSST11	HSST12	HSST13
Ult. Ld.(kN)	455	472	437	484	474	446
Punching Ld.(kN)	409	414	409	409	427	427

* - punching shear load was not recorded

The relationship between b_1/b_p and ultimate load of the connection is shown in Figure 4. The graph is quadratically increasing between b_1/b_p values of 0.45 and 0.9. The load appears to decrease linearly above b_1/b_p - 0.9 where the failure mode is by weld failure.

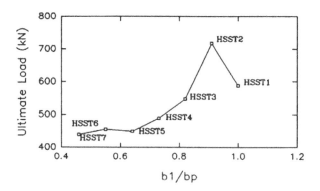

FIGURE 4. Ultimate tensile load vs width ratio

The thickness ratio t_1/t_p appears to have no significant effect on connection strength within the range of values tested. The effect of edge distance on ultimate load seems minimal from c - 75 mm to 372 mm (branch plate centered on reinforcing plate). In Figure 5, stress distribution across the branch plate from zero load to maximum load is shown for HSST1, HSST4, and HSST7, with b_1/b_p ratios of 1, 0.73, and 0.45, respectively. The graph indicates that the portion of the branch plate resisting tensile force decreases with decreasing b_1/b_p. It is also apparent that maximum difference in stress from the centre of the plate to its edges increases as the connection tends toward full width.

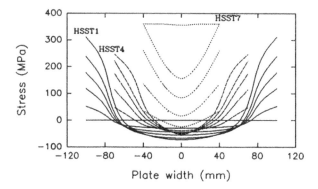

FIGURE 5. Stress distribution across branch plate

HSST13 was fabricated with a slightly narrower reinforcing plate width than the other 12 specimens. This modification appeared to have no effect on capacity.

Compression series: The ultimate load for each test specimen is given in Table 3. All but one specimen failed by RHS web buckling followed by yielding of the reinforcing plate and RHS flange. The branch plate of HSSC11 failed by buckling before the sidewalls and flange had reached maximum load.

TABLE 3
Compression series test results

Spec. no.	HSSC1	HSSC2	HSSC3	HSSC4	HSSC5	HSSC6	HSSC7
Ult. Ld.(kN)	601	688	710	380	421	681	416

Spec. no.	HSSC8	HSSC9	HSSC10	HSSC11	HSSC12	HSSC13
Ult. Ld.(kN)	588	799	501	630	409	603

Figure 6 shows the relationship between thickness ratios t_1/t_p and t_o/t_p and joint capacity. The joint capacity is improved when reinforcing plate thickness is increased and when the wall thickness of the RHS is increased.

A third parameter, edge distance, was examined for two different reinforcing plate thicknesses. The optimum edge distance for both sizes lies between 50 and 100 mm for connections of the dimensions tested. The capacity of the joint appears to improve with increasing b_1/b_p for the range of values tested. However the premature failure of HSSC11 does not permit a definitive analysis of the results.

The stress in the webs of the RHS is plotted against corresponding compressive load values for 4 different specimens in Figure 7. The point at which buckling of the web commences is different for each specimen because of varying parameters.

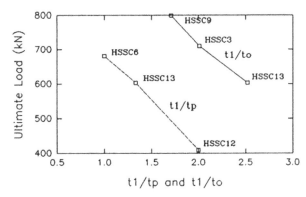

FIGURE 6. Ultimate compressive load vs thickness ratios

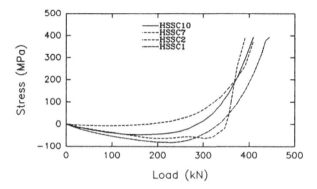

FIGURE 7. Average stress in RHS web at point of maximum bulging vs load

Discussion of Results

Tension series: Test results indicate that punching shear failure of the reinforcing plate is the governing criteria for b_1/b_p less than or equal to 0.82, for a particular value of t_1/t_p. This failure mode is related to reinforcing plate thickness, a parameter that was not explored in the tension series. There is some difficulty in interpreting the ultimate loads for HSST1 and HSST2, which were full width and 91% of full width, respectively. Additional specimens with similar dimensions would have to be tested before any conclusions could be made.

The optimum edge distance indicated for the tension zone of a moment plate connection is less than 75 mm for a reinforcing plate thickness of 12 mm. Recommended edge distance for a 10 mm plate from earlier research was 50 mm [3]. Preliminary analysis of stress distribution across the branch plate indicates that less efficient use is made of the plate section in resisting tensile force as b_1/b_p decreases.

Compression series: Buckling of the RHS webs appeared to be most sensitive to RHS wall thickness and reinforcing plate thickness. As either of these dimensions was increased, bulging became less pronounced and extended over a greater length. The reinforcment of the webs provided by additional beams framing in would reduce the tendency to buckle, depending on the similarity of loading from each direction.

Optimum edge distance for the compression zone indicated from the tests is between 50 mm and 100 mm for reinforcing plate thicknesses of 8 mm and 12 mm.

CONCLUSIONS

Preliminary recommendations for the design of moment plate connections between W beams and RHS columns are given below. Additional testing of branch plate-RHS connections and of W beam-RHS column connections is necessary to validate these guidelines.

Tension Zone
1. A parabolic relationship between b_1/b_p and punching shear resistance is indicated by test results. This relationship is valid for b_1/b_p values of between 0.45 and 0.91. Above 0.91 the capacity of the connection may be

assumed to be constant if adequate fillet weld is provided between the reinforcing plate and the RHS, adjacent to the tension plate.

2. A complete penetration groove weld at the base of the tension plate provides sufficient capacity to eliminate the possibility of failure at the weld.

3. The thickness of the tension plate within a range of realistic values will have no impact on resistance of the connection to punching shear for a constant reinforcing plate thickness.

4. End distance for a 12 mm reinforcing plate should be 75 mm.

Compression Zone

1. A linear relationship is appropriate for thickness ratios t_1/t_p and t_1/t_o and RHS web buckling resistance within the range of values tested.

2. End distance of 100 mm for 8 mm and 12 mm reinforcing plate will ensure that capacity of the connection is not affected by this parameter.

ACKNOWLEDGEMENTS

The authors appreciate the assistance of CIDECT and the Natural Sciences and Engineering Research Council of Canada.

REFERENCES

1. White, R.N. and Fang, P.J., Framing connections for square structural tubing, ASCE Journal of the Structural Division, 1966, 92(ST2), 175-193.

2. Stelco Inc., Hollow Structural Sections Design Manual for Connections, 2nd. ed., Stelco, Hamilton, Canada.

3. Dawe, J.L. and Grondin, G.Y., W-shape beam to RHS columns connections, Canadian Journal of Civil Engineering, 1990, 17, 788-797.

4. Pook, L.L., Strength of W-beam to HSS column connections, M.Sc. thesis, Department of Civil Engineering, University of New Brunswick, Fredericton, N.B., Canada.

5. Wardenier, J., Hollow Section Joints, Delft University Press, The Netherlands, 1982.

6. CIDECT Monograph No 6, British Steel Corporation, Tubes Division Technical Centre, Corby, Northants, Great Britain.

7. Picard, A. and Giroux, Y.M., Rigid connections for tubular columns, Canadian Journal of Civil Engineering, 1977, 3(2), 174-185.

8. Korol, R.M. and El-Zanaty, M., Unequal width connections of square hollow sections in Vierendeel trusses, Canadian Journal of Civil Engineering, 1977, 4, 190-201.

9. Wardenier, J., Davies, G. and Stolle, P., The effective width of branch plate to RHS chord connections in cross joints, Stevin Report 6-81-6, Department of Civil Engineering, Delft University of Technology, Delft, The Netherlands, 1981.

10. Davies, G. and Packer, J.A., Predicting the strength of branch plate - RHS connections for punching shear, Canadian Journal of Civil Engineering, 1982, 9, 458-467.

11. A370-91, Standard Specifications for Mechanical Testing of Steel Products, American Society of Testing and Materials, 1991.

THE DESIGN OF FLUSH AND EXTENDED END-PLATE CONNECTIONS WITH BACKING PLATES

D B MOORE
Building Research Establishment,
Garston, Watford, Herts,WD2 7JR, UK

and C GIBBONS
Ove Arup & Partners
13, Fitzroy Street, London W1P 6BQ, UK

ABSTRACT

This paper presents a step-by-step design procedure for flush and extended end-plate connections with backing plates. The design procedure is compared with a comprehensive range of published experimental data on T-stub connections and full-connections with and without backing plate stiffeners. The results of these comparisons show that the method is conservative and can be used for design purposes.

INTRODUCTION

In the design and detailing of a steel framed building, the normal sequence of events is to determine the layout of the framework and the loading to be carried, analyze the structure to determine the forces and moments in the individual members and then design the members. The final stage is to design and detail the connections between the members and to prepare information for the fabricator's shop.

When designing either flush or extended end-plate connections there is opportunity in the design to change the end-plate and the bolts but the size of the column flange is already established. If the column flange is not adequate the options available to the designer are:-

> Weld horizontal stiffeners into the web of the column between the tension bolts. This may make the connection of beams to the minor axis of the column difficult.

. Haunch the beam or use a deeper beam to increase the lever-arm and reduce the load in the tension region. This may not be economical or may not be possible because of the loss of headroom due to the increase in floor depth.

. Use a column section with a thicker flange.

. Add backing plates to increase the strength of the column flange.

One or more of the above options may be used to obtain an acceptable detail. However, it appears that the use of backing plates would be the simplest and most economic solution. Figure 1 shows a typical bolted extended end-plate connection with a backing plate stiffener. Each backing plate comprises a single plate, with either pre-punched or pre-drilled bolt holes. The backing plates can be either tack welded to the column flange in the fabrication shop or bolted in position during erection. Although they are not in common use, backing plates unlike the other options require minimal fabrication and can be fitted quickly and easily either in the fabrication shop or on site.

BACKGROUND

Early pioneering work on backing plates was undertaken by Zoetemeijer [1]. He studied a number of different moment connections and concluded that the stiffness and strength of a connection is increased with the addition of backing plates. Zoetemeijer also used yield line analysis to develop expressions for two possible modes of failure. The first mode consists of a hinge adjacent to the column web which produces a prying force leading to bolt failure. In the second mode, the prying force reaches a maximum and plastic hinges are formed at the bolt line and adjacent to the column web.

From a consideration of these modes, Zoetemeijer produced a single expression for the moment capacity of a connection having unstiffened flanges. The addition of backing plates increases the moment capacity of the connection since the total length of yield lines in the column flange may be increased and yield lines may form in the backing plates. However, Zoetemeijer did not carry out any thorough analytical study for connections with backing plates. Instead he proposed the following two empirical formulae, based on his test results:-

. Where bolt failure is the determining factor:
$$2F.b - 2(U-F).n \leq (2s + 4b_z + 1.25a)M_{pf} \tag{1}$$

. Where collapse of the column flange is the determining factor
$$2F.b_z \leq (2s + 4b_z + 1.25a)(2M_{pf} + M_{pp}) \tag{2}$$

From these formulae the minimum capacity of either the column flange or the bolts can be calculated. By increasing bolt diameters or backing plate thickness it is possible to increase the tensile capacity of the connection so that it is equal to, or approaches, the compressive capacity.

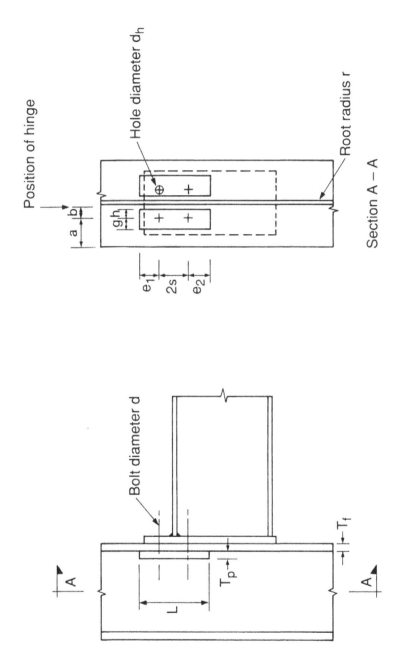

Figure 1 Typical extended end-plate connection with backing plate stiffeners

Work undertaken by Moore and Sims [2] on backing plates having a thickness equal to, or less than, the column flange thickness showed that backing plates increased the connection strength by an average of only 10-15%. This relatively small increase was partly due to the use of thin backing plates in which the potential for increased flexural action is reduced.

Tests carried out by Pynnonen and Granstrom [3] on symmetrical connections with backing plates concluded that Zoetemeijer's work overestimated the strength increase with thin backing plates. However, using thick backing plates and large bolts on columns with thin flanges, a strength increase of approximately 50% over the unstiffened flange may be obtained. From their test data Pynnomen and Granstrom proposed an empirically based design method. The test data were plotted on a graph of B/U against t_f/t_c and a design line was produced which accorded well with Zoetemeijer's test results.

RECENT EXPERIMENTAL WORK

The proposed backing plate procedure reported herein has been validated against previous experimental studies. These comprise tests performed by Moore and Sims [2], Zoetemeijer [1] and those conducted by Buller [4]. In the process of developing the design procedure it became apparent that further experimental tests were necessary in order to fully validate the method. Specifically, experimental data was required on the effect of bolt-preload; 'thick' backing plates (T_p/T_f approximately equal to 2); higher strength backing plates ($p_{yp} > p_{yf}$); lateral position of the plate (a/b ratio) and confirmation of the maximum projection of the plate for which an increase in connection resistance can be gained. A series of tests were therefore commissioned specifically to address these deficiencies in the existing experimental data. A total of 27 additional connection tests were performed at the Building Research Establishment, England. These comprised 21 extended end-plate (19 of which were stiffened by backing plates) and six flush end-plate connections (4 of which were stiffened by backing plates). A full review of these tests and the results is presented in reference [5].

THE PROPOSED DESIGN PROCEDURE

The proposed design method is based on a yield-line appraisal of the column flange and backing plate [5]. A number of previous comparisons of yield line appraisals with experimental data have shown this to be a valid approach [1,2]. However, steelwork designers are not normally familiar with the direct application of yield line principals. This is due to potential geometrical complexities when considering intersecting planes and hence the increased scope for calculation errors. In addition, for the approach to be valid, it is essential that the correct yield line pattern is identified - i.e. that which corresponds to the minimum load capacity. This can lead to lengthy trial and error calculations, thus further dissuading designers. The solution is to develop a design procedure based on an 'equivalent length' of yield line containing easily derived parameters which is a conservative lower bound to the actual length derived from a rigorous yield line appraisal.

In developing the 'equivalent length' design procedure, a rigorous appraisal of four yield line failure mechanisms (Mode A, B, C and D) was performed for connections of practical proportions. Figure 2 shows Mode A and C which are indicative of failure mechanisms where backing plates are used. Mode A, in which the yield lines in the column

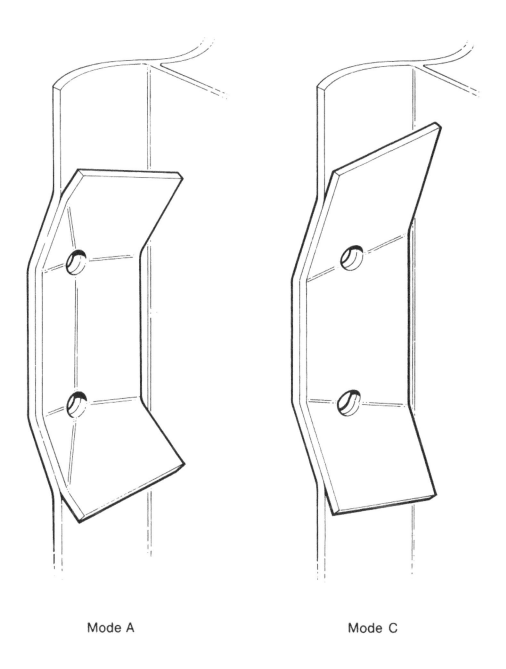

Mode A   Mode C

Figure 2 Indicative failure mechanisms where backing plates are used

flange and the backing plate are coexistent, is likely where the backing plates are relatively thin compared to the column flange. In Mode B, there are no diagonal yield lines present in the backing plate. This form of failure is typical where the backing plate is relatively thick. In modes C and D, there are no yield lines between the bolts. Consequently, these modes assume only limited prying action, a limited flexural contribution from the backing plates and are therefore appropriate for a column flange appraisal in the absence of backing plates.

The proposed expressions for the equivalent yield line length and bolt force are as follows:-

Column flange equivalent length $=$ s+a+1.4b (3)

Backing plate equivalent length $=$ s+g+1.4h (4)

$$\text{Bolt Force (F)} = \frac{p_{yf}T_f^2}{4b}2(s+a+1.4b)+\frac{p_{yp}T_p^2}{4b}(s+g+1.4h)$$ (5)

Figure 3 shows the proposed expression for bolt force F (eqn. 6), plotted with respect to a/b and Φ, in which Φ is the ratio of backing plate strength to that of the column flange. It should be noted that the condition Φ=0 corresponds to a situation in which backing plates are not used. It is evident that the 'equivalent length' expression generally provides a conservative approximation to the rigorously calculated values for both failure Modes A and B. Where the 'equivalent length' prediction of bolt force exceeds that obtained from a rigorous analysis, the overestimation is only 2.7%. Comparisons which have been made between the above expressions and C and D failure modes has shown that the above simple expressions are conservative for all practical values of Φ and a/b.

As described previously, prying forces must exist if a yield line is assumed to form along the line of bolts. Due to the enhanced flexural resistance, these prying forces tend to be significantly larger where backing plates are used. It is important therefore that the above simplified 'equivalent length' approach can conservatively predict the additional bolt forces due to the effects of prying.

Using a 'T' stub approach it has been shown that, in terms of predicting the bolt load, the above equivalent yield line method is 8 - 13% conservative for a mode B failure and 28 - 40% conservative for mode A. It is evident therefore that two separate formulae could be developed to cater for these conditions, thereby reducing the conservatism in the case of mode A. However, it is considered preferable to use a single formula to determine the prying force for both cases. There would then be no problem if collapse was actually by mode B when mode A was assumed in the design. It should be noted that this would occur in the likely event of a thicker backing plate being used to fabricate the connection, compared to that assumed in the calculations.

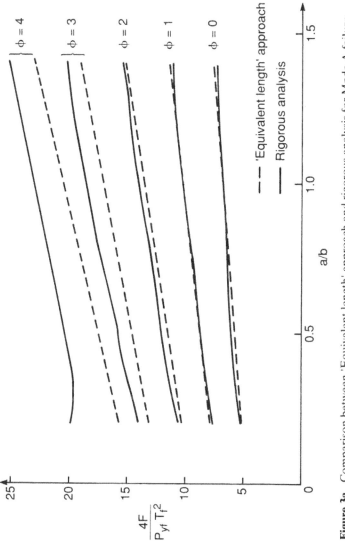

Figure 3a Comparison between 'Equivalent length' approach and rigorous analysis for Mode A failure

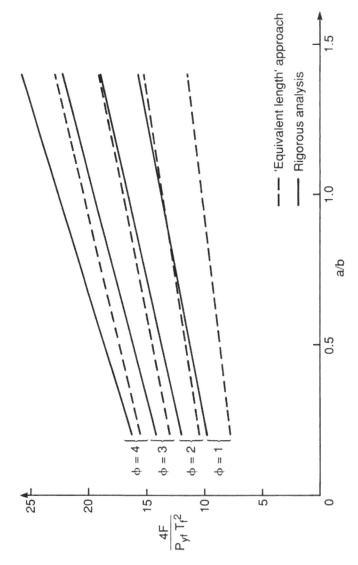

Figure 3b Comparison between 'Equivalent length' approach and rigorous analysis for Mode B failure

In addition to considering the flexural modes of failure of the column flange and backing plate, it is possible that the capacity of the connection could be dominated by the tensile resistance of the bolts. The following expression is proposed for this condition:

$$\text{Bolt Force (F)} = \frac{p_{yf}T_f^2}{4(b + n)}(s + a + 1.4b) + \frac{B_{tn}}{b + n} \tag{6}$$

The simplified approach effectively assumes that yield lines do not develop in the backing plate, but are present at the root of the column flange.

The following design procedure is based on the above 'equivalent yield line' equations for flexural failure of the backing plate/column flange (eqn. 5) and that governed by the tensile resistance of the bolts (eqn. 6). It is evident from these two equations that the moment resistance of a given connection detail can be determined with relative ease. However, the procedure is specifically presented in a step-by-step format which will enable the required connection detail to be designed from a known set of applied loads and moments. The procedure adopts the general limit state recommendations in the new European steelwork code, EC3[6].

Step 1:   Determine the basic connection detail by applying standard rules to determine the location of the bolts, thickness and size of beam end-plate.

Step 2:   Determine the permissible bolt tension, B_t.

For grade 4.6 and grade 8.8 bolts:
If $V_{net} \leq 0.127 f_{ub}A_s$ then $B_t = 0.67 f_{ub}A_s$ (7)
If $V_{net} > 0.127 f_{ub}A_s$ then $B_t = 0.93 f_{ub}A_s - 2.1V_{net}$ (8)

where:-
V_{net} $= (V - m_cP_s) / m_t$ (9)
V $=$ total applied shear force
m_c $=$ number of bolts in the compression zone
m_t $=$ number of bolts in the tension zone
P_s $=$ ultimate shear resistance of the bolts in the compression zone
f_{ub} $=$ ultimate tensile strength of the bolt
A_s $=$ tensile area of the bolt.

This check effectively reduces the tensile resistance of the bolts in the tension zone to allow for the presence of large shear forces. It should be noted that a reduction on the value of B_t is often not necessary for practical connection arrangements.

Step 3:   Calculate the moment resistance of the column flange, M_{pf}.

$$M_{pf} = 0.25 \, k_r . w_f T_f^2 . p_{yf} \tag{10}$$

where:-

k_r is a reduction factor to allow for the axial stress, f_n, in the column.

k_r = 1.0   when $f_n \leq 180$ N/mm^2

k_r $= \dfrac{2p_{yf} - 180 - f_n}{2p_{yf} - 360}$ when $f_n > 180$ N/mm^2

a =   distance from the centreline of the bolt to the edge of the effective width of the column flange. To guard against failure due to a local yield line pattern forming around the bolt.

'a' should be taken as the lesser of:-
           (i)   distance from the centreline of the bolt to the edge of the flange.
           (ii)   1.4b

b =   distance from the bolt centreline to 0.95 x root radius from the face of the web

w_f = $u_{fl} + u_{fr}$ i.e. the sum of the effective length of the yield line on either side of the bolt.

u_{fl} and u_{fr} should be taken as the least of :-
           i)   distance from the centreline of the bolt to the column end
           ii)   0.5 x bolt pitch to an adjacent bolt
           iii   a + 1.4b

Step 4:   Calculate the load capacity per bolt (F_t)

$$F_{t1} = \frac{1}{b+n}(M_{pf} + B_t n) \quad \text{(bolt failure)} \tag{11}$$

$$F_{t2} = \frac{2M_{pf}}{b} \quad \text{(column flange yielding)} \tag{12}$$

n =   distance from the centreline of the bolt to the point where the prying force is assumed to act.

'n' should be taken as the least of:-

i)   distance to edge of either the end-plate or the column flange.

ii)

	without backing plates	with backing plates
pre-loaded bolts	$1.5T_f$	$2.0T_f$
non pre-loaded bolts	$2.0T_f$	$3.0T_f$

This particular limitation is intended to limit the loss of pre-load in pre-loaded bolts after stressing to working load, and to limit the possibility of bolts being loosened in the case of non pre-loaded bolts.

Step 5:   Appraise the capacity of the connection without backing plates

If F_{f1} and $F_{f2} > F_{reqd}$ backing plates are not required.
If $F_{f1} < F_{reqd}$ increase bolt size - go to step 2.
If $F_{f2} < F_{reqd}$ backing plates are required - go to step 6.

Step 6:   Check for bolt failure when backing plates are provided.

Recalculate 'n' from the Table in step 4, appropriate for when backing plates are used. Determine F_n from equation (11). If the allowable bolt force is less than required then increase bolt size and go to step 2.

Step 7:   Calculate the bolt load (F_p) and moment resistance (M_{pp}) to be contributed by the backing plate.

$$F_p = F_{reqd} - F_f \qquad (13)$$
$$M_{pp} = F_p b \qquad (14)$$

Step 8:   Decide on the backing plate dimensions.

The following rules should be applied.

i) $T_p \le 2 T_f$ and $T_p \ge 0.5$ x bolt diameter

ii) $g \le 0.85a$ and $h \ge 0.75b$ to ensure that additional hinges do not form in column flange adjacent to the plate

iii) $h < b$ to ensure that the plate clears the root radius of the column flange.

Step 9:   Calculate the minimum thickness of backing plate required (T_p)

$$T_p \not< \sqrt{\frac{4 M_{pp}}{w_p P_{yp}}} \qquad (15)$$

where:-

P_{yp} = design strength of backing plate steel

$w_p = u_{pl} + u_{pr}$ i.e. the sum of the effective length of the yield line on either side of the bolt.

u_{pl} and u_{pr} should be taken as the least of:-
(i)   distance from the centreline of the bolt to the end of the backing plate
(ii)   0.5 x bolt pitch to an adjacent bolt
(iii)   g + 1.4h

COMPARISON OF PROPOSAL DESIGN PROCEDURE WITH EXPERIMENTAL RESULTS

In Table 1, the design procedure is compared with the T-stub tests reported by Zoetemeijer[1], Moore and Sims[2] and Buller[4]. These correlations were achieved using the measured values of yield stress. It is clear that for all the tests considered the design procedure gives conservative results with safety margins varying from 1.12 to 4.65.

Table 2 compares the design procedure with published experimental data on flush and extended end-plate connections with backing plates. The design method gives an adequate predication of strength for all tests with the safety margin varying from 1.96 to 2.38 for flush end-plates and from 1.48 to 2.71 for extended end-plates.

A comparison between the predicted capacities of T-stubs with and without backing plates indicates that backing plates can increase the tensile capacity of a column flange by up to 103%. However, in a full unstiffened connection this increase is limited to the failure load of the column web in the compression zone. For the tests studied in reference [5] it was shown that the maximum increase in the predicted moment capacity when using an unstiffened column was only 7.4%. Obviously the performance of such a connection would be enhanced if welded web stiffeners were provided in the compression zone of the column. However, these stiffeners are expensive to install, particularly when fixed to both sides of the column, and would detract from the advantages which can be gained by using backing plate stiffeners.

CONCLUSIONS

A design procedure for backing plate stiffeners has been presented and compared with published experimental data from which the following conclusions may be drawn:-

(1)   Comparisons between predicted and experimentally observed results shows that the proposed design method is conservative.

(2)   It is evident that in some cases the increase in strength which can be achieved using backing plates will be limited by the buckling resistance of the unstiffened column web in compression. However, instances have been highlighted where this effect can be minimised.

TABLE 1

Correlation between experimental and predicted capacity for
T-stub tests with backing plates.

Reference No	Specimen No	Experimental Capacity P_E (kN)	Predicted Capacity P_P (kN)	P_E/P_P
1	7*	300	126	2.38
1	15	350	207	1.69
1	16	410	236	1.74
2	T1*	266	114	2.33
2	T5	318	150	2.12
4	1a 23/17	358	121	2.96
4	1a 23/38	340	126	2.70
4	1a 23/29	333	126	2.64
4	1a 23/30	361	148	2.44
4	1a 23/7	350	163	2.15
4	1a 23/19	358	163	2.20
4	1a 23/39	376	167	2.25
4	1a 23/15	357	189	1.89
4	1a 23/16	371	189	1.96
4	1a23/11*	365	110	3.32
4	1b 30/23	410	222	1.85
4	1b 30/27	401	237	1.69
4	1b 30/44	424	239	1.77
4	1b 30/24	410	273	1.50
4	1b 30/28	414	308	1.34
4	1b 30/38	414	313	1.32
4	1b 30/33	413	313	1.32
4	1b 30/41	414	363	1.14
4	1b 30/45	415	371	1.12
4	1b30/26*	220*	183	1.20
4	1d 23/27	480	131	3.66
4	1d 23/36	450	143	3.15
4	1d 23/23	440	146	3.01
4	1d 23/26	446	164	2.72
4	1d 23/2*	460	99	4.65

* No backing plate

TABLE 2

Correlation between experimental and predicted capacities for
flush and extended end-plate connections with backing plates

Reference No	Specimen No	Experimental Capacity M_E (kNm)	Predicted Capacity M_P (kNm)	M_E/M_P
2	J3*	76	28	2.71
2	J4	86	37	2.32
5	E1*	141	87	1.62
5	E2*	146	87	1.68
5	E3	145	87	1.67
5	E4	197	126	1.56
5	E5	195	126	1.55
5	E6	186	126	1.48
5	E7	215	118	1.82
5	E8	207	127	1.63
5	E9	218	109	2.00
5	E10	200	135	1.48
5	E11	230	135	1.70
5	E12	237	135	1.76
5	E13	215	135	1.59
5	E14	226	135	1.67
5	E15	242	152	1.59
5	E16	199	106	1.88
5	E17	246	152	1.62
5	E18	207	115	1.80
5	E19	258	172	1.50
5	E20	228	113	2.02
5	E21	240	113	2.12
5	F1*	99	48	2.06
5	F2*	94	48	1.96
5	F3	158	74	2.14
5	F4	151	74	2.04
5	F5	164	69	2.38
5	F6	160	76	2.11

* No backing plates

E or J - Extended end-plate connection
F   - Flush end-plate connection

ACKNOWLEDGEMENTS

The contribution to this work from Mr Abdul Malik and Dr Adrian Dier of the Steel Construction Institute, Mr David White of the Building Research Establishment and of Mr Brian Cheal is greatly appreciated.

REFERENCES

1. Zoetemeijer, P., A design method for the tension side of statically loaded, bolted beam-to-column connections. Heron 20, No.1, Delft University, Delft, Holland, 1974.

2. Moore, D.B. and Sims, P.A.C., Preliminary investigations into the behaviour of extended end-plate steel connections with backing plates. Journal of Constructional Steel Research Vol.6, No.2, 1986.

3. Pynnonen, J. and Granstrom, A., Beam-to-column connections with backing plates. Swedish Institute of Steel Construction, Report No.86.6, April 1968.

4. Buller, P.S.J., The influence of backing plate dimensions on the behaviour of extended end-plate steelwork connections - interim report 1. Building Research Establishment internal note, March 1985.

5. Rules for the design of backing plate connections - calibrations with additional experimental structures. Building Research Establishment Contractors Report, DoE No SCI-RT-216, December 1991.

6. ENV 1993-1-1 Eurocode 3:Design of steel structures Part 1.1:General rules and rules for buildings, CEN, February 1992.

EXPERIMENTAL BEHAVIOR
OF ENCASED STEEL SQUARE TUBULAR COLUMN-BASE CONNECTIONS

SHIGETOSHI NAKASHIMA
Associate Professor/Department of Architecture
Osaka Institute of Technology
5-16-1 Ohmiya, Asahi-ku, Osaka 535, JAPAN

ABSTRACT

In order to increase the degree of fixation of the column bases of steel
structures and to facilitate reinforcing and repairing of such column
bases when they are damaged, steel column bases are often encased in
reinforced concrete. This report describes the results of the
experimental studies on the mechanical properties of the models of the
concrete-encased column bases supporting steel square tubular columns.
The values obtained by using the assumed mechanical models were compared
with the experimental values, and on the basis of such comparison, the
design method for such column bases is discussed.

INTRODUCTION

Japan's Aseismatic Design Standard provides that the story drift of the
frame subjected to lateral force should not exceed 1/200 of each story
height. Therefore, it is a common practice to encase the column bases in
reinforced concrete in order to increase the degree of fixity of the
column bases. On the other hand, exposed type column bases are prone to
damage when the lateral force is imposed on them. In order to enable
easy, economical and safe repair of such damaged column bases when minor
damage has been caused to them, additional reinforcing bars are often
welded to the vertical bars embedded around the anchor bolts, and by using
such additional bars as main bars reinforced concrete stubs are formed to
surround the column bases. This method enables stresses produced in steel
columns to be transferred smoothly to the reinforced concrete elements or
to transfer such stresses by the composite effects of steel column bases
and reinforced concrete elements. However, the mechanical properties of
these column bases have not yet been fully clarified. The principal
objective of this paper is to clarify the mechanical properties of the
encased column bases by taking up for consideration the part of columns
below their inflection points which are assumedly located at a certain
fixed height and by subjecting them to bending and shear forces under the
predetermined compressive force.

EXPERIMENT PLANNING

Table 1 shows the experiment plan. The experiment elements included the presence or absence of anchor bolts, amount of hoop bars for shear reinforcement, height and section of the encasement, presence or absence of main chord bar hooks and column axial force ratio. The height of inflection point varies depending on the details of the encasement stub, the magnitude of the load applied, etc., but for the purpose of the present experiments, a fixed value was assumed.

Specimens
Representative specimens out of the 30 specimens produced by concreting performed in 3 times will be shown in Table 1. The specimens are models scaled down to approximately 1/2.5 of the actual column base as shown in Fig. 1. Cold-formed square section steel tube of □-150 X 150 X 6.0 was used for columns. The encasement stub reinforcement was varied while the standard height and section of the encasement stub were assumed to be 3D (D: Outside dimension of steel square tube) and 2D X 2D, respectively.

Methods of loading and measurement
As shown in Fig. 1, a pin jig was installed at the position 75 cm above the lower surface of the base plate, and positive/negative alternating lateral force by deflection controls was applied. Prior to the application of lateral force, fixed axial pressures corresponding to 0, 1/6, 1/3

Table 1
Experiment Plan

No.		Specimen name	h	b×d (cm×cm)	N/Ny	Chord bar	Hoop	Top hoop	Anchor bolt
1)	A	MⅢ5-13A-0R	3D	30×30	0.0	10−D13	D6 @35	3−D10@17.5	4−M16
2)	A	MⅢ5-130-0R	3D	30×30	0.0	10−D13	D6 @35	3−D10@17.5	-------
3)	A	MⅢ5-330-0R	3D	30×30	0.0	10−D13	D6 @35	2−D10@35.0	-------
4)	A	MⅢ5-N30-0R	3D	30×30	0.0	10−D13	D6 @35	------------	-------
5)	B	MⅢ5-13A-0R	3D	30×30	0.0	10−D13	D6 @35	3−D10@20.0	4−M16
6)	B	MⅢ5-17A-0R	3D	30×30	0.0	10−D13	D6 @70	3−D10@20.0	4−M16
7)	B	MⅢ5-1NA-0R	3D	30×30	0.0	10−D13	-------	3−D10@20.0	4−M16
8)	B	MⅢ5-13A-0R*	3D	30×30	0.0	10−D13	D6 @35	3−D10@20.0	4−M16
9)	C	MⅡ5-25A-0R	2D	30×30	0.0	10−D13	D6 @50	3−D10@25.0	4−M16
10)	C	LⅢ5-25A-0R	3D	40×40	0.0	10−D13	D6 @50	3−D10@25.0	4−M16
11)	C	MⅢ5-25A-0R	3D	30×30	0.0	10−D13	D6 @50	3−D10@25.0	4−M16
12)	C	MⅢ5-25A-1R	3D	30×30	1/6	10−D13	D6 @50	3−D10@25.0	4−M16
13)	C	MⅢ5-25A-2R	3D	30×30	1/3	10−D13	D6 @50	3−D10@25.0	4−M16
14)	C	MⅢ5-25A-5R	3D	30×30	2/3	10−D13	D6 @50	3−D10@25.0	4−M16

※ ' No hook

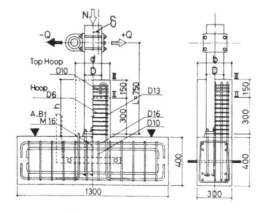

Fig. 1 Test specimens

Table 2
Mechanical properties of materials

Steel				σ_y (N/mm²)	σ_t (N/mm²)	E (×10⁵N/mm²)	ε (%)
Column	AB	□-150×150×6	STKR400	348	421	1.90	34.5
	C	□-150×150×6	STKR400	388	433	2.10	42.0
Anchor bolt	AB	M16	SR235	322	439	2.20	30.6
	C	M16	SR235	290	433	2.00	27.0
Chord bar	AB	D13	SD345	375	560	1.87	20.6
	C	D13	SD345	356	531	1.90	18.5
Top hoop	AB	D10	SD345	371	518	1.90	21.0
	C	D10	SD345	385	555	1.90	18.4
Hoop	AB	D 6	SD345	305	445	2.00	25.1
	C	D 6	SD345	479	533	1.90	12.3

Concrete		σ_c (N/mm²)	σ_t (N/mm²)	E_c (×10⁴N/mm²)		σ_c (N/mm²)	σ_t (N/mm²)	E_c (×10⁴N/mm²)
	A	21.8	1.80	2.01		21.8	1.90	2.01
Beam	B	23.9	1.80	2.01	Encasement	24.2	1.90	2.01
	C	25.8	1.95	2.54		26.0	1.91	2.53

and 2/3 times as much as the yield load of the column were applied. The frame holding the measuring instrument was installed to the threaded steel bar which had been fixed to the foundation beam and was assumed to be a stationary point, and the deflection of each point was measured with the use of this setup as shown in Fig. 1.

RESULTS OF THE EXPERIMENTS AND DISCUSSION

Hysteretic characteristics

The relations between lateral load (Q) and displacement at the loading point (δ) are shown in Fig. 2. Their envelope curves on the positive load side at each cycle are shown in Fig. 3. The straight lines in the figure show the calculated value where the column bases are fixed perfectly.

Effects of anchor bolts: Test specimen 1) is provided with anchor bolts while Test specimen 2) is not. Fig. 3 a) shows the envelopes of the hysteretic curves on the positive side for these specimens. The maximum strength of Specimen 1) is 10 - 15% greater than that of Specimen 2). It follows from this that the ultimate capacity of such column bases may be obtained as a sum of the capacity of the anchor bolts and that of the reinforced concrete encasing them.

Effects of top reinforcing bars: Effects of the top reinforcement on the mechanical properties of the column bases will be studied on the basis of the results of the experiments using Test specimens 2), 3) and 4). The envelopes on the positive side of the hysteretic curves are shown in Fig. 3 b). In each specimen, shear bars are placed at uniform spacings under the top reinforcing bars. The strength of Specimen 4) is 9% and 21% lower than that of Specimen 2) on the positive side and the negative side respectively. It is also noted that the less substantial the top rein- forcement is, the greater the lowering of the capacity due to repeated loading is.

Effects of shear reinforcing bars: Specimens 5), 6) and 7) are all reinforced at the top with an identical reinforcement ratio; however, spacings of hoop bars placed under the top bars differ among the specimens. Fig. 3 c) shows the envelopes on the positive side of the hysteretic curves for these specimens. These envelopes indicate that the

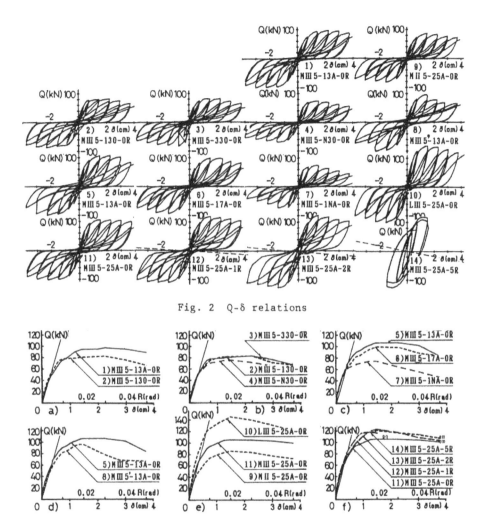

Fig. 2 Q-δ relations

Fig. 3 Envelope curves of Q-δ relations

larger the spacings of hoop bars in the specimens are, the larger the
lowering of the capacity beyond the maximum capacity as well as the lowering of the capacity due to the repeated loadings is. These phenomena are
particularly pronounced in the case of Specimen 7), which is not hooped.
However, the capacity of this specimen was 70 - 75% of Specimen 5), which
is reinforced at the top and hooped. These test results indicate the
necessity to provide column bases with top reinforcement which is
effective against shears and also to add adequate hoop bars. Also, it is
known from the above test results that the initial rigidity of the column
bases is not affected by such top reinforcement and hoop bars for shear.

Effects of hooks of chord bars: The vertical chord bars in the
concrete encasement of Specimen 5) is hooked whereas those of Specimen 8)
is without such hooks. Fig. 3 d) indicates the envelopes on the positive
side of the hysteretic curves for these specimens. The maximum capacity

of the specimen whose vertical chord bars are not end-hooked is somewhat lower than the specimen having hooked bars, and the pronounced lowering of the capacity beyond the maximum capacity is seen in the specimen without hooked bars.

Effects of the encasement height and sectional area: Based on the results of the tests of Specimen 9), 10) and 11), the influence of the encasement size on the hysteretic properties of the column bases will be discussed. Fig. 3 e) shows the envelopes on the positive side of the hysteretic curves for these test specimens. Specimen 11) has an encasement height of 3D and an encasement sectional area of 300 mm × 300 mm. Specimen 9) is identical with Specimen 11) with respect to the encasement sectional area and shear bar spacings but its encasement height is 2D. The capacity of Specimen 9) is ca. 20% lower on both the positive and the negative sides, and the lowering of its capacity beyond its ultimate capacity is slightly larger than in the case of Specimen 11). Further, the initial rigidity of Specimen 9) is small compared with the initial rigidity computed by assuming that the column base are fully fixed.

Specimen 10) is the same as Specimen 11) with respect to the encasement height and the shear bar spacings but its cross sectional area of the encasement is 400 mm × 400 mm. The maximum capacity of Specimen 10) is 38% and 22% higher on the positive and the negative sides respectively than that of Specimen 11); however, Specimen 10) shows great lowering of the capacity beyond its ultimate capacity and also great lowering of the capacity under repeated loadings compared with Specimen 11). In both of these two specimens, the initial rigidity proven by the tests is higher than the initial rigidity computed by assuming that the column bases are fully fixed. Thus, it is reasonable to presume that the ultimate capacity and the elastic rigidity of the concrete-encased column bases are substantially affected by the height and cross sectional area of the encasement and are also directly affected by the method by which the encasement is reinforced.

Effects of column axial force: The effects of the column axial force on the hysteretic properties of the column bases will be considered on the basis of the test results obtained for Specimen 11), 12), 13) and 14). Fig. 3 f) shows the envelopes on the positive side of the hysteretic curves obtained for these test specimens. Specimen 11) was not subjected to the column axial force. Specimen 12), 13) and 14) were subjected to the column axial forces equal to 1/6, 1/3 and 2/3 respectively of the yielding loads for the columns. The test results clearly indicate that the larger is the axial compressive force, the larger is the maximum capacity and the greater is the lowering of the capacity beyond the maximum capacity. In this case, the hysteresis loops show bulging when the axial compression is involved whereas such bulging is insignificant if the axial compression is absent. For Specimen 14), the yielding of the steel tubular columns due to bending at the top of the encasement turned out to be a governing factor to determine the capacity.

Rigidity
Fig. 4 shows examples of the envelope curves of the Q-δ relations in initial cycles on the positive force application side. The calculated value I in the figure represents the elastic rigidity of the encasement concrete before generation of initial bending cracks, and the calculated value II represents the rigidity after generation of cracks. The rigidity before and after crack generation was sought as follows:

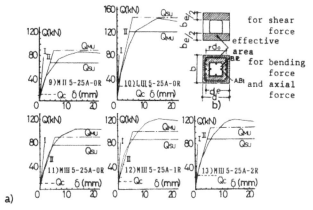

Fig. 4   Initial stiffnesses and effective area

Rigidity before generation of bending cracks: The bending rigidity of the encasement reinforced concrete portion before generation of bending cracks was sought by adding the rigidity of the structural steel element, that of the reinforcing bar element and that of the concrete element. The deformation of column was obtained by taking bending and shear deformation into consideration.

Bending deformation:

$$\delta_M = \frac{Q}{3} \left(\frac{\ell^3}{{}_sK} + \frac{L^3 - \ell^3}{{}_RK} \right) \tag{1}$$

where, ${}_sK$: Bending rigidity of steel column ($= {}_sE \cdot {}_sI$)
   ${}_RK$: Integrated bending rigidity of structural steel element, reinforcing steel element and concrete element in column base encasement ($= {}_sE \cdot {}_sI + {}_rE \cdot {}_rI + {}_cE \cdot {}_cI$)

where, ${}_sE$, ${}_rE$, ${}_cE$: Modulus of elasticity of respective elements of structural steel, reinforcing bar and concrete
   ${}_sI$, ${}_rI$, ${}_cI$: Moment of inertia of respective elements of structural steel, reinforcing bar and concrete
   L, ℓ, h   : See Fig. 6.

Shear deformation:

$$\delta_S = Q \left(\frac{{}_sk \cdot \ell}{{}_sG \cdot A_s} + \frac{{}_ck \cdot h}{{}_cG \cdot A_c} \right) \tag{2}$$

where, ${}_sk$, ${}_ck$: Shape factor of structural steel and concrete by respective section shapes
   ${}_sG$, ${}_cG$: Shear modulus of elasticity of structural steel and concrete
   A_s, A_c: Section area of structural steel and concrete

Rigidity after generation of bending cracks: The bending rigidity of the encasement portion after generation of bending cracks was obtained by assuming the bending rigidity of the concrete element to be zero. The deformation of the column was obtained by adding said deformation assuming the rigidity of the concrete element to be zero and the deformation of the chord bars drawn out. Further, the initial bending crack strength(Qc) was

sought assuming that the effective section of the reinforced concrete stub
is as shown in Fig. 4 b) and using AIJ standard formula for reinforced
concrete structure (1).

The rigidity values thus calculated tend to exceed their respective
experimental values but are generally indicative of the values before and
after crack generation.

Stress transmission mechanism and load bearing mechanism

Fig. 5 a) shows the representative examples of the relations between the
lateral force (Q) and the strain (rε) of the chord bars at the base level
of the encasement. The typical examples of bending moment borne by the
steel column and that borne by the encasement reinforced concrete as
obtained using these Q-rε relations are shown in Fig. 5 b) respectively.
In this figure, the moment borne by the encasement concrete is shown as
added to the moment borne by the steel column. As is clear from the
figure, the moment generated in the column is in part directly transmitted
from the steel column base to the foundation and is in part transmitted by
way of the encasement reinforced concrete. The former is transmitted
through the anchor bolts and the base plate and the latter is transmitted
through the bend resistance of the encasement stub. A large shear force
($_r$Q) is generated in the vicinity of the encasement stub top and the
bearing pressure by said shear force must be transmitted by way of the
shear reinforcing bars to the foundation concrete. Fig. 6 shows the load
bearing mechanism in a simplified form as clarified by the experiment. It
could be understood that the yield strength of the chord bar under a fixed
axial compression, when subjected to bending/shear force, increases as the
axial compression is increased. In case the encasement stub height is 3D
or over, the unit strain will be close to the value for a structural
steel/reinforced concrete composite structure. Accordingly, the axial

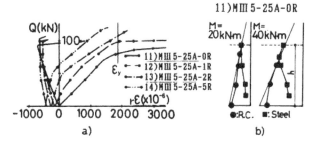

Fig. 5 Q-rε relations of chord bars and moment distributions of steel
column & encased reinforced concrete

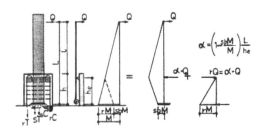

Fig. 6 Load bearing mechanism

force sharing ratios should be investigated in greater detail. In the present investigation, however, an assumption was made that the percentage of axial force shared by the encasement reinforced concrete portion was 1/2 of the axial force obtainable when a structural steel/reinforced concrete composite column is assumed.

Ultimate strength
The ultimate strength of the encasement stub is sought based on the load bearing mechanism as shown in Fig. 6 by adding the respective strengths of the encasement reinforced concrete element and the steel column base element.

<u>Ultimate bending strength determined by the steel column:</u>

$$_{sc}M_u = \{L/(L - h_e)\} \cdot M_p \tag{3}$$

where,
M_p: Plastic moment, taking axial force into consideration
L: Height from base plate lower surface to loading point
h_e: Effective height of encasement (= h-6 cm, h: encasement height)

<u>Ultimate bending strength of the encased column base:</u>

$$M_u = _{sb}M_u + _rM_u \tag{4}$$

where,
$_{sb}M_u$: Ultimate bending moment of steel column base element
$_rM_u$: Ultimate bending moment of reinforced concrete portion of encasement

In each case, the calculation will be made applying the ultimate strength theory of the reinforced concrete structure calculation standard on the assumption that the base plate and the encasement stub section are sections of reinforced concrete respectively.

<u>Ultimate shear strength of an encased column base:</u>

$$Q_u = _sQ_u + _rQ_u \tag{5}$$

where,
$_sQ_u$: Ultimate shear strength of steel column element (= $_{sb}M_u/L$)
$_rQ_u$: Ultimate shear strength of encasement reinforced concrete element

The application of the formula based on correspondingly the empirical formula of Ohno-Arakawa can be considered:
Using the load bearing mechanism as shown in Fig. 6, the shear force acting upon the encasement top was sought, and the relations of shear span ratios with ultimate shear stresses are shown in Fig. 7.

$$_rQ_u = \{\tau_c + 2.7 \sqrt{P_{wt} \cdot _{wt}\sigma_y + P_w \cdot _w\sigma_y}\} \cdot b_e \cdot _rj \cdot h_e/L \tag{6}$$

$$\tau_c = 0.080 \frac{k_u \cdot k_p \cdot (180 + F_c)}{h_e/d_e + 0.115} \tag{7}$$

where,
P_{wt}: Top portion shear reinforcing bar ratio (= $\Sigma\, a_{wt}/b_e \cdot h$, a_{wt}: sectional area of a group of top hoops, b_e: effective width of encasement reinforced concrete portion)

p_w: Hoop ratio (= $a_w/b_e \cdot X$, a_w: sectional area of a group of hoops, X: hoop bar interval)

$_{wt}\sigma_y$, $_w\sigma_y$: Yield stress of top hoops and hoops, respectively.

$_rj$: Stress center to center distance in section of encasement reinforced concrete

k_u: Compensation coefficient due to encasement section size

k_p: Compensation coefficient due to tensile chord bar ratio p_t (%)

d_e: Effective depth of encasement

F_c: Compressive strength of concrete

However, the following conditions must be satisfied:

$$2.7 \sqrt{P_{wt} \cdot {}_{wt}\sigma_y + P_w \cdot {}_w\sigma_y} \cdot b_e \cdot {}_rj \cdot h_e \leq {}_ra_t \cdot {}_r\sigma_y \cdot {}_rd_o \tag{8}$$

where, $_ra_t$: Total area of tensile side chord reinforcement of encasement

$_r\sigma_y$: Yield stress of tensile side reinforcement

$_rd_o$: Chord bar interval on tensile side and compressive side

Table 3
Experiment results

No.	Specimen name	scQu (kN)	Qwu (kN)	Qsu (kN)	cQu (kN)	Qmax(kN) +	Qmax(kN) −	Qmax/cQu +	Qmax/cQu −
1)	MⅢ5-13A-OR	175.2	99.3	89.8	89.8	97.4	98.3	1.08	1.09
2)	MⅢ5-130-OR	175.2	78.9	69.4	69.4	82.9	87.6	1.19	1.26
3)	MⅢ5-330-OR	167.0	78.9	61.3	61.3	82.9	82.5	1.35	1.35
4)	MⅢ5-N30-OR	167.0	78.9	55.7	55.7	75.5	68.9	1.36	1.24
5)	MⅢ5-13A-OR	169.3	99.9	86.2	86.2	106.6	96.4	1.24	1.12
6)	MⅢ5-17A-OR	169.3	99.9	80.1	80.1	99.0	94.3	1.24	1.18
7)	MⅢ5-1NA-OR	169.3	99.9	72.1	72.1	74.5	72.8	1.03	1.01
8)	MⅢ5-13A-OR*	169.3	99.9	86.2	86.2	96.0	101.1	1.11	1.17
9)	MⅡ5-25A-OR	133.0	89.3	67.6	67.6	86.6	82.3	1.28	1.22
10)	LⅢ5-25A-OR	186.2	120.7	145.7	120.7	145.5	124.3	1.21	1.03
11)	MⅢ5-25A-OR	186.2	89.3	76.7	76.7	105.5	102.0	1.38	1.33
12)	MⅢ5-25A-1R	186.2	99.2	85.1	85.1	120.5	111.7	1.42	1.31
13)	MⅢ5-25A-2R	158.6	108.4	88.4	88.4	122.7	127.4	1.39	1.44
14)	MⅢ5-25A-5R	104.9	111.3	83.4	83.4	116.3	120.8	1.39	1.45

※ ' No hook

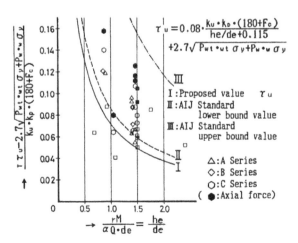

Fig. 7 Relations between shear span ratios and ultimate shear stresses

249

Ultimate strength of encased column base: The ultimate strength of the encased column base can be sought by the following formula:

$$_cQ_u = \min \{_{sc}Q_u, Q_{MU}, Q_{SU}\} \tag{9}$$

where, $_{sc}Q_u = _{sc}M_u/L$, $Q_{MU} = M_u/L$, $Q_{SU} = Q_u$

Table 3 shows the comparison between empirical values and calculated values. The ratio of the empirical value to the calculated value is approximately 1.2 on the average meaning that the empirical values are on the safe side.

CONCLUSIONS

From this experiment, the following facts were clarified:
1) Encased column bases such as those used in the present experiments show inverted "S" shape hysteretic behaviors accompanying the slip phenomenon.
2) The elastic rigidity of column base is governed by the height and section area of the encasement and can be sought by the method described in this paper.
3) If the height of the encasement is approximately 3 times the outside dimension of the steel tube, its rigidity will exceed that of the fixed column base.
4) Top reinforcing and shear reinforcing steel bars do not affect the initial rigidity but do affect the maximum strength and the strength beyond the maximum strength.
5) The ultimate strength of the encased column base can be sought by adding the strengths of the steel column base element and the encasement concrete element. However, the former should be ignored in case of repair of the damaged column base.
6) In case the vertical chord bars in the encasement are not hooked at ends, pronounced lowering of the strength beyond the ultimate strength is liable to occur.

REFERENCES

1. AIJ, Standard for Structural Calculation of Reinforced Concrete Structures, Architectural Institute of Japan, Tokyo, 1988, pp. 177-185. (In Japanese)
2. Arakawa, T., Experimental studies on shear strength of reinforced concrete beams, Trans. of AIJ, No. 66, 1960, pp. 437-440. (In Japanese)
3. Nakashima, S. and Igarashi, S., Influence of details of column bases on structural behaviour, Prereport 13th congress, IABSE, Seminar sessions, Helsinki, 1988, pp. 705-710.
4. Nakashima, S., Suzuki, T. and Igarashi, S., Mechanical characteristics of exposed type steel column bases, Proceedings of the international colloquium, Bolted and special structural joints, Vol. 2, USSR National Committee, IABSE, Moscow, 1989, pp. 148-152.
5. Nakashima, S., Igarashi, S., Kadoya, H., Suzuki, M. and Noda, N., Experimental study on exposed steel square tubular column bases using a special grout method, Tubular structures, 4th International symposium, Delft, 1991, pp. 109-118.

STRENGTH OF MULTIPLANAR TUBULAR JOINTS WITH SEMI-FLATTENED BRACING MEMBERS

Jacques RONDAL
Associate Professor,
Institute of Civil Engineering,
University of Liège,
Quai Banning, 6, B-4000 Liège, Belgium

ABSTRACT

For statically loaded CHS lattice girders, the conventional profiling of the ends of the bracing members can be, in some cases favourably replaced by a partial flattening of these ends. The aim of the contribution is to present the results of an experimental research on the static strength of multiplanar tubular K-joints in welded lattice girders. The ends of the bracing tubular members used in the tested specimens have been prepared by means of three different technologies: classical profiling, partial flattening and cropping.
On base of the test results, design strength formulae, modifying the IIW formulae to take account of the multiplanar effect, are proposed for the three types of technologies.

INTRODUCTION

Circular hollow sections combines aesthetic and strength in a very attractive manner. For this reason, they are widely used in steel lattice structures. In these applications, welding represents the favourite joining method but, unfortunately, it leads to a relatively high cost of the connections.

For statically loaded structures, a simplification of the end preparation of the bracing members, leading to a reduction of the cost, can be obtained by using cropped members (cropping is a technique in which flattening and shearing of a round tube is made by the same tool so that straight fillet welds can be used to weld the braces to the chords). Unfortunately this technique can affect significantly the performance of the joint. However, very often, it will be sufficient to produce a semi-flattening of the end of the bracing members. This technique allows not only for a simplification of the end preparation of the tubes but also for a reduction of the weld gap between the brace - and the chord member up to a limit quite acceptable for a correct welding, i.e. 3 mm [1]. With semi-flattened bracing members, the performance of the joint is better than that of joints with cropped members and is in relation with the required level of flattening which is necessary to ensure an allowable weld gap.

The objective of this contribution is to summarize the results of an experimental research on joints in multiplanar lattice girders using cropped and semi-flattened bracing members.

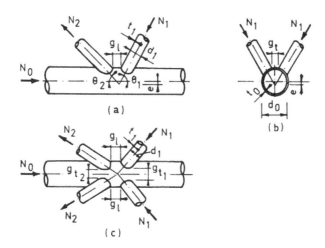

(a)

(b)

(c)

Figure 1 - Main notations for a multiplanar K-joint

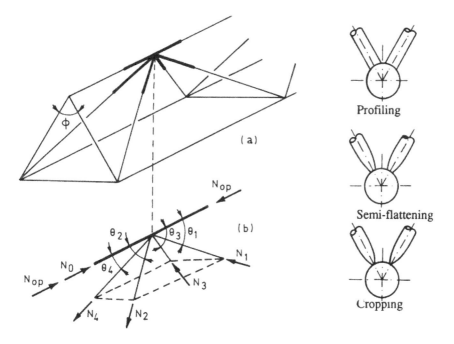

(a)

Profiling

Semi-flattening

(b)

Cropping

Figure 2 - Main parameters of the test specimens

However, due to the present lack of knowledge, tests have also been performed on joints prepared with the classical profiling technique. Figures 1 and 2 give the main parameters governing the joint behaviour. It must be stressed that, both, the longitudinal joint gap g_l and the transverse joint gap g_t play an important role in the behaviour of the joint. These two factors are in direct relation with the level of flattening of the bracing members (Figure 3).

EXPERIMENTAL PROGRAMME

96 tests have been performed on multiplanar K-joint specimens up to collapse. However, in order to obtain a precise knowledge of the geometry of the joints, a series of measurements has been performed on 72 specimens with different levels of flattening. These geometry measurements have yielded to the following relations (Figure 3) [2] :

- longitudinal joint gap:

$$g_l = \sum_{i=1}^{2} g_i \tag{1}$$

with:

$$g_i = \frac{d_o}{2 \ tg \ \theta_i} - \frac{d_{imax}}{2} \tag{2}$$

- transverse joint gap :

$$g_t = \frac{\pi \ d_o}{360} \ \phi^o - d_{imin} \tag{3}$$

where :
- profiling :

$$d_{imax} = d_{imin} = d_i \tag{4}$$

- semi-flattening :

$$d_{imax} = \frac{0.65}{\sin \theta_i} \ (2.54 \ d_i - 2\sqrt{j(d_o - j)} \ ) \tag{5}$$

$$d_{imin} = 2 \ \sqrt{j(d_o - j)} \tag{6}$$

$$j \ (weld \ gap) = 3 \ mm \tag{7}$$

253

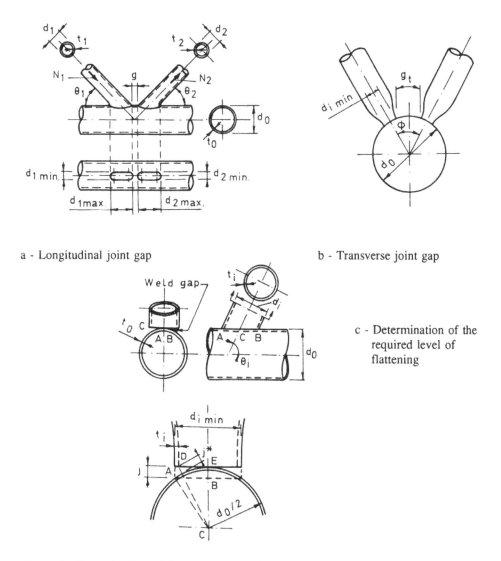

a - Longitudinal joint gap

b - Transverse joint gap

c - Determination of the required level of flattening

Figure 3 - Determination of the longitudinal joint gap, of the transverse joint gap and of the level flattening

- cropping :

$$d_{imax} = \frac{1.57}{\sin \theta_i} (d_i - 1.52 \, t_i) \tag{8}$$

$$d_{imin} = 3 \, t_i \tag{9}$$

ANALYSIS OF THE RESULTS

Two types of failure mode have been observed during the experimental tests :
- a chort plastification, practically without deformation between the compression braces - i.e. the two compression braces act as one member ;

- a plastic buckling close to the ends of the compression braces.

The first failure mode is influenced by the transverse gap as shown by some authors [3, 4, 5]. This influence may be taken into account by means of a correction factor k_{gt} multiplying the strength of the equivalent planar joint, given, for example, in the annex K of the Eurocode 3 [6] or in the IIW recommendations :

$$k_{gt} = 1 - 0.49 \, \frac{g_{tl}}{d_o} \tag{10}$$

The second failure mode is influenced by the non-uniform distribution of the stresses in the compression braces close to the joint weld (a shell plastic buckling for joints with profiling or small flattenings gradually leading to a plate buckling for large flattenings or cropping). This influence may be taken into account by a correction factor given by :

$$k_s = 0.7 + 0.2 \, \frac{d_{1min}}{d_1} \tag{11}$$

STATISTICAL EVALUATION

A statistical evaluation of the calculated strengths against the experimental results has been performed on base of the procedure recommended in the annex Z of the Eurocode 3 [7]. Table 1 gives the main results of the evaluation, where the ratio Δ_k and the partial safety factor γ_M^* are defined by means of the relations :

$$\Delta_k = \frac{nominal\ strength}{characteristic\ strength} \leq 1 \tag{12}$$

and

$$\gamma_M^* = \frac{nominal\ strength}{design\ strength} \tag{13}$$

The partial safety factor has to be compared with the level of safety required, which can varies from country to country.

Figure 4 shows the correlation between calculated and experimental joint strengths.

TABLE 1
Statistical evaluation

Type of end preparation	profiling	semi-flattening	cropping
Comparison between calculated and experimental joint strength (N_{exp}/N_{th}):			
- number of tests n =	16	19	20
- coefficient of correlation ρ =	0.99	0.98	0.98
- mean m =	1.106	1.104	1.453
- standard variation s =	0.068	0.090	0.099
- coefficient of variation s/m=	6.14 %	8,18 %	6.84 %
Statistical evaluation :			
- Δ_k minimum :	0.96	0.98	0.76
maximum :	0.99	1.02	0.79
mean :	0.975	1.005	0.777
- γ_M^* minimum :	1.13	1.20	0.94
maximum :	1.17	1.23	0.98
mean :	1.151	1.214	0.950

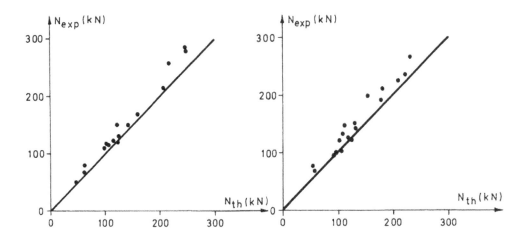

a - Profiling

b - Semi-flattening

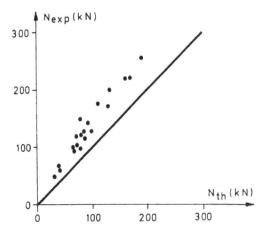

c- Cropping

Figure 4 - Correlation between calculated and experimental joint strengths

DESIGN EQUATIONS

Table 2 gives all the design equations allowing the calculation of the design strength of multiplanar K-joints for the three technologies of preparation of the end sections of the bracing members. These formulae have been calibrated for a partial safety factor $\gamma_M = 1.10$ [6].

CONCLUSIONS

As mentioned in the introduction, the aim of the contribution was to propose, on base of experimental tests, design strength relations for multiplanar K-joints with three different types of preparation of the end sections of the bracing members. The other factor of the cost-to-performance ratio - i.e. the fabrication cost of the joints - is in relation with the unit cost of steel, the man-hour cost and, last but not least, the available equipments. These factors can differ from country-to-country. As a consequence, the technology leading to the minimum cost-to-performance ratio can vary around the world.

REFERENCES

1. Wardenier, J., Kurobane, Y., Packer, J.A., Dutta, D. and Yeomans, N., Design guide for circular hollow section joints under predominantly static loading, Verlag Tüv Rheinland, Köln, 1991.

2. Mouty, J. and Rondal, J., Etude du comportement sous charge statique des assemblages de profils creux circulaires dans les poutres de sections triangulaires et quadrangulaires. Final Report, Commission of the European Communities, Research 7210 SA 310, 1991.

3. Makino, Y, Kurobane, Y. and Ochi, K., Ultimate capacity of tubular double K-joints. Proceedings of the 2nd International Conference of IIW, Boston, 1984.

4. Scola, S., Redwood, R.G. and Mitri, H.S., Behaviour of axially loaded tubular V-joints. J. Constr. Steel Res., 1990, 16, pp. 89 - 109.

5. Paul, J.C., Ueno, T., Makino, Y. and Kurobane, Y., The ultimate behaviour of circular multiplanar TT-joints. Proceedings of the International Symposium on Tubular Structures, Delft, 1991, pp. 448 - 460.

6. Eurocode 3, Annex K, Hollow section lattice girder connections. Eurocode n° 3, Design of Steel Structures, 1990.

7. Eurocode 3, Annex Z, Procedure for the determination of design resistance from tests. Eurocode n° 3, Design of Steel Structures, 1989.

TABLE 2

Design equations for the strength of multiplanar K-joints made with circular hollow sections

Braces (profiling)
$d_{imin} = d_1$
$d^* = d_1$
$g_1 = \sum\limits_{i=1}^{2} g_i \not< 0.0$ with $g_i = \dfrac{d_o}{2\,tg\theta_i} - \dfrac{d_i}{2\sin\theta_i}$
$g_{t1} = \dfrac{d_o \cdot \phi^o}{114.6} - d_1 \not< 0.0$
Braces (semi-flattening)
$d_{imin} = 2\sqrt{3(d_o - 3)} \not> d_i$ (with d_o in mm)
$d^* = \dfrac{d_1 + d_{1min}}{2}$
$g_1 = \sum\limits_{i=1}^{2} g_i \not< 0.0$ with $g_i = \dfrac{d_o}{2\,tg\theta_i} - \dfrac{0.65}{2\sin\theta_i}(2.54d_i - d_{imin})$
$g_{t1} = \dfrac{d_o \cdot \phi^o}{114.6} - d_{1min} \not< 0.0$
Braces (cropping)
$d_{imin} = 3\,t_i$
$d^* = \dfrac{d_1 + d_{1min}}{2}$
$g_1 = \sum\limits_{i=1}^{2} g_i \not< 0.0$ with $g_i = \dfrac{d_o}{2\,tg\theta_i} - \dfrac{1.57}{2\sin\theta_i}(d_i - 1.52t_i)$
$g_{t1} = \dfrac{d_o \cdot \phi^o}{114.6} - d_{1min} \not< 0.0$

Notations	Design strength (i = 1,2)

	Plastification of the chord

$$N_{1,Rd} = \frac{f_{yo} \cdot t_o^2}{\sin\theta_1}(1.8 + 10.2\frac{d^*}{d_o})$$

$$\cdot k_{g1} \cdot k_{gt} \cdot k_p \cdot k_\gamma$$

$$N_{2,Rd} = \frac{\sin\theta_1}{\sin\theta_2} N_{1,Rd}$$

Punching shear

$$N_{i,Rd} = \frac{f_{yo}}{\sqrt{3}} \cdot t_o \cdot \pi \cdot d_i \cdot \frac{1+\sin\theta_i}{2\sin^2\theta_i}$$

Functions	Plastification of the braces

$$k_{g1} = \gamma^{0.2}[1 + \frac{0.024\gamma^{1.2}}{e^{(0.5g'-1.33)}+1}]$$

$$k_{gt} = 1 - 0.49\frac{g_{t1}}{d_o}$$

$k_p = 1.0$ *for* $n_p \geq 0$ *(tension)*

$= 1 + 0.3n_p - 0.3n_p^2$ *for* $n_p < 0$

(compression)

$$N_{1,Rd} = \pi \cdot t_1 \cdot (d_1 - t_1) \cdot f_{y1}$$

$$\cdot (0.7 + 0.2\frac{d_1\min}{d_1})$$

$$N_{2,Rd} = \pi \cdot t_2 \cdot (d_2 - t_2) \cdot f_{y2}$$

Special notations

$k_\gamma = 0.96$ *(profiling)*

   0.91

   (flattening)

   1.00 *(cropping)*

$$d^* = \frac{d_1 + d_{1\min}}{2}$$

$$\gamma = \frac{d_o}{2t_o} \quad ; \quad g' = \frac{g_1}{t_o}$$

$$n_p = \frac{N_{op}}{A_o \cdot f_{yo}} : \geq 0 \text{ in tension;}$$

   < 0 *in compression.*

Validity ranges

$$0.2 < \frac{d_i}{d_o} < 1.0 \; ; \; 5 \leq \frac{d_i}{2t_i} \leq 25 \; ; \; 30° \leq \theta_i \leq 90°$$

DEVELOPMENT, ANALYSIS AND EXPERIMENTATION OF ATLSS CONNECTIONS

ROBERT B. FLEISCHMAN, Research Scholar
B. VINCENT VISCOMI[1], Research Professor
LE-WU LU, Professor of Civil Engineering
MARK R. KACZINSKI, Research Engineer
ATLSS Center, Lehigh University, Bethlehem PA 18015

ABSTRACT

A series of new beam-to-column connections known as ATLSS connections (ACs) are currently under development. The emphasis of these new designs is on a self-guiding feature for use in automated construction. This feature will minimize human assistance during construction and will result in quicker, safer, and less expensive erection procedures. The AC concept is based on using a tapered male piece on the beam which slips into a female guide mounted on the column. The AC concept can cover a large range of structural needs, including shear, partial- and full-moment connections. The analytical analysis of this connection is performed using plane strain finite element procedures. Experimentation is being conducted concurrently with the theoretical work. It is anticipated that this phase of the program will provide additional modifications to the AC. Then, scale-model plane frames with shear and moment ACs will be tested, and a 3-D building will be evaluated.

INTRODUCTION

A series of new beam-to-column connections known as ATLSS connections (ACs) are currently under development. The emphasis of these designs is on geometric configurations which provide a self-guided erection feature to greatly facilitate initial placement. This feature will minimize human assistance during construction, resulting in quicker, less expensive erection procedures in which workers are less susceptible to injury or fatalities.

ATLSS Integrated Building System Program
The ACs are being developed at the Center for Advanced Technology for Large Structural Systems (ATLSS) at Lehigh University. The center was created in 1986 by a grant from the National Science Foundation. The primary mission of the ATLSS Center is to serve as a focal point for research and education that will lead to technological developments which increase the competitiveness of the U.S. construction industry. Within ATLSS, the AIBS (ATLSS Integrated Building Systems) program was developed to coordinate ongoing research projects in automated construction and connection systems. Its objective is to

---

[1]Professor of Civil Engineering, Lafayette College, Easton, PA 18042

provide a means to design, fabricate, erect and evaluate cost-effective building systems with a focus on automation and computer integration [1]. An important component of this system is the family of ACs. These connections, in both concrete and steel, possess the capability of being erected by automated construction techniques.

The approach to automated construction of building components at ATLSS incorporates recent advances in crane technology, which are related to the development of the Stewart platform. A Stewart platform uses six individually controlled cables to move a lower platform and payload relative to the upper platform position. Orientation of the cables is such that the system has properties of a space frame, and the lower platform can move with six degrees-of-freedom. This configuration not only provides excellent translational and rotational stiffness when compared to a boom crane, but can also adjust the position and orientation of the lower platform with precision to make insertion of ACs possible [2]. A scale-model Stewart platform has been constructed to test the feasibility and limitations of automated construction with the ACs. Future plans include improving manual control to incorporate force and position feedback; development of a vision system; and extending the platform's duties to on-site material storage and inventory management (See Fig. 1).

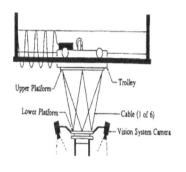

Figure 1. Stewart Platform.

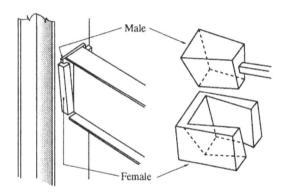

Figure 2.　a: 2-D AC　　　　b: 3-D AC

ATLSS Connections

The AC concept is based on using a tapered male piece on the beam which slips into a female guide mounted on the column [3]. This is referred to as the keystone coupling and has been designed as 2-D or 3-D units as illustrated in Figure 2a,b. To accomplish its intended purpose, the AC must contain the following features : The connection must be able to guide the beam toward the proper location once the male and female pieces make contact; The male piece cannot jam or catch [4] on the female guide, nor can it pull out horizontally once it engages; Due to the tolerances allowed for each piece involved in the construction process, the connection must allow for misalignment or out-of-plumbness, and also have the ability to be adjusted easily when the building is being plumbed; The connection must be stable during erection and, in its final form, it must be able to carry the design loads. The ultimate goal of this concept is to have a limited assortment of mass-produced connections with a standard shop fitting operation and quick, automatic erection capabilities.

The 2-D AC, Figure 2a, consists of guide plates on the column and dove-tailed angles connected to the beam web in a manner similar to conventional framing angles. However, since the shear is carried by contact, this AC requires only securing the beam to the column. Test results for the 2-D AC indicate that the stiffness, strength and ductility performance is equal to that of a conventional connection. Details can be found elsewhere [5].

Analytical studies on the insertion characteristics of the 3-D AC have indicated that it is well suited for automated erection because of its three sloping surfaces. This chamfered piece's three slopes are : in the vertical direction at the column face; in the vertical direction on the sides; and in the horizontal direction on the sides (See Fig. 2b). The reasons for the chamfer are to proportion the insertion forces so that the male part slides easily into the female; and, to provide a means for alignment during the insertion phase [6]. This connection piece is a solid volume, and its higher costs could be partially defrayed by casting and mass-producing. This 3-D connection is more accurate and easier to place than its 2-D counterpart, and also requires little or no securing. The tapered male seat finding the large female target is forgiving to tolerances; slotted holes on the beam web allow for adjustment.

While the concepts associated with the AC can be applied to all types of building connections (column splices, bracing, beam-to-girder, etc.) and materials, the scope of this paper is limited to the steel beam-to-column 3-D AC. Design calculations based on finite element analysis results and an experimental program have been completed on the first generation of ACs and the second generation is under study.

Background Theory

To develop models for the behavior of the AC, traditional structural analysis must be combined with several sciences not usually associated with structural engineering. These include tribology, solid body contact (elastic and plastic), wedging, and extrusion analysis.

Tribology is the science and technology of the interface between two bodies in relative motion. It incorporates factors such as energy consumption, physical and chemical changes, and altering or removal of surface topography [7]. Basically, surfaces come together only at their extreme points, or asperities. As the normal force increases, the surface areas at the tips of the asperities cannot support the force without violating plasticity, so they compress and increase in area. This crushing of the asperities is dependant on the surface quality, and is responsible for the seating behavior of the AC at the beginning of loading.

While the stiffness of the connection is determined in part by the deformations due to shear and moment, a significant role is played by the solid body deformations at the contact surfaces. Classical solutions by Hertz and Flamant to related problems are being used to develop the model of behavior.

Wedging is an important statical property involved in the behavior of the AC. When an object is wedged between two surfaces, self-equilibrating normal forces develop. These high forces remain even in the absence of the applied force, and the friction associated with them oppose the removal of the object. This provides a beneficial lock for the AC.

It was neccessary to calculate bounds on the limit states of different designs. A sufficiently strong male would cause a mechanism in the female; likewise with a strong female the male would tend to extrude. Bounds were set up for the steady plastic flow of the male (extrusion) and the unrestricted plastic opening of the female (mechanism).

AC DEVELOPMENT AND DESIGN

The AC was developed by not restricting imagination or limiting thinking to current technology or practice, while at the same time involving industry experts at an early stage and throughout the genesis to keep the product practical and functional. The industry panel assembled by ATLSS stressed three main points, one of which was the tolerance/adjustment issue discussed in the previous chapter. The other two points were that the connection be a structural memeber, not an erection aid; and that the AC's acceptance would occur only if field fastening was, for the most part, eliminated. With this in mind, the emphasis of the research turned to the structural aspects of the sloped contact surfaces.

Structural Role

The AC concept can cover a large range of structural needs. These include the connection as an erection aid only, as a shear connection, as a partial-moment connection, and as a full-moment connection. After review by an industry panel, the AC as an erection aid was given a low priority due to its additional fabrication expense with no structural contribution. The shear AC requires a smooth contact surface to facilitate uniform transfer of force. Consideration was given to where the coupling device should go in relation to the beam cross section. The preferred location is above the beam centroid because supporting a member above its center of gravity is stable. On the other hand, placing the keystone near the bottom of the web and using the composite action of the floor slab would give partial moment resistance. Web buckling in the beam at the keystone would have to be investigated.

To develop substantial moment capacity, a connection must engage the beam flanges. There are various ways to accomplish this with ACs : 1) The AC can be attached to the beam web and used in conjunction with flange tees. Because high accuracy is required for the welds, a 3-D AC is envisioned (See Fig. 3a). For clearance, the lower tee is shop attached to the column, while the upper tee is shop attached to the beam. Backing bar emulators are built into the tees. 2) The AC itself can make up the flange connections. 3) A full-depth, tight-fitting, end-plate AC can be used (See Fig. 3b). These connections were deemed possible by the excellent tensile resistance determined in the analytical work.

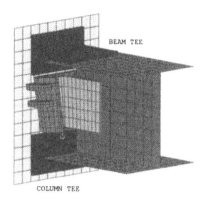
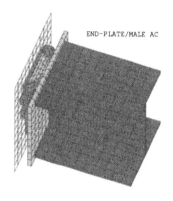

Figure 3: a. Flange-Tee Moment AC.   b. End-Plate Moment AC.

Erection Techniques

The ACs will be erected by leaving the beam web bolts loose. The object then would be to get the beam in the general position safely, but not in a fixed manner. Then, during plumbing, the beam could be brought to its final location and the male piece tightened. This would eliminate performing strenuous tasks in a potentially dangerous environment while the beam is unstable. The future approach may be to tighten the male piece initially and by applying a downward force, use the corrective slopes to bring the columns to their proper orientation. A tradeoff between erection expense due to adjusting and fastening and fabrication expense due to tolerance is established.

Because the AC uses gravity to its advantage, most designs involve dropping the beam in place. This is not feasible in situations where there are overhead obstructions such as in beam-to-girder connections. With this in mind, some ACs have been designed to swing into place from the horizontal [8]. Another approach to the beam-to-girder connection would be to use the standard AC with a tubular girder, or to extend the female AC beyond the projection of the flange. In either case, the girder's upper flange ceases to be an obstruction, and it may be possible to drop multiple floor beams into place simultaneously.

Geometric Considerations

The vertical slope of the contact planes is a key factor in both the erection and the structural behavior of the connection. A steep slope benefits both the erection procedure and the wedging action (See *Background Theory*). However, a steep slope imparts high horizontal forces. Examining a simple planar model of the contact surfaces, and assuming impending motion, as can be seen in Fig. 4a the effect of the slope on the horizontal force component is quite pronounced. Of course the limiting case is a seat: a one-to-one relationship with no horizontal force, but self-erection and stable wedging are completely lost. Therefore, the steep slope required for erection creates a large horizontal component of the contact force. These self-equilibrating, outward components tend to open the female.

There are certain aspects that are controlled by the combination of the slope and the vertical length. Structurally, these dimensions determine the horizontal projection of the contact surface which resists the vertical force. For erection, the slope and vertical length determine the ratio of object area to target area. Also, from a practical standpoint, the width limitation imposed by the column face dimension keeps the slope from becoming too flat. The vertical length must be sufficient to fit, thereby fully utilizing, the required bolts on the beam web, and must provide a sufficient runway for automatic erection.

Once the male piece enters the female guide, it is desired that the male piece should not be able to pull out horizontally. This geometric constraint restricts the bottom horizontal dimension of the male piece to be larger than the upper inside dimension of the female guide. The strength of the female guide is dependent on the weld that attaches it to the column face. The outer perimeter of the female piece defines the weld pattern, and hence the moment of inertia of the weld. This property is of importance because of the large, outward opening moments created by the large lateral forces discussed above. The moments attempt to rotate the female off the column face, so a line weld is unacceptable [8]. From this standpoint, and to make shop layout easier, it would be beneficial to make the female guide one piece.

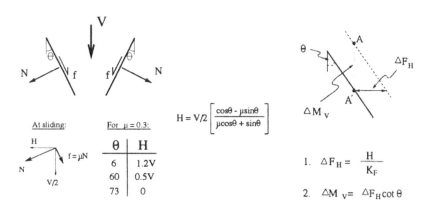

Figure 4: a. Planar Equilibrium Model.   b. Kinematic Compatibility Model.

Phase I Prototype Design
The AC has one feature that separates it from traditional connections. In its purest form there exists no mechanical or welded connection, and the keystone units depend on contact, wedging, and friction from gravity loads and beam end rotation to form a load transferring contact surface between the male and female connection parts. The location of the contact surfaces can vary due to the intial lack-of-fit, the inelastic history, and the current loading.

The deformation of the connection is comprised of two separate entities: First are the deformations of the continua, but since there are no traditional fasteners, rigid body motion (RBM) is possible between the parts. Figure 4b shows the kinematic compatibility of the planar slice model. Using the equilibrium equations from Figure 4a, combined with the compatibility equation and stiffness equation, a relationship between vertical RBM and vertical load is acheived. Clearly, the vertical RBM is dependant on the horizontal force component and the horizontal stiffness of the female. Furthermore, the RBM from tensile or flexural loading would be directly affected by the female's horizontal stiffness. Therefore, high horizontal stiffness of the female results in a connection with good moment resistance and little vertical RBM. Of course, for connections subjected to reversed loading, the RBM becomes a heightened problem due to play in the connection from permanent set of reversal plasticity. A traditional fastening procedure may be required. The more load the connection transfers with its contact surface, the less expensive it will be.

The prototype design evolved into a tapered male solid piece, and a female whose section is stiffer and stronger at planes closer to the bottom (See Fig. 5a). The ACs were manufactured at a local foundry to ASTM A27 cast steel Grade 70-36 specifications and exhibited excellent tensile strength, 76 ksi, and elongation properties, 32%. Radiographic analysis was used to examine material integrity. The contact surface angle was designed to optimize connection strength and ease of erection.

Phase II Modification
In an attempt to limit the RBM, two modifications were made to the female casting (Fig. 5b). Flutes, or stiffeners which prevent the female from prying open, and a seat which closes the bottom section of the female were added. The idea to combine a seat with the side slopes was initially avoided because the seat would attract a disproportionate amount of the vertical force, thereby causing an uneven shear distribution on the beam web bolts.

To study the effect of surface roughness on the seating, the connection's contact surfaces were machined. These machined connections represent an upper bound of initial fit for connections manufactured using more sophisticated casting methods. A practical method of improving the seating of the connection is to design a mechanism or installation procedure which preloads the connection above the seating threshold. This preload will enhance the AC's service load performance by eliminating the initial slip. Another desirable feature of a preloading mechanism is the opportunity to use it as a lock between the male and female pieces. The lock reduces the possibility of construction accidents and also allows ACs to be used in industrial applications subjected to vibrating loads.

Figure 5: a. Phase I Prototype.   b. Phase II Prototype.

ANALYTICAL WORK

The analytical analysis of this connection is complicated by a number of factors that include: unusual geometry; material nonlinearity of the solid body contact aspects; and the geometrical nonlinear nature of the contact surface. The approach taken here is to break the problem into simpler parts by forming plane strain solutions for slices of the connection and then stacking them to obtain the overall behavior.

Finite Element Analyses
Finite element meshes have been developed for the connections. The force paths through the sloped bearing surfaces, the connection stiffness, and the effect of out-of-fit are assessed. ANSYS[2], a general purpose finite element program, is being used. The contact surfaces are being modelled by interface elements capable of bearing and large sliding displacements. A gap option of these elements will be used in order to assess lack-of-fit effects. Analyses will be performed to obtain the proper configuration and stiffness of the 3-D piece so as to carry the required forces while still exhibiting ductile behavior.

Shear Loading
The analysis for shear loading was performed in two steps. First, an analysis examined the whole depth of the female AC, loaded in shear. The results show that the highest stress concentration occurs at the fillet of the lower region (See Fig. 6a). Then, a plane strain slice was examined to find the stress distribution across a section. Since the shear loading effectively forces a larger cross-section of the male into a smaller female opening, the effect was achieved by providing a thermal coefficient of expansion and then heating the male piece. The stress distribution (See Fig. 6b) indicates high stresses at the neck (fillet).

---

[2]Swanson Analysis Systems, Inc, Houston, PA

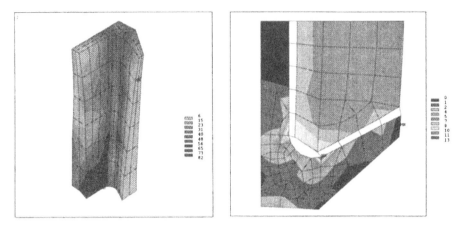

Figure 6. Stress Intensity, Shear Loading, 1/2 Symm: a. Female. b. Slice.

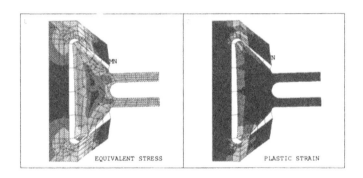

Figure 7. Tensile Loading, Slice: a. Stress Intensity. b. Plastic Strain.

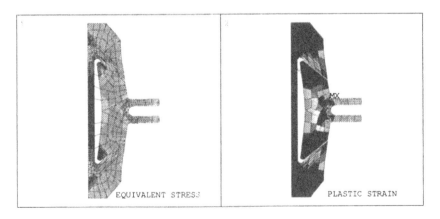

Figure 8. Tensile Loading, Flutes: a. Stress Intensity. b. Plastic Strain.

Tensile Loading
The finite element analysis for tension used solid plane strain elements to model slices of the AC. Figure 7 shows the deflected shape, the stress intensity contour, and the regions of plastic deformation for a representative slice. A plastic hinge has formed at the neck, and the outward displacement is a combination of 1) RBM as the female's geometry changes; 2) Elastic elongation of the male plates and deflection of the female; 3) Plastic deformation from solid body compressive contact. Figure 8 shows the same results for a section through the flutes. Note the much different stress distribution. The flutes have provided a force path (similar to the chords of a truss) and redistributed the high stress and deformation to the center of the male. The results of these analyses are encouraging: The slice with the flutes resisted double that of the plain section. Even ignoring the stiffness of the seat and the securing bolt, the stacked tension resistance fo the Phase II prototype would be easily sufficient for the three moment connection designs associated with Figure 3.

EXPERIMENTATION

Third-scale experimentation is being conducted concurrently with the theoretical work. This is especially useful with the AC because of the connection's dependence on a contact surface for force transfer. Unlike welds or bolts, which are defined by a fixed geometry, the contact surface can vary in location. This creates a challenging modelling situation for structural analyses, making experimentation essential. The experiments are being used to check the theoretical model before a full range of theoretical analyses are performed.

The experimental work is planned in four phases. Phase I involved testing the initial prototype in 'pure shear'. Phase II involves modifying the original prototype and augmenting the shear experiments with tension and rotation tests. Phase III involves cantilever tests of the AC, and the associated moment ACs to determine their moment resistance and performance under cyclic loading. Phase IV involves 3-D testing of a structure.

Phase I
Shear connections on 8" beam sections were tested to evaluate structural and erection behavior. The test span was only 2' 9", loaded at the third points to make the loading almost purely shear. Figure 9a shows the results from a typical test, a plot of RBM vs. load. Four things to note from the results: 1) The connection exhibited excessive vertical deflections as the male section pushed through the female casting; 2) The connection did carry a high shear (80% beam ultimate shear); 3) The initial portion of the curve shows a seating period where the male deflected 1/8" under low loading. This was typical of all tests, and is a result of the flattening of the asperities (See *Background Theory*). This is unacceptable behavior; 4) The unload-load curves show that the wedging action prevents the connection from deflecting until a higher load than previously experienced is attained. This is important because the connection is capable of being pre-loaded (See *Phase II Modifications*).

Test results of the AC with flutes (Fig. 9b) show that the RBM is arrested in one case but not in another. This erratic behavior is a result of the initial fit of the connection. If the male, through dimensional variance, can slip below the upper level of flutes, the remaining flutes provide insufficient resistance to stop the RBM. For this reason, and because it provides a ply for a securing bolt, a seat was tried. A seat was originally rejected because it would unload the sides and cause a stress concentration at the bottom. The AC with the seat behaved similarly to the AC with flutes except, since the seat provides a positive stop, there is no uncertainty involved (Fig. 9b). However, as expected, the lower bolt and the beam web surrounding it fractured from carrying a disproportionate amount of the load. The flutes were retained, partly to create an equitable force distribution between the sides and the seat, and also to provide tensile and flexural resistance.

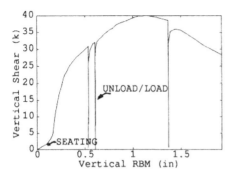

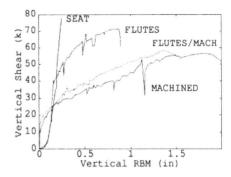

Figure 9: a. Original AC.   b. Modified ACs.

The machining of the contact surfaces completely eliminated the seating effect (See Fig. 9b). This process is impractical from an economic standpoint, but indicates an upper bound on surface quality. A more feasible way of eliminating the seating is to preload the AC using a small bolt (See *Phase II Modifications*). This would also place beneficial wedging forces on the male piece. A phenomenon called flexure lock also keeps the male piece inside the female . Two experimental results verified this. First, during a shear span test, after a small downward load had been applied, an attempt was made to remove the beam by reversing the load. Even on this short span (2'9"), slight asymmetry in the setup caused one side to be pulled up slightly before the other. The lagging side's male piece caught in the female from the rotation and did not allow the beam to be pulled out. A second instance can be seen in Figure 10, which shows the results for a short span vs long span test. Clearly, for the same loading, the span with the higher moment/shear ratio has less RBM.

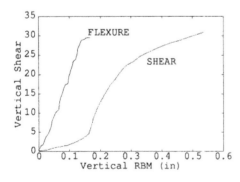

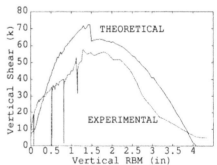

Figure 10: Flexure Lock.   11. Theoretical Model Verification.

Comparison with Analytical : Using the equilibrium/compatibility model (Refer to Fig. 4a,b), a step-by-step incremental solution was obtained. As can be seen in Figure 11, the results were in good agreement with experimental results. The stress contours from the FEA correctly predicted the location of the failure to be the radius of the lowest slice.

Phase II,III

From the results of the Phase I program, the AC's details and geometry were revised and new connections have been manufactured for the next phases of testing. The slopes have been softened to lower the horizontal force. Flutes and a seat have become permanent additions. The horizontal return angles of the sides have been made sharper to promote tension resistance. The new AC will be tested in tension, seating, and rotational capacity. Plane frames with a composite concrete deck will be tested to evaluate connection performance when subjected to gravity and lateral loads. The behavior of a full-moment connection which uses an AC on the web and tees on the flanges will be studied.

FUTURE WORK

From the results of the analytical and experimental program, the AC will undergo additional modifications. Scale-model plane frames with and without a composite concrete deck will be tested to evaluate connection performance when subjected to gravity and lateral loads. Analytical work has recently begun on the development of an AC system for precast concrete beams and columns. A variation of the 3-D steel AC will be used as a model for this system.

Once the examination of the ACs themselves is complete, the scope of the project will move to a systems level as the structure will be evaluated from planning, design, fabrication, erection, and monitoring. An experimental structure will then be built in ATLSS's 3-D testing laboratory. It will be a low-story, steel frame incorporating full- and partial-moment ACs in one direction and shear ACs in the other, and be erected using the Stewart platform. After performance evaluations, as well as a comprehensive economic study, it is anticipated that these connections will be field tested in actual structures.

REFERENCES

1. ATLSS, FIFTH-YEAR ANNUAL REPORT TO NATIONAL SCIENCE FOUNDATION, Volume I: Programs and Plans, Center for Advanced Technology for Large Structural Systems, Bethlehem, PA, February 1990.

2. Kaczinski, M.R., Viscomi, B.V., Lu, L.W., and Fleischman, R.B., An Approach to Integrated Building Systems. Proceedings of the Tenth Structures Congress, ASCE, San Antonio, TX, April 1992, pp.

3. Fleischman, R.B., Viscomi, B.V., and Lu, L.W., ATLSS Connections: Concept, Development and Study. Proceedings of the Ninth Structures Congress, ASCE, Indianapolis, IN, May 1991, pp. 426-429.

4. Whitney, D.E., Quasi-Static Assembly of Compliantly Supported Rigid Parts. Journal of Dynamic Systems, Measurement and Control, 1982, **104**, pp. 65-77.

5. Croshaw, B.A., 2D ATLSS Connections: Experimental and Theoretical Study. ATLSS REU Program Report, Lehigh University, Bethlehem, PA, August 1991.

6. Nguyen, V.X., and Perreira, N.D., A Chamfered Connection for Use in Automated Framing of Buildings. ATLSS Report 88-06, ATLSS Center, Lehigh University, Bethlehem, Pennsylvania, June 1988.

7. Suh, N.P., Tribophysics, Prentice-Hall, Englewood Cliffs, NJ, 1986.

8. Fleischman, R.B., Viscomi, B.V., and Lu, L.W., ATLSS Connections: Concept, Development and Experimental Investigation. ATLSS Report 91-02, ATLSS Center, Lehigh University, Bethlehem, Pennsylvania, June 1988.

INNOVATIVE CONNECTIONS FOR DUCTILE STEEL FRAMES

MARIUS B.WECHSLER
Professional Member AISC, Fellow ASCE
Consulting Engineer, Downey, California 90240-2966 USA

ABSTRACT

Based on theoretical and practical considerations, a research was developed for improving the non-linear behavior of steel ductile braced frames. The innovative connections resulting from the performed research, might develop the required ductility for the subject frames. They are herein described and illustrated with details and examples of calculations. The ductility of such frames is mostly achieved by strengthening steel plates welded on the end portions of the braces. The research deals with bolted brace to column connections only.The results of the performed research might be applied for the design of large steel structures, such as turbine buildings, elevated tanks and towers, in areas of high seismicity. The basic philosophy used for the design of connections described in this paper, can be applied for anykind of steel braces in ductile frames.

INTRODUCTION

Ductile behavior of bracing systems is assured if the brace tension capability is governed by yielding of the gross cross-sectional area of the member rather than by its net cross-sectional area. Three principles are used in order to develop full-member ductility:

1. To oversize connections such that they do not yield before the member. An increase of 1/3 of the axial load is considered as adequate to account for potential capacity variabilities resulting from nonuniform distribution of the applied load.

2. To reinforce the brace cross-section before the joint region by welded steel elements such that the stress, anywhere in the vicinity of the connection is less than or equal to that in the gross cross-sectional area of the member.

3. To lengthen the reinforcing elements both sides of the joint such that concentration of stresses in the connection is avoided.

Braces composed of steel pipes and structural tubing are not covered in this paper since they are not suitable to be connected by bolts. The research was focused on two kind of braces, namely: the double angle strut, and the I-shaped beam connected through sets of four angles.

The procedure applied for improving the non-linear behavior of the double strut brace can be applied as well for braces composed of channels, structural tees, or I-shaped beams when they are in the plane of the column flanges. In this case, the connection is made by the aid of double gusset-plates, while the brace presents two important advantages: they have the larger radius of gyration in the direction of the larger buckling length, and, where cross-braces are required, the middle connection can be made in a simple fashion with two gusset-plates in the outside of the braces (see Example #1).

Nevertheless, there are difficulties to achieve these latter advantages where the braces are in the plane of the column web, usually requiring either fork-arrangements on simple gusset-plates, or double gusset plates spanning the flanges of the columns. These difficulties can be avoided by making the connections through sets of four angles. And the disadvantage of this type of brace to bear the connection on the thin webs of I-shaped beams, can be avoided by welding additional plates on the web, which are anyhow required in non-linear design conditions, for ductility reasons (see Example #2).

METHODS

In the light of the above stated principles, let an axially loaded tension brace transfer its load, P, to a connection element, like gusset-plate, set of cover plates, or set of cover angles. Consider the brace as working for its full tension capacity, thus:

$$P = A \times F_t \tag{1}$$

where A is the gross-sectional area of the brace; and F_t, the allowable axial tensile stress [1]. In accordance with the principle 1., the required number of bolts for the connection is:

$$N = (1 + 1/3) \, P \, / \, P_b \tag{2}$$

where P_b is the allowable load for one bolt to be carried by shear [1]. The required thickness for the member portion penetrated by bolts, should be:

$$t_b = P_b \, / \, (d \times F_p) \tag{3}$$

where d is the diameter of the bolt, and F_p, the allowable bearing stress on steel [1].

When applying the principle 2., consider a more accurate expression for the effective net area of the brace than the conservative one actually in use [1]. On purpose, and taking into account the stress funneling in the member, assume that the effective width of an outstanding portion, b_o wide, of brace, start to diminish from its center line, like a high degree parabola shown in Figure 1., with the quantity:

$$b = b_o \, (x \, / \, L)^k \tag{4}$$

where x is the distance from the center line of the brace to any of its cross-sections; L is a half of the brace length, which is considered between the outermost bolts of the connection; and k, an unknown exponent. For the determination of k, assume that the curve b is similar to a trajectory of stress [2]. Consequently, its derivative :

$$b' = k \times b_o \times x^{k-1} \, / \, L^k \tag{5}$$

272

should be tan 45° = 1, at x = L, thus:

$$k \times b_0 / L = 1, \quad \text{or:} \quad k = L / b_0 \tag{6}$$

Implementing Eq.6 in Eq.4, it follows:

$$b = b_0 (x / L)^{L / b_0} \tag{7}$$

It should be stressed that, in the spite of the fact that b_0 is considered, theoretically, as starting to diminish from the centerline of the brace, a practical vue of the phenomenon should consider a full tensile capacity of the brace until b exceed 0.01% of b_0. Consequently, where:

$$(x / L)^{L / b_0} = 0.0001, \quad \text{or:} \quad x = L \times 0.0001^{b_0 / L} \tag{8}$$

The related cross-section to the afore shown value of x, is marked with the letter C in Figure 1. and x_C is its abscissa.

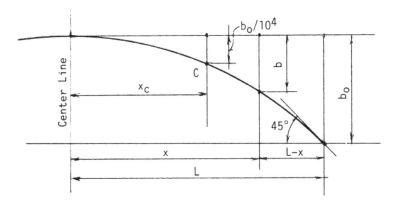

Figure 1. Plot of the variable ineffective width of an outstanding portion of an axially loaded tension brace

Taking into account the afore shown considerations, it follows that the effective net area of the brace, in the region of its connections, is:

$$A_e = A - r \times t \times D - n \times t_0 \times b + A_{sn}(\text{if any}) \tag{9}$$

where r is the number of bolts in a row; t, the total thickness of the brace portions penetrated by bolts; D, the width of a bolt hole used in calculating net areas [1]; n, the number of the outstanding portions of the brace; t_0, their thickness; and A_{sn}, the net area of the strengthening plates, which is:

$$A_{sn} = p (w - r \times D) t_s \tag{10}$$

where p is the number of the strengthening plates; w, their width; and t_s, their thickness. Implementing Eq.7 and Eq.10 in Eq.9, it follows:

$$A_e = A + p (w - r \times D) t_s - r \times t \times D - n \times t_0 \times b_0 (x / L)^{L / b_0} \tag{11}$$

From the condition stated in the principle 2., namely that A_e be larger than A, and solving the Equation 11 with respect to t_s, it follows:

$$t_s > [rtD + nt_ob_o \, (x/L)^{L/b_o}]/[(w - rD)p] \tag{12}$$

To comply with the principle 3., the strengthening plates shall be extended towards the brace such that the cross-section C, in Figure 1, is exceeded
by the length:

$$s_1 = pwt_sF_t \, / \, (m \, \Delta \, F_f) \tag{13}$$

where Δ is the size of the fillet weld; F_f, its allowable stress [1]; and m,
the number of the fillet welds. While towards the gusset-plate, the strengthening plates shall be extended such that the cross-section where they are
theoretically no more utile therefore where the reduced load and the resistance match, be exceeded by the length:

$$s_2 = A_{sn}F_t \, / \, (m \, \Delta \, F_f) \tag{14}$$

CONCLUSIONS

In order to obtain non-linear response of bolted axially loaded steel braces,
it is common practice to provide strengthening elements in their connection
region. This is necessary because the lost tensile capacity of the brace created by the bolt holes, and the partially stressed outstanding flanges or legs,
must be compensated for. By the method presented herein, one can determine
the cross-sectional area, as well as the length and the adequate location of
these strengthening elements. By the recommended detailing of the connection
it is assured that the brace member would yield first and hence provide stable hysteresis loops under reversed loadings. The ultimate adequacy of the recommendations should be verified by laboratory testing.

ACKNOWLEDGEMENT

The author would like to thank Asadour H.Hadjian, Bechtel Fellow, for his
constructive comments concerning the presented paper.

REFERENCES

1. American Institute of Steel Construction (AISC), Manual of Steel Construction, Allow. Stress Design, Ninth Edition, Chicago, Illinois, pp. 4-5 & 6;
and 5-33, 34, 70, 71, 117 & 118.

2. Timoshenco, S. and Macbullough, G.H., Elements of Strength of Materials,
Third Edition, D.Van Nostrand Company, Princeton, New Jersey, 1956. p.149

NUMERICAL EXAMPLES

Two types of braces are considered, namely a double angle strut, and a I=sha-
ped beam with its web perpendicular to the plane of the single gusset-plates,
on which the brace end connections are made through sets of four angles.
Both types are of A 36 Steel, with strengthening plates fillet welded
by XX 70 electrodes, and connected to single gusset plates 3/4 in. thick by
the aid of A 325, d=1-1/8 in. bolts, fully tightened in standard round holes.
It follows [1]: F_t= 22 ksi; F_p=69.6 ksi; F_f=21 ksi; P_b=33.8 kips; D=1.25 in.

274

Example #1. Improve the non-linear behavior of a 56 ft - 3 in.long brace com-
posed of two $9 \times 4 \times 5/8$ angles, reinforced by two steel plates $7\frac{1}{2}$ in.wide fillet welded both sides on the larger legs of the angles, which are used also
for the connections of the brace with rows of two bolts (Figure 1). It follows:
$L=(56\cdot12+3)/2=337.5$ in; $m=4$; $n=p=r=2$; $t=2\cdot5/8=1.25$; $t_o=5/8=0.625$; $b_o=4-5/8=$
3.375 in; $L/b_o=337.5/3.375=100$; $b_o/L=0.01$; $A=2\cdot7.73=15.46$ in^2 [1]; and from:
Eq.1: $P=15.46\cdot22=340$ kips; Eq.2: $N=(1+1/3)340/33.8=13.4$ say 14 bolts; Eq.3;
$t_b=33.8/(1.125\cdot69.6)=0.432 < 3/4=0.75$ in: OK; Eq.12: $t_s=[2\cdot1.25\cdot1.25+2\cdot0.625\cdot$
$3.375(313.5/337.5)^{100}]/[(7.5-2\cdot1.25)2]=0.338$ use $9/16=0.562$ in, and fillet
welds of $\Delta=\frac{1}{4}$ in; Eq.10: $A_{sn}=2(7.5-2\cdot1.125)0.562=5.62$ in^2; Eq.13: $s_1=2\cdot7.5\cdot$
$0.562\cdot22/(4\cdot0.25\cdot21)=8.83$ in; Eq.8: $x_c=337.5\cdot0.0001^{0.01}=307.8$ in; $L-x_c=29.7$ in;
$\ell_1=29.7+8.83-(24+2)=12.53$ say $\overline{13}$ in; Eq.14: $s_2=5.62\cdot22/(4\cdot0.25\cdot21)=5.89$ in;
$\ell_2=2+4+5.89=11.89$ say 12 in*.

*Note: The determination of this length is based on the fact that A_e computed
in accordance with Equation 9, but without the strengthening plates, for the
cross-section corresponding to the third row of bolts, where $x=321.5$ in.,is
12.30 in^2, therefore sufficient to carry the related load which requires an
$A_e=15.46\cdot5/7=11.04$ in^2, while, for the second row of bolts, $A_e=12.22$ in^2,which
does not satisfy the required $A_e=15.46\cdot6/7=13.25$ in^2 and needs strengthning.

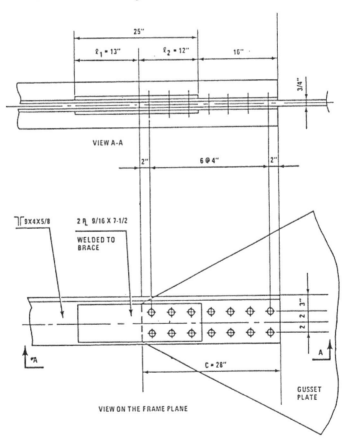

Figure 2. Ductile connection for a double angle axially loaded brace. Fillet
welds of the strengthening plates are 1/4 inch.

Example #2. Improve the non-linear behavior of a I-shaped axially loaded brace with its web perpendicular to a single gusset plate. The shape is W 12×53,
64 ft - 4 in.long; and it is reinforced by steel plates 8-3/4 in.wide, welded
on the both sides of the web which is used also for the connection of the brace with rows of two bolts (Figure 3). It follows: $L=(64\cdot12+4)/2=386$ in; $m=n=4$
$p=r=2$; $t=0.345$ in: $t_o=0.575$ -in; $b_o=(9.995-0.345)/2=6.825$ in; $L/b_o=386/4.825=80$;
$b_o/L=0.0125$; $A=15.6$ in^2 [1]; and from: Eq.1: $P=15.6\cdot22=343$ kips; Eq.2: $N=(1+$
$1/3)343/33.8=13.5$ say 14 bolts; Eq.3: $t_b= 33.8/(1.125\cdot69.6)=0.432>t=0.345$ in:
no good, hence the strengthening plates have to take care about; Eq.12: $t_s=$
$[2\cdot1.25\cdot0.345+4\cdot0.575\cdot4.825(362/386)^{80}]/[(8.75-2\cdot1.25)2]=0.074$ use $3/16=$
0.187 in; check bearing $33.8/1.125(0.345+2\cdot0.187)=41.79<69.6$: OK; use $\Delta3/16$ for
the fillet welds;Eq.10: $A_{sn}=2(8.75-2\cdot1.25)0.187=2.34$ in^2; Eq.13: $s_1=2\cdot8.75\cdot$
$0.187\cdot22/(4\cdot0.187\cdot21)=4.58$ in; Eq.8: $x_c=386\cdot0.0001^{0.0125}=344$ in; $L-x_c=42$ in;
$\ell_1= 42+4.58-(24+2)=20.58$ use 20 in; Eq.14 should not be applied since, for
constructive reasons, ℓ_2 shall be extended over the entire bold region of 28
inches. The comparison between the required net areas and the actual ones computed with Equation 9, are shown in the table below.

TABLE

Section on bolt row x =	362	366	370	374	378	382	386	in
Required effective area	15.60	13.37	11.14	8.91	6.69	4.46	2.23	in^2
Actual effective area	17.01	16.92	16.70	16.19	15.00	12.26	5.98	in^2

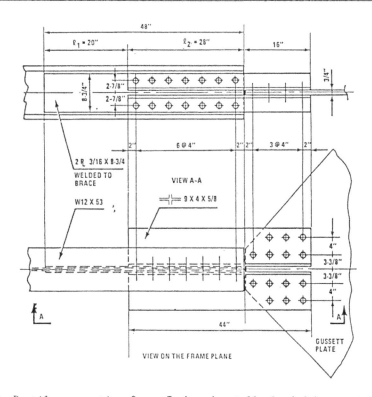

Figure 3. Ductile connection for a I-shaped axially loaded brace with its web
perpendicular to a single gusset-plate. Fillet welds of the strengthening plates are 3/16 inch.

MECHANICAL MODELS FOR SEMIRIGID CONNECTIONS

MARIO DE STEFANO
Istituto di Tecnica delle Costruzioni, Università "Federico II"
P.le Tecchio, 80125 Napoli, Italy.

ANTONELLO DE LUCA
Istituto di Ingegneria Civile ed Energetica, Università di Reggio Calabria
Via Cuzzocrea 48, 89100 Reggio Calabria, Italy.

ABSTRACT

In this paper a mechanical model which is able to simulate the cyclic moment-rotation response of flexible and semirigid connections is presented. After describing the main features of the physical behavior incorporated into the model and the adopted solution algorithm, the numerical loops obtained when simulating cyclic behavior of double angle and top-and-bottom angle connections are compared.

INTRODUCTION

Mechanical models of semirigid connections have been so far developed for simulating only their monotonic response [1,2]. These models idealize the joint with two rigid bars connected by nonlinear springs representing the angle segments which the joint can be divided into. Both double angle connections and top and bottom angle connections can be modelled this way [3].

Regarding the force-displacement relationship assumed for the springs, which accounts for the nonlinear behavior of the connection, different laws are usually considered depending on whether a spring is subjected to compression or to tension. Multilinear functions, whose parameters were defined according to geometrical and mechanical properties of the connection, or mathematical expressions, which were calibrated by curve fitting against experimental results, have been used elsewhere [1,2].

In this paper, a mechanical model which is able to predict cyclic moment-rotation response of semirigid connections is presented. The model represents the nonlinear axial behavior of the angle segments, which is mainly dependent on bending of the legs connected to the column over a short span the bolts edges and the angle fillet, by introducing beam elements with distributed plasticity. When simulating cyclic flexural behavior of the connection, another source of nonlinearity (boundary nonlinearity) influences the response due to the contact between the angle leg and the column flange, which occurs as the angle segment is compressed. This phenomenon is taken into account by introducing an elastic gap element whose stiffness can be defined by following an approximate procedure shown in [1]. This model has been demonstrated to closely predict the actual cyclic response of double angle connections in [4].

Aim of this paper is to carry out a comparison between the numerical responses provided by this model when simulating flexural behavior of a flexible connection (double angle connection) and a semirigid connection (flange and web cleated connection).

MECHANICAL MODELS FOR FLEXIBLE AND SEMIRIGID CONNECTIONS

In this section, the mechanical models used for simulating the response of two types of beam-to-column connections are described. Both models are constituted by two rigid bars which represent the webs of the column and of the beam.

In the case of the flexible connection (figure 1) the two rigid bars are connected by inelastic springs which model the axial response of the angle segments which the web cleats are divided into. In the case of the semirigid connection, in addition to the springs representing the web angle segments, two springs are located at the top and the bottom of the rigid bars for modeling the axial response of the flange cleats.

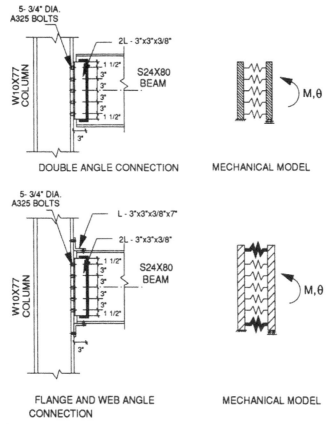

Figure 1. *Flexible and semirigid connections under examination with their mechanical models*

It is worth noting that each spring actually represents the mechanical system shown in figure 2 for the double angle connection, which is composed by an inelastic beam element and a linear gap element. The coupling of the axial responses of the springs is due to the rigid bar CD (figure 3), which imposes a linear pattern of the axial displacements δ_i of the angle segments along the height of the connection.

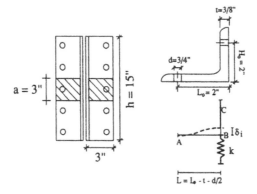

Figure 2. *Mechanical system representing the behavior of an angle segment*

Since the problem of the moment-rotation response of the mechanical model is nonlinear, an incremental solution has to be pursued. If the solution is obtained by imposing increments $\Delta\theta$ of the applied rotation, the incremental compatibility condition due to the presence of the rigid bar CD is expressed by:

$$\Delta\delta_i = (y_i - x_c)\Delta\theta \tag{1}$$

where y_i is the position of the i-th spring and x_c is the distance of the istantaneous center of rotation CR from the tip C.

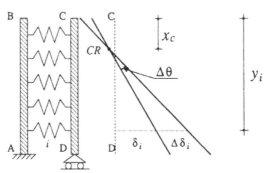

Figure 3. *Coupling of axial forces of the angle segments*

The increment of the axial displacement for the i-th spring, $\Delta\delta_i$, can be related to the increment of the axial force sustained by the i-th spring, ΔN_i, through the equation:

$$\Delta N_i = K_{si}\Delta\delta_i \tag{2}$$

where K_{si} is the secant stiffness of the i-th spring. Observe that K_{si} is given by the sum of the secant stiffness of the beam element and the stiffness k of the gap element if the axial displacement δ_i is less than zero, whereas only the secant stiffness of the beam element is considered if δ_i is greater than zero.

Since only the bending moment M is applied to the connection, the position of CR, defined by x_c, derives from the equilibrium condition that the resultant of the increments ΔN_i of the axial forces acting on the springs is equal to zero:

$$\sum_i \Delta N_i = \sum_i K_{si}\Delta\delta_i = \Delta\theta \sum_i K_{si}(y_i - x_c) = 0 \tag{3}$$

This nonlinear equation can be solved for x_c iteratively. Once x_c is determined, equations (1) and (2) allow to obtain the increments ΔN_i, while the increment of the applied moment ΔM can be expressed as:

$$\Delta M = \sum_i \Delta N_i y_i \tag{4}$$

RESULTS PROVIDED BY THE MECHANICAL MODEL

A comparison between the numerical simulation obtained by the mechanical model and the test response has been carried out in [4] for the web cleated connection shown in figure 1. Plots of the experimental loops and the simulated ones are contained in figure 4. It can be clearly seen that a close prediction of the strength of the connection is achieved, while the stiffening of the test curve, due to the contact alternatively occurring at the top and the bottom angles is taken into account in a suitable way. Results provided by the model when predicting the cyclic bending behavior of the flange and web cleated connection of figure 1 are represented in figure 5. From this representation it is clear the capability of the model to simulate also this semirigid connection. In particular, the addition of top and bottom angles leads to a dramatic increase of strength but still a considerable pinching, due to contact phenomena, affect the cyclic behavior.

CONCLUSIONS

The applications reported in this paper of the mechanical model presented in [4] evidenced the capability of such a model to simulate the cyclic response of connections characterized by different degree of flexibility. The model appears to be promising and, therefore, further comparisons with experimental data are needed in order to evaluate its reliability.

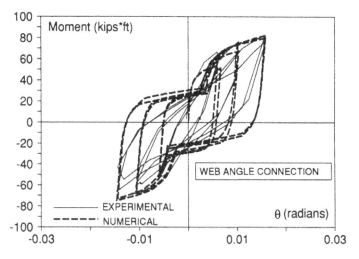

Figure 4. Comparison between experimental and numerical loops for the flexible connection

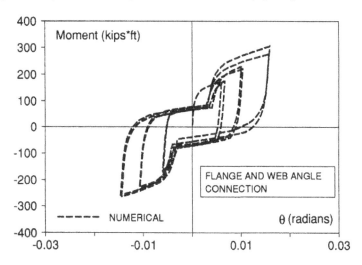

Figure 5. Numerical loops for the semirigid connection

REFERENCES

[1]   **Wales, M.W., Rossow, E.C.,** "Coupled Moment-Axial Force Behavior in Bolted Joints", Journal of Structural Engineering, ASCE, vol.109, 1983.
[2]   **Richard, R.M., Hsia, W.K., Chmielowiec, M.,** "Moment Rotation Curves for Double Framing Angles", Materials and Members Behavior, Proceedings of the Session at ASCE Structures Congress '87, Orlando, Florida, 1987.
[3]   **Nethercot, D.A., Zandonini, R.,** "Methods of Prediction of Joint Behavior", Chap.2 in Structural Connections: Stability and Strength, ed. R. Narayanan, Elsevier Applied Science Publishers, London, 1989.
[4]   **De Stefano, M., De Luca, A., Astaneh, A.,** "Prediction of Double Angle Connections Cyclic Moment-Rotation Behavior by a Mechanical Model", Proceedings of the SSRC Annual Technical Session, Pittsburgh, April, 1992.

RESTRAINT CHARACTERISTICS OF FLEXIBLE BEAM-TO-COLUMN JOINTS

ANIL K. AGGARWAL
Associate Professor
Department of Civil Engineering
University of Technology
Lae, Papua New Guinea

ABSTRACT

The paper describes an experimental investigation into the behaviour and moment-rotation characteristics of flexible beam-to-column connections. In the experimental programme, twelve specimens - six each of angle cleat and web-side plate joints have been tested under gradually increasing 'static' loads.

From the experimental moment-rotation curves, an assessment of end-restraint offered by the joints has been done. The magnitude of restraint observed is large to allow for the distribution of bending moments assumed in the analysis of frames with flexible joints.

INTRODUCTION

All flexible beam-to-column joints are capable of transmitting shear from the beam into the column. The important difference between different types of flexible connections in the context of structural behaviour lies in their rotational stiffness i.e. their ability to transmit the restraining effects of the beam to the column and thus to transmit moment. This facility will vary from about zero for the most flexible connections upto almost 100% for rigid connections. This property of connection behaviour can be most appropriately described in terms of the connection's moment-rotation characteristics. In this study, restraint characteristics of two different types of flexible beam-to-column connections i.e angle cleat and web side plate are determined experimentally.

DESCRIPTION OF TEST SPECIMENS

To study the behaviour and moment-rotation characteristics, twelve specimens were tested under gradually increasing 'static' loads. Of these, six were angle cleat joints (Table 1(a)) while the other six had a web-side plate to make the beam-column connection (Table 1(b)). In addition, all joints had a seat angle to connect a beam flange to a column flange. The beam section used in all test specimens was 200UB 25.4 kg/m and the column 200UC 46.2 kg/m. All specimens were made from steel having a nominal yield stress of 250 MPa. The design method suggested by Hogan and Thomas (1) was used in calculations. A sketch of a typical angle cleat beam-to-column connection used for testing is shown in Fig 1.

In four angle cleat specimens, the web cleats were welded to the beam web and bolted to the column flange (welded-bolted construction (type 1 joint)) while in the other two specimens, the web cleat angles were bolted both to the beam web and column flange (bolted-bolted joint (type 2)).

TABLE 1(a)

Experimental programme

Specimen No.	Web cleat size	Flange Cleat size/location	Bolt Dia. web cleats	Joint type
AC1	102x102x10	102x102x10(T)	20	1
AC2*	102x102x10	102x102x10(T)	20	1
AC3	102x102x10	102x102x10(T)	24	1
AC4	102x102x10	102x102x10(C)	20	1
AC5	102x102x10	102x102x10(T)	20	2
AC6	102x102x10	102x102x10(C)	20	2

(*) means single web cleat connection
(T) cleat on tension flange of beam and (C) cleat on compression flange

TABLE 1(b)

Experimental programme

Specimen No.	Web Bolts Diameter (mm)	Flange Bolts Diameter (mm)	Plate thickness (mm)	Seat angle size
WS1	24	20	10	76x76x8
WS2	20	20	10	76x76x8
WS3	20	20	16	76x76x8
WS4	24	20	16	76x76x8
WS5	24	20	12	76x76x8
WS6	20	20	12	76x76x8

MEASUREMENTS

To measure the rotation of the joint ($\Theta - \emptyset$), it was necessary to measure the rotation of the beam 'Θ' and the rotation of the column '$\emptyset$'. An optical technique was adopted using a pair of ordinary plane mirrors and a pair of theodolites.

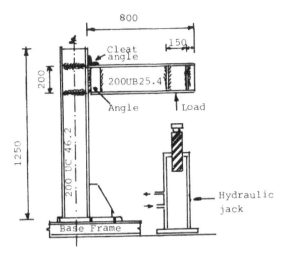

Figure 1   Details of test specimen and loading arrangement

TEST RESULTS AND DISCUSSION

From the experimental data, moment-rotation curves for all connections were obtained. An assessment of end restraint (connection stiffness) provided by the joints was also made using these curves. The end restraint of a connection has been defined as a proportionate complement of 'β' where 'β' is the slope of the moment-rotation curve at zero rotation and is represented by:

$$\beta = \tan^{-1} (\Theta - \emptyset)/(M) \qquad (1)$$

If the joint rotates by 14.0 x 10^{-3} radians under a moment of 60 kNm, then $\beta = \tan^{-1} (14.0/60.0) = 13.1$ degrees and the end restraint provided is $(90 - 13.1)/90 = 85.4\%$.

To determine the average end restraint provided by the joints, plots of moment versus total cumulative rotation (Θ -$\emptyset$) were obtained for all angle cleat joints (Fig. 2) and web-side plate joints (Fig.3).

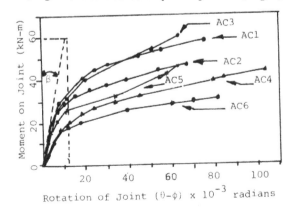

Figure 2. Variation of total cumulative rotation with moment for angle cleat joints.

It is observed from the plot that maximum restraint is provided by joint AC1 (85.4%) with welded-bolted construction and the minimum by joint AC6 (76.9%) with bolted-bolted construction and having a cleat on the compression flange of the beam. Larger rotations were expected in type 2 joints, because in bolted construction, relative movement between the beam web and angle cleats cannot be eliminated eventhough bolts connecting the two faying surfaces may be fully tensioned.

In the case of web-side plate joints, it is observed that the maximum end restraint (97%) is provided by specimen WS2 and the minimum restraint (76%) is shown by joint WS3. The average end restraint (86.5%) is certainly far too greater that what is normally assumed for this type of connection.

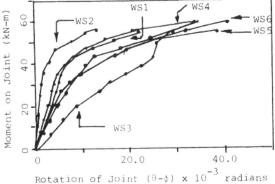

Figure 3. Variation of total cumulative rotation with moment for web-side plate joints.

CONCLUSIONS

Moment-rotation curves of both angle cleat and web-side plate joints have been obtained experimentally under gradually increasing 'static' loads.

The curves indicate non-linearity over the complete range of loading. The rotation of angle cleat joints results primarily from the deformation of the angles, while in the case of web-side plate joints, the rotation was caused by slip and elongation of bolt holes

The average end restraint shown by angle cleat (81.1%) and web-side plate (86.5%) joints is too large to allow for the moment distribution assumed in the analysis of frames with flexible joints.

ACKNOWLEDGEMENTS

The financial support given by the Papua New Guinea University of Technology for this project is gratefully acknowledged.

REFERENCES

1. Hogan, T.J. and Thomas, I.R., Design of Structural Connections (Standardised Connections Manual - Part B), 2nd edn 1981 (first published 1978), Australian Institute of Steel Construction, 1981

Composite Structures

NON LINEAR ANALYSIS
OF CONTINUOUS COMPOSITE BEAMS

EDOARDO COSENZA
Istituto di Ingegneria Civile, Università di Salerno
Fisciano, 84080, Salerno, Italy

STEFANO MAZZOLANI, MARISA PECCE
Istituto di Tecnica delle Costruzioni, Università Federico II
P.le V. Tecchio, 80125, Napoli, Italy

ABSTRACT

In case of steel-concrete composite sections, the behaviour of continuous beams in the hogging moment regions is characterized by two phenomena; one is the cracking of the concrete slab and the other is the possible local buckling of the lower flange or of the web. The cracking produces a redistribution of bending moment, before yielding of materials, reducing the hogging moment and increasing the sagging moment; the local buckling of flange or web of the structural steel section can substantially reduce the rotation capacity of section and the following redistribution of moment. In this paper the non linear behaviour of continuous composite beams taking into account cracking and local buckling is analyzed. In particular the "required" plastic rotation to obtain the design degree of redistribution is evaluated; then an evaluation of "allowable" plastic rotation is proposed as function of the local buckling of flange or web in compression. An example relative to a beam designed according to Eurocode 4 concludes the paper.

INTRODUCTION

The behaviour of simple supported beam is more convenient for steel-concrete composite beam from many points of view; in fact every section is subjected to sagging moment so that the concrete slab is in compression while the steel section is in tension: both the materials are used by the best way.

However continuous beams are often considered more appropriate, both structural aims (i.e. displacement reduction and vibration problem) and technical problems (i.e. for eliminating joints). Using this kind of beams the section at the intermediate supports are subjected to hogging moments, inducing different effects to the concrete slab and to the steel section.

Firstly the concrete of the slab is not properly used, because it is in tension; the cracking so obtained reduces the structural stiffness increasing deformations in service conditions. Beside the lower flange and the web of steel section are in compression so that local buckling can occur; this phenomenon can be an important limitation for plastic rotations, as a consequence it can limit or sometimes eliminate the convenient aspects of "limit design" with the redistribution of the bending moment for statically indeterminate structures.

From the design point of view, the verification of simple supported beams can be effectuated assuming the materials fully plastified and evaluating if the ultimate bending moment of the section is greater than the one due the design load. On the contrary the problem is theoretically very difficult for continuous beams; in fact a redistribution degree of moments has to be chosen: this redistribution is due both to cracking of the sections at the hogging moments and to non linear behaviour of materials. The verification is satisfied if the "required" plastic rotation at the design load, that needs for the compatibility condition, is lower than the "allowable" one.

The evaluation of the required and allowable plastic rotation is very difficult. For the required rotation the structural behaviour until failure condition must be modelled while for the allowable one the local buckling has to be taken into account; the following two paragraph examine these two problems.

The modern codes (i.e. [3]) allow to consider a redistribution of the bending moments as a function of the ductility classification of sections. The composite sections are divided in 4 classes on the grounds of the flange and web slenderness. The class 1 sections allow the plastic hinges introduction and the development of rotation capacities predicted by the plastic analysis of the structures; on the contrary for the class 4 sections local buckling can occur in the elastic range of behaviour, so that this phenomenon has to be clearly taken into account [14]. As regards class 2 sections, the full plasticity can be achieved but rotation capacities are lower while for class 3 the yielding of the flange in compression occurs but the full plasticity of the section is impossible. The plastic analysis is allowable only for class 1 and 2 while the only elastic analysis can be effectuated for the 3 and 4 class. The slenderness limitations of EC 4 are reported in the Appendix.

In the last section of this paper a beam designed according to EC 4 is analysed and verified by comparing that the "required" plastic rotation is lower than the "allowable" one; sections of class 1, that is "compact", are considered.

THE EVALUATION OF ALLOWABLE PLASTIC ROTATIONS

The numerous experimental studies on rotation capacities proposed in the technical literature are mostly referred to steel elements, beams or beam-columns; on the contrary the experimental results about rotations of composite beams are a few number.

It can be observed that in hogging moment regions the slab provides a compressive action to the steel section, and so such a part of section has a behaviour of a beam-column even if the whole section is subjected to bending moment. Thus, the experimental results regarding rotation capacities of steel beam-column elements could be directly applied to composite sections too if compression is rightly evaluated. This is confirmed by the experimental data reported in [5].

According to this idea, the calculation of the composite continuous beams needs of a reliable procedure to evaluate the rotation capacities of steel beam-column elements. As regards this problem a model has been recently proposed by Kato [10,11,12] based on theoretical studies and experimental results.

In particular the ratio between yield stress σ_y and the stress σ_{cr} for which the local buckling occurs is defined by the following equation:

$$\frac{1}{s} = \frac{\sigma_y}{\sigma_{cr}} = c_1 + c_2 \frac{\sigma_y}{E} \left(\frac{c}{t_f} \right)^2 + c_3 \frac{\sigma_y}{E} \left(\frac{d_e}{t_w} \right)^2 \tag{1}$$

being E the steel elasticity modulus and c/t_f, d_e/t_w the slenderness of flanges and web (cfr. Tab A1); the coefficients c_1,c_2,c_3 are obtained for different steel types by means experimental data. Considering the mean values of the coefficients referred to the different steel types analysed in [12] they result:

$$c_1 = 0.74 \quad ; \quad c_2 = 0.51 \quad ; \quad c_3 = 0.050 \tag{1'}$$

For taking into account the axial force, the "efficient" value d_e of the web height is introduced varying from a minimum of 0.5, for the case of beams, to a maximum of 1 for web completely in compression; d_e is calculated by means of the following formula:

$$\frac{d_e}{d}=\frac{1}{2}\left[1+\frac{A}{A_w}\rho\right] \quad ; \quad \rho=\frac{N}{A\sigma_y} \tag{2}$$

N being the axial force and A, A_w being the area of steel section and of steel web; in particular for the composite beams the compression force at hogging sections is the tensile force of the reinforcement steel, so that $N=A_f\sigma_f$, where A_f and σ_f are area and yield stress of reinforcement.

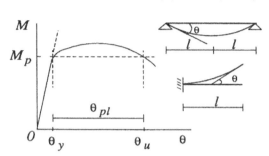

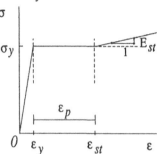

Fig. 1. Rotational yielding and at failure.   Fig. 2. Constitutive law of the steel.

The expression (1) means that if $s<1$ the buckling occurs in the elastic range of behaviour so any plastic rotation is possible. On the contrary if $s>1$ local buckling occurs after yielding and the structure can have a redistribution capacity. The rotation capacity R_θ is defined as:

$$R_\theta=\frac{\theta_{pl}}{\theta_y}=\frac{\theta_u-\theta_y}{\theta_y} \tag{3}$$

where the meaning of θ_{pl}, θ_y and θ_u is explained in Fig.1; if the condition $\rho>(s-1)/2$ is verified, as it usually happens for the composite section at hogging moment, this rotation capacity can be evaluated by the following formula [12] (cfr. Fig.2):

$$R_\theta=\frac{s-1}{2(1-\rho)(s-\rho)}\left[\frac{E}{E_{st}}(s-1)+2\frac{\varepsilon_{st}-\varepsilon_y}{\varepsilon_y}\right] \tag{4}$$

The reliability of this formula for composite sections is evaluated comparing the theoretical results and the experimental ones presented in [5]; the values of c_1, c_2, c_3 are the same that in (1') and $\varepsilon_p/\varepsilon_y=10$, $E/E_{st}=50$ is assumed.

More in detail the comparison is reported in Tab.1. For the experimental data when the maximum rotation capacity is non clearly defined, because of no local buckling, the lowest experimental limit of 10 is reported; in such cases even the theoretical values are very high.

Beam	Experimental Rotational Capacity	Theoretical Rotational Capacity
SB 3	>10	7.8
SB 9	>10	7.8
HB 40	>10	7.1
SB 2	6.45	5.2
SB 8	6.83	5.2
HB 41	4.12	4.4

Tab. 1. Comparison between theoretical and experimental data
of the rotation capacities of composite beams.

The analysis of the results clearly show the good reliability of the theoretical evaluation. The proposed formulation seems to develop useful results for forecasting allowable rotations of composite beams.

THE EVALUATION OF REQUIRED PLASTIC ROTATIONS

The evaluation of plastic rotations required to restore the compatibility with loads, whenever a design with redistribution is effectuated, needs the structural analysis taking into account the non-linear behaviour of materials by suitable models.

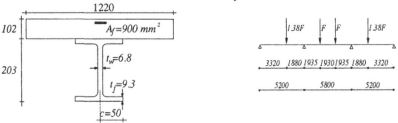

Fig. 3. Geometric characteristics, in mm, of the beams analysed in the theoretical-experimental comparison.

In order to analyze this problem the comparison between the theoretical results obtained by means of different analysis methods and experimental results [9] is effectuated for the beam named CB12; it is a continuous beam on four supports, as is drawn in Fig. 3. In particular the following models are applied:

a- a monodimensional finite elements model, adopting the constitutive laws of concrete in tension (including tension stiffening) and compression, of structural steel (including the residual tensions) and of the reinforcement steel suggested by EC2, EC3, EC4 and CEB [1÷4]; the analysis is effectuated integrating the non-linear moment-curvature diagram along the beam.

b- a monodimensional finite elements model [6], considering the tension stiffening effects of concrete by means of the simplified procedure suggested by EC2 [1] and CEB [4] and used in [6], that is assuming a medium curvature $1/r_m$ depending on the cracked stage curvature $(1/r_2)$ and the uncracked one $(1/r_1)$, on the ratio between the actual moment of section and the cracking one (M_{cr}/M) and on the quality of bond by the parameter $\beta_1 \beta_2$ (equal to 1 for high bond reinforcement bars and short term load):

$$\frac{1}{r_m} = \frac{1}{r_1}\beta_1\beta_2\left(\frac{M_{cr}}{M}\right)^2 + \frac{1}{r_2}\left[1 - \beta_1\beta_2\left(\frac{M_{cr}}{M}\right)^2\right] \qquad (5)$$

The plastic rotations are concentrated in the critical sections;

c- the simplified models of problem according to EC4 that is uncracked (c1) and cracked (c2) analysis; in the first case an elastic analysis has to be developed assuming the uncracked section while in the second one the section si considered completely cracked, without "tension stiffening" effect, for 15% of length of the span from each side of internal supports. Also in these models, the plastic rotations are concentrated in the critical sections.

In Fig.4 the diagram moment M - rotation Θ is drawn for the different models and for the experimental results, and the rotation is referred to the controflexure point near the two internal supports. The two experimental curves, lightly different each other, have a descending part due to the local buckling of the flange.

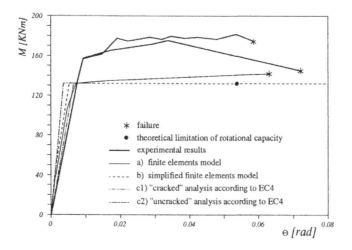

Fig. 4. Hogging moment-rotation diagram.

About the maximum moment value, all the proposed methods are clearly conservative compared to the experimental result. On the contrary about deformations it needs to consider failure criteria since the proposed models do not consider local buckling; in particular for the finite elements model (a) the failure is defined by the maximum strength of materials experimentally evaluated, while for the plastic hinges models the limit to the rotation capacity suggested by Kato is introduced as shown in the picture. All the curves stay completely down the experimental ones, as in terms of strength as of ultimate rotation.

All the methods introduced are different in a significant way only in the first part of the curves, where the simplified method of EC 4 with cracking analysis provides the smaller tangent at the origin, while the other methods have the same beginning tangent. When the plasticity gets on the simplified "uncracked analysis" gives good results. In this case the rotation in a section can be evaluated by the following formula:

$$\theta = \frac{ql^3}{24E_cI_1}(1 - 6\xi^2 + 4\xi^3) + \frac{M_p^{(-)}l}{6E_cI_1}(-1 + 3\xi^2) \qquad (6)$$

where q is the uniform load, l is the span length, E_cI_1 is the flexural stiffness of the section, evaluated by the "modular ratio" method, at the uncracked stage, $M_p^{(-)}$ the plastic hogging moment of the section and ξ the ratio between the distance of the analysed section from the extreme support and the span length.

When the mechanism occurs the following relation occurs:

$$M_p^{(+)} + \frac{1}{2}M_p^{(-)} \cong \frac{ql^2}{8} \qquad (7)$$

being $M_p^{(+)}$ the sagging plastic moment of the section; in particular, in the elastic controflexure point ($\xi=0.75$), the following rotation is obtained for the design load and the mechanism respectively:

$$\theta_d = \frac{0.6875q_dl^3}{24E_cI_1}\left(1 + \frac{4M_p^{(-)}}{q_dl^2}\right) \quad ; \quad \theta_{mec} = \frac{1.375M_p^{(+)}l}{6E_cI_1} \qquad (8)$$

For evaluating the rotation when the first hinge occurs the redistribution due to cracking cannot be neglected; so the cracked analysis has to be applied. Assuming the inertia I_2 for the sections of the beam subjected to hogging moment and introducing the parameter $i=I_1/I_2$, it can be shown that [8]:

$$M^{(-)} = -\frac{ql^2}{8} \cdot \frac{0.890+0.110i}{0.614+0.386i}$$

(9)

and the rotation which correponds to the full plastification on the support, for $\xi=0.75$, is:

$$\theta_y = \frac{M_p^{(-)}l}{6E_cI_1} \cdot \frac{0.193+0.497i-0.002i^2}{0.890+0.110i}$$

(10)

The reliability of the proposed formulas θ_y, θ_d, θ_{mec} will be analysed in the following numerical example.

DESIGN APPLICATION: THE CASE OF "COMPACT" SECTIONS

An example of design according to procedures suggested by EC 4 is developed and then the verify by means the proposed method is effectuated. In particular, a continuous beam on three supports, with symmetric spans of 8 m is considered with a design load of 52 KN/m; the structure is fully propped during the construction, full shear connectors are considered, and the only bending problem is taken into account.

The design is related to class 1 sections, assuming the uncracked analysis and plastic design of the sections; a redistribution from the support to the span of 40% is considered, that is the maximum value suggested by EC 4 (cfr. Appendix).

The design of the structure is not easy. In fact the width of the slab and a steel section that satisfies the conditions of slenderness of flange and web at limit of class 1 can be chosen. If the redistribution degree is fixed in the hogging moment section the reinforcement area of slab is directly defined; this reinforcement provides values of ρ and α calculated by (2) that reduce the slenderness limit of the web for the class 1, according to the expression in table A2. As a consequence the section is no more in class 1 so the section has to be overdimensioned for the moment in the span in order to be in class 1 and an iterative procedure occurs.

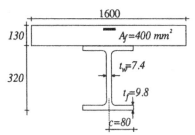

Fig. 5. Section designed according to EC 4.

For this example the section of Fig. 5 is analysed, that is characterized by the following parameters for the steel section:

$$\frac{c}{t_f} = 8.1; \quad \frac{d}{t_w} = 40.5 \quad ; \quad \rho = \frac{A_f\sigma_f}{A\sigma_y} = 0.09; \quad \alpha = 0.61$$

and the limits of the class 1 ($\sigma_y = 355$ MPa by which $\varepsilon = 0.81$) are (cfr. Tab. A1, A2 of Appendix):

$$10\varepsilon = 8.1; \quad \frac{396\varepsilon}{13\alpha-1} = 46$$

Thus, the section is in class 1 of EC4; for this designed beam the plastic hogging and sagging moments values are 250 KNm and 447 KNm respectively.

About the structural analysis the difference between the results of the already introduced models is clearly shown by the diagrams of Fig. 6 where the redistribution on the support is presented by means the ratio of the actual moment M and the elastic one M_e as a function of the applied load q.

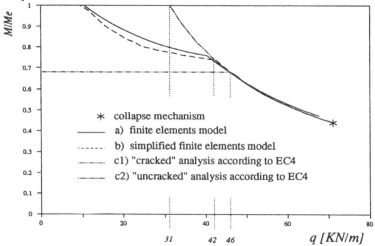

Fig. 6. Redistribution of bending moment for the analysed beam.

In the curves obtained by the finite elements methods (a,b) the singular points correspond to the beginning of cracking and to the first plastic hinge after; in the two methods suggested by EC4 there is not singular point when cracks occurs because the cracking is not taken into account (uncracked analysis) or it is considered from beginning (cracked analysis). All the methods in which plastic hinges are introduced, by the way, reach the same maximum load corresponding to the mechanism.

The redistribution of the stresses is similar for the two different finite elements methods, while it is different for the two models suggested by EC4; in particular the cracked analysis provides a redistribution too large for low load levels, since the cracking is considered from beginning, on the contrary the uncracked analysis neglects the redistribution due to cracking.

More in details, it can be observed that adopting the uncracked analysis of EC4 the plastic hinge on the central support develops for a load of 31 KN/m, that is 60% of design load, according to the assumed redistribution of 40%. In the cracked analysis, the plastic hogging moment occurs for a load of 46 KN/m; the ratio of this value and the redistribution coefficient 0.75, that EC 4 allows in this case, is equal to 61 KN/m, that is higher than the value of the design load: so the uncracked analysis provides a result conservative than the more complex cracked analysis.

It is interesting to observe that the cracked analysis is not conservative respect to a more sophisticated analysis; in fact the increase of flexibility due to cracking is generally overevaluated, because tension stiffening effect is completely neglected. However this model is conservative for the evaluation of deflections since it provides displacements larger than the actual ones, but the redistribution of hogging moment is overstimated.

As already said, the reliability of the design can be evaluated only comparing the required rotation at the design load to the allowable one. In order to this aim in fig. 7 the required rotation, that is the rotation at the controflexure point of the span, and the moment on the support both as functions of the load are drawn.

The plastic hogging moment, the design load and the mechanism can be reached with rotations at the controflexure points, evaluated by the method b), respectively of 0.0051, 0.0070 and 0.0105 rad. The use of equations (8), (10) provides (I_1=243800 cm^4, I_2=78833 cm^4, E_c=32000 MPa =3200 KN/cm^2) θ_y=0.0059 Rad, θ_d=0.0068 Rad and θ_{mec}=0.0105 Rad, that is a good approximation.

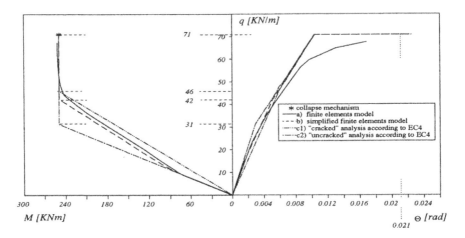

Fig. 7. Rotation-load-hogging moment diagram for the analysed beam.

The evaluation of the rotation capacity is effectuated by the methods previously shown; being $\rho=0.09$, it is $s=1.18$ by eqn. (1) and $R_\theta=3.1$ by eqn. (4). Because fo rotation at yielding is 0.0051 Rad the allowable plastic rotation is 0.021 Rad, almost twice the required one. For this example it is possible to reach both the design load and the structural mechanism; it should be noted that the simplified rules provided by EC 4 to use limit design, based on [15], are, in this design case, satisfied.

CONCLUSION

The analysis effectuated has shown that it is possible to evaluate allowable and required rotational capacities of the continuous composite beams, and the results are in reasonable agreement with available experimental data. Therefore a rational procedure to verify the continuous composite beams, based on comparison between required plastic rotation and allowable ones, could be developed.

In particular the procedure has been applied to a continuous beam designed as a class 1 sections, according to EC 4; so the safety of these rules has been analysed and some remarks as starting points for an extensive research work can be take.

However, many research and design problems are completely open; for example it seems interesting to observe that the coefficients to evaluate characteristic values and design values of allowable plastic rotation does not seem to exist in the literature, and this subject, together with the accurate verification of the proposed procedure to evaluate average allowable plastic rotations, needs of further theoretical and experimental studies.

REFERENCES

1. Commission of the European Communities: Eurocode n.2, Common Unified Rules for Concrete Structures, Final Draft, October, 1989.

2. Commission of the European Communities: Eurocode n.3, Common Unified Rules for Steel Structures, Edited Draft, November, 1990.

3. Commission of the European Communities: Eurocode n.4, Common Unified Rules for Composite Steel and Concrete Structures, Revised Draft, October, 1990.

4. CEB Model Code 90, First Draft, Bulletin d'Information n. 195, March, 1990.

5. Climenhaga J.J., Johnson R.P., Local Buckling in Continuous Composite Beams, The Structural Engineer, Vol. 50, No. 9, 367-374, September, 1972.

6. Cosenza E., Finite Element Analysis of Reinforced Concrete Elements in a Cracked State, Computers & Structures, Vol. 36, n.1, 71-79, 1990.

7. Cosenza E., Pecce M., Deflections and Redistribution of Moments Due to the Cracking in Steel-Concrete Composite Continuous Beams Designed to Eurocode 4, Proceedings of ICSAS 91, Elsevier Science Applied, Vol. 3, 52-61, Singapore, May, 1991.

8. Cosenza E., Pecce M., The Evaluation of Short Term Deflections of Continuous Composite Beams Designed According to Eurocode 4 (in Italian), Proceedings XIII C.T.A. Congress, October, 1991.

9. Hope-Gill M.C., Johnson R.P., Tests on Three Three-Span Continuous Beams, Proc. Instn. Civ. Engrs., Part 2, Vol. 61, 367-381, June, 1976.

10. Kato B., Rotational Capacity of Steel Members Subject to Local Buckling, Proceedings IX WCEE, Tokyo-Kyoto, Vol. IV, 115-120, 1988.

11. Kato B., Rotational Capacity of H-Section Members as Determined by Local Buckling, J. Construct. Steel Research, Vol. 13, 95-109, 1989.

12. Kato B., Deformation Capacity of Steel Structures, J. Construct. Steel Research, Vol. 13, 33-94, 1991.

13. Kidmann R., Composite Girders, IABSE Short Course "Composite Steel-Concrete, Construction and Eurocode 4", 117-145, Brussels, 1990.

14. Johnson R.P., Fan C.K.R., Strength of Continuous Composite Beams Designed to Eurocode 4, IABSE Proceedings P-125/88, 33-44, May 1988.

15. Johnson R.P., Hope-Gill M.C., Applicability of Simple Plastic Theory to Continuous Composite Beams, Proc. Instn. Civ. Engrs., Part 2, Vol. 61, 127-143, March, 1976.

APPENDIX: EC 4 CLASSES OF SECTIONS

In EC4 four classes of sections of continuous beams are defined, in the same way of EC 3 [2] for steel beams.

In particular the classes of the sections are defined by means of the compression flange and web slenderness. The slenderness limitations of the flange proposed by EC4 are shown in Tab. A1, and include beams encased into concrete. Rolled and welded steel sections are considered, so that four different cases are reported for every class.

	Web not encased		Web encased	
CLASS	ROLLED	WELDED	ROLLED	WELDED
1	$c/t_f \leq 10\varepsilon$	$c/t_f \leq 9\varepsilon$	$c/t_f \leq 10\varepsilon$	$c/t_f \leq 9\varepsilon$
2	$c/t_f \leq 11\varepsilon$	$c/t_f \leq 10\varepsilon$	$c/t_f \leq 15\varepsilon$	$c/t_f \leq 14\varepsilon$
3	$c/t_f \leq 15\varepsilon$	$c/t_f \leq 14\varepsilon$	$c/t_f \leq 21\varepsilon$	$c/t_f \leq 20\varepsilon$
4	$c/t_f > 15\varepsilon$	$c/t_f > 14\varepsilon$	$c/t_f > 21\varepsilon$	$c/t_f > 10\varepsilon$

Tab. A1. Characterization of classes of sections; flange slenderness; $\varepsilon = \sqrt{\sigma_y/235}$ [MPa].

The classification based on web characteristics is presented in Tab. A2, where the elastic analysis is represented by triangular stress diagram and the plastic one by constant block stress diagram; beside the stress condition of the web is important.

With reference to classes 1 and 2, the evaluation of the parameter α, presented in Tab. A2, can be effectuated by the translation equilibrium of the sections when the plastic hogging moment is arise, by the following way:

$$[\alpha d - (d - \alpha d)]t_w \sigma_y = A_f \sigma_f \tag{A1}$$

that leads to:

$$\alpha = \frac{1}{2}\left[1 + \frac{A}{A_w}\rho\right] \quad ; \quad \rho = \frac{A_f \sigma_f}{A \sigma_y} \tag{A2}$$

In the comparison between eqns. (2) and (A2), being $N = A_f \sigma_f$, it can tried that the parameter α defined by EC 3, EC 4 coincides with the value defined by Kato.

CLASS	Web subjected to bending	Web subjected to compression	Web subjected to bending and compression	
Stress distribution (Compr. +)				
1	$d/t_w \leq 72\varepsilon$	$d/t_w \leq 33\varepsilon$	$\alpha > 0.5$: $d/t_w \leq 396\varepsilon/(13\alpha-1)$	$\alpha < 0.5$: $d/t_w \leq 36\varepsilon/\alpha$
2	$d/t_w \leq 83\varepsilon$	$d/t_w \leq 38\varepsilon$	$\alpha > 0.5$: $d/t_w \leq 456\varepsilon/(13\alpha-1)$	$\alpha < 0.5$: $d/t_w \leq 41.5\varepsilon/\alpha$
Stress distribution (Compr. +)				
3	$d/t_w \leq 124\varepsilon$	$d/t_w \leq 42\varepsilon$	$\psi > -1$: $d/t_w \leq 42\varepsilon/(0.67+0.33\psi)$	$\psi \leq -1$: $d/t_w \leq 62\varepsilon/(1-\psi)\sqrt{-\psi}$
4	$d/t_w > 124\varepsilon$	$d/t_w > 42\varepsilon$	$\psi > -1$: $d/t_w > 42\varepsilon/(0.67+0.33\psi)$	$\psi \leq -1$: $d/t_w > 62\varepsilon/(1-\psi)\sqrt{-\psi}$

Tab. A2 Characterization of classes of sections; web slenderness.

The structural analysis to evaluate the internal forces for the design load can be effectuated using the redistribution degrees of table A3. The limitations to redistribution depend on the class of sections and they are different for uncracked and cracked analysis; the difference between the redistribution degrees of the two models is only due to cracking. A parametric analysis developed in [7] has shown these values are generally conservative and strongly variable, in fact redistribution due to cracking can be even higher than 25%.

Class of cross section	1	2	3	4
For "uncracked" elastic analysis	40%	30%	20%	10%
For "cracked" elastic" analysis	25%	15%	10%	-

Tab. A3. Allowable redistribution degree.

NEW COMPOSITE BEAMS AND SLABS IN FINLAND

TIMO INHA
Division of Structural Engineering
Tampere University of Technology
P.O. Box 600, SF-33101 Tampere, Finland

ABSTRACT

The use of composite steel-concrete structures has increased considerably in Finland during the last two years. At the same time, composite structures have been developed strongly.

The discussed new composite beams are Ekobalk beam, Hava beam and Delta beam. Moreover, there is discussed slightly a new composite slab developed from Super Holorib slab.

Ekobalk beam is manufactured by profiling steel plate. The inside of steel section is concreted upside down. Before concreting, short reinforcing bars are placed through the webs of steel section for interaction.

Hava beam consists of three different steel parts welded together. The parts are structural steel profile, bottom section and between theirs diagonal ties or burning cut steel sheet. The bottom section is either straight or U-shaped.

Delta beam consists of steel plates welded together. The webs have round punched holes for concreting inside the steel section.

The discussed beams have complete interaction between steel and concrete section. They have the official approval even for fire resistance of 120 minutes without fire protection.

Super Holorib slab has been developed during the years 1991 - 92 by testing in Tampere University of Technology different modifications of base slab with or without end anchorages.

In this paper there are discussed tests, constructions and structural design included structural fire design.

INTRODUCTION

The prestressed hollow core slabs have been the most typical Finnish floor system during the last two decades. At first, reinforced concrete wall elements were the most typical vertical bearing construction of the Finnish multi-storey buildings. Beam-column system as bearing constructions became an usual system not until during the last decade. At the same time, steel construction is been increased strongly in Finland.

The beam-column system introduced slim flooring systems into building constructions. Concrete has been traditionally the most important material in the Finnish building constructions. However, steel or composite beams are more suitable for the slim flooring system than reinforced or prestressed concrete beams. Thus, HQ-beam, Fig. 1, was developed for the slim flooring system a few years ago.

HQ-beam is not a composite beam actually. However, its action as the structural part of floor is nearly like a composite beam's, although HQ-beam is designed as a steel beam without interaction between structural steel and concrete sections.

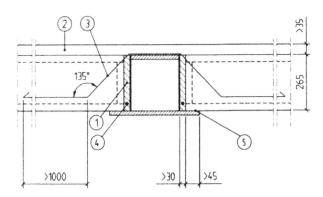

Figure 1. HQ-beam as the part of slim flooring system [1].
(1) Joint concrete
(2) Concrete topping (thickness ≥ 35 mm)
(3) Reinforcement in the joints of slab elements
(4) Reinforcement around the slab field to ensure its behaviour as a shear panel
(5) Neopren pad for flushing (if required).

HQ-beam is the most usual Finnish steel beam yet. The new composite beams as Eko, Hava or Delta have become usually used within the slim flooring systems of steel built multi-storey buildings. Reinforced or prestressed concrete beams are the most used within the concrete built buildings according to the long tradition. The Finnish concrete industry clearly tries to avoid using steel or composite members to the concrete built buildings.

The use of composite slabs has been very minor to the Finnish multi-storey buildings so far. However, composite slabs have been quite usual in single-family houses for instance. The use of composite slabs probably increases considerably to multi-storey buildings while steel construction is increasing.

Three different composite slab systems have been used. According to the official Finnish approval only Super Holorib has no complete interaction between steel sheet and concrete section. Thus, Super Holorib has been developed by testing old slab with end anchorages and new slab whose thin steel sheet has been profiled slightly unlike the steel sheet of old slab.

NEW COMPOSITE BEAMS

Ekobalk

Ekobalk is a low steel concrete composite beam which is suitable for use as primary or secondary beam. The sructural principle is based on complete interaction between structural steel profile, reinforcement, concrete section inside the steel profile and concrete slab.

The development work of Ekobalk began in 1989. The sructural behaviour has been verified by tests. The tests of normal temperature have been carried out in Tampere University of Technology and the furnace test in Borås in Sweden.

The development work contained the following aims: 1) Beams should be joined either to concrete columns or to steel columns, 2) Costs should be so low as possible, 3) Fire class of 60 minutes at least without a fire protection, 4) Small total height, 5) Working methods should be traditional if possible.

Ekobalk's typical cross-section is shown in Fig. 2, where two different slab types have been joined to Ekobalk simultaneously. The optimization of beam is carried out case by case. Variables are the height of cantilevers, the thickness of steel plate and the amount of reinforcement among others.

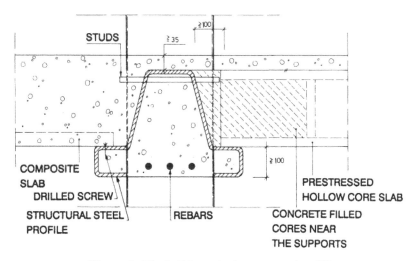

Figure 2. Ekobalk's typical cross-section [2].

The steel plate is cold-formed. The holes for short reinforcing bars (studs in Fig. 2) either are drilled or punched. The usually used steel class is quite high (yield point 355 MPa). The thickness of profiling steel plate is from 3 to 12 millimetres.

The cubic strength of concrete should be at least 25 Mpa and the maximum size of aggregate is 16 mm.

The profiled steel section is as concrete form at first and as structural steel later. Moreover, the rebars are usually necessary for fire resistance. Because the inside of steel section is concreted upside down, it is easy to cause camber which is taken into account in determination of total deflection.

The floor can be consisted of composite slabs, concrete elements or cast-in-situ concrete slab. In most cases, the prestressed hollow core concrete slabs are used in Finland.

Ekobalk's structural design can be divided into three steps: 1) Concreting of beam when it is possible to design camber for beam, 2) Erection of element slabs or concreting of floor when beam and slab together is not yet a composite structure, 3) Beam and slab together as a composite beam. The necessity of shores are determined case by case.

The structural design as composite beam is based on the tests which verified complete interaction between steel and concrete section by using short reinforcing bars as studs shown in Fig. 2.

The composite element of furnace test consisted of Ekobalk whose length was 1,7 m and of composite slab whose area was $1,7 \cdot 1,7$ m^2 and heigth 250 mm, the cross-section is shown in Fig 3. The composite element was unloaded. During the furnace test of 120 minutes, spalling was observed and temperatures of 35 points were measured. The test specimen remained unbroken without spalling of concrete shown after fire exposure in Fig 4. The temperature of inside corner of steel cantilevers was only 500 °C after the standard fire of 120 minutes.

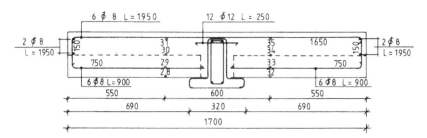

Figure 3. Ekobalk's cross-section in furnace test.

The structural fire design can be carried out by using ideal plastic stress distribution of the cross-section. The exposed parts of steel section are ignored. Moreover, the steel parts whose temperatures have increased over 500 °C during the fire exposure are ignored.

The temperatures of rebars can be determined by using the tabulated temperature distribution of concrete slab.

The beam to column connections can be designed freely. Details should be designed so that the connections can be easily protected against the fire exposure if necessary. Brackets of concrete column for the beam to column connections are shown in Fig. 5. An erected Ekobalk is shown in Fig. 6.

Ekobalk has the official Finnish approval as a composite beam.

Figure 4. Unbroken Ekobalk after furnace test.

Figure 5. Brackets of concrete columns for Ekobalk beam to column connections.

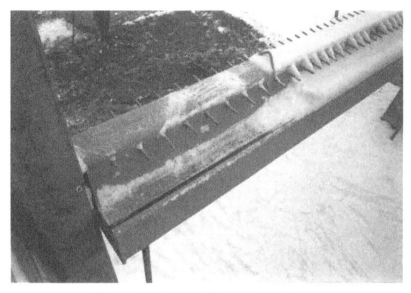

Figure 6. An Ekobalk erected in winter.

Hava beam

The Hava beam is designed for use of intermediate floors of multi-storey buildings alongside conventional concrete and steel beams. It consists of two steel sections and joined ties between theirs. The bottom section, which remains partly visible, can be a straight or formed steel plate. The top section is an I-profile. The two sections are joined by means of multifunction ties or burning cut steel sheet.

The development work of Hava beam began in 1990. Its structural bahaviour has been verified by tests in Tallinn Technical University in Estonia and in Tampere University of Technology. Furnace tests were not necessary for the official approval.

The two typical cross-sections of the Hava beam are shown in Fig. 7. Of course, the slabs of intermediate floor can be composite slabs, too. However, the prestressed hollow core slabs shown in Fig. 7 are the most used as intermediate floor in Finland.

Structural design of Hava beam can be divided into two steps: 1) Design for erection of element slabs or concreting of floor. 2) Beam and slab together as a composite beam. The necessity of shores are determined case by case.

The structural design as composite beam is based on tests. They verified complete interaction between steel and concrete section by using diagonal welded ties between bottom and top section. The burning cut sheet shown in Fig. 7 is corresponding to ties of test beams.

The structural fire design can be carried out by using ideal plastic stress distribution of the cross-section. The exposed bottom section of beam is ignored. The temperature of the parts of I-profile can be determined by using the tabulated temperature distribution of concrete slab.

The beam to column connections can be designed freely. Two erected Hava beams connected to round steel column is shown in Fig. 8.

The Hava beam has the official Finnish approval as a composite beam.

Figure 7. Typical cross-sections of Hava beam.

Figure 8. Two erected Hava beams and their connections to round column.

Delta beam

The Delta beam is designed for use of intermediate floors of multi-storey commercial and office buildings. It consists of four straight steel plates welded together to box girder. The webs of steel box are welded to slope for improvment of resistance of floor to beam connection. Moreover, the webs have round punched holes.

The development work of Delta beam began in 1988. Its structural behaviour has been verified by tests at normal temperature and by furnace tests in the Technical Research Centre of Finland.

The typical cross-section of the Delta beam is shown in Fig. 9. Rebars inside the box are designed for fire resistance. Thus, fire protection below the bottom plate is unnecessary.

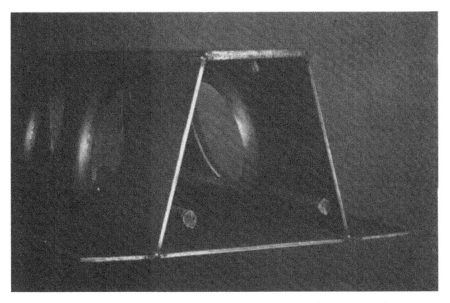

Figure 9. Typical cross-section of Delta beam before concreting.

Structural design of Delta beam can be divided into two steps as that of Hava beam. The structural design as composite beam is based on tests. They verified complete interaction between steel box and concrete section by using round punched holes as studs. Moreover, the inside of box is concreted through the holes and reinforcing bars can be designed for joints of element slab through the steel box by using holes.

The structural fire design can be carried out as that of Hava beam. The unprotected bottom plate is ignored, the webs are reduced and the rebars are taken into account according to their temperature.

Connection between prestressed hollow core slabs and Delta beam is shown in Fig. 10.

Also the Delta beam has the official Finnish approval as a composite beam.

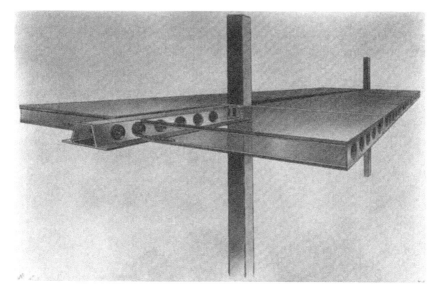

Figure 10. Connection between prestressed hollow core slabs and Delta beam.

A NEW COMPOSITE SLAB

Super holorib slabs have been tested in Tampere University of Technology. The aim of development work was to develop composite slab where complete interaction is between thin steel sheet and concrete. The normal Super Holorib slab has not complete interaction.

At first, the effect of end anchorages was tested. The dimensions of test specimens are shown in Fig. 11.

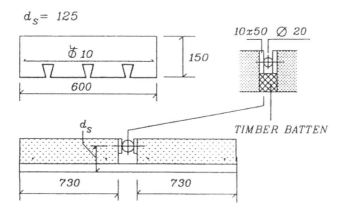

Figure 11. Dimensions of composite test slabs.

The simple supported composed test slabs shown in Fig. 11 were loaded by a central line load. The end anchorages shown in Fig. 12 improved the flexural resistance of the Super Holorib slab over 50 %.

New profile forms without end anchorages have been tested, too. The results are very promising. Unfortunately, the patent protection of new form is not yet world-wide. Thus, pictures can't be shown in this paper.

Figure 12. End anchorages of tested Holorib slab.

CONCLUSIONS

The use of the discussed new composite beams increases in Finland strongly by decreasing the use of concrete beams at the same time. Development work of discussed beams is continuing and also new are developed.

The use of composite slabs is quite insignificant yet, but their use is increasing probably strongly in a near future.

REFERENCES

1. Leskelä, M., State-of-the-Art Report on Composite Flooring Systems in Finland, Presented in ECCS TC11 workshop in December 1991, unpublished.

2. Inha, T. and Kallio, K., Ekobalk - a New Finnish Composite Beam, Teräsrakenne 4/90, The Finnish Constructional Steelwork Association Ltd, pp. 40-42, (in Finnish).

COMPOSITE FLOOR SYSTEM DESIGN

W. SAMUEL EASTERLING
Assistant Professor
The Charles E. Via, Jr. Department of Civil Engineering
Virginia Polytechnic Institute and State University
Blacksburg, Virginia, 24061-0105, USA

ABSTRACT

Two aspects of the design of composite floor systems will be presented. These are slab design for gravity loading and composite beam design. Each topic will be presented in the context of recently conducted or ongoing research with which the author has been involved. The impact of the research on design specifications in the United States (U.S.) will be discussed. Recommendations based on the composite slab study will be considered by Steel Deck with Concrete Standards Committee of the American Society of Civil Engineers (ASCE). Recommended specification changes that come about as a result of the composite beam studies will be submitted to the American Institute of Steel Construction (AISC) Specification Committee.

INTRODUCTION

The economy that is inherent in steel framed buildings is in large part due to the efficiency of the composite floor system. Composite floor systems in the U.S. typically consist of a concrete slab placed on a steel deck that is supported by beams or joists. Horizontal shear stress at the steel-concrete interfaces must be resisted for composite action to be developed. Shear resistance in composite slabs is provided by embossments or the geometry of the deck cross section. Headed shear connectors are used to resist the horizontal shear forces between the beam and slab. Reinforcement for the control of shrinkage and temperature induced cracking will typically be the only additional steel placed in the slab. The efficient use of material, as well as the relative ease of construction, make composite floor systems economically attractive.

While engineers may generally agree that composite floor systems are economic, design procedures for the various components of the system are not as readily agreed upon.

This is true within the U.S. alone and certainly so when the various international design procedures are considered. Although making comparisons between all available design specifications would be an interesting topic for discussion, that is not the objective of this paper. The focus of this paper will be on recently completed or ongoing research projects at Virginia Tech and how those might impact U.S. design specifications. The discussions contained here are general in nature and the reader is referred to the cited literature for detailed quantitative information.

COMPOSITE SLABS

In the U.S., composite slab design for gravity loads is primarily based on the results of extensive experimental testing programs from which design load tables are developed. Historically the methods by which load tables have been developed varied between steel deck manufacturers. This was primarily due to the lack of an accepted, standardized test procedure and data reduction technique. This shortcoming was eliminated by the acceptance of an ASCE Standard entitled *Specifications for the Design and Construction of Composite Slabs* [1]. The test procedure and methods specified for converting the test strengths to design strengths that are described in Ref. [1] were primarily developed in research studies conducted at Iowa State University (ISU) [2, 3, 4].

The procedure developed at ISU, commonly known as the "m and k" method, has been incorporated in several international design specifications. The method is based on a series of simply supported, single span tests, with the specimen being the width of one sheet of deck. Two line loads are placed symmetrically along the span at approximately the third points. In virtually all such tests on typical U.S. manufactured deck sections, a shear bond mode of failure will result. This failure is characterized by a degradation of the deck-concrete interface within the shear span. Noticeable end-slip between the deck and concrete results and composite action is lost.

Design methods based on experimental testing are necessary due to the difficulty in accounting for the partial composite behavior that is characteristic of composite slabs. While the ASCE Standard provides a uniform way by which composite slabs constructed with various types of deck may be evaluated, the test method is severe and does not represent the service conditions of a typical floor slab. Field details that are not commonly incorporated in composite slab test programs include adjacent spans, shear studs and stay-in-place pour stops. Studies conducted at Virginia Tech [5] indicate that the performance of composite slabs is improved if such details are included in the test program and that the shear bond approach is usually not appropriate. More importantly, the studies have shown that simple calculations procedures may be used to determine design load capacities based on the flexural strength. It should be noted that the work reported in Ref. [5] represents a preliminary study, in that a limited number of tests were conducted, and additional confirmatory tests are needed to evaluate a wider range of slab parameters.

Steel Deck Institute Design Procedure

The results of Ref. [5], and other related studies [6, 7, 8, 9, 10, 11] have been synthesized and a design handbook published by the Steel Deck Institute (SDI) for common U.S. composite deck sections [12]. The design handbook provides information for the evaluation of flexural strength, vertical shear strength, effects of concentrated loads and deflection criteria. Design tables and example problems are also contained in the handbook.

Two flexural strength calculation procedures are given; one applies to designs without shear studs and the other to designs with shear studs. For designs that do not use shear studs the limit state is yielding of the bottom flange of the steel deck section. Stresses due to casting are added to live load stresses. Vertical separation and end-slip in the shear span, which is associated with a shear bond failure, is prevented by web deformations, adjacent spans, pour stops or other end restraints. If a true single span condition without end restraints is part of the design then the procedure in the SDI handbook is not applicable and the strength should be based on the shear bond method. If shear studs are present, additional end anchorage is provided and the slab capacity is based on an under-reinforced concrete strength approach.

Little attention has been paid in past research studies to the vertical shear strength of composite slabs, therefore design calculations must be based on conservative assumptions. The method in the SDI handbook assumes that only a portion of the concrete is effective in resisting vertical shear forces. The steel deck is not counted on to resist any of the shear. This is similar to the approach used in Ref. [11], with the only difference being a slight variation in the definition of the area that is effective in resisting the shear. This is most likely not an accurate representation of the true vertical shear strength. However, until research data is available that shows that a more liberal strength model is appropriate, the approach taken in the handbook should be followed.

Two approaches for evaluating concentrated loads are given in the handbook; one for loads less than or equal to the typical 8.9 kN load specified in U.S. building codes and one for loads larger than this load. The approaches are similar with the only difference being in the definition of the effective resisting slab width. The definitions given in the 1985 draft of Eurocode 4 [11] are used for the large load case. Distribution steel transverse to the deck ribs is required in the amount of 0.2% of the concrete area above the top flange of the deck.

Deflection checks are made using the average of the transformed (cracked) section and the uncracked section. This is the recommendation given in the ASCE Standard [1].

ASCE Standard

At the time of this writing an update to the ASCE Standard [1] had recently been completed. The document passed the public review process and should be published in late 1992. The new document contains several revisions but the fundamental reliance on the shear bond approach remains in place. The methods outlined above that are contained in the SDI handbook are not part of the revised document, but will be submitted for consideration by the ASCE Standard [1] committee for possible inclusion in the next revision. The flexural strength determination procedure is fundamentally different in the SDI handbook and the ASCE Standard, thus a resolution of the differences will be required prior to the inclusion of the SDI approach. The revised ASCE Standard does not address concentrated loads or vertical shear strength. The information in the SDI handbook should therefore be a welcome addition for designers. Deflection checks are the same in both documents.

Composite Diaphragms

In addition to the gravity load design, a composite slab will often be part of the lateral load resisting system in a multi-story building. At present there is no U.S. design specification that addresses the design of composite diaphragms. Design information is available from several sources [13, 14, 15], but not in a authoritative design specification. However, the ASCE Steel Deck with Concrete Standard Committee is presently developing such a specification. Information used in the development of the diaphragm standard will be drawn from existing design guidelines [13, 14] as well as recent and ongoing research with which the author has been associated [15, 16]. Development of the diaphragm design standard has not progressed to the point that meaningful discussion regarding it can be provided here. It is anticipated that a draft of the diaphragm standard will be completed by the conference date, and therefore a presentation of the document can be made at that time.

COMPOSITE BEAMS

Design provisions for composite beams constructed with ribbed steel deck consider the strength of headed shear connectors. Variables typically taken into account in the strength determination include concrete material properties, shear stud material properties and the geometry of the steel deck. Recent research results, from a study conducted at Virginia Tech [17] as well as by other researchers [18-22], indicate that the position of the stud relative to the stiffener in the bottom flange of the steel deck is also a critical design parameter. Additionally, the Virginia Tech studies indicate that the strength of the steel deck influences the shear connector strength in certain cases. More importantly, all of the cited studies have reported that the commonly used strength reduction equation, which accounts for the geometry of the steel deck section, is too liberal in many cases.

The stud reduction factor, SRF, which accounts for the influence of steel deck geometry, was developed by Grant, et al. [24] and is given by

$$SRF = \frac{0.85}{\sqrt{N_r}} \left(\frac{w_r}{h_r}\right) \left(\frac{H_s}{h_r} - 1.0\right) \leq 1.0 \tag{1}$$

where N_r = number of studs in one rib at a beam intersection, w_r = average width of concrete rib, h_r = nominal rib height and H_s = length of stud after welding. This reduction factor applies to cases in which the deck ribs are perpendicular to the steel beam and is applied to the nominal strength of a shear stud placed in a solid slab [25, 26] which is given by

$$Q_n = 0.5 A_{sc} \sqrt{f'_c E_c} \leq A_{sc} F_u \tag{2}$$

where A_{sc} = cross-sectional area of a stud shear connector, f'_c = specified compressive cylinder strength of concrete, E_c = modulus of elasticity of concrete and F_u = minimum specified tensile stress of a stud shear connector. Equation (1) was developed empirically by comparing experimental moment capacities of full-size beam tests to predicted moment capacities using Eq. (2) and a flexural model presented by Slutter and Driscoll [27]. The Slutter and Driscoll model is essentially the same as used in current U. S. practice [26].

The precise reason for the discrepancy between recent experimental results with those predicted using Eqs. 1 and 2 is not clear. The researchers reporting on recent studies [18, 19, 22, 23] have suggested various revised models. However, none of the studies have reported reasons for the discrepancy with Eqs. 1 and 2.

Results from full-size beam tests and push-out tests conducted at Virginia Tech [17] also indicate that composite beam strengths predicted using Eqs. 1 and 2 are in need of modification. The results of the 4 full-size beam tests and the 7 push-out tests are described briefly here.

The full-size beam tests were constructed with W410x46 steel sections on a simply supported span of 9150 mm. The composite slab for each of the tests consisted of 76 mm of normal weight concrete cover placed on a 76 mm deep steel deck, giving a total slab depth of 152 mm, with a 2050 mm width. Shear connectors consisted of 19 mm diameter x 125 mm (after welding) headed shear studs. A total of 12 shear studs, 6 per half span, were used for each test, resulting in an approximately 40% partial composite design. Welded wire fabric (WWF152x152-W25.4xW25.4) was placed on the top of the deck. Four equally space concentrated loads were applied directly over the beam. A variety of strain, displacement and slip measurements were recorded for each test. One test had all the shear studs in the strong position, one test had them all in the weak position and two tests had the studs in alternating strong and weak positions. A stud placed in the strong position will be on the side of the bottom deck flange stiffener that is closest to the near end support. A stud placed on the side of the bottom flange deck stiffener nearest midspan of the beam is in the weak position.

The push-out tests were constructed by casting slab segments, using the same deck and slab thickness as in the beam tests, with a width of 610 mm and a length (parallel to the steel section) of 915 mm. Each half of the push-out specimen was cast on a WT125x16.5 section. After the concrete cured, the two halves were bolted through the stem of the tee, creating a W shape with a slightly offset web. This procedure permits the same batch of concrete to be used for both halves of the push-out specimen. The specimens were placed on elastomeric bearing pads and tested in a universal testing machine. Additionally, a yoke mechanism was used to apply normal load to the slab surface, thus simulating the vertical load present on the composite beam and restraining premature separation of the concrete and steel deck. The quantity of applied normal load was 10% of the applied shear load. Three of the push-out specimens had the studs in the strong position and four had the studs in the weak positions.

Material properties test results for the composite beams are given in Table 1. Key results for the full-size and push-out tests are summarized in Table 2., all presented in terms of either calculated or measured shear stud values. The Q_{ne} values were calculated by substituting the maximum composite beam test moment, along with measured material properties, into the flexural moment equations given in Ref. 26. Therefore, assuming the plastic analysis model to be correct, Q_{ne} represents a measured shear connector capacity in the composite beam. The Q_n values were calculated using Eq. 2. The SRF, given by Eq. 1 is equal to 1.0 for the steel deck profile used in all the tests. The Q_{po} values were determined from the push-out tests with the adjustments described in the following paragraphs.

TABLE 1
Material properties for composite beam specimens

TEST	FLANGE f_y (N/mm^2)	FLANGE f_u (N/mm^2)	WEB f_y (N/mm^2)	WEB f_u (N/mm^2)	f_c (N/mm^2)
1	289	474	315	496	33.2
2	307	485	298	509	22.1
3	315	483	332	522	15.7
4	301	437	311	434	34.4

The shear studs in the weak positions, in both the full-size and push-out tests, failed by bulging or punching through the web of the deck. Very little concrete was present between the shear stud and deck web for these cases, therefore it is hypothesized that the strength of the shear stud in the weak position is not primarily dependent on the concrete strength. Rather the stud strength is primarily a function of the steel deck strength. Certainly some interaction between the concrete and steel occurs, but the dominant component is assumed to be the steel deck. Based on this hypothesis the weak position push-out test results were averaged and used for all the weak position studs in the beam tests, regardless of the concrete strengths. The average value obtained from the push-out tests was 60.3 kN.

The strength of the shear studs in the strong position was assumed to be a function of the concrete strength. Therefore the Q_{po} values for the strong position studs in the beam tests were obtained by normalization using the concrete strengths. The average concrete strength in the push-off tests was 31.5 N/mm^2 and the average push-out test strength was 83.7 kN. Therefore the values of Q_{po} were calculated by

$$Q_{po} = 83.7\,kN\sqrt{\frac{f_c'}{31.5}} \qquad (3)$$

where f_c' is the concrete compressive strength for the beam test.

TABLE 2
Experimental results

TEST	Q_{ne} (kN)	Q_n (kN)	Q_{po} (kN)	Q_{ne}/Q_n	Q_{po}/Q_n	Q_{po}/Q_{ne}
1(strong)	85.4	127.7	85.9	0.67	0.67	1.01
2(weak)	59.6	100.5	60.3	0.59	0.60	1.01
3(alt.)	64.5	77.8	59.1	0.83	0.76	0.92
4(alt.)	75.6	127.7	73.9	0.59	0.58	0.98

The results in Table 2 clearly indicate two things; the strength predicted by Eqs. 1 and 2 are too liberal and that push-out tests, if properly compared to beam tests, are good indicators of shear connector strength. Additionally, while a comparison between strong and weak position shear studs indicates some difference, the more pronounced and significant difference is between the predicted values, Q_n, and the beam and push-out test results. Specific modifications to Eq. 1 are not proposed here, but will be following an evaluation of existing procedures [11, 18, 19, 22, 23]. The hypothesis regarding the influence of the steel deck material properties must be evaluated at the same time and perhaps included as a modification to one of the existing methods.

SUMMARY

Two aspects of composite floor systems were discussed in the context of recently completed or ongoing research studies with which the author has been involved. The main thrust of the composite slab discussion was the influence of end restraints on the flexural strength. Also, a recently published composite slab design handbook [12] was outlined. The procedures given in the handbook provided the structural engineer with the necessary information to evaluate composite slab designs for the common limit states associated with flexure, vertical shear, concentrated loads and deflection. Once the design procedures are put in a specification format they will be submitted to the ASCE composite slab standards committee for consideration in future revisions of Ref. 1.

A brief review of an experimental composite beam project was presented. Results of the research program support those of other recent studies. A hypothesis was put forth that the strength of shear studs in the so-called weak position is primarily dependent on the steel deck strength. This hypothesis, while not completely proven, was supported by the results of the research program. Further validation of the idea is required. The results clearly indicate that a revision to the strength reduction equation, accounting for the presence of steel deck, is required.

ACKNOWLEDGMENTS

The composite slab studies were sponsored by the Steel Deck Institute, Inc. Richard B. Heagler of United Steel Deck, Inc. and Professor Larry D. Luttrell of West Virginia University have contributed significantly to the research and were largely responsible for the development of the design provisions described herein. The composite beam studies were sponsored by the American Institute of Steel Construction and the Virginia Tech Civil Engineering Department.

REFERENCES

1.   *Specifications for the Design and Construction of Composite Slabs,* 1984, ASCE Standard, New York.

2.   Schuster, R. M., Strength and Behavior of Cold-Rolled Steel-Deck-Reinforced Concrete Floor Slabs. 1970, thesis presented to Iowa State University, Ames, Iowa, in partial fulfillment of the requirements for the Degree of Doctor of Philosophy.

3.   Porter, M. L. and Ekberg, C. E., Jr., Design Recommendations for Steel Deck Floor Slabs. *ASCE J. Struct. Div.*, 1976, **102**(11), 2121-2136.

4.   Porter, M. L. and Ekberg, C. E., Jr., Compendium of ISU Research Conducted on Cold-Formed Steel-Deck-Reinforced Slab Systems. Iowa State University Engineering Research Institute Bulletin No. 200, Ames, Iowa, 1978.

5.   Easterling, W. S. and Young, C. S., Strength of Composite Slabs. *ASCE J. Struct. Engr.*, 1992, **118**(9), *in press.*

6.   Slutter, R. G., Load Tests of Composite Slabs with 1.5, 2 and 3 in. Decking for United Steel Deck, Inc., Lehigh University, Fritz Engineering Laboratory Report No. 200.74.571.1., Bethlehem, Pennsylvania, 1975.

7.   Roeder, C. W., Point Loads on Composite Deck Reinforced Slabs. *ASCE J. Struct. Div.*, 1981, **107**(12), 2421-2429.

8.   Porter, M. L. and Greimann, L. F., Shear Bond Strength of Studded Steel Deck Slabs. *Proc. 7th International Specialty Conf. on Cold-Formed Steel Structures,* 1984, Univ. of Missouri-Rolla, 285-306.

9.   Luttrell, L. D. and Prassanan, S., Strength Formulations for Composite Slabs, *Proc. 7th International Specialty Conf. on Cold-Formed Steel Structures*, 1984, Univ. of Missouri-Rolla, 307-326.

10.   Luttrell, L. D. and Prassanan, S., Method for Predicting Strengths in Composite Slabs. *Proc. 8th International Specialty Conf. on Cold-Formed Steel Structures*, 1986, Univ. of Missouri-Rolla, 419-431.

11.   *Eurocode 4: Common Unified Rules for Composite Steel and Concrete Structures* (DRAFT)", 1985, Commission of the European Communities, Rep. EUR 9886.

12.   Heagler, R. B., Luttrell, L. D. and Easterling, W. S., *Composite Deck Design Handbook*, Steel Deck Institute, Canton, Ohio, 1991.

13.   Departments of the Army, Navy and Air Force, *Seismic Design of Buildings*, Army TM 5-809-10, Washington, D.C., 1982.

14.   Luttrell, L. D., *Diaphragm Design Manual*, Second Edition, Steel Deck Institute, Canton, Ohio, 1987.

15.   Porter, M. L. and Easterling, W. S., Behavior, Analysis and Design of Steel-Deck-Reinforced Concrete Diaphragms. Final Report ISU-ERI-AMES-78263, Engineering Research Institute, Iowa State University, Ames, Iowa, March 1988.

16. Widjaja, B. and Easterling, W. S., Analysis of Composite Diaphragms, Report No. CE/VPI_. Virginia Polytechnic Institute and State University, Blacksburg, VA, *in preparation*

17. Gibbings, D. R., Easterling, W. S. and Murray, T. M., Composite Beam Strength as Influenced by the Shear Stud Position Relative to the Stiffener in the Steel Deck Bottom Flange, Report No. CE/VPI_. Virginia Polytechnic Institute and State University, Blacksburg, VA, *in preparation*

18. Hawkins, N. M. and Mitchell, D., Seismic Response of Composite Shear Connections. *ASCE J. Struct. Engr.*, 1984, , **110**(9), 2120-2136.

19. Jayes, B. S. and Hosain, M. U., Behaviour of Headed Studs in Composite Beams: Push-out Tests. *Can. J. Civ. Engr.*, 1988, **15**, 240-253.

20. Jayes, B. S. and Hosain, M. U., Behaviour of Headed Studs in Composite Beams: Full-Size Tests. *Can. J. Civ. Engr.*, 1989, **16**, 712-724.

21. Robinson, H., Multiple Stud Shear Connections in Deep Ribbed Metal Deck. *Can. J. Civ. Engr.*, 1988, **15**, 553-569.

22. Mottram, J. T. and Johnson, R. P., Push Tests on Studs Welded Through Profiled Steel Sheeting. *The Structural Engineer*, 1990, **68**(10), 187-193.

23. Lloyd, R. M. and Wright, H. D., Shear Connection between Composite Slabs and Steel Beams. *J. Construct. Steel Research*, 1990, **15**, 255-285.

24. Grant, J. A., Fisher, J. W. and Slutter, R. G., Composite Beams with Formed Steel Deck. *AISC Engr. J.*, 1977, **14**(1), 24-43.

25. Ollgaard, J. G., Slutter, R. G. and Fisher, J. W., Shear Strength of Stud Connectors in Lightweight and Normal Weight Concrete. *AISC Engr. J.*, 1971, **8**(2), 55-64.

26. *Load and Resistance Factor Design Specification for Structural Steel Buildings*, American Institute of Steel Construction, Chicago, Illinois, 1986.

27. Slutter, R. G. and Driscoll, G. C., Flexural Strength of Steel-Concrete Composite Beams. *ASCE J. Struct. Div.*, 1965, **91**(ST2), 71-99.

SLIM FLOOR CONSTRUCTION IN THE UK

R MARK LAWSON
GRAHAM W OWENS
DEREK L MULLETT
Steel Construction Institute
Silwood Park, Ascot, SL5 7QN, UK

ABSTRACT

The design of 'Slim Floor' beams contained within the slab depth is reviewed. This form of construction is common in Scandinavia and has been adapted for use in the UK. Moment capacity calculations are based on plastic section analysis principles for both non-composite and composite forms of construction. Design tables are presented for common cases. Other design aspects such as torsion due to out-of-balance loading, and use of deep decks are also considered. Fire resistance has been justified by 8 fire tests, leading to achievement of 60 minutes fire resistance for the unprotected beam in most cases.

INTRODUCTION

An innovative and economic form of steel construction in buildings has been developed in Scandinavia and has become the most popular form of constructing medium rise buildings there. This is the so-called 'Slim floor' where a steel beam is fabricated so that it can support precast concrete slabs, and the beam occupies the same depth as the slab. These developments have been closely followed in the UK where steel construction has achieved 60% of the market for multi-storey commercial buildings. However, for buildings up to 5 storeys height, reinforced concrete is still popular, and it was with the aim of increasing the market share for steel in this sector that the slim floor concept was adapted to suit British construction practice.

The slim floor beam can use fabricated sections as shown in Figure 1, but the simplest system is to weld a support plate to the lower flange of a Universal Column (or HEB or W) section. This means that the p.c. units can be easily lowered into position. Three forms of construction are recognised.

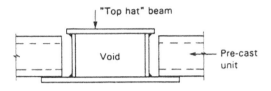

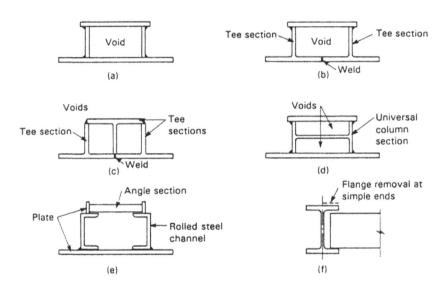

Figure 1   Example of slim floor beams using welded sections

Type 1:   Dry construction with grout only for sound insulation purposes (Figure 2(a)).

Type 2:   Encased construction, offering better fire resistance and stiffness properties (Figure 2(b)).

Type 3:   Composite construction with an in-situ concrete slab and welded shear connectors (Figure 2(c)).

The slim floor beams and columns form a 2-dimensional frame. In British practice it would be normal to 'tie' the columns together in the third dimension using a tie member. This tie can be contained within the slab depth (see Figure 3).

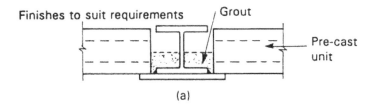

(a)

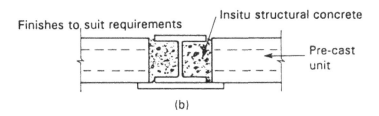

(b)

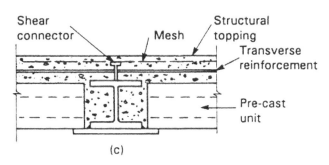

(c)

Figure 2   Types 1, 2 and 3.   Slim floor construction

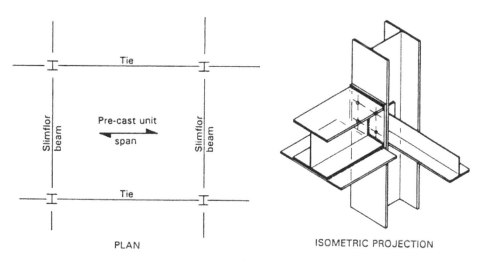

PLAN   ISOMETRIC PROJECTION

Figure 3   Typical framing arrangement

Slim floor construction provides a number of benefits:

- The floors have a flat soffit.

- The overall floor construction depth can be reduced. This will reduce cladding costs.

- It improves the fire resistance of the section. The concrete that surrounds the beam partially insulates the section. This can lead to the elimination of the fire protection giving up to 60 minutes fire resistance.

- The concrete that surrounds the beam produces an increase in the second moment of area of the section. This enhancement is helpful in reducing deflections.

- It offers unhindered passage for services

- In the case of local element instability the concrete will improve the load carrying characteristics of the beam. For the future this could prove an asset for continuous construction.

- In certain circumstances, "dry construction" can be employed, thus saving time before the building is occupied.

The modified slim floor beam shown in Figure 2 has the additional benefits:

- It uses standard steel sections.

- The beam is easy to fabricate with full depth end plate connections. Only two fillet welds are required to attach the longitudinal plate which can be automatically welded without turning the section.

- The system provides relatively long spans with minimum construction depths. This will have the effect of reducing cladding costs.

- No internal voids for sound or heat transfer in fire are created. This reduces the amount of fire protection.

- The system is inherently versatile to suit the requirements of a given building. This is emphasized by the three forms of construction shown in Figure 3 which range from dry to composite construction.

Note: Torsional effects in the construction stage and possibly at the edges of the building may occur under eccentric loading. This can be reduced by design, or by propping during construction.

DESIGN

The design procedures are based on the use of BS5950: Parts $1^{(1)}$ and $3^{(2)}$ (or to the forthcoming Eurocodes 3 and 4). The cross-sections will be limited to plastic (Class 1) or compact (Class 2) sections. Semi-compact (Class 3) sections limit the design of the section to the elastic moment capacity. This complicates the design procedure but, more importantly, is an uneconomical use of steel.

The basic design assumptions are summarised as follows:

a.   Unpropped simply supported beams subject to uniformly distributed loading.

b.   Use of plastic or compact cross-sections.

c.   Plastic analysis of the cross-section based on rectangular stress blocks.

d.   Serviceability checks use elastic analysis with unfactored loads. To ensure that irreversible deformation (under normal service loads) does not occur in the steel, the extreme fibre stress is limited to p_y, and in situ concrete stress is likewise limited to $0.5f_{cu}$.

e.   Deflections of beams are limited to span/360 under imposed loads, which applies to buildings of general usage; allowances should be made where deflections under serviceability loads could cause damage to the finishings. The total deflection of the beam is limited to span/200.

Only some of the salient aspects of slim floor design can be discussed here; full guidance is given in the SCI guide[3]. Design assumptions for assessing the moment capacity and torsional effects for the three forms of construction are as follows:

Type 1: Non-composite

a.   Slim floor beam is unrestrained in the construction stage.
b.   Slim floor beam is restrained under imposed load.
c.   Out-of-balance loads to be considered for construction and imposed loads.

Type 2: Semi-composite

a.   Slim floor beam is unrestrained in the construction stage.
b.   Slim floor beam is restrained under imposed load.
c.   Out-of-balance loads to be considered for the construction stage only.

Type 3: Composite. As for Type 2, and in addition

a.   Self weight and construction loads resisted by steel section (if unpropped).
b.   Imposed loads resisted by composite section.

Construction Stage

In the construction stage it is not always possible to ensure that the precast units are placed in such a way as to eliminate out-of-balance loads. The complexities introduced by torsion are best avoided in building structures. However, in this instance this is not easily achieved unless strict erection procedures are adhered to. Out-of-balance loads will also have an undesirable influence on the lateral torsional buckling (LTB) of the section.

The combination of these two effects is highly complex and requires simplification for general use.[4] This has been achieved by developing an expression similar in form to that used in BS5950: Part 1 for the combination of stresses.

$$\frac{M_x}{M_b} + \frac{M_y}{M_{cy}} \leq 1.0 \qquad (1)$$

where M_x is the applied moment about the x-x axis; M_b is the buckling resistance moment about the x-x axis, determined using BS5950: Part 1 ($M_b \leq M_c = S_x p_y$). M_y is the applied moment to the top flange of the steel section about the y-y axis, which is obtained by considering the torsional moment as two opposing forces in the flanges as shown in Figure 4. M_{cy} is the plastic moment capacity of the top flange of the steel section about the y-y axis.

An alternative method of eliminating torsion is to prop the beam during construction. This can have an adverse influence on construction times, and generally has a nuisance factor in site operations. However it may be appropriate on small projects, or for long span beams.

Moment capacity (non- and semi-composite sections)

In the construction stage, LTB and pure torsion in slim floor beams are treated in the same manner as in a non-composite beam. The concrete that surrounds the beam is used for stiffness purposes only and is assumed to provide adequate lateral restraint to the member for ultimate load conditions. It is difficult to show that the moment capacity of the steel beam is enhanced by composite action unless additional shear connection is provided. It is for these reasons that the composite action is neglected at the ultimate limit state. Figure 5 shows the method of deriving the equation for moment capacity (steel member only) using plastic analysis for the design of the cross-section at ultimate loading.

Figure 5(a) shows the position of the plastic neutral axis (PNA) in the web, below the centreline of the steel section at a distance y. In order to simplify calculation of the moment capacity, M_c, of the section, Figures 5(b) and (c) show a standard method of rearranging the rectangular stress blocks.

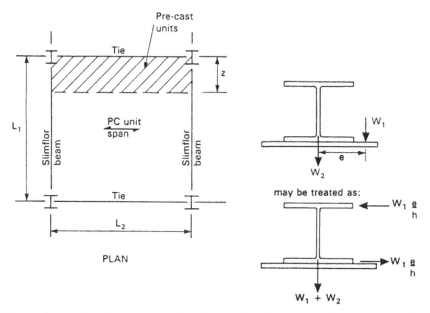

Figure 4   Simple treatment of torsion on slim floor beams during construction

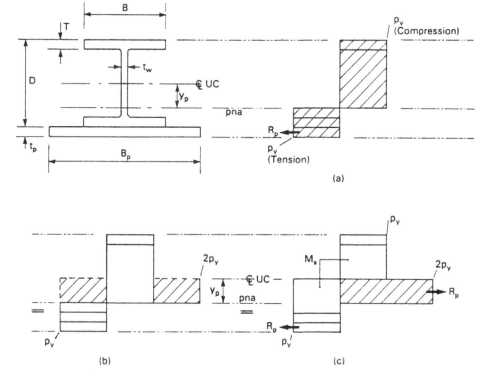

Figure 5   Plastic analysis of non-symmetric steel section

If moments are taken about the centreline of the section:

$$M_c = M_s + R_p \left[\frac{D}{2} + \frac{t_p}{2} \right] - R_p \frac{y}{2}$$

where

$$y = \frac{R_p}{2p_y t_w}$$

$$M_c = M_s + \frac{R_p}{2} (D + t_p) - \frac{R_p^2}{4p_y t_w} \tag{2}$$

where M_s is the moment capacity of the steel section (ignoring the plate) $= S_x p_y$

R_p is the tensile resistance of the plate of thickness, t_p

D is the depth of the steel section

p_y is the design strength of steel

t_w is the web thickness.

Moment capacity (composite section)

The moment capacity of composite slim floor beams is dependent on the area of concrete in compression and on the degree of shear connection between the concrete and steel beam. The effective breadth of the slab, B_e, is taken as beam span/4, as for a normal composite beam.[2,5,6]

Full shear connection. Figure 6 shows a typical cross-section through a composite beam. Also shown is the plastic stress distribution across the section. Full shear connection occurs where the number of shear connectors provided is sufficient so that the force they transfer is greater than the resistance of the concrete or steel member. This will generate the maximum moment capacity of the cross-section. In the majority of cases, the concrete resistance R_c is less than the resistance of the steel member, ie.

$$R_c < (R_s + R_p)$$

where R_c $=$ $0.45 f_{cu} B_e D_s$ and $R_s = A p_y$.

and D_s the slab depth above the section (assuming $D_{pc} < D$)
    f_{cu} the cube strength of concrete ($\approx$ 1.2 x cylinder strength).

In cases where the plastic neutral axis (PNA) for full shear connection lies within the precast units, higher forces may be generated in the shear connectors, potentially leading to an overload in longitudinal shear. Therefore, to account for this effect the number of shear connectors is determined for the maximum possible force developed in the concrete (including the precast units), or in the steel section. Conversely, when considering the moment capacity it would appear prudent to assume that this bonding of the concrete no longer exists. In these circumstances a conservative approach has to be adopted, otherwise, the degree of shear connection could be lower than the minimum required value. The moment capacity calculation is best illustrated by referring to the following case:

Partial shear connection. Partial shear connection design is attractive where the moment capacity is much greater than the applied factored moment. When this situation occurs BS5950: Part 3 and Eurocode 4 permit a reduction (to a maximum of 40%) in the number of shear connectors relative to full shear connection. The principle is that the number of shear connectors is reduced so that the longitudinal force, R_q, transferred by the shear connectors is sufficient to provide an adequate moment capacity.

The moment capacity, M_c, is given by the following expression based on the stress block approach as illustrated in Figure 6:

$$M_c = M_s + R_q \left[D_s + \frac{D}{2} - \frac{R_q}{0.9 f_{cu} B_e} \right] + \frac{R_p}{2} (D + t_p) - \frac{(R_q - R_p)^2}{4 p_y t_w} \quad (3)$$

In all cases M_c is less than the value for full shear connection. The degree of shear connection is defined as R_q/R_c for $R_c < R_s + R_p$, which is usually the relevant case.

For full shear connection $R_q = R_c$ in equation (3).

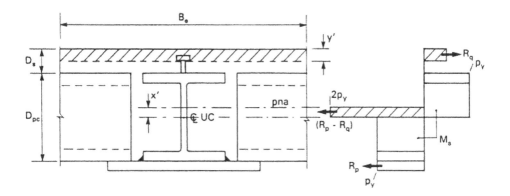

Figure 6   Plastic analysis of composite section for partial shear connection

FURTHER DESIGN CONSIDERATIONS

Biaxial stress effects in the flange plate

Biaxial stresses in the flange plate have to be considered as a direct result of the way the loads are applied to the flange plate. The plate is subject to longitudinal and transverse effects as shown in Figure 7. The longitudinal stress due to overall bending of the section, σ_1, has an influence in reducing the resistance of the plate when also subject to a transverse bending stress, σ_2. This is irrespective of whether the stresses are plastic or elastic. Plastic analysis of the section may be continued but using a reduced strength of the plate taking account of σ_2.

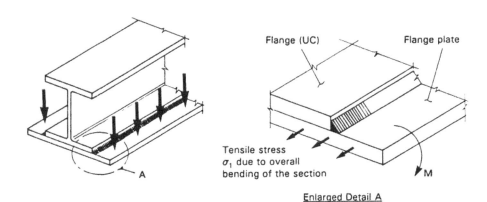

Figure 7   Biaxial stresses in flange plate

Design for Torsion

It is usually not possible to eliminate out of balance loads which cause torsion of the section. A simple way of taking this into account in design is to treat the beam as subject to transverse forces in the flanges.[4] Longitudinal and transverse bending stresses are then linearly combined using equation (1). Torsional moments are resisted at the ends of the beams by four bolted end plate connections (see Figure 3).

Edge Beams

The design of the edge beams is more problematical because torsional effects on an I section cannot be eliminated in this case. However, use of a C section as an edge beam can reduce the torsional effect because the shear centre of the section lies away from the web. Alternatively, a downstand beam may be used if clear height is not needed at the cladding line.

Use of Deep Steel Decking

A UK manufacturer of profiled sheeting is currently developing a steel deck which can span 6m unpropped and has a depth of 210mm. The deck when acting compositely with an in-situ concrete slab is an ideal replacement for the pc units as it will reduce the dead weight of the floor. This has the effect of reducing steel weight and the size of foundations. Also, it lends itself for passing minor services through the slab depth between the ribs of the deck (Figure 8).

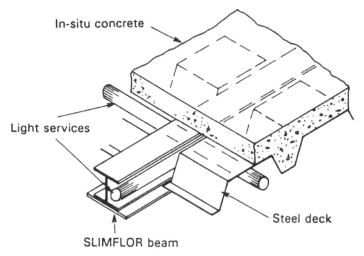

Figure 8　　Slim floor beam using deep steel decking.

FIRE RESISTANCE

The fire resistance of slim floor beams has been investigated both experimentally, by a number of fire resistance tests, and analytically using computer modelling techniques. As a result of these studies recommendations on the use of unprotected sections have been prepared for up to 60 minutes fire resistance[8]. For more than 60 minutes fire resistance additional fire protection is generally recommended. This may take the form of a board pinned to the underside of the flange plate.

The fire resistance of a slim floor beam is inherently good because most of the section is shielded from the fire by the floor units. For 60 minutes fire resistance, with minor exceptions, Type 1 and Type 2 systems can be used unprotected. However, the composite system, (Type 3), or those using the deep deck option may require additional protection or may have restrictions on design loading.

Summary of fire tests

The recommendations are based on the analysis of 8 fire resistance tests on loaded beams[3]. These are summarised in Table 1. In tests 1 and 2 the floor units were built directly into the beam web and no supporting plate was used. In tests 3 to 8, 15 mm bottom plates were used together with 8 mm fillet welds.

Table 1   Summary of fire resistance tests

Test	Section UC size x kg/m	Type	Fire Resistance (mins)	Load Ratio	Plate (†) Temperature (°C)	Flange (†) Temperature (°C)
1	254 x 254 x 73	-	44	0.56	-	746
2	254 x 254 x 89	-	93	0.42	-	783
3	254 x 254 x 107	1	60	0.55	799	661
4	203 x 203 x 86	3	68	0.44	727	558
5	203 x 203 x 60	1	82*	0.51	812	691
6	254 x 254 x 73	2	78	0.47	778	578
7	152 x 152 x 30	1	75*	0.48	788	731
8	305 x 305 x 283	1	115*	0.17	728	411

Note: * Test discontinued before failure.
        † at 60 minutes.

For tests 7 and 8 it was decided to test a Type 1 design with the section only half filled with concrete as temperatures recorded would then be conservative when applied to a Type 2 or 3 design. In tests 3 and 5 sand infill was used which resulted in a lower fire resistance compared to concrete infill.

It can be seen from Table 1 that there is always an appreciable temperature difference between the plate and the bottom flange. This difference is very important and is due to an interface resistance between the two surfaces. However, in test 7 the interface resistance between the 152 × 152 × 30 UC and the plate was appreciably lower than in the other tests.

The behaviour of the beams can be modelled mathematically using the measured and predicted temperatures and correlation with the test results was good.

DESIGN TABLES

In order carry out the scheme design of slim floor beams with p.c. concrete slabs various Design Tables have been prepared and examples are presented in Tables 2a. and b. for grade 50 steel (355N/mm^2 yield strength) in Type 1 (non-composite) and Type 3 (composite) slabs. It can be seen that composite construction is only economic for relatively long span beams with heavy imposed loads. These tables assume that the p.c. units are placed on the beams in such a way as to cause the most onerous case for torsion during construction. Symmetric placing of the units or propping during construction will reduce torsion, potentially leading to the use of a lighter steel section, if torsion is critical to the design. A diskette offering generalised design of Slim Floors has been prepared.

Table 2a:Non-Composite Beam

TYPE 1	UNIVERSAL COLUMN SIZES IN GRADE 50 STEEL		
	IMPOSED LOADING = 3.5 kN/m^2		
Beam Span m	PRECAST UNIT SPAN, m		
	4.5	6.0	7.5
4.5	152 x 152 x 37 kg/m	203 x 203 x 52 kg/m	203 x 203 x 71 kg/m
6.0	203 x 203 x 60 kg/m	203 x 203 x 71 kg/m	254 x 254 x 107 kg/m
7.5	254 X 254 X 89 kg/m	254 x 254 x 107 kg/m	254 x 254 x 132 kg/m
pc unit depth mm (and wt)	150 (2.2 kg/m^2)	150 (2.2 kg/m^2)	200 (2.7 kg/m^2)

Table 2b:Composite Beam

TYPE 3	UNIVERSAL COLUMN SIZES IN GRADE 50 STEEL		
	IMPOSED LOADING = 6 kN/m^2		
Beam Span m	PRECAST UNIT SPAN, m		
	6.0	7.5	9.0
6.0	203 x 203 x 71 kg/m	254 x 254 x 107 kg/m	305 x 305 x 118 kg/m
7.5	254 x 254 x 107 kg/m	254 x 254 x 167 kg/m	305 x 305 x 158 kg/m
9.0	305 x 305 x 158 kg/m	305 x 305 x 198 kg/m	305 x 305 x 240 kg/m

Overall slab depth = Beam depth + 80mm

TABLE 2: SCHEME DESIGN OF SLIM FLOOR BEAMS

ACKNOWLEDGEMENT

The work leading to the preparation of design guidance on slim floor construction was funded by British Steel (General Steels).

The work leading to the preparation of design guidance on slim floor construction was funded by British Steel (General Steels).

REFERENCES

1. BRITISH STANDARDS INSTITUTION
   BS 5950: Structural use of steelwork in building
   Part 1: Code of practice for design in simple and continuous
   construction: hot rolled sections
   BSI, 1990

2. BRITISH STANDARDS INSTITUTION
   BS 5950: Structural use of steelwork in building
   Part 3: Codes of practice for design in composite construction
   Section 3.1: Design of simple and continuous composite beams
   BSI, 1990

3. MULLETT, D.L.
   Slim floor design and construction
   The Steel Construction Institute, 1991

4. NETHERCOT, D.A., SALTER, P.R. and MALIK, A.S.
   Design of members subject to combined bending and torsion
   The Steel Construction Institute, 1989

5. LAWSON, R.M.
   Design of composite slabs and beams with steel decking
   The Steel Construction Institute, 1989

6. LAWSON, R. M.
   Commentary to BS 5950: Part 3: Section 3.1 "Composite beams"
   The Steel Construction Institute, 1990

7. WYATT, T.A.
   Design guide on the vibration of floors
   The Steel Construction Institute, 1989

8. BRITISH STANDARDS INSTITUTION
   BS 5950: The structural use of steel in building
   Part 8: Code of practice for fire resistant design
   BSI, 1990

Moment Resistance of Composite Connections in Steel And Concrete

Y Xiao, D A Nethercot and B S Choo
Department of Civil Engineering,
University of Nottingham, University Park Nottingham, UK

ABSTRACT

A series of tests designed to investigate the interaction of a variety of different steel beam to column details with a composite metal deck floor is described. The main emphasis is on assessing the connections' moment capacity, rotational stiffness and rotation capacity. Based on the different types of failure observed, a simple method for calculating moment capacity is proposed.

INTRODUCTION

Efficient and economical lightweight floor systems may be created by integrating the structural properties of concrete and cold formed steel decking by ensuring composite action through the use of through deck welded shear studs. This type of construction has now became a common feature in multistory steel frame buildings in several countries[1,2,3]. Traditional methods of steel frame design disregard the actual behaviour of the beam to column joints, assuming one of the ideal models of perfectly rigid or perfectly pinned. In addition to the inherent stiffness and moment capacity of the steel detail, neglect of the contribution to joint properties from composite action with the floor slab may result in either an over conservative or an unconservative assessment of the actual behaviour of the composite frame. Inclusion of more realistic joint effects does, however, require that a proper understanding of the structural behaviour of composite joints be available. A number of researchers have studied this topic [4,5]. However, a general basis for the design of composite joints has not yet been established and the large number of variables present means that the existing database of test results provides very patchy coverage of the subject. No attempt has so far been made to quantify requirements for rotational capacity, nor to predict rotational stiffness.

Recent tests conducted at Nottingham University are reported herein. The main emphasis is on assessing the three key indications of performance: moment resistance, rotational stiffness and rotation capacity. Knowledge of these fundamental properties is necessary before the true semi-rigid and partial strength nature of the connections can be properly incorporated into more realistic methods of frame design. Specimens were therefore selected to include several currently used joint types as well as to investigate slab parameters expected to influence the connection behaviour directly.

Test specimens and set-up
Specimen description

The conventional metal deck flooring system comprises concrete topping on profiled metal decking on top of the steel beams, with shear connection provided by through deck welded shear studs. The test specimens were configured as 'cruciform' type, as shown in Fig.1. Four different types of steel joint have been used: flush end plates, partial depth end plates, web side plate and cleats as shown in Fig.2. The members used in the construction of the specimens were 305x165x40 UB and 203x203x52 UC in grade 43 steel. Reinforcement was A142 mesh, supplemented in some cases by T10 or T12 rebars. PMF CF46 deformed metal decking was used as bottom shuttering.

The bare steel joint was first assembled in position on the baseplate, the decking was added and then the slab was cast. All bolts (M20, Gr. 8.8) were tightened to 150 N·m torque by a torque wrench to maintain consistency and comparability between specimens.

The steel decking was filled with ready mixed normal weight concrete. This had a design cube strength of 30 N/mm^2, maximum aggregate size of 20 mm, and a slump of 45 mm. In practice the slump values were found to be in the range 45-75mm, and the concrete strength on the day of testing in the range 30-54 N/mm^2.

Concreting work was carried out inside the laboratory with the specimen in the test position. The metal decking was used as bottom formwork with non-absorbent plywood as side shuttering. Some curing oil was brushed on the surface of the side shuttering for easy demolding. Casting of cubes and cylinders for strength tests was carried out at the same time. After the surface finishing, the specimen was covered by a polythene sheet and left to cure. The cylinders and cubes were cured in water at an average temperature of 20^0 C. These samples were tested at 7 days, 14 days and on the test day according to the BSI test standard. Usually, the connections were tested when the concrete strength reached 30 N/mm^2 or after 28 days.

Figure 1. Test set-up and instrumentation

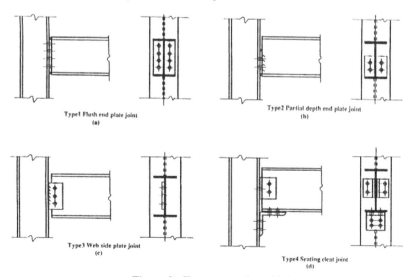

Figure 2. Four types of steel joint

TABLE 1

Specimen details and test results

Specimen	Joint Type	Web Stiffening	Reinforcement Ratio(%)	First Crack Moment (kN· m)	Ultimate Moment (kN· m)	Ultimate Rotation (mRad)*	Maximum Rotation (mRad)	Failure mode*
SJ1	Seating cleat	None	None	—	39.5	> 59.8	—	A
SCJ1	Seating cleat	None	A142 mesh (0.2%)	15	43.1	14.3	21.7	B
SCJ2	Web side plate	None	A142 mesh (0.2%)	12	29.6	16.4	30.4	B
SCJ3	Flush end plate	None	A142 mesh (0.2%)	30	85.7	7.2	26.6	B
SCJ4	Flush end plate	None	T12 Rebar (1.0%)	40	202.9	23.4	41.1	C,D
SCJ5	Flush end plate	Web stiffeners	T12 Rebar (1.0%)	45	240.8	26	35	G,E
SCJ6	Flush end plate	None	T10 Rebar &A142 Mesh (1.0%)	36	157.6	11.5	23	C,D
SCJ7	Flush end plate	Plate stiffening	T12 Rebar & A142 Mesh (1.2%)	37.5	204.5	26.5	46.9	G,D
SCJ8	Partial depth end plate	None	T10 Rebar &A142 Mesh (0.8%)	22.5	84	29	44.5	F
SCJ9	Partial depth end plate	None	T10 Rebar &A142 Mesh (0.8%)	31.5	107.5	27	43.9	F
SCJ10	Partial depth end plate	None	T10 Rebar &A142 Mesh (0.8%)	35	147.8	16.5	30	C,D
SCJ11	Seating cleat	None	T10 Rebar &A142 Mesh (0.7%)	33.8	169.5	14.3	21.7	C,D
SCJ12	Web side plate	None	T10 Rebar &A142 Mesh (0.7%)	22.5	101.3	39	79	H

*Note: 1mRad = 0.05729 Degree
A --- Excessive deflection of beams
B --- Fracture of the mesh reinforcement
C --- Excessive deformation of column flange
D --- Buckling of column web
E --- Buckling of beam flange
F --- Buckling of beam web
G --- Shear studs failure
H --- Web side plate twisting

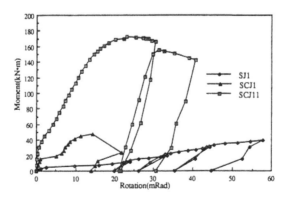

Figure 3. Moment-rotation relations of cleated connections

Test set-up and instrumentation

The column of each specimen was fixed to the laboratory floor through a base plate and supporting angles. This arrangement allowed for accurate adjustment of the specimen. The load was applied to the specimen using hand operated hydraulic jacks mounted on the test rig, which was itself attached to the strong floor of the laboratory. Two loads were applied symmetrically at 1.5 m from the column flange through two hollow section load spreaders. Measurement of the load was obtained through load cells positioned between the jack and a spherical seat. The arrangement allowed in-plane movement so as to ensure that the load was always applied to the same point on the slab.

Strains in the reinforcement and on the steel beam were measured using strain gauges. The resistance and insulation were checked after the strain gauge coating to ensure proper operation after concreting. Having assembled the test specimen, all the instruments were mounted on the specimen. Inclinometers were used to measure the rotation of the column and the relative rotation of the beam so as to provide full information from which the moment-rotation relationship could be derived. Displacement transducers were used to measure the deflection of the specimen on the bottom flange of the beam and also to monitor slip of the seat cleat. The relative rotation of the beam to column can also be obtained from the deflection values within the calibrated horizontal length of the beam. Dial gauges were used to measure the deformation of the concrete slab and horizontal movement of the specimens.

Load was applied to the specimen by a manual pump. The load increment was one twentieth of the calculated capacity of the specimen, usually 5-10 kN each step, with half increments as failure was approached. Load cells were also connected to the load output chart recorder to ensure the load history and maxima were graphed simultaneously. Loading was continued for several steps after the peak load value so as to get information of the ductility of the specimen and the post-failure situation. All the test data were monitored through a Data-Logging system using Axis software to display selected test information on line.

After the test, all the specimens were carefully inspected, including taking the joint apart, removal of the bolts, checking the slip and the hole deformation. When significant large longitudinal cracking appeared in the slab, it was cut and the shear studs were checked.

Test results and analysis

It is not possible to properly present and discuss all of the test results herein. Thus attention will be focussed on the completed tests on the cleat and web side plate specimens; an equivalent consideration of the end plate specimens is available elsewhere (6). The details and the main results of all the specimens are summarized in Table1.

Specimen SJ1 was a bare steel joint, as shown in Fig.2(d) with a seating cleat and double web stability cleats. The test result was required for a direct comparison with composite joint SCJ1 which had the same steel detail. The moment-rotation curve shown in Fig.3 indicates the early onset of slip, with increasing growth of deflection and rotation as the load was increased. Horizontal slip occured at the connection as the top of the beam moved out and the bottom of the beam pushed towards the column, causing opening of the web angles. The failure mode was excessive deflection of the beam. The last recorded beam rotation value was greater than 58 mRad. The measurements show that there was no vertical slip and only small deformations of the seating cleat. This test confirmed that bare steel joints do posses some degree of moment resistance and rotational stiffness.

Specimen SCJ1 had the same steel detail as SJ1 and was cast with a 120 mm deep concrete slab onto the profiled decking. Only the mesh reinforcement (A142), which is used to control shrinkage cracks in current construction practice, was included in the concrete floor. SCJ11 had the same detail as SCJ1 but with an increased reinforcement ratio to 0.7%. As expected, both specimens exhibited increased stiffness in the initial loading stage compared with SJ1 as shown in Fig.3. The first crack for SCJ1 was found parallel to the column flange at 15 kN·m. This later formed into the main cracks of the slab. The stiffness of the joint decreased after the concrete cracked. This part of the path to failure is similar to a 'yield plateau'. The specimen regained stiffness later as the load was approaching its maximum value. Debonding of the concrete to the metal decking took place due to the horizontal cracking formed at the surface between concrete and decking and mesh fracture occured eventually. The loading value fell rapidly after this stage. The ultimate bending strength of the specimen was increased by about 40% as compared with bare steel joint SJ1. But the ultimate rotational of the such joint was only 14.3 mRad.

A mixture of rebar and mesh was introduced in specimen SCJ11. The decrease in the initial stiffness caused by the onset of slip was not severe because of the reinforcement preventing the large cracks from forming in the concrete. The cracks were smeared over the whole length of the specimen with the width of the main cracks being much less than for SCJ1. The specimen failed in a more ductile way after the maximum loading was reached and still possessed quite high residual strength. The rotational capacity and moment capacity were both increased as compared with SCJ1. The ultimate moment resistance was almost four times that of the bare steel joint and three times that of the connection with only mesh reinforcement. Prior calculation had identified the column web as the weakest element. Buckling of the column web in compression as illustrated in Fig.4 appeared after

Figure 4. Failure mode of SCJ11

the reinforcement had reached yield as indicated by the strain gauge readings. The presence of additional reinforcement changed the form of the eventual failure to one in which the steel members were the controlling factor, leading to a substantial increase in the moment resistance as shown in Fig.3.

The web side plate joint shown in Fig.2(c) was also included in the first phase work. The steel detail derives its flexibility from bolt deformation in shear, bolt hole distortions in bearing in the web side plate and/or the beam web and out of plane bending of the web side plate. SCJ2 with mesh reinforcement was tested to check the behaviour of such joints when functioning in a composite fashion. Fig.5 shows how the stiffness decreased after the first crack appeared in a similar way to SCJ1. After the peak load, the moment resistance dropped sharply as the cracks became enlarged and the mesh fractured. The ultimate rotation was higher than SCJ1, reaching 16.4 mRad as shown in Fig.5. This type of joint was more flexible than the cleated joint. An increased ratio of reinforcement

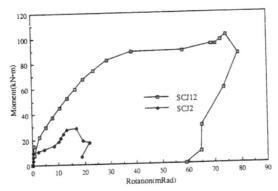

Figure 5. Moment-rotation relations of web side plate connections

was used in SCJ12 to improve the moment resistance. The moment-rotation relation for SCJ12 given in Fig.5 shows significantly greater ductility in the region of maximum load. The beam bottom flange moved towards the column flange when the loading continued. After contact of the flanges, the connection stiffness increased and the moment resistance was further increased. The ultimate rotation of the beam reached 39 mRad, with eventual failure being caused by the bearing strength of the beam web and excessive twisting of the web plate. Permanent deformation had taken place in the web bolt holes as revealed in the check after the test. The moment resistance of this type of connection was clearly governed by the tensile force of the reinforcement and the bearing in the beam web or web side plate. The performance of both types of connections has clearly been enhanced by the additional reinforcement. The change of the failure mode away from the connection into the members makes it possible to fully utilize the strength of the surrounding members.

A full discussion of the performance of the composite end plate connections details is provided in ref.6. Comparison of the behaviour of three different joints with the same type of mesh reinforcement(0.2%) is shown in Fig.6, from which it can be seen that the flush end plate connection has the highest moment capacity. Decrease of stiffness caused by the cracks in the slab and onset of slip is apparent for both cleated and web side plate connections. Failure in each case was initiated by mesh fracture. Tensile tests on samples of the mesh showed a low elongation of 3-5%. With additional reinforcement, the stiffness and moment capacity were greatly increased due to the direct enhancement of the tensile zones as shown in Fig.7. These composite connections also exhibited superior rotational capacities.

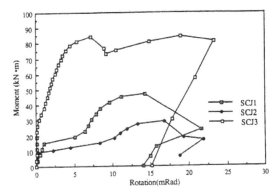

Figure 6. Moment-rotation relations of different steel details with A142 Mesh (0.2%)

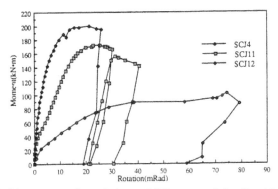

Figure 7. Moment-rotation relations of different steel details with extra rebar

Simplified calculation of moment capacity

At present no generally applicable methods exist for the design of composite connections. One previous suggestion was to determine the value $M_p = A_s f_y d_f$ to roughly represent the connection capacity [7]. This represents a conservative assessment of the joint ultimate strength since it assumes a simple fixed lever arm value d_f for the internal force of the connection. It therefore neglects the variation of the neutral axis position caused by the reinforcement ratio and steel detail.

Separate calculation procedures have been developed for web side plate, flush end plate and partial depth end plate composite connections. These are consistent with the current BS 5950; Part 3.1 approach [8] for composite beams and composite slabs. The plastic analysis concept has been adopted within the framework of ultimate limit state design. Tensile resistance of the cracked concrete is neglected and strain hardening of the reinforcement is not considered in the calculation.

The basic concept for calculating the moment capacity of a composite web side plate connection is presented below as an illustration. The calculation is based on failure being controlled by yielding of the reinforcement, acting in conjunction with the bearing capacity of the plate or beam web, whichever is the thinner. The shear and compression zone of the column web should also be checked using EC3 [9]. In this method only the contribution of the beam section, reinforcement and connecting components are included. Comparisons between the calculated and test values for the different connections are listed in Table2.

TABLE 2

Comparison of calculated and test moment values

Specimen	Test Moment (kN· m)Mt	Calculated Moment (kN·m) Mc	Comparison Ratio Mt/Mc
SCJ1	43.1	49.4	1.15
SCJ2	29.6	30.0	1.01
SCJ11	169.5	162.0	1.05
SCJ12	101.3	110.0	0.92

Case one   Neutral axis within the concrete section

Initially the reinforcement in the top of the section is assumed to yield. On this basis, a calculation is performed to determine the neutral axis depth. If the calculated neutral axis depth is too close to the top of the concrete section (because the reinforcement ratio is excessive), it is then assumed that the reinforcement has not reached yield and a new equilibrium condition is established to determine the neutral axis position again. If the neutral axis is within the concrete section, the following calculations are then performed.

Tensile force in the top reinforcement is

$$F_s = f_y \, A_f \tag{1}$$

in which $A_f = \rho \, A_c$

f_y is the yield strength of the reinforcement

ρ is the ratio of reinforcement

A_c is the effective composite floor section area

Compressive force in the bottom component is

$$F_b = P_{bs} \text{ or } P_S \text{ whichever is smaller} \tag{2}$$

Bearing capacity P_{bs} is determined by the plate or beam web whichever is the thinner

$$P_{bs} = n\, d\, t_p\, p_{bs} \quad \text{or} \quad \frac{1}{2} P_1\, t_{bw}\, p_{bs}. \tag{3}$$

in which   n is the number of bolts

                d is the diameter of bolt hole

                t is the thickness of plate or web

                p_{bs} is the bearing strength of the bolts

                P_1 is the top edge distance of the plate

Shear capacity of the bolts T_S is

$$T_S = n\, P_s\, A_t \tag{4}$$

in which P_s is the shear capacity of a single bolt

                A_t is the net area of one bolt

This case is usually associated with the condition $F_s > F_b$, hence, the neutral axis is within the concrete area and the lower portion of the concrete is in compression.

Compressive force F_c of the concrete is

$$F_c = F_s - F_b \tag{5}$$

Depth of the compressive concrete section is

$$x_p = \frac{F_s - F_b}{0.4\, b_c\, f_{cu}} \tag{6}$$

in which b_c is the effective calculated composite floor width

                f_{cu} is the cube strength of the concrete

The depth of the web or plate in compression must also be checked using a suitable shear buckling formula.

The ultimate moment capacity of the connection can be obtained by taking moments about the centroid of the reinforcement.

$$M_p = F_c(d_{cc} - \frac{x_p}{2}) + F_b(d_{cc} + t_{bf} + t_1 + P_1 + (n-1)P) \tag{7}$$

in which   P is the bolt pitch

                t_1 is the distance from top bolt to the top flange of the beam

Effective depth of concrete slab is

$$d_{cc} = d_c - c \tag{8}$$

                in which c is the thickness of the concrete cover

Case two   Neutral axis in the steel section

This case is usually associated with the condition $F_s < F_b$

The number of bolt rows m above the neutral axis is derived as follows:

$$m = \frac{F_b - F_s}{P_{bs}} \quad \text{or} \quad m = \frac{F_b - F_s}{T_s} \text{ whichever is}$$

the larger $\qquad$ (9)

Moment capacity M_p of the connection can be derived by taking moments about the centroid of the compressive stress block.

$$M_p = F_s((n-1)P - \frac{(n-m-1)P}{2} + d_{cc} + t_1 + P_2 + t_{bf}) + td\frac{(n-1)P}{2} \qquad (10)$$

Reinforcement tensile force F_s and number m of bolts acting in shear are already available from formulae [1] and [9].

CONCLUSIONS

A test programme designed to systematically study the moment resistance and rotational capacity of composite connections has been described. Several different types of joint were chosen as a way of investigating changes to the major parameters which affect joint behaviour. The following conclusions can be drawn from the analyses of the tests:

(i)   A low ratio of reinforcement consisting only of mesh cannot be used for structural purposes. In every test on this type of specimen failure was caused by fracture of the mesh and was associated with low rotational capacity and relatively little enhanced strength.

(ii)   Provision of additional reinforcement in the form of loose rebars dramatically increased the moment resistance of the connections. The rotational capacity was also improved.

The influence of slip of the bolts on the loss of the stiffness was minimized.

(iii) Different failure modes were observed for the different steel details and the different reinforcement arrangement. The failure was usually through a combination of yielding of the reinforcement and buckling of the lower part of the connection, either the flange or web of the column or the lower portion of the beam. The presence of column stiffening further increased the moment capacity and changed the failure modes.

(iv)   Utilization of composite action in enhancing the strength and stiffness of the connection will require the availability of simple, safe calculation methods; the basis of such a method has been suggested.

The work described in this paper covers the first and second phases of the project, the third phase of the experimental work covers tests on minor axis connections and edge joints.

ACKNOWLEDGEMENT

The project was funded through a grant from the Building Research Establishment as part of their role in calibrating the forthcoming Eurocode 4 for use in the U.K. This financial support is gratefully acknowledged.

REFERENCES

1. Patrick, M., Hogan,T. J.and Firkins, A. Composite Floor Systems in Commercial Buildings, Proceeding of 3rd Conference in Nuclear Development , AISC, Melbourne, Australia, May 1988, pp54- 66.

2. Mathys, J. H., Multistory Steel Buildings-a New Generation: the Current Scene, The Structural Engineer, Vol. 65A, No.2, February 1987, pp 47-51.

3. Wright, H. D., Evans, H. R., and Harding, P. W., The Use of Profiled Steel Sheeting in Floor Construction, Journal of Constructional Steel Research, 1987, pp279-295.

4. Zandonini, R., Semi-Rigid Composite Joints, Structural Connections-Stability and Strength, ed. R. Narayanan, Elsevier Applied Science, 1989, pp 63-120.

5. Nethercot, D. A., Tests on Composite Connections Second International Workshop: Connections in Steel Structures, Behaviour, Strength and Design, Pittsburgh, Pennsylvania, U.S.A, April 1991.

6. Xiao, Y., Nethercot, D. A., Choo, B. S., The influence of Composite Metal Deck Floorings on Beam-Column Connection Performance Paper submitted to the 11th International Speciality Conference in Cold Formed Steel Structures, Missouri-Rolla, U.S.A.

7. Johnson, R. P., and Hope-Gill, Semi-rigid Joints in Composite Frames Proceedings of IABSE Ninth Congress, Preliminary Report, Amsterdam, Holland, May 1972 pp 133-144

8. BSI, BS5950 Structural Use of Steelwork in Building Part 3. British Standard Institution, London. 1985.

9. Eurocode 3 Design of Steel Structures, Part1, General Rules and Rules for Buildings, Prepared for the Commission of the European Communities, Edited draft, Issue 3, April 1990.

STRUCTURAL ANALYSIS OF COMPOSITE BRIDGES
UNDER THERMAL AND MECHANICAL LOADS

ENRIQUE MIRAMBELL and PERE ROCA

Universitat Politècnica de Catalunya
Department of Constructional Engineering
Gran Capitán s/n. 08034 Barcelona. Spain.

ABSTRACT

Design engineers evaluate accurately dead and live load acting on the superstructure of bridges. However, in the case of the thermal loads, the designer of bridges bases the evaluation of such loads on his experience and on recommendations of existing codes of practice, which usually consider temperature effects in a trivial manner. The present paper is focussed on the behaviour of composite bridges under temperature effects. Results related to the thermal response and stress distributions in composite bridges are presented. In addition, an assessment of the temperature distribution to be considered in the design of composite bridges is carried out. Lastly, the influence of the thermal stresses on the behaviour of composite bridge decks in serviceability conditions is analysed.

INTRODUCTION

Environmental thermal actions introduce imposed deformations on steel–concrete structures which can influence significantly their structural behaviour. In order to evaluate, in an accurate way, the structural effects of such thermal loads on the thermal and mechanical response of composite structures, several aspects must be considered in the analysis.

First, temperature distributions within the cross–sections of composite decks are, in general, nonlinear and are dependent on the environmental conditions, the location of the bridge and the thermal material properties. Moreover, due to the nonlinearity of temperature distributions, self–equilibrated stresses can be induced. Therefore, it is necessary to determine the magnitude of the thermal actions and the stress distributions to be considered in the design of composite bridges. In second place, the structural effects caused by thermal actions are highly dependent on the materials state and on ductility conditions of the structure. In this paper, the structural analysis is carried out in service conditions, considering linear elastic relationships for concrete and steel. On the other hand, cracking of concrete may be considered by means of the reduction of the stiffness along the cracked length of the bridge.

THERMAL ANALYSIS OF COMPOSITE BRIDGES

Assuming that concrete hardening has finished and considering concrete and steel as isotropic homogeneous materials, the differential equation that governs the heat transfer problem is

$$K \cdot \nabla^2 T = \partial T / \partial t \tag{1}$$

where K is the thermal diffusivity of the material. The general boundary condition at the external surfaces of the steel–concrete deck is the Newmann condition of prescribed heat flow, which considers the heat transfer mechanisms between the bridge deck and the surrounding air: solar radiation, convection and thermal radiation. Due to the complexity of the contour of the deck and of the boundary condition, the differential equation (1) is solved by numerical methods, based on finite difference and finite element schemes ([1], [2], [3], [5]).

Numerical method of thermal analysis

In this study the numerical model developed in order to determine the time–dependent temperature distributions within the cross–section of composite bridges is based on the finite difference method. For the concrete cross–section, the thermal analysis is carried out on a two–dimensional domain, while for the steel cross–section the analysis is one–dimensional because of the assumed hypothesis of constant temperature along the thickness of the steel plates. The main characteristics of the numerical program developed and the convergence and numerical stability conditions can be found in [5].

Likewise, perfect interaction between concrete and steel is assumed. Therefore, the continuity of heat flow and temperature at the contact surface between both materials must be imposed in the thermal analysis. In any case, in order to obtain actual time–dependent temperature distributions within the cross–section of composite bridges, the incident solar radiation acting on the steel web of the deck and the evolution of the temperature inside the cell in box girder bridges must be considered ([3], [5]). In order to verify the convergence and stability conditions, the time step adopted in the thermal analysis is 6 min. Besides, such time step allows to determine the time in which the maximum imposed thermal deformations and the maximum stresses in concrete and steel appear. The period of time analyzed in this study is 4 days, therefore the actual temperature distribution does not depend on the assumed initial temperature distribution.

Self–equilibrated stress distributions

Temperature distributions within the cross–section of composite bridges are nonlinear. In order to obtain stress distributions at sectional level, the Navier–Bernouilli hypothesis is assumed. Due to this fact, self–equilibrated longitudinal stresses appear and are independent of the support conditions. Such stress distributions can be obtained superposing the results derived from two analyses: The loading case I with total restraints and the loading case II without restraints.

Assuming that there's no cracking in concrete, the self–equilibrated stress distributions can be determined by means of the following expressions

$$\sigma_{c,z}(x,y) = \sigma_{c,z}^I(x,y) + \sigma_{c,z}^{II}(x,y) = E_c[-\alpha_c T_c(x,y) + \varepsilon_m + \psi_x y + \psi_y x] \qquad (2.a)$$

$$\sigma_{s,z}(x,y) = \sigma_{s,z}^I(x,y) + \sigma_{s,z}^{II}(x,y) = E_s[-\alpha_s T_s(x,y) + \varepsilon_m + \psi_x y + \psi_y x] \qquad (2.b)$$

In these expressions, ε_m is the imposed thermal deformation and ψ_x and ψ_y are the imposed thermal curvatures to which the composite deck is subjected. They can be calculated according to the following expressions

$$\varepsilon_m = N_o/A^* = [\alpha_c E_c \sum_i (T_{c,i} A_{c,i}) + \alpha_s E_s \sum_j (T_{s,j} A_{s,j})]/A^* \qquad (3.a)$$

$$\psi_x = M_{ox}/I_x^* = [\alpha_c E_c \sum_i (T_{c,i} A_{c,i} y_i) + \alpha_s E_s \sum_j (T_{s,j} A_{s,j} y_j)]/I_x^* \qquad (3.b)$$

$$\psi_y = M_{oy}/I_y^* = [\alpha_c E_c \sum_i (T_{c,i} A_{c,i} x_i) + \alpha_s E_s \sum_j (T_{s,j} A_{s,j} x_j)]/I_y^* \qquad (3.c)$$

where A^* is the fictitious area (axial stiffness) and I_x^* and I_y^* are the fictitious moments of inertia (bending stiffnesses) of the composite cross–section of the deck.

Assessment of the design temperature distribution in composite bridges

In order to assess the temperature distribution to be considered in the design of composite bridges, the numerical program developed allows to determine two types of temperature distributions which approximate the actual nonlinear temperature distributions existing within the cross–section of composite decks. These temperature distributions are presented in figure 1.

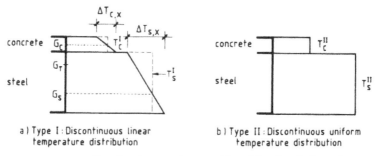

a) Type I: Discontinuous linear temperature distribution

b) Type II: Discontinuous uniform temperature distribution

Figure 1. Temperature distributions to be considered in the design of composite bridges.

Both types are suitable for the structural analysis of composite bridges in front of environmental thermal effects. Six parameters define the temperature distribution type I and they can be calculated through the determination of the linear temperature distribution equivalent to the actual nonlinear temperature distribution for both partial cross-sections (average temperature and two linear thermal gradients for each partial cross-section). Temperature distribution type II is only defined by two parameters, T_c^{II} and T_s^{II}. Such uniform temperatures do not represent the average temperature in concrete and steel, but they are defined imposing that the thermal deformation and the vertical thermal curvature obtained from this temperature distribution must be equal to the results of thermal deformation and vertical thermal curvature obtained from the thermal analysis of the composite cross-section.

This temperature distribution does not consider the effect of the imposed thermal curvature ψ_y. However, the influence of this curvature on the structural behaviour of the composite bridge is not significant, due to its small magnitude. In general, it can be stated that the horizontal thermal curvature only influences the values of the maximum compressive stress in steel and the maximum tensile stress in concrete. In addition, such influence is practically null when there is no incident solar radiation on the steel webs of the deck.

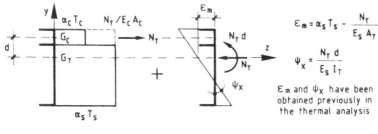

G_c = centroid of concrete cross-section
G_T = centroid of total cross-section
A_T = area of total cross-section
I_T = moment of inertia of total cross-section

Figure 2. Determination of temperature distribution type II.

NUMERICAL RESULTS OF THERMAL ANALYSES

The temperature distribution in a bridge can be determined starting from data related to the location and the azimuth of the bridge, geometrical and material properties of the structure, and environmental conditions existing in the placement of the bridge. Parametric studies have been carried out by several authors, in order to establish the influence of different variables on the thermal response and on self-equilibrated stress distributions in composite bridges ([3], [6]). On the basis of the results derived from these studies, some assumptions can be done in order to obtain the highest temperature difference and highest stresses caused by thermal effects. Such assumptions lead to fixed values of the different parameters which influence the thermal response of composite bridges. In any case, values must be in accordance with the environmental conditions existing in the placement of the structure. In this study, the bridge is located in Barcelona and the analysed cross–section is presented in figure 3. Table 1 shows data for the thermal analysis.

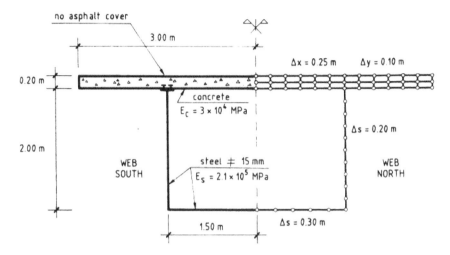

Figure 3. Analysed cross–section of composite bridge and finite difference mesh.

TABLE 1
Data for the thermal analysis

Thermal properties	Concrete	Steel
Thermal diffusivity (m²/h)	0.003	0.045
Solar absorptivity	0.5	0.685
Emissivity	0.88	0.80
Thermal expansion coef. (°C⁻¹)	$10 \cdot 10^{-6}$	$12 \cdot 10^{-6}$

Environmental conditions		
	Air temperature (°C)	$-2.0 \div 20.0$
	Wind speed (m/s)	0.0
	Turbidity factor	2.0
	Day of the year	December, 21

Location and orientation of the bridge		
	Latitude (°N)	43.0
	Altitude (m)	100.0
	Azimuth (°)	East–West, 0.0

Only selected results will be presented here. Figure 4 shows the evolutions of average temperatures for both partial cross–sections. In the same figure, daily variations of the air temperature inside the cell and the maximum temperature in steel are also shown. From that figure, one can observe a strong correlation between the temperature inside the cell and the average steel temperature. The maximum temperature in the steel is about 46 °C and occurs in several nodes of the web subjected to solar radiation.

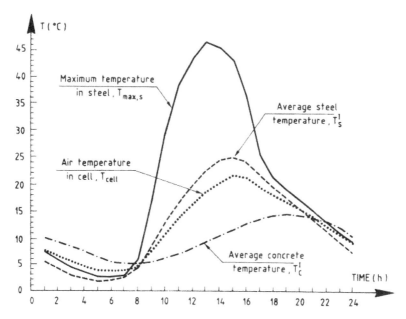

Figure 4. Daily variations of temperatures in a box girder composite bridge.

On the other hand, from the results related to the daily evolutions of equivalent linear thermal gradients for both partial cross–sections, only the thermal gradient $\Delta T_{s,y}$ in the steel box girder shows a significant variation during the day. These results lead to determine the design temperature distribution type I by means of the average temperatures in concrete and steel, T_c^I and T_s^I, and the linear thermal gradient $\Delta T_{s,y}$.

Figure 5 shows daily evolutions of the imposed thermal deformation, ε_m, and the imposed thermal curvatures, ψ_x and ψ_y, for the global composite cross–section. One should note that the structural response can be estimated considering only the thermal deformation ε_m, and the thermal curvature ψ_x. From this figure, it can be seen that the daily variation of the thermal curvature ψ_y is small and its magnitude is not significant. Consequently, from these results, a suitable temperature distribution to be considered in the design of composite bridges can be only expressed in terms of the average temperatures of the partial cross–sections, T_c^I and T_s^I (see figure 4).

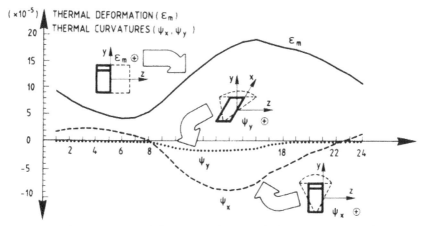

Figure 5. Daily variations of the imposed thermal deformations for the composite cross–section.

Now, we compare the evolution of the average temperatures in concrete and steel with the evolution of the temperatures T_c^{II} and T_s^{II} (temperature distribution type II). The results related to these daily variations, presented in figure 6, show that temperature differences between concrete and steel are very similar for both design temperature distributions. Therefore, in order to take into account the environmental thermal effects in the design process of composite bridges, a discontinuous uniform temperature distribution may be suggested. This will be useful for bridge designers to carry out the structural analysis under thermal loads, in a simple and realistic way. Of course, accurate values of T_c^{II} and T_s^{II} must be in accordance with the environmental conditions existing in the location of the bridge and with the geometry of the composite cross–section (web depth, overhanging length).

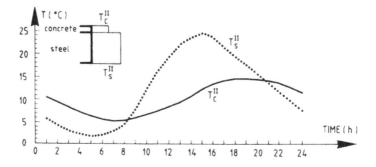

Figure 6. Daily variations of T_c^{II} and T_s^{II}. Discontinuous uniform temperature distribution.

STRESS DISTRIBUTIONS

In the case of simply supported composite bridges, only self–equilibrated stress distributions appear due to the nonlinearity of temperature distribution within the composite cross–section. Figure 7 shows the self–equilibrated stress distributions corresponding to the actual temperature distribution at the time when the maximum thermal curvature ψ_x is induced (14.10 hours). The cross–section of the hypothetical bridge analised is the one presented in figure 3.

In figure 8, stress distributions at several cross–sections of a hypothetical continuous composite bridge are presented. Such stress distributions are induced by dead and live loads plus the design temperature distribution (discontinuous uniform temperature distribution) at the time when the maximum thermal curvature ψ_x is reached. We must remark some aspects related to these results. The design temperature distribution does not allow to assess in an accurate way the magnitude of compressive self–equilibrated stresses in the steel cross–section. This effect tends to vanish in the case of continuous composite bridges due to the appearance of additional bending moments which induce continuity stresses opposite to the previously existing ones. However, for the partial concrete cross–section, the continuity thermal stresses must be added to the self–equilibrated thermal stresses.

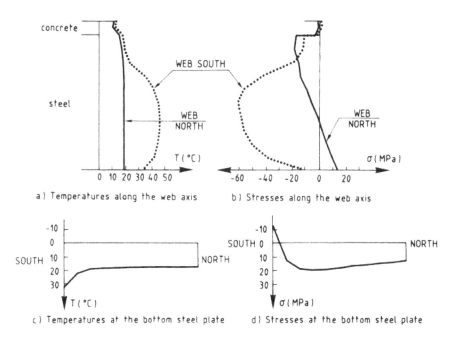

Figure 7. Self–equilibrated stresses corresponding to the actual temperature distribution.

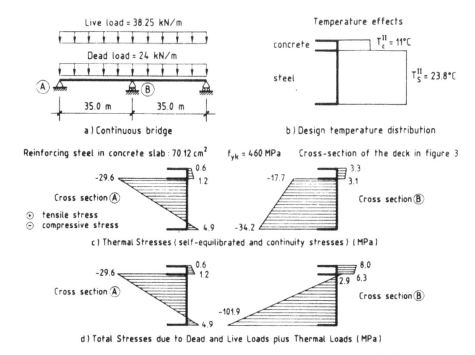

c) Thermal Stresses (self-equilibrated and continuity stresses) (MPa)

d) Total Stresses due to Dead and Live Loads plus Thermal Loads (MPa)

Figure 8. Stress distributions in a continuous composite bridge.

Figure 8 also shows that the total tensile stresses in concrete, due to dead and live loads plus temperature effects, cause cracking. Therefore, we must account for this effect in a nonlinear analysis of composite bridges. In our example, the cracked length of the bridge is about 10 m around the intermediate support. Assuming a reduction of the stiffness along this zone, the bending moment distribution and the vertical displacements vary significantly. Some selected results related to both structural analyses, considering dead and live loads plus thermal loads, are presented in table 2.

TABLE 2
Results of structural analyses

Structural analysis	No cracking	Cracking
Bending moment at intermediate support (kN·m)	14715.0	8540.0
Maximum deflection (cm)	1.55	2.45

CONCLUSIONS

A method of analysis is presented to estimate realistic temperature distributions and thermal stress distributions in composite bridges. In addition, in order to take into account temperature effects in the design process of composite bridges, a discontinuous uniform temperature distribution is suggested. This will be useful for bridge designers to carry out the structural analysis under thermal loads, in a simple and realistic way. On the order hand, the magnitude of the thermal stresses (self–equilibrated and continuity stresses) are very significant when compared to stresses due to dead and live loads. Likewise, temperature effects in composite bridges induce tensile stresses in the concrete which can exceed the tensile strength and generate cracking. In order to ensure controlled cracking in the concrete slab, required minimum area of reinforcement must be provided. Lastly, the effects of cracking on the global stiffness of the deck should be considered in order to calculate, in an accurate way, the maximum deflection and the bending moment distribution of composite bridges in serviceability conditions.

REFERENCES

1. Berwanger, C., "Transient Thermal Behaviour of Composite Bridges". Journal of Structural Engineering ASCE. Vol.109, No.10, October, 1983, pp. 2325–2339.

2. Chan, M.Y.T., Cheung, M.S., Beauchamp, J.C. and Hachem, H. M., "Thermal Stresses in Composite Box–Girder Bridges". Third International Conference on Short and Medium Span Bridges. Ed. B. Bakht, R.A. Dorton and L.G. Jaeger. Toronto. August, 1990, pp. 355–366.

3. Dilger, W.H., Ghali, A., Chan, M., Cheung, M.S. and Maes, M.A., "Temperature Stresses in Composite Box Girder Bridges". Journal of Structural Engineering ASCE. Vol.109, No.6, June, 1983, pp. 1460–1478.

4. Fu, H.C., Ng, S.F. and Cheung, M.S., "Thermal Behaviour of Composite Bridges". Journal of Structural Engineering ASCE. Vol.116, No.12, December, 1990, pp. 3302–3323.

5. Mirambell, E. and Aguado, A., "Temperature and Stress Distributions in Concrete Box Girder Bridges". Journal of Structural Engineering ASCE. Vol.116, No.9, September, 1990, pp. 2388–2409

6. Soliman, M. and Kennedy, J.B. "Simplified Method for Estimating Thermal Stresses in Composite Bridges". Transportation Research Record. 1072.

A FINITE BEAM ELEMENT FOR LAYERED STRUCTURES AND ITS USE WHEN ANALYSING STEEL-CONCRETE COMPOSITE FLEXURAL MEMBERS

MATTI V. LESKELÄ
Research Council for the Technical Sciences,
Academy of Finland,
Geologintie 1, SF 90570 OULU, FINLAND

ABSTRACT

Plane finite elements with in-plane degrees of freedom have frequently been used to solve beam problems employing interconnected members, e.g. layered beams and composite beams. Plane beam elements can also be applied, however, when special types of finite elements for layered beams are developed. The number of degrees of freedom will be the same as in the basic element from which they are derived. The stiffness coefficients of the transformed beam elements are easily manipulated to meet the requirements of the internal connection structure of the system. The theory and structure of the stiffness matrix are briefly explained and its possible uses are discussed.

INTRODUCTION

Plane beam elements having three degrees of freedom at the nodes, i.e. rotation and two displacements, can also be employed to solve problems involving layered structures, e.g. composite structures and sandwich structures. The degrees of freedom are then transformed to allow the elements to be interconnected on top of others. The stiffness matrix is formulated by deriving the changes required using the matrix of an ordinary beam element. Timoshenko's beam theory can be applied to cases having considerable shear deformations, but for most applications within composite beams and slabs it is satisfactory to employ the technical flexural theory with no shear deformations allowed for.

A beam element suitable for layered structures is such that its nodes can be connected flexibly to those of another element by applying proper spring elements between the appropriate nodes. The three degrees of freedom at the end nodes of the ordinary element are therefore separated into three nodes, each having now only one degree of freedom, axial or transverse displacement (Figure 1.). The top and bottom nodes, $'ti'$ and $'bi'$ allow for displacements in the direction of the element axis and the middle node, situated on the centroidal axis, deflects transversely. Different configurations can be produced for the axial nodes, depending on the structural system for which the element is intended. The coefficients of the stiffness matrix are always derived according to the location of these nodes.

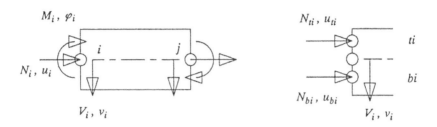

Figure 1. Principle of transforming an ordinary beam element by dividing the displacements among nodes having only one degree of freedom each.

DERIVATION OF STIFFNESS COEFFICIENTS

Stiffness coefficients can be constructed by introducing positive displacements one at a time at individual nodes and considering the nodal force components of the ordinary beam element required to produce one non-zero displacement. As an example, we will take a closer look at the displacement $u_{ti} > 0$ and introduce the forces required to impose the state of Figure 2. u_{ti} is generated using axial displacement $u_i = \xi_t u_{ti}$ and rotation $\varphi_i = u_{ti}/h$ in the ordinary element. These must be due to the forces $\{N_i \ V_i \ M_i \ N_j \ V_j \ M_j\}^T$, i.e.

$$N_i = (EA)u_i/L = -N_j,$$

$$V_i = 6(EI)\varphi_i/L^2,$$ (1)

$$M_i = 4(EI)\varphi_i/L \ ; \ M_j = 2(EI)\varphi_i/L \ .$$

The notations *(EA)* and *(EI)* are used for the axial and flexural stiffnesses of the element, respectively. The axial forces on the nodes *'ti'* and *'bi'* of the transformed element can be arranged so that they result in a total axial force N_i and a bending moment M_i. To produce N_i, the components N_{tn} and N_{bn} are required, while the moment M_i is composed of the force couple $N_{tm} = -N_{bm} = M_i/h$:

$$N_{tn} = \eta_b N_i/(\eta_t + \eta_b) = \xi_t N_i,$$

$$N_{bn} = \eta_t N_i/(\eta_t + \eta_b) = \xi_b N_i.$$ (2)

The resultant forces on the nodes of the 'i' end are then $N_{ti} = N_{tn} + N_{tm}$, and accordingly $N_{bi} = N_{bn} + N_{bm}$. The nodal vectors being arranged as $\{F_e\} = \{N_{ti}\ N_{bi}\ V_i\ N_{tj}\ N_{bj}\ V_j\}^T$ and $\{U_e\} = \{u_{ti}\ u_{bi}\ v_i\ u_{tj}\ u_{bj}\ v_j\}^T$, the stiffness matrix $[S_e]$ in the stiffness equation $\{F_e\} = [S_e]\{U_e\}$ for the element can be written in the form

$$[S_e] = \begin{bmatrix} [S_1] & [S_2] \\ [S_2]^T & [S_3] \end{bmatrix}$$

$$[S_1] = \begin{bmatrix} \xi_t^2 SCA + SCF1 & \xi_t\xi_b SCA - SCF1 & SCF3 \\ & \xi_b^2 SCA + SCF1 & -SCF3 \\ SYMM. & & SCF4 \end{bmatrix}$$

$$[S_2] = \begin{bmatrix} -\xi_t^2 SCA + SCF2 & -\xi_t\xi_b SCA - SCF2 & -SCF3 \\ -\xi_t\xi_b SCA - SCF2 & -\xi_b^2 SCA + SCF2 & SCF3 \\ SCF3 & -SCF3 & -SCF4 \end{bmatrix}$$

$$[S_3] = \begin{bmatrix} \xi_t^2 SCA + SCF1 & \xi_t\xi_b SCA - SCF1 & -SCF3 \\ & \xi_b^2 SCA + SCF1 & SCF3 \\ SYMM. & & SCF4 \end{bmatrix}$$

$$\xi_b = \eta_t/(\eta_t + \eta_b);\quad \xi_t = \eta_b/(\eta_t + \eta_b);\ h = (\eta_b + \eta_t)H$$

$$SCA = (EA)/L;\quad SCF1 = 4(EI)/(Lh^2);\quad SCF2 = 2(EI)/(Lh^2) \tag{3}$$

$$SCF3 = 6(EI)/(L^2 h);\quad SCF4 = 12(EI)/L^3.$$

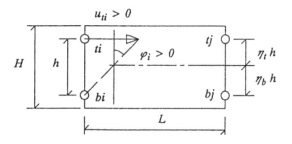

Figure 2. Correspondence between nodal displacement and rotation of the element end face

APPLICATION OF THE ELEMENT TO COMPOSITE BEAMS

The following set of parameters may be used for the flexural composite structures consisting of a concrete slab on top of a steel joist :

concrete elements	$\eta_t = 0$ and $\eta_b = 0.5$,
steel elements	$\eta_t = e_{sj}/h_s$ and $\eta_b = 0$,

e_{sj} being the distance from the connection interface to the centroid of the joist and h_s the depth of the joist. The selections presented imply that the nodes having longitudinal displacements are located on the centroidal axes and connection interface of the members.

The element layers are connected by introducing springs between the layers. The spring stiffnesses can be modelled relative to the shear-slip characteristics on the unit length of the connection. Where studs are concerned, the spring stiffnesses can be derived from the load-slip curves of a standard push test. When analysing beams having stud spacings 0.5 .. 1.0 times the depth of the steel member, the element lengths can be selected directly according to the spacing of the studs, and hence the spring stiffnesses match the shear stiffness of one row of connectors. For deeper beams, lumped spring stiffnesses can be arranged when element lengths greater than the stud spacing are employed.

Non-linearity effects can be included in the calculation by introducing reductions in the stiffnesses (EI) and (EA) whenever the softening stage in material behaviour has been entered. The steel stiffnesses should not be altered until yielding has started, but the overall non-linear material behaviour of the concrete suggests that the program should consider changes from the very beginning. The load-slip characteristics of the studs should also be considered to be non-linear if the true behaviour of the beam is to be traced.

An example is given by comparing the calculation with a test beam data described by Ansourian [1]. When the true material behaviour is followed, i.e. the strain hardening is included in the steel characteristics and the softening of the concrete is considered in addition to the load-slip curve obtained from a push test on the stud connectors, the measured load-deflection behaviour can be traced at least to 90 % of the observed failure load.

The shear force distribution of the connectors for three load levels presented in Figure 3. [2] is seen initially to follow approximately the complete interaction analysis (load level 100 kN). Due to the markedly high strength and stiffness of the shear connection, the beam is unable to develop the typical uniform plastic distribution of connector shear loads usually found in more flexible connections. At the highest load level presented, the maximum connector shears are slightly lower than 60 % of the resistance and are found around a plastic hinge on the section experiencing the concentrated load.

Although the example presents a case of full shear connection, all kinds of connection can be analysed. It is only a question of selecting the proper elements and connector characteristics (initial stiffness, strength, ductility).

DISCUSSION

The elements explained above have been found to behave well, due to their origin among two-noded beam elements. They are especially applicable to composite structures of steel and concrete when partial interaction problems are concerned. While beams designed according to the partial shear connection principle are quite common, they may include prominent slipping in the connection interface even in the service state, and thus the distribution of connection forces does not follow that given by the complete interaction analysis. Moreover, when true deflections are required, they cannot be estimated reliably unless the real behaviour of the connection is considered.

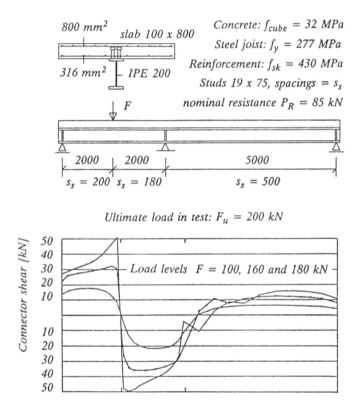

Calculated distribution of connector shear forces [2] in CTB1 [1]

Figure 3. An example of calculation results. Composite beam according to a test by Ansourian [1].

REFERENCES

1. Ansourian, P., Experiments on continuous composite beams. The Institution of Civil Engineers, Proceedings, Part 2, **73**, 25-51.

2. Leskelä, M., Calculation models for concrete-steel composite beams, considering partial interaction. University of Oulu, Department of Civil Engineering, Oulu 1986. Acta Univ. Oul. C 36. Artes Constr. 8.

LATEST DESIGN EQUATIONS, TESTING AND CRITERIA OF HEADED CONCRETE ANCHORS IN APPLICATIONS OF COMPOSITE CONSTRUCTION

M. SOMMERSTEIN P. ENG.
CHIEF ENGINEER & ENG'G MANAGER
VICWEST CORPORATE ENGINEERING
1296 SOUTH SERVICE ROAD WEST
OAKVILLE, ONTARIO
CANADA   L6L 5M7

ABSTRACT

An investigation and conclusion illustrating the basic strength mechanism of complex welded headed concrete studs when incorporated in close groups and trapezoidal profiled steel sheets.

INTRODUCTION

The efficiency and design strength of the shear connector in plain slab is easy to evaluate (see Fig. 1). However, the efficiency of the stud connector will be reduced substancially as a result of stud location in relation to the metal deck (see Fig. 2).

Full scale composite beam, push off, and push out tests at VicWest Test Lab and other test facilities have verified the pyramidal shear theory.

BASIC CONCRETE CONE THEORY

When the deck is placed with flutes perpendicular to beams, the ultimate strength of headed concrete anchors attached to structural steel shapes or plates and embedded in concrete is controlled by the following factors:

(a)   Boundary conditions
(b)   Beam flange or plate thickness and strength
(c)   Concrete shear strength
(d)   Shear friction
(e)   Centre to centre concrete anchor spacing
(f)   Concrete anchor spacing in relation to metal deck
(g)   Strength of concrete anchor
(h)   Full concrete shear cone
(i)   Partial concrete shear cone

A full concrete shear cone will develop only if:

(a)   Anchor itself is stronger than the encasing concrete
(b)   If beam flange or plate are stronger than the encasing concrete
(c)   If there is no break interference of metal deck profile
(d)   If adjacent anchor centre lines are a minimum distance of 2h−dh
      (see Fig. 3.)

PYRAMIDAL PULL OUT AREA METHOD OF DESIGN

When metal deck flutes are oriented parallel to steel beams, shear values of studs may be the full solid values for all cases when $W_r/t_d \geq 1.5$ (see Fig. 4). When the steel deck is placed with flutes perpendicular to the beams, the strength of the headed studs embedded within concrete ribs is a function of the pyramidal pull out area "Ap" governed by conditons (a) to (i) described in basic concrete cone theory.

Using limit state design with performance factor for shear connections of
  $\emptyset sc = .80$

The following equations are applicable:
1)   In solid slabs

$$q_{rs} = (.4)\,(A_{sc})\,\sqrt{f'_c\,E_c} \leq (.8)\,(A_{sc})\,(F_u) \leq (.8)\,(415)\,(A_{sc})$$

  (F_u) = tensile strength of stud 415 Mpa

2) In slabs with metal deck ribs parallel to beams

   (a) When $W_d/h_d \geq 1.5$
      $$q_{rr} = q_{rs}$$
   (b) When $W_d/h_d < 1.5$
      $$q_{rr} = \left[(.6)(W_r/t_d)(h/t_d-1) q_{rs}\right] \leq 1.0\, q_{rs}$$

3) In slabs with metal deck with ribs perpendicular to beams

   a) When $h_d = 75$ mm
      $$q_{rr} = (.28)(\rho)(A_p)\sqrt{f'_c} \leq q_{rs}$$
   (b) When $h_d = 38$ mm
      $$q_{rr} = (.488)(\rho)(A_p)\sqrt{f'_c} \leq q_{rs}$$

   Where:

   q_{rs} = Factored ultimate shear resistance of headed stud embedded in solid concrete (N)

   A_{sc} = Normal area of stud shear connector (mm^2)

   E_c = Modulus of elasticity of concrete (Mpa)

   $$E_c = (W_c^{1.5})(.043)(\sqrt{f'_c})$$

   W_c = Mass density of concrete (Kg/m^3)
   f'_c = specified concrete compressive strength at 28 days (Mpa)

   q_{rr} = Factored ultimate shear resistance of headed stud embedded in concrete inside a metal rib (N)

   ρ = 1.0 for normal concrete density (2150 to 2500 kg/m^3)

   ρ = .85 for semi−low concrete density (1850 to 2150 kg/m^3)

   A_p = Concrete pyramidal pull out area (mm^2) governed by factors as described

   W_r = Average rib width (mm)   see Fig. 4

   t_d = Metal deck rib height (mm)   see Fig. 4

   h = Actual headed stud height after welding (mm)   see Fig. 4

Often designers ignore checking the bending strength of the beam flange when subjected to tensile strength of the stud, this omission may lead to early failure of the beam flange and serious strength consequences. The method of checking beam flange thickness requirements is shown in Fig. 5.

Fig. 6, 7 and 8 show the method of determining pyramidal pull out areas Ap for a particular metal deck (HB 30V).

CONCLUSIONS

The ultimate resistance of the connection of the headed shear stud and the steel beam is less than the connection in solid slabs.

The failure mode can be predicted fairly accurately by using the pyramidal pull out area method of design as shown.

References:
   Can/CSA −S16.1−m89
   Composite beams   17.6 Interconnections

361

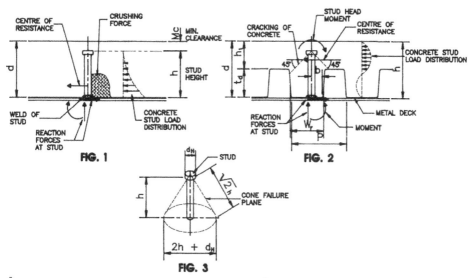

FIG. 1   **FIG. 2**

FIG. 3

 # Estimated Pyramidal Pull-out Areas, A_p (mm²)

Nominal Deck Profile			Stud Connectors				Pyramidal Pull-out Area		q_r (kN) per Connection		
t_d (mm)	p (mm)	W_r/t_d	No. per rib	length†	"s" (mm)	"x" (mm)	Area, A_p 10³ mm²	Sensitivity Ratio	Computed △	Current Rules ⊂	TEST ⊕ x $Ø_s$
75	300	≈ 2.0	1	115	/	*	41.6	1.03	52.1	67.3	58.5
			2	115	76	*	58.4	1.02	73.1	95.2	76.7
			1	115	/	65	35.2	1.01	44.1	N.A.	39.0
			1	115	/	38	29.2	1.02	36.6	N.A.	35.5
75	400	≈ 2.4	1	115	/	*	49.3	1.18	61.7	74.3	72.7
			2	115	76	*	64.0	1.17	80.1	105	85.5
			1	115	/	65	37.9	1.17	47.5	N.A.	42.1
			1	115	/	38	32.0	1.17	40.1	N.A.	38.4

For f'c = 20 MPa, density = 2300 kg/m³ . (Effect of stud placement tolerance assumed to be included)

* = Interior condition

† = Length after welding (burn off length ≈ 10mm is used for through-deck application)

△ = Based on minimum pull-out cone area.

⊂ = Current rules used by designers: S16.1-M84 and Grant-Fisher-Slutter formula (19mm dia. studs)

$Ø_s$ = .80

⊕ = Results of a total of 68 tests of pull out push variety.

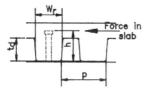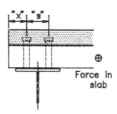

FIG. 4

362

$$t_{allow.} = \sqrt{\frac{6\,M}{f_b}}$$

Where:

$f = (.90)\,(F_y)$

$b = 150\ mm$

P_u = Utimate tensile force on stud (N)
F_y = Yield of steel beam (MPa)
x = Distance between studs (mm)
t = Beam flange thickness (mm)
k_1 = Distance from ℄ of beam to tangent of r

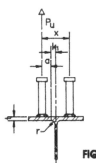

FIG. 5

$a = \dfrac{x}{2} - k_1$

$\left[\, M = (P_u)\,(a)\, \right]\ (N - mm)$

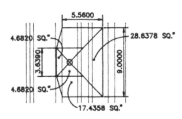

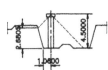

HB30V profile
1 = stud per rib
stud ht. = 4.5"
pull-out area
= 77.5554 sq."

FIG. 6

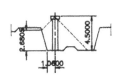

HB30V profile
2 = studs per rib
stud ht. = 4.5"
pull-out area
= 104.367 sq."

FIG. 7

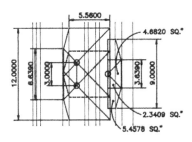

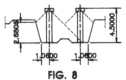

HB30V profile
3 = studs per rib
stud ht. = 4.5"
pull-out area
= 104.367 sq." (2-studs pull-out)
+ 20.2794 = 124.6464 sq."

FIG. 8

Dynamics and Fatigue

Design of Steel Beam-to-Box Column Connections for Seismic Load

Keh-Chyuan Tsai
Associate Professor of Civil Engineering
National Taiwan University
Taipei, Taiwan, Republic of China

ABSTRACT

In order to provide the needed stiffness for the beam-to-box column joint in moment resisting frames, column diaphragm plates are installed opposite the beam flanges. As a result, beam bending moments are transferred primarily through the beam flanges into the box column. In this paper, the flange strength requirements for the beam-to-column moment connections are critically reviewed. Experimental results confirmed that if the ultimate flexural capacity of the beam flanges alone is greater than the strain-hardened beam moment, the bolted web and welded flange beam-to-box column connection details performed very well and are well likely to be able to provide the ductility capacity needed to survive a severe earthquake. The paper concludes that the proposed beam flange strength requirements is suitable for the design of beam-to-column connections for moment resisting frames in regions of high seismic risk.

INTRODUCTION

Ductile moment resisting frames (MRFs) have been widely adopted as viable structural system for buildings in high seismic risk. In particular, steel MRFs using wide flange beams connected to box columns have gained wide acceptance in the constructions of typical three dimensional MRF and the corner columns in a framed tube system. As the panel zone in the box column joints is relatively strong, the beam-to-box column connections are required to develop the beam flexural capacity thereby dissipate seismic energy.

Past experimental research has extensively assessed the performance of beam-to-wide flange column connections (Popov et. al. 1985, Popov and Tsai 1989, Popov et. al 1989) and numerous field applications have provided a database for code provisions. Due to the cost benefits, the conventional bolted web and welded flange details commonly used for connecting the beam to wide flange column have been widely adopted in the construction of beam-to-box column connections. Nevertheless, research results on the behavior of beam-to-box column connections of realistic sizes are very limited (Linderman and Anderson 1990). Recent code provisions require that, in addition to the web bolting, supplementary welds are required to connect the beam web to the shear tab when the beam web moment capacity exceeds 30% of the entire beam flexural capacity (AISC 1990, UBC 1991). While this is one step forward toward providing a stronger connection, the design criteria for the supplementary welds remain to be developed and validated (Popov and Tsai 1989). Further more, the relatively flexible box column plate, adjacent to the shear tab and between the two column diaphragms, may not be able to transmit the beam web moment regardless whether the beam web is fully welded to the shear tab or not.

In this paper, results of analytical and experimental investigations (Tsai and Lin 1992, Tsai and Liu 1992) on the cyclic behavior of large steel beam-to-box column connections conducted recently at the National Taiwan University are discussed. From these studies, it is found that the beam web moment at the connection is primarily carried by the beam flanges due to the rigidity of the column diaphragm plates opposite the beam flanges. Accordingly, the beam flange flexural strength criterion for the construction of seismic beam-to-box column connection using conventional bolted web and welded flange details is proposed. Experimental test results confirmed that the proposed beam flange strength criterion is suitable for the design of beam-to-box column connections in severe seismic environment.

STRENGTH REQUIREMENTS

As noted in many experimental tests (Linderman and Anderson 1990, Popov et. al. 1985, Popov and Tsai 1989, Engelhardt 1991) for the severe service intended for beam-to-column connections, traditional bolted web and welded flange beam-to-column connections show inadequate ductility and are not reliable in providing the rotational capacity needed to survive a major earthquake. While the poor ductility is generally considered being attributed to the bolt slippage (Popov et. al. 1985, Popov and Tsai 1989) it is instructive to review the strength of this particular type of connection as it can provide further insights into the failure of connections.

Considering a conventional beam-to-column moment connection where the beam flanges are welded to the column flange using full penetration weld and the web is bolted to the shear tab to resist the design shear only, then the following strength criterion should be met in order to sustain the beam bending moment as inelastic rotation develops in the beam:

$$Z_F F_u \geq \alpha Z F_y \quad \text{or} \quad \frac{Z_F}{Z} \geq \alpha \frac{F_y}{F_u} \tag{1}$$

where Z_F and Z are plastic section moduli of the flanges and the entire beam section, respectively; F_y and F_u are the minimum yield strength and the ultimate tensile strength of the beam, respectively; α, depends on the magnitude of the beam inelastic rotation and represents the effect of strain hardening. Fig. 1 illustrates the relationships of the section moduli ratio, Z_F/Z, and the strength ratio, F_u/F_y, for various α in Eq. 1.

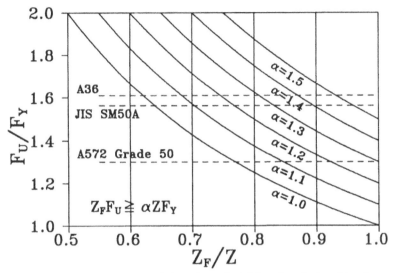

Figure 1 Sectional moduli ratio and yield ratio relationships for various strain hardening effects

An extensive parametric study on beam moment-rotation relationships has been conducted for cantilever beams using elastic-plastic-strain hardened trilinear constitutive model (Tsai and Liu 1992) for beams of various grade, length and sectional moduli ratio. It is found from the study that, depends on the beam length, grade and section moduli ratio, in general a bending moment of magnitude between $1.1ZF_y$ and $1.2ZF_y$ will be developed at the beam-column connection when a beam plastic rotational demand of 0.015 radian is reached.

Since a plastic rotational capacity in the order of 0.015 radian is generally required for a typical beam-column connection in MRFs with strong beam-column panel zones (Popov et. al. 1985), an α of 1.2 in Eq. 1 is therefore appropriate for evaluating the required strength for the beam-column connection. From Eq. 1 or Fig. 1, it is clear that for certain steel grades and beam sections where web flexural capacity is significant (i.e. small Z_F/Z ratio) and yield ratio (F_y/F_u) is relatively large, beam-to-column connection using the conventional bolted web and welded flange details may not be able to sustain the beam flexural demand. In such cases, i.e. $Z_F F_u < \alpha Z F_y$ beam web should be welded to the column flange or shear tab in a beam-to-wide flange column connection and the design strength of the welds should at least be $\alpha Z F_y - Z_F F_u$ from the flexural strength point of view. Additional bolting and web welds should be provided to develop the beam shear force.

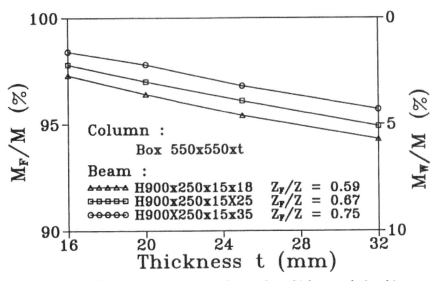

Figure 2 Beam flange moment versus column plate thickness relationships

However, for a typical beam-to-box column connection where column interior diaphragm plates are provided opposite the beam flanges, beam flanges carry most of the beam bending moment regardless whether the beam web is fully welded to the column or not. This is illustrated in Fig. 2 for three different cantilever beams connected to box columns using finite element analyses (Tsai and Lin 1992). The stiffness of the shear tabs were included in the finite element models while the beam webs were assumed to be fully attached to the column. From the ratios of the bending moment carried by beam flanges and the entire beam moment, M_F/M, shown in the figure, it can be found that beam moments are transferred primarily through the beam flanges into the column. Since the stiffness at the beam web-to-column region is not pronounced as demonstrated in these finite element analyses, it appears that even for box columns with relatively thick plates, the beam flange reinforcing plates may be required for the beam-to-box column connection in MRFs when the connection flexural strength requirement is violated. The stiffening requirement for the beam flange should be based on Eq. 1 in which Z_F includes the effects of the beam flange stiffeners. The beam web should be adequately bolted or welded in order to develop the beam ultimate shear.

Using the strength criterion Eq. 1 and the corresponding α of 1.2 for the design of test specimens, ductile behavior in recent tests of several beam-to-box column subassemblages discussed later in this paper have been achieved. The strength requirements for both the beam-to-wide flange column and beam-to-box column connections described in this section, not adequately addressed in the current model steel seismic building code (AISC 1990, UBC 1991), have been incorporated into the draft of the building code of Republic of China (CSSC 1991).

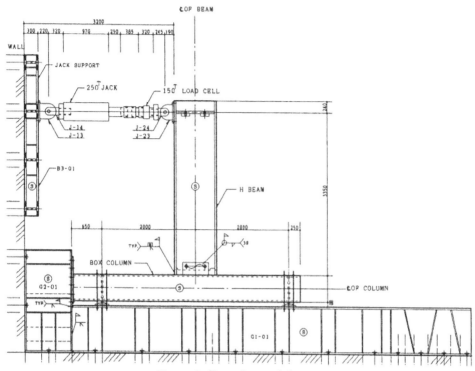

Figure 3   Experimental Set-up

EXPERIMENTAL PROGRAM

All specimens were fabricated as cantilevers attached to column stubs. As shown in the experimental set-up in Fig. 3, the column is place horizontally and the cantilever beam loads are applied laterally through a actuator mounted between the reaction wall and the beam end. The beam and column sizes used are given in Table 1, where the ratios of the plastic beam flange moduli Z_F to the respective plastic beam moduli Z are given.

All columns are built-up box sections. Columns in Specimens TB1 through TB6 were made from ASTM A572 Grade 50 steel while columns in Specimens SB2 through SB5 were made from JIS SM50A grade. All beams are built-up wide flange sections. Beams in Specimens TB1 through TB6 were made from ASTM A36 material while beams in Specimens SB2 through SB5 were made from JIS SM50A grade. In Table 1, values of $F_{Y F}$ and $F_{U F}$ indicate the actual tensile yield strength and ultimate strength of beam flanges, respectively, while values of $F_{Y W}$ and $F_{U W}$ give actual yield strength and ultimate strength of beam webs. Since the Z_F/Z ratios of the beam section for Specimens SB2 through SB5 are smaller than that required from Eq. 1 for a corresponding α of 1.2 (see Fig. 1), Specimens SB2 and SB4 were reinforced with cover plates at each exterior side of the beam flanges as shown in Fig. 4 while in Specimens SB3 and SB5, triangular wing plates were added to the side edges of the beam flanges. Ratios of Z_F/Z considering the effect of flange reinforcements are also given in parentheses in Table 1.

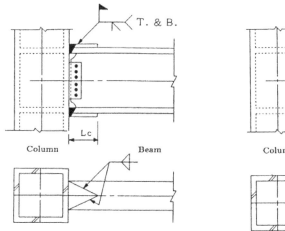
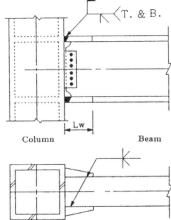

Improved Connection by Cover Plates   **Improved Connection by Wing Plates**

Figure 4   Stiffened beam-to-box column connection details

Table 1   Schedule of specimens

Specimen (1)	Column	Beam				Beam Strength (t/cm^2)			
	Size (mm) (2)	Size (mm) (3)	Length L_b (4)(mm)	Z_F/Z^1 (5)		F_{YF} (6)	F_{UF} (7)	F_{YW} (8)	F_{UW} (9)
TB1	□ 550 × 550 × 24 × 24	H690 × 240 × 14 × 24	2305	0.73		2.67	4.55	2.88	4.59
TB2	□ 550 × 550 × 24 × 24	H772 × 250 × 16 × 27	2305	0.71		2.76	4.89	2.93	4.80
TB3	Same as TB1	H690 × 320 × 14 × 24	2305	0.78		2.64	4.52	2.86	4.55
TB4	Same as TB2	H772 × 380 × 16 × 27	2305	0.79		2.85	4.85	2.93	4.80
TB5	□ 900 × 900 × 32 × 32	H700 × 350 × 15 × 38	3000	0.86		3.03	4.88	3.10	4.76
TB6	Same as TB5	H700 × 350 × 15 × 38	3000	0.86		3.03	4.91	3.10	4.76
SB2	□ 550 × 550 × 16 × 16	H900 × 250 × 15 × 18	3175	$0.59 (0.85)^2$		3.93	5.30	4.64	5.73
SB3	Same as SB2	H900 × 250 × 15 × 18	3175	$0.59 (0.82)^3$		3.90	5.32	4.72	5.78
SB4	□ 550 × 550 × 25 × 25	H900 × 250 × 15 × 25	3175	$0.67 (0.95)^2$		4.22	5.68	4.37	5.71
SB5	Same as SB4	H900 × 250 × 15 × 25	3175	$0.67 (0.94)^3$		4.07	5.69	4.53	5.70

Notes: 1. Values in parentheses give ratio of Z_F/Z including flange stiffeners.
      2. With cover plates.
      3. With wing plates.

All the beam-to-column connections were made in the structural testing laboratory of the Department of Civil Engineering of National Taiwan University by certified welders using bolted web and full penetration welded flange details. All web copes for the flange welds were ground smooth before welding. The beam flange welding for Specimens TB1 through TB4 were fabricated using flux-cored arc welding while the remainders were made using shielded metal arc welding. All full penetration flange welds were ultrasonically tested. All beam flange welds satisfy the specified requirements except the one in the beam bottom flange of Specimen SB2. The defect was removed and rewelded twice before passing the requirements. During the test of each specimen, increasing cyclic displacements were applied at the cantilever beam end.

EXPERIMENTAL RESULTS

The resulting cantilever beam load versus beam plastic rotation hysteresis diagrams for the specimens are illustrated in Fig. 5, in which a circle in the diagram indicates a fracture occurred in the heat affected zone near the flange weld whereas a square indicates a fracture occurred near the diaphragm electroslag weld. The Specimen SB1 (not shown here), consists of a wide flange built-up beam section same as the one used in Specimens SB2 and SB3 but without any flange stiffeners, failed abruptly at both flanges with no ductility. A summary of the test results for the specimens is given in Table 2.

It can be seen from Fig. 5 and Table 2 that except in the tests of Specimens SB3 and SB5 in which premature fractures of diaphragm weld occurred and except in Specimen TB4, all of the remaining beam-to-box column connections sustained significant cyclic plastic deformations before fracture occurred in the beam flange. In Specimens SB3 and SB5 where fractures of diaphragm weld occurred, lacks of fusion of the associated electroslag weldings were found after disassembling the beam-column joints. It appears that if these diaphragm weldings had been made properly, the beam-column connections in Specimens SB3 and SB5 might have well been able to sustain larger inelastic rotations. For Specimens SB2 through SB5 where flange stiffeners were installed, values of P_F as defined in Table 2 were computed assuming the stiffeners are fully effective and the stiffeners are of the same material as the associated beam flanges. As indicated in columns (6) and (7) of Table 2, it is evident that the effects of strain hardening on most of the beam-column connections are pronounced. Moreover, it can be found in columns (8) and (9) for each specimen that, except premature failures of diaphragm weld occurred in Specimen SB3 and SB5, the ratio of the maximum cantilever load, P_u, attained before the failure to the ultimate capacity, P_F, computed from the beam flange strength alone is very close to unity. This suggests that the ultimate flexural capacity of a beam-to-box column connection can be adequately predicted considering the flexural strength of the beam flanges only.

CONCLUSIONS

Based on these limited experimental tests, the following conclusions can be reached:

- Due to the in-plane stiffness of the column diaphragm plate opposite the beam

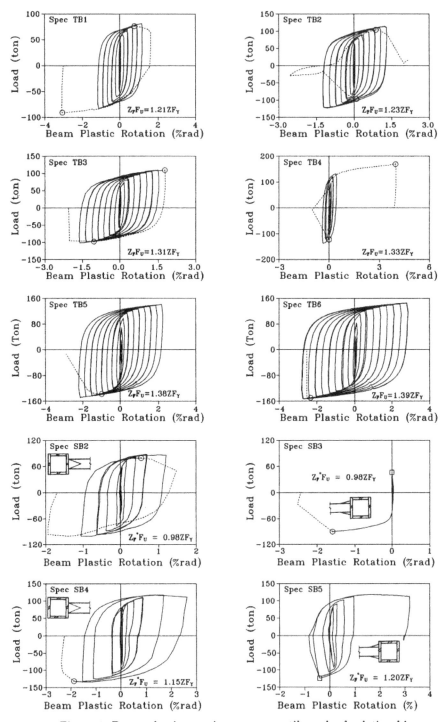

Figure 5 Beam plastic rotation versus cantilever load relationships

Table 2   Summary of experimental results

Specimen (1)	θ_{bp} (% rad)		P_u (ton)		P_u / P_p		P_u / P_F		α (10)
	+ (2)	− (3)	+ (4)	− (5)	+ (6)	− (7)	+ (8)	− (9)	
TB1	1.16	3.07	81.4	91.1	1.30	1.46	1.08	1.20	1.21
TB2	1.29	1.20	114.3	123.0	1.32	1.42	1.07	1.15	1.23
TB3	1.76	1.60	109.6	101.1	1.43	1.32	1.09	1.01	1.31
TB4	3.97	0.37	168.1	136.9	1.39	1.13	1.05	0.85	1.33
TB5	2.17	2.14	141.3	147.8	1.36	1.42	0.99	1.03	1.38
TB6	2.80	2.75	145.2	153.4	1.40	1.48	1.01	1.07	1.39
SB2	1.20	1.90	88.3	97.8	0.90	0.99	0.91	1.01	0.98
SB3	---	1.60	51.0	91.0	0.54	0.96	0.55	0.98	0.98
SB4	2.60	1.90	116.7	134.0	0.97	1.11	0.84	0.97	1.15
SB5	3.20	0.40	118.8	125.2	1.04	1.10	0.87	0.91	1.20

θ_{bp} : max. beam plastic rotation attained before failure

P_u : max. cantilever load attained before failure

"+" indicates tension in top flange

$P_p = (Z_F F_{YF} + Z_W F_{YW}) / L_b$

$P_F = (Z_F F_{UF}) / L_b$

$\alpha = (Z_F F_{UF}) / (Z_F F_{YF} + Z_W F_{YW})$

flanges, the beam bending moment is transferred primarily by the beam flanges into a beam-to-box column joint. It is confirmed that the variation of the bending moment carried by the beam flanges at the joint is not sensitive to the thickness of the column plate.

- If the ultimate flexural capacity of the beam flanges alone is greater than the strain-hardened beam moment, the conventional beam-to-box column connection details are very likely to provide the beam rotational capacity needed to survive a severe earthquake.

- For beam sections violate the proposed strength criterion, the strength and ductility of the beam-to-box column connections can be enhanced by properly detailed cover plates or stiffeners at the beam flanges.

- The proposed beam flange strength criterion with a corresponding strain hardening factor of 1.2, not adequately addressed in the current model seismic steel building codes, appears to be appropriate for the design of beam-to-column connections intended for severe seismic service.

- Since the column diaphragm plates are highly stressed under the beam moment, the quality of the electroslag welds is very essential to the success of a earthquake resistant beam-to-box column joint.

ACKNOWLEDGEMENTS

The authors gratefully acknowledge the supports of National Science Council of Republic of China (NSC80-0414-P002-15B) and China Steel Structures Company, Ltd. The valuable suggestions provided by Prof. Egor P. Popov of University of California at Berkeley are very much appreciated.

REFERENCES

1 AISC, 1990, "Specifications of Seismic Provisions for Structural Steel Buildings," American Institute of Steel Constructions.

2 CSSC, 1991, Chinese Society of Structural Engineering, "Recommended Ductility Requirements and Commentary for Building Structures", A Report to Building Research Institute, Ministry of Interior of Republic of China.

3 Engelhardt, M.D., 1991, Private Communications, Department of Civil Engineering, University of Texas at Austin.

4 Linderman, Robert R. and Anderson, James C., 1990, "Steel Beam to Box Column Connections", Proceedings of Fourth U.S. National Conference on Earthquake Engineering.

5 Popov, E. P., Amin, N. R., Louis, J. J. C. and Stephen, R. M., 1985 "Cyclic Behavior of Large Beam-Column Assemblies", *Earthquake Spectra*, 1, No.2.

6 Popov, E.P., and Tsai, K.C., 1989, "Performance of Large Seismic Steel Moment Connections Under Cyclic Loads," *Engineering Journal*, American Institute of Steel Construction, Chicago, Vol. 26, No. 2.

7 Tsai, K.C. and Lin, K.C., 1992, "Effects of Flange Stiffeners on Seismic Response of Beam-to-Box Column Connections", Report No. CEER/NTU R81-1, Center for Earthquake Engineering Research, National Taiwan University.

8 Tsai, K.C. and Liu, M.C., 1992, "A study of Ductility Requirements for Steel Beam-Column Connections", Center for Earthquake Engineering Research, National Taiwan University.

9 UBC, 1991, "Uniform Building Code", International Conference of Building Officials, Whittier. California.

SEMI-RIGID STEEL CONNECTIONS UNDER CYCLIC LOADS

CLAUDIO BERNUZZI, RICCARDO ZANDONINI, PAOLO ZANON
Department of Structural Mechanics and Design Automation
University of Trento, Mesiano - 38050 Trento, ITALY

ABSTRACT

The seismic efficiency of steel frames depends on the ability of dissipating energy by undergoing large plastic deformations without formation of brittle mechanisms. Use of semi-rigid joints can be envisaged as part of the dissipation system. This implies that an appropriate design philosophy be established and validated. To this aim a research project is currently carried on at the University of Trento.
As a vital prerequisite, prediction models for steel beam-to-column connections must be developed for the response under both monotonic and cyclic loading. This paper presents and discusses the main results of a first series of cyclic tests on semi-rigid connections.

INTRODUCTION

The ability to dissipate energy by undergoing large plastic deformations without formation of "brittle" mechanisms, i.e. the ductility, is a key parameter to assess the efficiency of steel frames in seismic zones. The ductility of the framework as a whole depends as well on the stability of the hysteretic behaviour of individual components.
Traditional design models concentrate energy dissipation in the bracing cantilever system of simple non sway frames, or in the beam-to-column joints of rigid sway frames. In the former case the bracing system does possess high stiffness but the cyclic response is not satisfactory in terms of ductility and of shape of the response cycles; in the latter the higher ductility of the nodes is combined with a quite low overall stiffness and high fabrication costs.
The semi-rigid design concept was recently developed with the aim to achieve an engineering optimisation by incorporating the actual joint response into the design analysis. Experimen-

tal analyses [1,2,3] and numerical studies [4] indicate that this philosophy may be applicable also in seismic zones: i.e. the cost-effectiveness and the seismic performance of steel frames can be improved if semi-rigid joints are part of the dissipation system.

The European Code for steel structures (Eurocode 3,[5]) includes semi-continuous frame models among design options. Basic concepts and requirements as well as general guidance and limited specific recommendations are provided. This Code clearly states that design must be based on a reliable assessment of the joint behaviour (including rotational capacity). The lack of data banks, wherefrom designers could readily obtain information, as well as of simple methods permitting approximation of the joint response, actually prevents these methods from being used in current practice. The research project in progress at the University of Trento aims at developing models for steel beam-to-column connections under both monotonic and cyclic loading suitable for use in design practice. The preliminary phase of the study focussed on the experimental analysis of the monotonic behaviour of several forms of connections commonly used in practice (extended as well as flush and header end plates, and cleated connections) subject to monotonic loading [6,7]. The key behavioural parameters were identified, and methods were proposed for determining the ultimate moment capacity. This paper reports on a second experimental phase, devoted to the cyclic response of the same forms of connection. The main results of the tests are presented, and the parameters defining the joint rotational cyclic performance are determined, including the hysteretic behaviour and the dissipation capability. Finally, an appraisal of the test data enables some considerations to be conducted on the cyclic test procedures, with reference also to their significance as to seismic design.

THE EXPERIMENTAL ANALYSIS

The first series of cyclic tests comprises six connections different in type and /or significant geometric parameters. As in the previous analysis under monotonic loading, the study is based on the general prediction philosophy by components. The attention is focussed here on the connection only. Therefore, specimens consist of a long beam stub (IPE 300 section), attached through the connection to be tested to a "rigid" counterbeam, as shown in figure 1a. The loads are applied to the free end of the specimen by means of a device that transfers horizontal forces only. Testing conditions approximate quite closely the case of beam-to-column joints with negligible column deformability.

The measuring system was set up so as to allow both the rotation of the connection as a whole and the contributions of the various components to be determined. If reference is made to figure 1b, displacement transducers (LVDTs') enabled determination of: (1) the rotation at a distance of 300mm from the beam end (LVDTs' A), (2) the total connection rotation (LVDTs' B), (3) the contribution of bolt deformation (LVDTs'

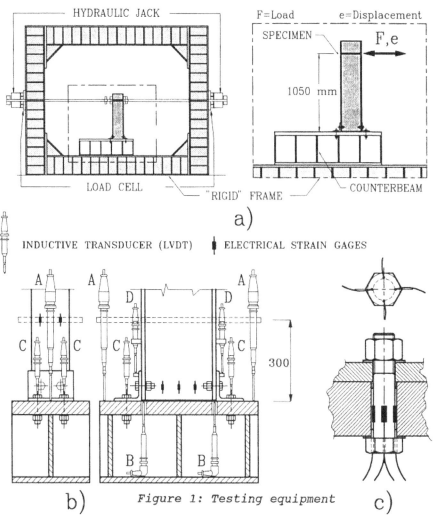

a)

INDUCTIVE TRANSDUCER (LVDT)   ELECTRICAL STRAIN GAGES

b)   *Figure 1: Testing equipment*   c)

C), and, when relevant, (4) the slip contribution (LVDTs' D); electrical strain gages were used in order to have an apprais- al of the strain state in the beam in the vicinity of the connection, as well as of the evolution of the extensional and flexural deformation of the bolts (fig. 1c).

The following forms of connections were investigated (fig. 2): (a) top and seat angles connection (1 test: TSC-1); (b) flush end plate connection (2 tests, differing for number and size of the bolts: FPC-1 and FPC-2); (c) end plate connection extended on both sides of the beam (2 tests differing for the plate thickness: EPBC-1 and EPBC-2); (d) end plate connection extended only on one side of the beam (1 test: EPC-1).

In the first four specimens the thickness of the connection elements (end plate or angles) was 12mm, whilst in specimens EPBC-2 and EPC-1 it was 18mm.

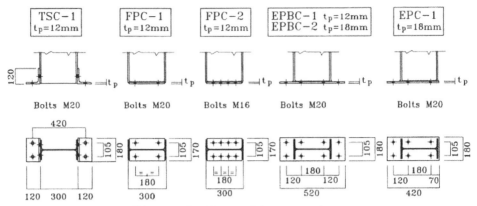

Figure 2: Connection specimens

All bolts were grade 8.8 bolts, preloaded to 40% of the
nominal yield strength; such a pretension level corresponds
approximately to the pretension induced by hand tightening up
to the snug tight condition. Beams were made of steel grade
FE 510; tension coupon tests indicated that material properties have a very low scatter, thus allowing an average value
to be adopted for the elastic and plastic moments of resistance, i.e. $M_e=243$kNm and $M_p=284$kNm respectively.
The testing procedure was based on the ECCS Recommendations
[8], assuming the deflection, e, at the beam stub free end as
the controlling displacement parameter. In the elastic stage,
the amplitude of the cycles was selected to be small enough to
ensure a satisfactory approximation in detecting the onset of
the inelastic phase (e_y). When this phase was entered, two
cycles were performed for each integer multiplier of the yield
displacement limit until the collapse was achieved. The quasi-
static test procedure required a very low rate of load
application.

THE CYCLIC RESPONSE OF THE JOINTS

The evaluation of the experimental data, correlated to the
main phenomena observed during the tests, permitted identification of the key behavioural facets as well as determination
of the parameters characterizing joint responses. The behaviour of each specimen is discussed here with reference to the
moment-rotation curves M-Φ: features specific to the different
joint forms are singled out for both the overall and the
component responses. Furthermore, the problems of the deformation capacity and of the failure mode are addressed and
related. An assessment of the joint response with reference to
seismic design will be presented in the next section.
Figure 3 shows some of the M - Φ hysteresis loops for the top
and seat angle connection (TSC-1). The contribution of the
slippage of the angles relative to the beam is substantial; in
particular this is the major contribution for cycles associated to low and intermediate partial ductility ratios (up to
$e/e_y=6$). When the imposed deformation increases further bending

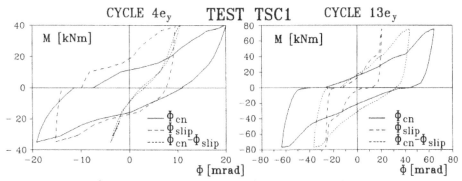

Figure 3: Moment-rotation hysteresis loops

of the angles tends to become the most important factor. The amplitude of the "slip loops" is in fact related, and limited, by local ovalization of the holes. First slip was observed at connection moments close to 20kNm (i.e. at about 45% in average of the elastic limit moments of the connection). The onset of slippage tends then to decrease in the subsequent cycles. The significant pinching of the global response is also associated to the stiffness deterioration of the "angle component". The permanent deformation of the angles affects noticeably the behaviour of these elements: e.g. the residual bent of the angle legs beyond the bolt row induces prying forces to develop at increasing rotation levels. Collapse occurred by fracture of the column side leg of the angle, which was first stressed beyond the elastic limit. A partial ductility ratio $e/e_y = 13$ was attained.

The two flush end plate connections tested differ for number and diameter of bolts (4 M20 in specimen FPC-1, 8 M16 in the FPC-2 one). The consequent difference in the ratio between the end plate bending stiffness and the axial stiffness of the bolts (higher in FPC-2) appears to be an important factor affecting connection behaviour. Connection FPC-2 (Fig. 4) showed higher initial stiffness and resistance capacity. Bolts were, however, more strained and more significantly elongated entering the plastic range of behaviour. As a result, the bolts closer to the beam flange lost contact with the plate while unloading and, which is more influential, in the first phase of the reloading. Hysteresis loops exhibit hence a greater pinching and a more significant stiffness deteriora-tion than for connection FPC-1 (Fig. 5), in which axial deformation of the bolts has been lower. On the other hand, being the plastic deformation of FPC-1 basically concentrated in the end plate and the failure mode associated in both connections to the weld fracture in correspondence of the internal part of the beam flange, this connection showed a lower partial ductility ratio. However, this parameter remains as high as 14 (it was 16 for FPC-2). The above mentioned difference in the main behavioural features affects also the value of the bending moment achieved in the two subsequent cycles performed at the same displacement: a non negligible decrease of moment capacity was observed for specimen FPC-2, whilst for connection FPC-1 the resistance drop was modest.

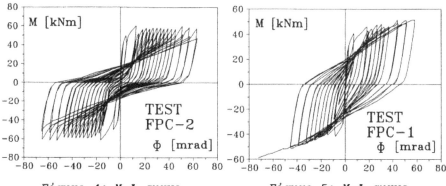

Figure 4: M-Φ curve   Figure 5: M-Φ curve

Similar considerations can be drawn for connections EPB with
end plates extended on both sides of the beam (Figs. 6 and 7).
The plate extension ensures a noticeable increase in stiffness
and strength of connection EPBC-1 with respect to the corresponding flush end plate connection (test FPC-1). The key
features of the behaviour are nonetheless the same, with the
plate contributing the most to the response in both the
elastic and the inelastic range. Connection EPBC-2, with the
end plate thickness increased to 18mm, showed a substantially
different behaviour. This was remarkably affected by bolt
inelastic elongation due to the higher bolt forces consequent
to the increase of the plate to bolt stiffness and strength
ratios. When the most stressed bolts (the ones   internal   to

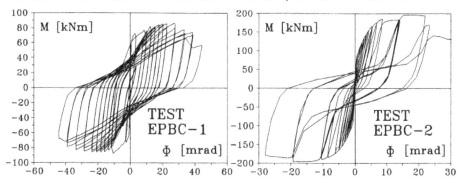

Figure 6: M-Φ curve   Figure 7: M-Φ curve

the beam) enter the inelastic range in tension (in the cycle
with amplitude $2e_y$), because of the residual deformation there
is no more contact between these bolts and the plate in the
reloading branch of the next cycle. The hysteresis loops
present, hence, a noticeable pinch that becomes more and more
significant as the loop amplitude increases. The failure mode
and the rotational capacity vary as well: (1) during the
cycles at $18\phi_y$, a crack started in specimen EPBC-1 at the weld
in the plate extension. This crack further expanded in the
following cycles until collapse was achieved by fracturing of

the plate over its full width. Connection EPBC-2 failed by
bolt rupture. (2) It is not surprising that connection EPBC-2
showed a lower rotation capacity, being this typical of
failure modes involving bolt fracture. Specimen EPBC-2,
however, had significantly higher stiffness and ultimate
moment capacity.

Connection EPC-1 with the end
plate extended only on one
side showed a response very
similar to that of connection
EPBC-2 (Fig. 8). Collapse was
due, also in this case, to
fracture of the internal
bolts in the cycle performed
at $4e_y$. The test was then continued without inverting the
load direction, in order to
assess more completely the
cyclic performance of the
connection. The collapse was
again associated to bolt fracturing during the cycle at
$6e_y$. The lack of symmetry of

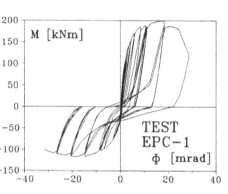

Figure 8: M-Φ curve

the behaviour clearly reflects the asymmetry of the connection
geometry.

EVALUATION OF THE RESULTS

The key features of the experimental response illustrated in
the previous section generally indicate that the connections
have a satisfactory cyclic behaviour. Yet the significant
differences observed in the tests stress the importance of an
appropriate design selection of the strength and stiffness of
the end plate relative to the bolt group, in order to reduce
the pinching and improve the stability of the hysteresis
loops.
Table 1 reports the main parameters characterizing the
connection performance in terms of initial stiffness, K_e,
ultimate moment capacity, M_{max}, and rotational ductility, Φ_{max};
the values of the parameters considered were obtained from the
envelopes of the M-ϕ curves in the positive and negative range
of the applied moments.

TABLE 1

TEST	K_e^+	M_{max}^+	$\dfrac{M_{max}^+}{M_{p,b}^+}$	Φ_{max}^+	$\dfrac{\Phi_{max}^+}{\Phi_y^+}$	K_e^-	M_{max}^-	$\dfrac{M_{max}^-}{M_{p,b}^-}$	Φ_{max}^-	$\dfrac{\Phi_{max}^-}{\Phi_y^-}$
TSC-1	11.0	75.5	0.26	67.0	14.0	16.0	74.3	0.27	64.1	25.8
FPC-1	9.7	51.5	0.18	58.0	12.9	14.4	63.0	0.22	75.6	30.3
FPC-2	30.0	60.3	0.21	63.2	31.6	30.0	61.2	0.22	65.3	32.6
EPBC-1	78.0	86.0	0.34	22.5	22.5	97.5	86.2	0.31	21.0	26.2
EPBC-2	35.4	196.3	0.69	18.6	3.9	37.8	195.8	0.68	19.3	4.3
EPC-1	40.5	199.4	0.70	16.5	3.9	77.3	116.3	0.40	17.6	16.0

K_e is expressed in kNm/mrad, M_{max} in kNm and Φ_{max} in mrad.

Figure 9 compares the M-ϕ envelopes and the Eurocode boundaries for joint classification [5] for a beam span equal to 6
meters. Extended end plate connections with rather thick
plates (EPC-1 and EPBC-2) lie in the upper part of the semi-
rigid zone, whilst all the other connections are close to the
lower boundary of this region. Connection EPBC-1 shows a
fairly stiff elastic response
associated to a quite low
ultimate resistance. All connections possess a significant rotation capacity (in
terms of both maximum rotation and partial ductility
ratio), which seems satisfactory even for the high demand
typical of seismic design. It
is interesting to note that a
comparison with the response
of nominally identical connections tested under
monotonic loading indicated
that cycling affects remarkably the moment capacity only
for the connections where
plate deformation is the dom-

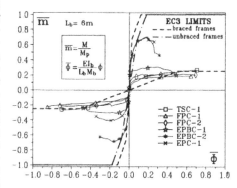

*Figure 9: M-Φ envelopes in
Eurocode 3 domains*

inant factor whilst bolt deformation gives a negligible
contribution to the overall connection rotation. The increasing residual plate deformation reduced the stiffness of the
plate at each increase of the cycle amplitude. This is clear
in figure 10, where the EPBC-1 M-ϕ hemicycles are shifted in
order to make all of them to start at the axes origin. Higher
moments could have presumably
been attained, had the applied force been used as the
controlling parameter, instead of the free end displacement: i.e. the testing
procedure affects the
"measured" performance of a
joint. No noticeable reduction in the rotation capacity
was observed with respect to
the monotonic tests.
The cumulated energy is
plotted in figure 11 versus
the connection maximum rotation attained in the positive
hemicycles. The extended end
plate connections, in particular connection EPBC-1, present the highest energy ab-

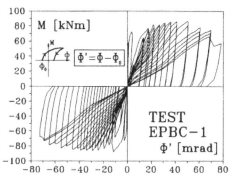

*Figure 10: M-Φ curve without
residual plate deformations*

sorption capacity. The other forms of connection show comparable dissipation capabilities, although the flange cleated
connection curve is remarkably higher in the range of rotations up to 20 mrad, which is of great practical interest.
The more flexible connections tested dissipate energy mainly
in the range of rather high rotational deformations (ϕ > 20

CUMULATED ABSORBED ENERGY [kJ]

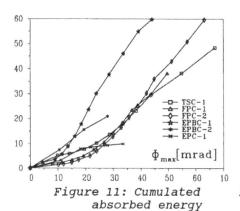

Figure 11: Cumulated
absorbed energy

CUMULATED ABSORBED ENERGY CONTRIBUTION:
END PLATE (ANGLES)/TOTAL

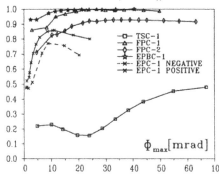

Figure 12: Energy contributions

mrad), which is usually of lower importance in seismic design. The following figure (Fig. 12) intends to provide an appraisal of the different contributions to the cumulated absorbed energy. It should be noted that: (1) the cleated connection TSC-1 mainly dissipates through the slip mechanism (angle and bolt deformation contributed less than 50%); (2) the energy absorbed by bolt deformation may be practically neglected in connections FPC-1 and EPBC-1 with higher ratio of bolt diameter to end plate thickness, whilst it increases for connection FPC-2 and becomes significant (of about 20% and 30% for positive and negative hemicycles respectively) for specimen EPC-1.

CONCLUDING REMARKS

The main points of an experimental investigation of the cyclic behaviour of semi-rigid connections were shortly presented in this paper. A first appraisal of the observed performance and the evaluation of the test data indicate that:
(1) The cyclic response of semi-rigid connections is quite satisfactory in terms of stiffness, strength and rotational ductility. A wide range of responses can be obtained by varying the form of the connection and the end plate (cleat) stiffness and strength relative to the bolts. If inelasticity is concentrated in the end plate (cleat), the hysteresis loops show a reduced pinching and a higher deformation capacity. The performance of cleated connections is substantially affected by slippage, which should be accounted for and possibly controlled at the design stage.
(2) When bolt contribution to connection rotation can be neglected, simple response models can be used in order to approximate the M-φ cyclic curve, which basically must recognize the stiffness deterioration (see fig. 10).
(3) The energy dissipation capacity seems adequate, although it becomes significant mainly for high rotations, with associated frame sway outside the range of interest in seismic design. It should hence be concluded that on the one hand semi-rigid connections confirm their suitability to be used as

a part of a system which combines different sources of energy dissipation in order to "optimize" the seismic performance; on the other hand, connection design criteria should be developed to appropriately balance stiffness and ductility under cyclic loading (the connections tested were designed in accordance to static design criteria).

At present, the research is tackling the problem of the influence of the testing procedure on the experimental assessment of the cyclic behaviour of joints. Besides, the attention is focussed both on the development of a suitable model to approximate the hysteretic response, and on the definition of connection response requirements to be met in seismic design of braced frames.

ACKNOWLEDGEMENTS

This research was supported by a grant of the Italian Ministry of the University and Scientific and Technological Research (M.U.R.S.T.). The authors greatly appreciate the skillful work of the technical staff of the laboratory of the Department of Structural Mechanics and Design Automation, and express their thanks to Mr. Paolo Rosa for his assistance in the evaluation of experimental data.

REFERENCES

1. Ballio G., Calado L., De Martino A., Faella C., Mazzolani F.M., "Cyclic behaviour of steel beam-to-column joints: experimental research", Costruzioni Metalliche (in English), No. 2, 1987, pp 160-181.
2. Popov E.P., Amin N.A., Louie C., Stephem M., "Cyclic Behaviour of Large Beam-to-Column Assemblies", Engineering Journal, American Institute of Steel Construction, First Quarter, 1986, pp 9-23.
3. Korol R.M., Ghobarah A., Osman A., "Extended End-Plate Connections Under Cyclic Loading: Behaviour and Design", Journal of Constructional Steel Research, Vol. 16, 1990, pp 253-280.
4. De Martino A., Faella C., Mazzolani F.M., "Simulation of beam-to-column behaviour under cyclic loads", Costruzioni Metalliche (in English), No. 6, 1984, pp 346-356.
5. Commission of the European Communities, "Eurocode 3 - Design of Steel Structures - Part 1: General Rules EC3 and Rules for Buildings", June 1992.
6. C. Bernuzzi, R. Zandonini, P. Zanon, "Rotational behaviour of end plate connections", Costruzioni Metalliche (in English), No. 2, 1991, pp 74-103.
7. C. Bernuzzi, R. Zandonini, "Rotational behaviour of flexible steel beam-to-column joints: experimental analysis and methods of prediction", Intenational Conference on Steel Constructions (in Italian), Trieste, October 1987.
8. European Convention for Constructional Steelwork, "Recommended Testing Procedures for Assessing the BEhaviour of Structural Elements under Cyclic Loads", Technical Committee 1, TWG 1.3 - Seismic Design, Publ. No. 45, 1986.

SIMPLE EXPRESSIONS FOR PREDICTING FUNDAMENTAL NATURAL PERIODS OF HIGH-RISE BUILDINGS

MASAYOSHI NAKASHIMA
Disaster Prevention Research Institute, Kyoto University,
Gokasho, Uji, Kyoto, 611 JAPAN

HISATAKA YANAGI
Graduate School of Faculty of Engineering, Kobe University,
Rokkodai, Nada, Kobe, 657 JAPAN

&

JUNKO HOSOTSUJI
Graduate School of Faculty of Engineering, Kobe University,
Rokkodai, Nada, Kobe, 657 JAPAN

ABSTRACT

This paper presents an investigation into the fundamental natural period of high-rise steel buildings. Based on a survey for high-rise steel buildings designed previously in Japan, it is found that their fundamental natural periods (T_1) are given approximately by $T_1=0.026L$, with T_1 in sec and L as the building height in meter. Simple procedures to estimate the fundamental natural period of high-rise steel buildings are formulated using the Ritz method. Using the procedures, it is found that the fundamental natural period estimated by $T_1=0.026L$ is the period needed to ensure 1/200 in story drift angle under the base shear coefficient given by $1/(3T_1)$.

INTRODUCTION

In designing a high-rise building in a region with high seismicity, its fundamental natural period is a very important design variable since the design earthquake force is given as a function of its fundamental natural period. The period can be computed accurately by eigenvalue analysis if all data on the geometrical, material, and weight properties of the building are given. We should remember, however, that those properties vary in accordance with the design earthquake force, which in turn is a function of the fundamental natural period.

In Japan, any building over 60 m in height need be designed specially, and the design should be reviewed and authorized by a panel consisting of experts. For evaluating the seismic performance of such a building, the panel usually requires two phase of analysis. In the first phase, it should be verified that the building designed would remain elastic, with all stresses smaller than the short-term allowable stresses, and its story drift angles less than 1/200, under several representative ground input motion records whose maximum velocity is normalized to 200 mm/sec. In the second phase, the building should be checked against the same records, but with the maximum velocity enlarged to 400 to 500 mm/sec, and the building should neither collapse, nor its maximum story drift angle should exceed 1/100. In both phases, time history analysis, elastic for the first phase and inelastic for the second phase, is employed. Therefore, as long as the seismic performance of the building is verified in this way, the design earthquake forces used for proportioning and sizing the building do not necessarily follow the forces stipulated in the Japanese seismic design code. As the time history analysis can be made only after all material, geometrical and weight properties of the building are provided, in Japanese design of a high-rise building, it is particularly crucial to predict the fundamental natural period of the building in the preliminary design stage, in which only limited information on associated structural properties is available.

Considering the importance of predicting the fundamental natural period of high-rise buildings only with limited information, the study presented herein examines how to predict it for high-rise steel buildings to be designed in Japan. To this end, first, characteristics of the fundamental natural periods that had been used for designing Japanese high-rise steel buildings are surveyed. Next, simple formulas, using the Ritz method, are derived for predicting the fundamental natural period of high-rise buildings, and, in the final part, physical background on the fundamental natural periods used for the design of Japanese high-rise steel buildings is investigated in reference to the formulas derived. Although the study focuses on Japanese high-rise steel buildings, the writers believe that the discussion in the final part can provide a guideline with respect to the preliminary prediction of the fundamental natural period of high-rise buildings designed in other countries and regions.

SURVEY OF FUNDAMENTAL NATURAL PERIODS OF JAPANESE HIGH-RISE STEEL BUILDINGS

Design fundamental natural periods of high-rise steel buildings designed in Japan in the period of 1968 to 1988 were investigated in reference to their design summary sheets, which are documented in "Building Letters" published by the Building Center of Japan. A total of 239 high-rise steel buildings were selected from the sheets, and their fundamental natural periods were stored in a data base. Here, all of the periods were obtained by eigenvalue analysis, in which the building was modelled as a discrete spring-mass system, with the weight per story lumped as one discrete mass and the story stiffness represented by one elastic spring. Figure 1 shows the relationship between the design fundamental natural period (T_1) in sec and the height of the building in meter, denoting that the majority of the data plotted are enclosed between the two straight lines: $T_1 = 0.02L$ and $T_1 = 0.03L$. The linear regression line of the data was found as $T_1 = 0.026L$, with the coefficient of variation as 0.18. The data were classified in accordance with the location, structural type, and usage, whose statistics are listed in Table 1. Observations from this table are as follows. (1) Buildings constructed in the Kanto region (Tokyo and its vicinity) are slightly shorter in the averaged fundamental natural period than those constructed in the Kansai region (Osaka and its vicinity). This has something to do with the difference in seismicity between the two areas. Often, buildings in the Kanto region are designed with 500 mm/sec in the maximum velocity for the second phase of analysis, whereas those in the Kansai region with 400 mm/sec in the maximum velocity. (2) The averaged fundamental natural period of braced frames, given by $T_1 = 0.025L$, is about 10 % shorter than that of unbraced frames: $T_1 = 0.028L$. (3) The averaged fundamental natural period of apartment buildings: $T_1 = 0.031L$, is about 20 % shorter than that of office buildings: $T_1 = 0.026L$. (4) The averaged fundamental natural periods of the two principal axes of the building are nearly identical, which has been found true regardless of the width-to-depth ratio of the plan.

All of the fundamental natural periods discussed above were those obtained by eigenvalue analysis, so they are not necessarily the periods that the buildings possess in their service condition. There have been studies looking into the correlation between the design and actual (obtained by vibration tests) natural periods [1-4], which indicate that the actual natural period is

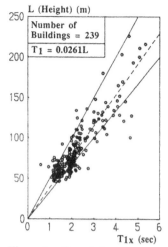

Figure 1. Correlation Between Fundamental Natural Period and Building Height

Table 1. Statistics on Correlation Between Fundamental Natural Period and Building Height in High-Rise Steel Buildings

Classification		Number	γ	[T (sec) = γL(m)]	
			Average	Standard Deviation	Coef. of Variation
Total	X-Direction	239	0.0261	0.00477	0.183
Total	Y-Direction	239	0.0267	0.00493	0.185
Location	Tohoku	18	0.0269	0.00771	0.295
	Kanto	142	0.0257	0.00392	0.153
	Tokai	19	0.0253	0.00483	0.191
	Kansai	60	0.0272	0.00535	0.197
Type of Structure	Braced (steel)	121	0.0250	0.00415	0.166
	Unbraced	89	0.0280	0.00506	0.181
	Braced (RC)	29	0.0253	0.00488	0.193
Usage	Office	157	0.0262	0.00383	0.146
	Hotel	41	0.0265	0.00466	0.176
	Apartment	11	0.0308	0.00655	0.213
	Tower	7	0.0235	0.01033	0.440

usually longer than the design natural period (by about 25 % on the average according to [1]). Further, Reference [2] notes that, not in a few cases, the difference between the two periods is conspicuous and that the most reliable prediction is to use the one given by regression analysis of experimental natural periods. The major objective of the study presented herein is to examine how the design natural period can be predicted in the preliminary design stage, and emphasis is given onto the correlation between the period provided by simple prediction and the period given by more rigorous eigenvalue analysis. Although the correlation between the design and actual natural periods is also important for evaluating the safety of buildings against earthquakes, this subject is beyond the scope of this paper.

PREDICTION OF FUNDAMENTAL NATURAL PERIOD

To develop a simple expression for predicting the fundamental natural period of a building, an analysis was carried out using the Ritz method. In this analysis, a building was simplified as a cantilevered shear beam. In applying the Ritz method, the vibrational mode (profile) was assumed to be the deflected shape of the shear beam, with the weight of the beam assumed as the distributed shear force. Using the vibrational mode thus computed, the maximum strain and kinematic energies were equated, and its natural period was obtained [5].

Natural Period of Building Having Uniform Mass and Stiffness
To begin with, considered was the simplest case, in which both the mass and stiffness were assumed to distribute uniformly along the height (L) as shown in Figure 2. Applying the Ritz method leads us to the following expression for its natural period (T_r):

$$T_r = 3.97\sqrt{\frac{m}{G}}L \qquad (1)$$

In which m and G are the mass per unit length in ton sec^2/m^2 and the stiffness per unit length in ton/m. For information, the exact period is given by:

$$T = 4\sqrt{\frac{m}{G}}L \qquad (2)$$

When an actual building is simplified to a shear beam, its weight, normally more concentrated on each floor level, need be smeared throughout the height. To evaluate the accuracy of such smearing, the period given by Equation 1 was compared with the period obtained from eigenvalue analysis. Figure 3 shows the relationship between the number of stories (the abscissa) and the ratio of the period given by Equation 1 (T_r) to the exact period given by eigenvalue analysis (T). To plot the data, considered was a building whose story height was 4 m, weight per story 500 ton, and story stiffness 70 ton/mm for all stories. Further, eigenvalue analysis was made by representing the building as a discrete spring-mass system. This figure shows that the ratio: T_r/T, approaches unity with the increase in the number of stories and that an accuracy of over 95% can be achieved if the number of stories exceeds 12. Similar plots were obtained for buildings having different mass and stiffness properties, and the same results were obtained.

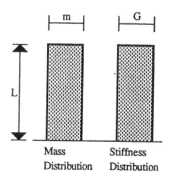

Figure 2. Shear Beam Representation
(Uniform Mass and Stiffness)

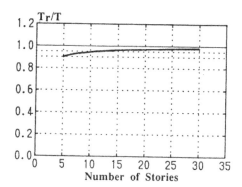

Figure 3. Accuracy of Natural Period
Predicted by Ritz Method   (Uniform
Mass and Stiffness)

Natural Period of Building With Linear Stiffness Distribution

In actual buildings, their stiffness distribution along the height (story) is hardly uniform, but rather decreases for upper stories. To consider the fundamental natural period of such buildings, considered was a shear beam, in which the mass was uniform throughout the height but the stiffness decreases linearly from the bottom to the top (Figure 4). Applying again the Ritz method for such a shear beam gives us the following natural period (T_k):

$$T_k = 2\pi\sqrt{\frac{mA}{\theta^3 B}} \qquad (3)$$

$$A = -G_N^3(\log R)^2 + (G_1^3 - 5G_1^2\theta L + 7G_1\theta^2 L^2 - \theta^3 L^3)\log R + (17/6)\theta^3 L^3 - (7/2)G_1\theta^2 L^2 + G_1^2\theta L$$

$$B = (3/2)\theta L^2 - G_1 L - (G_N^2)/\theta)\log R$$

In which, G_1, G_N, θ, and R are the stiffness per unit length at the bottom of the beam, the stiffness per unit length at the top of the beam, the angle of inclination in the stiffness distribution, and the ratio of the stiffness at the top to that at the bottom, respectively. Analysis was made for a building in which its story height and story weight were 4 m and 500 ton for all stories, and its story stiffness (k) was assumed as $70 - 2(i-1)$ (ton/mm), with

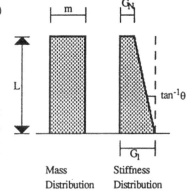

Figure 4. Shear Beam Representation
(Uniform Mass and Linear Stiffness)

i as the number of stories counted from the lowest story. The period of the building, estimated by employing Equation 4, was compared with the exact period obtained from eigenvalue analysis, and it was found out that the period thus estimated has an accuracy of over 95 % if the number of stories is greater than 12.

Natural Period of Building With Additional Concentrated Mass

In actual buildings, their mass distribution is not uniform, either. To account for this effect of the natural period, considered was a shear beam shown in Figure 5, in which both the mass (m) and stiffness (G) are uniformly distributed, and a concentrated mass (m') is added at a height of αL, measured from the bottom of the beam. The Ritz method provides us with the following expression for the natural period of this beam:

$$T_m = 2\pi\sqrt{\frac{C}{GD}} \tag{4}$$

$$C = m'm^2(\alpha L)^4 - (5m'^2m + 4m'm^2L)(\alpha L)^3 + 3(m'^3 + 3m'^2mL + m'm^2L^2)(\alpha L)^2 + 2m'm^2L^3(\alpha L) + (2/5)m^3L^5$$

$$D = -3m'm(\alpha L)^2 + 3(m'^2 + 2m'mL)(\alpha L) + m^2L^3$$

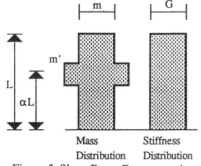

Mass Distribution — Stiffness Distribution

Figure 5. Shear Beam Representation (Uniform Mass and Stiffness, with Additional Concentrated Mass)

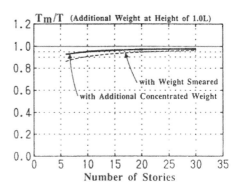

(a) Concentrated Mass at Top

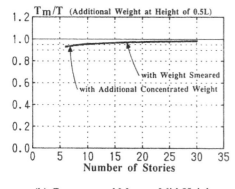

(b) Concentrated Mass at Mid-Height

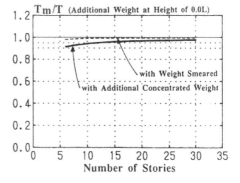

(c) Concentrated Mass at Bottom

Figure 6. Accuracy of Natural Period Predicted by Ritz Method (Uniform Mass and Stiffness, with Additional Concentrated Mass)

As an example, analyzed was a building having the story stiffness of 70 ton/mm, story weight of 500 ton, and an additional concentrated weight of 500 ton at a height of αL. The natural period obtained from Equation 4 (T_m) was compared with the exact natural period (T) in Figures 6(a), (b), and (c). In these figures, the solid line shows the ratio of T_m to T, and the broken line shows the natural period, in which the concentrated weight was smeared along the height of the beam and Equation 1 (a formula for the beam with uniform mass and stiffness distribution) was employed, relative to the exact natural period (T). Further, Figures 6(a), (b), and (c) respectively show the results when the concentrated weight was added at the top of the beam (α=1.0), at the middle of the beam (α=0.5), and at the bottom of the beam (actually, at the height of the lowest story) (α=0.0). Those figures indicate that, if the concentrated weight is added at the top of the beam, Equation 4 gives an accuracy of over 95 % for buildings with more than 10 stories, whereas, the same accuracy cannot be achieved until the number of stories reaches 22 if Equation 1 is employed with the concentrated mass smeared (Figure 6(a)). When the concentrated mass is added at the middle of the height, Equation 4 provides an accuracy of over 95 % for buildings with more than 10 stories, and Equation 1 with the smearing treatment also gives the same level of accuracy for buildings with more than 12 stories.(Figure 6(b)). When the concentrated weight is added to the lowest story, Equation 4 gives an accuracy of over 95% for buildings with more than 12 stories, whereas Equation 1 with the smearing treatment gives a 97% accuracy for the building with only 6 stories (Figure 6(c)). As shown in this figure, when the concentrated weight is added to the lowest story, Equation 1 with the smearing treatment is more accurate than Equation 1. This can be understood because the Ritz method gives a natural period smaller than the exact period, whereas smearing of the additional weight at the lowest story adds mass to the upper stories, which makes the natural period longer.

Simplification of Natural Period of Building With Linear Stiffness Distribution
When the stiffness linearly changes along the height (Figure 4), Equation 3, which is more complex in form than Equation 1, should be employed. To predict the natural period of such a beam in a simpler manner, it was considered to select the stiffness at a particular height as a representative stiffness and employ Equation 1 with this representative stiffness assumed to distribute uniformly along the height. Provided that the height to be selected is taken as βL, measured from the bottom of the beam, β can be obtained by equating Equation 4 and Equation 1 with the uniform stiffness of ($G_1 - \theta\beta L$) and is given as:

$$\beta = - (2/5)(B/A)\theta^2 L + G/(\theta L) \tag{5}$$

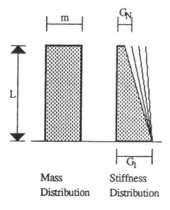

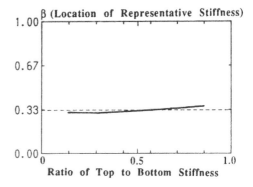

Figure 7. Effect of Stiffness
Distribution

Figure 8. Location of Representative Stiffness
in Linear Distribution of Stiffness

As an example, analyzed was a building whose story height was 4 m, story weight 500 ton, and story stiffness given as $(70 - 2(i-1))$ ton/mm. Applying Equation 5 leads β to 0.303 for the number of stories of 6, 0.325 for the number of stories of 20, and 0.358 for the number of stories of 30. All of those values are in the vicinity of 1/3. To examine the effect of the angle of inclination in the stiffness distribution on β, the stiffness at the top (G_N) was varied, with the stiffness at the bottom (G_1) left unchanged, as shown in Figure 7, and it was found that the same β value is obtained regardless of the number of stories if the ratio of the stiffness at the top (G_N) to the stiffness at the bottom (G_1) is the same (Figure 8). This figure clearly shows that β varies very little with respect to the ratio of G_N to G_1 and stays in the proximity of 1/3.

Effect of Smearing Additional Concentrated Mass

Further investigation was made about the applicability of Equation 1 with the smearing treatment, in the case where a concentrated weight is added at a particular height. A building whose story height was 4 m, story stiffness 70 ton/mm, and story weight 500 ton was considered again, with the height of the concentrated weight as a parameter. The magnitude of the concentrated weight was also varied from 0.1 to 1.0 of the total weight (not including the concentrated weight itself). Table 2 lists the ratio of the period obtained from Equation 1 with the smearing treatment (T_r) to the period obtained from Equation 4 (T_m). This table indicates that the smearing treatment makes the prediction less accurate with the increase in the magnitude of the concentrated weight, which seems quite natural. This table also reveals that the same accuracy can be achieved regardless of the number of stories as long as the ratio of the concentrated weight to the total weight

Table 2. Accuracy of Natural Period
with Smearing Treatment

Location of Additional Weight	Number of Stories	Amount of Additional Weight			
		0.1W T_r/T_m	0.4W T_r/T_m	0.7W T_r/T_m	1.0W T_r/T_m
1.0L	10	1.000	0.885	0.821	0.782
	20	0.976	0.872	0.813	0.777
	30	0.968	0.867	0.810	0.775
0.75L	10	1.000	0.920	0.876	0.848
	20	0.983	0.911	0.870	0.844
	30	0.978	0.908	0.869	0.843
0.5L	10	1.000	0.990	0.978	0.967
	20	0.999	0.988	0.976	0.976
	30	0.999	0.987	0.975	0.965
0.25L	10	1.000	1.092	1.152	1.191
	20	1.018	1.092	1.160	1.196
	30	1.024	1.108	1.162	1.197
0.0L	10	1.000	1.140	1.265	1.378
	20	1.025	1.162	1.285	1.396
	30	1.033	1.169	1.291	1.042

L=Height of Building, W=Total Weight of Building,
T_r=Natural Period by Ritz Method with Smeared Weight Distribution (sec),
T_m=Natural Period by Ritz Method with Concentrated Additional Weight (sec).

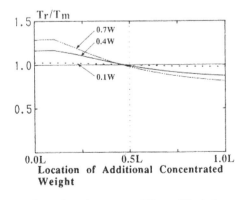

Figure 9. Accuracy of Natural Period with Smearing Treatment With Respect to Location

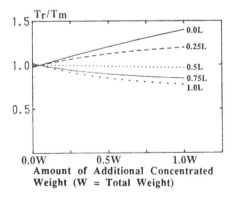

Figure 10. Accuracy of Natural Period with Smearing Treatment With Respect to Magnitude

remains the same.

Figure 9 shows the ratio of T_r to T_m, with respect to the height of the concentrated weight, for the number of stories of 20, indicating that a good accuracy is obtained if the concentrated weight is located at the mid-height and that the accuracy is aggravated more when the location is shifted further away from the mid-height. According to Figure 10, in which the ratio of T_r to T_m is plotted against the magnitude of the concentrated weight, the smearing treatment can guarantee an accuracy of over 95 % regardless of the height of the concentrated weight as long as its magnitude is less than 20% of the total weight.

ACCURACY OF SIMPLE PREDICTION WITH RESPECT TO HIGH-RISE STEEL BUILDINGS DESIGNED

Major findings obtained from the examination above are as follows. (1) If a building has more than 12 stories, Equation 1 ensure an accuracy of over 95 %. (2) When a building has a linearly decreasing story stiffness, Equation 1 is still applicable if the stiffness located at 1/3 of the total height of the building (measured from the ground) is taken as the equivalent uniform stiffness. (3) Equation 1 can be used even if a building has a concentrated weight at a particular story as long as the magnitude of the weight is less than 20 % of the total weight (not including the concentrated weight itself). To evaluate the capacity of such a simple procedure to predict the fundamental natural period of high-rise steel buildings, 11 buildings whose detailed properties on materials and geometries were available were tested for their fundamental natural periods. Figure 11 shows the ratio of the height at the i-th story to the total height of the building in the abscissa, with respect to the ratio of the sum of the story weights of the i-th story and above to the total weight of the building in the ordinate. All of the 11 buildings considered have straight lines in this relationship, which demonstrates that the story weight is more or less uniformly distributed along the height.

Table 3. Accuracy of Natural Period for Designed Buildings

Building	Height L(m)	T_r (sec)		T_r/T
(1)	47.6	X	1.421	0.975
		Y	1.434	0.972
(2)	69.4		2.037	0.998
			2.026	0.992
(3)	63.85		1.831	0.989
			1.981	0.997
(4)	66.6		1.763	0.956
			1.784	0.969
(5)	62.04		1.917	1.000
			1.954	0.977
(6)	92.7		2.516	0.986
			2.426	0.995
(7)	90.9		2.467	1.050
			2.704	1.038
(8)	87.01		3.347	0.995
			3.366	1.000
(9)	120.87		3.004	1.06
			2.882	1.05
(10)	133.1		3.920	0.971
			3.955	0.988
(11)	140.05		4.073	0.985
			4.094	0.973

X=X-direction, Y=Y-direction,

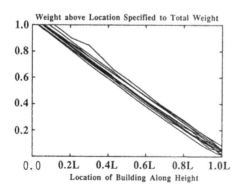

Figure 11. Relationship Between Story Weight and Building Height

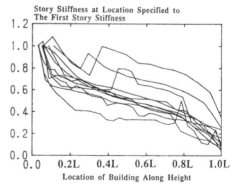

Figure 12. Relationship Between Story Stiffness and Building Height

Figure 12 indicates the ratio of the height at the i-th story to the total height in the abscissa, with respect to the ratio of the story stiffness at the i-th story to the stiffness at the lowest story. Not like the relationship between the height and weight (Figure 11), this relationship is significantly different from one building to another, and furthermore, these lines are by no means linear The fundamental natural period of those 11 buildings were predicted using Equation 1, in which the weight distribution was considered to be uniform and the stiffness at a story, located at 1/3 of the total height from the ground, was taken as the equivalent uniform stiffness. Table 3 shows the periods thus obtained (T_r) as well as the fundamental natural periods obtained from eigenvalue analysis (T). The ratio: T_r/T, ranges from 0.96 to 1.06, which suggests that this simple prediction is accurate although a very bold assumption was made with respect to the stiffness. As shown in Figure 12, the stiffness distribution was very complex in all of the 11 buildings.

EXAMINATION OF EXPRESSION: $T_1 = 0.026L$

Finally, examination was given for the reason why the fundamental natural periods (T_1) of the high-rise steel buildings introduced in the first section are expressed approximately as 0.026L. In Japan, under earthquake loading condition, the design story shear force applied to the i-th story is given as:

$$Q_i = A_i \times \sum_{j=i}^{N} W_j \times V \qquad (6)$$

In which W_j and V are the weight at the j-th story and the base shear coefficient, N the number of stories, and A_i a factor to consider the amplification of the design story shear force at the i-th story. Here, considered was an N-story building having uniform story weight distribution. Considering the previous observation that the stiffness located at 1/3 of the total height of the building represents the equivalent uniform stiffness, the story stiffness at this location is estimated using Equation 6. The design story shear force at this location ($Q_{1/3}$ in ton) is given as:

$$Q_{1/3} = 2/3 \times A_i \times W \times V \qquad (7)$$

In which W is the total weight of the building in ton. The coefficient A_i varies in accordance with the natural period of the building, taking a value of 1.0 (for T =0) to 1.37 (for T = ∞) at the location of 1/3 of the total height. As we look into high-rise buildings, 1.35 is taken as an approximate value for A_i. Then:

$$Q_{1/3} = 0.9 \times W \times V \qquad (8)$$

Further, considering that, in the design of high-rise steel buildings, their members are normally sized by story drift requirement rather than by stress requirement, the stiffness required at the location of 1/3 of the total height ($k_{1/3}$) is assumed to be the one that provides a story drift angle of 1/200 under the story shear force $Q_{1/3}$. This leads us to:

$$Q_{1/3}/k_{1/3} = 1/200 \times h \qquad (9)$$

In which h is the story height in meter. Provided that this stiffness ($k_{1/3}$) is taken to be distributed uniformly along the height, the stiffness per unit length (G) is obtained as:

$$G = k_{1/3} \times h = 200 \times Q_{1/3} = 200 \times 0.9 \times W \times V \qquad (10)$$

Substituting this stiffness (G) and the mass per unit length (m = W/(9.8L)) into Equation 1 leads us to:

$$T_1 = 0.095 \times \sqrt{1/(VL)} \times L \qquad (11)$$

Figure 13 shows the design base shear coefficients used for the 239 high-rise steel buildings. The coefficients decrease with the increase in the fundamental natural period and are reasonably approximately by:

$$V = 1/(3T_1) \qquad (12)$$

Supposing that $T_1 = \gamma L$ and substituting this into the left side of Equation 11 and the right side of Equation 12, and further inserting Equation 12 into V of the right side of Equation 11, we obtain an equation whose unknown is γ. Solving this equation provides us with 0.027 for γ, which is very close to the experimentally obtained coefficient of 0.026. From this observation, it can reasonably be interpreted that the natural period (T_1) given as 0.026L is the natural period required to make the building remain less than 1/200 in the story drift angle for the base shear coefficient given by Equation 12.

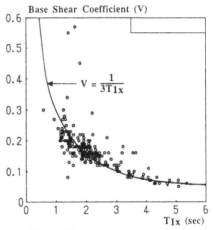

Figure 13. Relationship Between Fundamental Natural Period and Design Base Shear Coefficient

CONCLUSIONS

A summary and major findings obtained from this study are as follows:

(1) Characteristics of the fundamental natural period of high-rise steel buildings were examined based on the data for 239 high-rise steel buildings previously constructed in Japan. The averaged fundamental natural period (T_1 in sec) was found to be given as 0.026L, with L as the building height in meter.

(2) Buildings represented as discrete spring-mass systems were modelled as shear beams, and simple formulas were obtained for predicting their natural periods.

(3) An interpretation was made as to the meaning of the expression: $T_1 = 0.026L$. It was found that the natural period given by this expression is the natural period required to make the building remain less than 1/200 in the story drift angle for the base shear coefficient given by $1/(3T_1)$.

REFERENCES

1. Data for seismic design of building structures. The Architectural Institute of Japan, 1981 (in Japanese).
2. Ellis, B. R., An assessment of the accuracy of predicting the fundamental natural frequencies of buildings and the implications concerning the dynamic analysis of structures. Proc. Ins. Civ. Engrs., Part 2, 69, 1980, pp.763-776.
3. Takeuchi, Y and Watanabe, M., A study on brief formula of evaluating fundamental natural period of high rise buildings by multivariable statistical methods. Journal of Struct. and Const. Engrg., the Architural Institute of Japan, 360, 1986, pp.84-93 (in Japanese).
4. Housner, G. W. and Brady, A. G., Natural periods of vibration of buildings. J. Engrg. Mech. Div., ASCE, 89(EM4), 1963, pp.31-65.
5. Clough, R. W. and Penzien, J., Dynamics of Structures, McGraw-Hill, New York, 1975.

DIFFICULTIES WITH DESIGN FOR MOTION GENERATED LOADS

G.J.Krige
Department of Civil Engineering
University of the Witwatersrand
P.O.Box 3, Wits 2050, Johannesburg, South Africa

ABSTRACT

The paper argues that in general the loading codes used in steel design assume well defined conditions and structural responses. This leads to certain difficulties arising in both the definition of loads and the manner in which design codes are used for structures supporting such loads.

The difficulties are examined with regard to the loads from overhead cranes and the loads between conveyances in mine shafts and their guiding steelwork. It is shown that there is commonly a direct dependence of these loads on travelling velocities, on the structural stiffness at different positions within a structure, and on the maintenance and operational conditions in the plant. The empirical nature of much current quantification of the loads leads to certain difficulties with the scope of situations for which load formulations are applicable. Finally, the possibility of accident or emergency conditions arising has to be dealt with in some rational manner.

Methods of dealing with these difficulties are discussed, and it is shown how a new code for the design of conveyances in mine shafts has been drafted to accommodate them.

INTRODUCTION

Many modern structures are required to withstand the effects of loads which are generated by the motion of vehicles or machinery of some type. Such loading types are clearly not static, as their origin must lead to some form of variation. In some cases, where the loads are repeated at reasonably high frequencies, these motion generated loads can correctly be treated as dynamic loads, and the response of the structure analyzed accordingly. In many other cases, the loads are repeated infrequently, and vary sufficiently slowly that a dynamic analysis is not appropriate. Typically, such loads have been dealt with as quasi-static loads, and loading codes have defined them as some static proportion of the weight of the moving equipment.

This quasi-static approach can give quite acceptable results, but there are several inherent difficulties. These difficulties become apparent when undertaking a careful consideration of the origin and definition of the loads. Two specific motion generated load cases are considered in this paper, the difficulties are discussed, and suggestions are made as to how they can be adequately specified.

TYPICAL DIFFICULTIES FACED IN DESIGN

The author has specific recent experience of two situations of motion generated loads. The one case is of overhead travelling cranes, and the other concerns conveyances travelling vertically in mine shafts. Neither of these can appropriately be considered to be typical dynamic loads, yet in both cases a purely static idealisation may lead to either dangerous under-design on the one hand, or uneconomic over-design on the other. These two situations highlight many of the difficulties which are likely to arise due to the influence of various factors.

Lateral Loads from Overhead Cranes

Many industrial structures, most of which are constructed of steel, carry overhead travelling cranes. The South African loading code, SABS0160 [1], in common with most national loading codes, specifies all the loads generated by the operation of these cranes as proportions of the weight of the crane and its components and the load lifted. The actual proportion used is dependent on geometrical factors, such as the crane span and the wheel centres, and also usage factors which are incorporated into a "class" of crane. Thus, for example, the lateral load at each wheel, generated by the skewing of the crane, is given by:

$$L = a.W \qquad . \quad . \quad . \quad . \quad . \quad . \quad 1$$

where: L = the lateral load at the wheel

   a = an impact coefficient, as given in table 1

   W = the static vertical load on the wheel

TABLE 1 : Lateral Impact Coefficients

Crane Class	Lateral Load Coeff.
Hand operated	0.075
Maintenance	0.180
Low production	0.225
High production	0.300

Design Loads during Loading and Travelling in Mine Shafts

The loads between conveyances being hoisted in mine shafts and the guides along which they travel are also motion generated loads which cannot readily be dealt with as normal dynamic loads. Traditionally, it was assumed that these loads had a magnitude of 5% of the total weight of the fully loaded conveyance. During the 1970's this was increased to an assumption that a load of 5% of the total weight acted both at the top of the conveyance and at the bottom. In the early 1980's the assumption was adjusted again to a load of 10% of the total weight, acting either at the bottom or at the top [2]. These increases in the design loads were made because experience showed that they were necessary as a result of increasing winding velocities, ageing shafts where the alignment of the guides was no longer very good, and changes in the stiffness of the guide steelwork. This empirical approach did not, however, prevent very severe problems developing in some mine shafts [3].

The loads applied to cages during loading or unloading of materials are implicitly of an impact nature, and have traditionally been dealt with by empirically specifying minimum size members [4]. These members have been proven over many years of mining experience. New devices for holding cages are currently being fairly widely introduced, but little experience has been built up with the use of these devices, so it is not yet known whether the same floor beam sizes will still suffice. This problem highlights one of the difficulties with any empirically based design procedure.

Influence of Various Factors

In the following section, the influence of various factors on the magnitude of the loads generated by moving bodies will be discussed. The brief summary of design codes and approaches above shows that many of these factors are not considered at all, implying that the procedures are insufficient, and that they assume well defined conditions which do not typically exist in practice.

Velocity. In an evaluation of the actions causing lateral loads to develop between overhead cranes and the gantries on which they run, Krige [5] has shown that some of the actions are strongly velocity dependent. Figure 1 shows the relationship between the lateral loads caused by dynamic assymetry, or skewing during acceleration or braking,

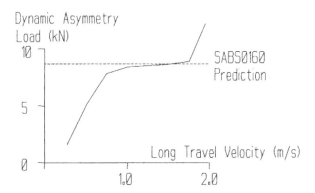

Figure 1: Crane Load Variation due to Velocity

of a typical crane. The figure also shows the load specified by SABS0160 [1] for this case. This figure clearly shows the over-prediction of load for slow cranes with velocities less than 0.7 m/s, and the under-prediction of load for very high speed cranes with velocities over 1.7 m/s. The author is not aware of cranes operating at such a high velocity in South Africa at present, but this is presumably not an unrealistic possibility in the future. The empirically based requirements of SABS0160 thus have the potential to significantly under-predict the loads on fast cranes.

Where loads are caused by a moving mass impacting buffers, such as a crab hitting its end stops or a material car hitting buffers at the back of a cage, it can again be shown that the loads are velocity dependent. The kinetic energy is:

$$KE = mv^2 \qquad . \quad . \quad . \quad . \quad . \quad . \quad 2$$

and it can be shown that the load applied to the buffers is:

$$F = v\,km \qquad . \quad . \quad . \quad . \quad . \quad . \quad 3$$

where: KE = the kinetic energy

m = mass of the travelling body

v = velocity of travel

F = load on buffers in cage

k = stiffness of the buffer

Where the velocity is known, these equations can be used to determine the loads induced by this impact. However, in cases such as material cars hitting cage buffers, the velocity is not controlled, so it is unknown. In fact, material cars are typically manually loaded, and it is known by the people doing this work, that the faster the cars can be moved the less likely they are to jam. Hence an attempt to achieve maximum possible velocity, which results in maximum possible force.

COMRO [6] defines two loads for guiding conveyances travelling in mine shafts, a roller load and a slipper load. In both cases, the dynamic behaviour of the conveyance

Figure 2: Slipper and Roller Load Variation with Velocity

travelling on the guides is considered, so a fairly complex dependence of loads on velocity emerges. This is shown in figure 2 for a typical 15 ton skip.

<u>Operator Skills and Abuse.</u> Moving equipment generates loads, firstly because of the very nature of the operational motion. An overhead crane must accelerate and brake along the gantry, and the crab must move across the bridge, in performing its function. The severity of the loads generated by these motions are dependent upon motors used in the crane, and how they are operated. A major South African steel producer uses woman crane drivers almost exclusively, because it has been found that their gentler operation of the cranes increases productivity by significantly reducing maintenance downtime. Presumably this indicates that smaller motion generated loads are induced. This influence probably cannot be quantified, but in the case of loading of equipment into cages in mine shafts, there is a quantifiable influence. When material cars are loaded into cages, they constitute a suddenly applied load, if the cage is properly docked, ie if the cage deck is exactly aligned vertically with the level of the loading station. This gives an impact factor of 2.0 on the weight at each axle. If the cage deck is too high, it is impossible to load material cars, so operators have learnt to ensure that the cage is at, or slightly below, the loading level. The material cars are thus likely to fall a small distance as they are loaded, increasing the impact load. A survey [4] has shown that some operators are happy for the material cars to drop as much as 700 mm leading to impact factors of as much 5 or 6.

It is known that cranes and mine shaft conveyances are subject to certain practices which design engineers would consider as abuse. For example, the practice, apparently fairly common in South Africa, of driving cranes at quite high speed into the end stops in order to straighten them. The 700 mm drop of material cars quoted above would also be considered as abuse. Improved training of operators may be of assistance in addressing this difficulty, but the engineer should be thoroughly aware of typical operating procedures when any design is undertaken.

<u>Installation and Maintenance Accuracy.</u> Moving equipment generates loads, secondly because of misalignment of the structure on which it moves. This dependence has been

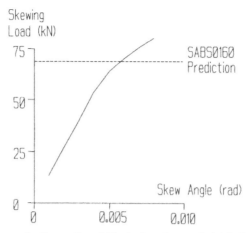

Figure 3: Crane Load Variation due to Initial Skew

shown by Krige [5], for cranes where the loads are generated by skewing of the crane. Figure 3 shows the variation of skewing loads for varying initial skew, which depends directly on the maximum longitudinal misalignment of the crane rails and wear of the rails and the wheels. COMRO [6] demonstrates that there is a linear relationship between the lateral slipper loads on conveyances, and the misalignment of their guide rails.

The level of misalignment is thus directly related, initially to the installation tolerances specified, and later to the quality of maintenance carried out. By law, mine shafts in South Africa must be regularly inspected, and some mines have regular programmes of measuring the lateral acceleration of conveyances in their shafts, which gives some idea of the alignment accuracy. The major difficulty in the case of this influence on motion generated loads is to find an acceptable manner in which it can be specified at the design stage. Current loading codes implicitly assume tolerances which constitute "normal good practice".

Structural Stiffness. When considering motion generated loads, the stiffness of the structure has two important influences. The first relates directly to the absolute stiffness at any point, and the second relates to the ratio of stiffnesses at different points along the structure.

If a moving body is travelling skew, or if its support is misaligned, there is a changing lateral displacement of the moving body relative to its support. Where the absolute stiffness of the support is low, the moving body will tend to move straight, and push the structure aside. This will tend to lead to a load:

$$F = ke \qquad . \quad . \quad . \quad . \quad . \quad . \quad 4$$

Where the stiffness of the support is high, the moving body will accelerate laterally to minimise the relative displacement. This action will tend to lead to an acceleration load:

$$F = ma \qquad . \quad . \quad . \quad . \quad . \quad . \quad 5$$

where: F = the lateral load induced

    k = stiffness of the structure

    e = maximum relative lateral displacement

    m = mass of the moving body

    a = lateral accelaration of the moving body

For cases of intermediate stiffness, there is an interaction between these two actions. A typical relationship between lateral loads on a crane, and the stiffness of the gantry on which it is supported, is shown together with the SABS0160 prediction for this crane, in figure 4.

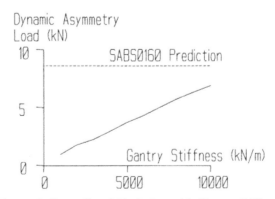

Figure 4: Crane Load Variation with Gantry Stiffness

A further difficulty arises where there is a difference in the structural stiffness along the direction of motion. This is commonly the case, as crane beams and mine shaft guide

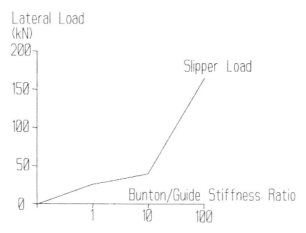

Figure 5: Slipper Load variation with Stiffness Ratio

rails are discreetly supported by columns and buntons respectively. Where the crane beams and guide rails are flexible relative to the supports, the moving body can move laterally quite easily along the flexible mid-span portion of its path. Very high loads can then be induced as the moving body approaches the more rigid supports, and is accelerated laterally around them. COMRO [6] describes this action carefully for skips in mine shafts, referring to it as "slamming". Figure 5 shows the variation of loads on mine shaft conveyances for differing ratios of guide stiffness to bunton stiffness. Current loading codes ignore this interdependence between the structural stiffness and the loads induced.

Emergency Conditions

The final difficulty which will be considered relates to the possibility of emergency conditions arising. The author is aware of one crane building where cranes have been known to jump off the rails due to misalignment and wear of the rails and wheels. The loads required for this to happen are shown in figure 6, and it can be seen that they must be of the order of 5, or more, times the static vertical wheel loads. Krige [5] also shows forces of a very high magnitude which may develop if one wheel of a crane jams while the crane is travelling at high velocity. In the case of conveyances in mine shafts, operator or plant errors may result in conveyances hitting crash beams at the top of the shaft or hitting the shaft bottom leading to extremely loads, which cannot be calculated.

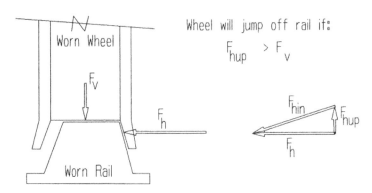

Figure 6: Crane Wheel Loads on Worn Wheels

Some new mines have installed buffer systems, with known stiffness and damping values, to limit these loads, but these are not widely used. It is also possible for the winder rope to break under certain conditions, which applies loads to the conveyance of the order 10 to 15 times its weight. Under these conditions limited damage to equipment is usually acceptable, provided a major disaster is avoided.

A NEW LOADING AND DESIGN CODE FOR MINE CAGES

The South African Institute of Steel Construction has a committee which is currently working on the drafting of a new code which defines the loading and design procedures for conveyances in mine shafts [7]. In drafting this code, careful attention has been given to the problems associated with motion generated loads, as many of the loads which must be considered are directly attributable to some or other motion. This code thus forms a useful example of a rational approach to dealing with motion generated loads in design codes. Design loads are specified in terms of the motions by which they are generated. It has not been possible to comprehensively eliminate all assumptions and empirical requirements, but an attempt has been made to minimise such clauses.

Loading and Unloading of Conveyances

These loads are attributable to the motion of material cars being loaded into, or unloaded from cages, or to the motion of ore being loaded into skips. The loads due to the loading of ore have not yet been formallised in the code, so material car loading will be discussed with reference to the horizontal and vertical loads which are specified. A quasi-static approach is taken to the definition of these loads, but cognizance is taken of certain variations in common practice.

Horizontal Loads. Observation of the manual loading procedures used at mines led to the conclusion that it was not possible to accurately define the speed at which material cars are travelling when they enter cages. Speeds varied between less than 0.3 m/s and over 1.0 m/s. However, the buffers used at the back of cages, have been developed over many years, so that a small range of steel section sizes is currently used as sacrificial members. These members typically require loads of about 25 kN to yield them, and typical material cars have a mass of about 5.0 tons. In order to develop this load, a loading speed of about 0.5 m/s is required. Observations of loading speeds and assessment of the members typically used thus agree roughly. The code thus specifies that a loading speed of 0.5 m/s should be used, and that the loads applied should be calculated in accordance with the stiffness of the buffers used.

Vertical Loads. The vertical loads are induced by the application of the weight of the material cars onto the floor of the cage. In defining these loads, the code had to recognise that there are two distinct possibilities. Either the cage is held by some form of holding device, or it is free to move downwards or upwards with rope stretch, as material is loaded or unloaded. In the former case, it is assumed that there is no misalignment of levels. A maximum impact factor of 1.5 is defined for this case, assuming a typical material car having two equally loaded axles. In the latter case, it is assumed that the initial level misalignment is equal to the amount of rope stretch when one axle of the material car, ie half its weight, is loaded. This leads to the definition of an impact factor of 2.5.

Conveyance Travelling in the Shaft

The definition of loads occurring whilst the conveyance is being hoisted takes a novel approach in two different ways.

The first is in following a procedure for the assessment of the loads which has been developed by COMRO [6]. This procedure makes specific allowance for several different factors which influence the magnitude of the loads. The influence of the winding velocity, guide and bunton stiffnesses, the geometry of the shaft steelwork, and the magnitude of misalignments are all included. This enables the engineer to design the shaft steelwork and the conveyances in a manner which minimises the loads induced by motion of the conveyance.

The second innovation is that the engineer must determine the installation and maintenance tolerances in the shaft. Shafts are classified as "good" if the maximum local guide misalignment does not exceed 5 mm, "average" if the misalignment does not exceed 10 mm, and "poor" if it does not exceed 20 mm. The code specifically allows the engineer to extrapolate outside of these limits if he believes a shaft to be in better condition than "good" or worse than "poor". The onus is thus on the engineer to ensure that his design is properly linked with the maintenance in the shaft. There is typically close liaison between the mines and design engineers, so that this is a realistic approach. The approach to assessing the fatigue life of the conveyances is also linked to this shaft classification.

Emergency Load Conditions

The draft code specifies two emergency loads which must be applied to the conveyance, both recognising the possibility of the rope breaking. The one load is the rope break load, which is assumed to be applied at the bottom of the conveyance, and must thus be transferred throughout the main load carrying frame of the conveyance. The other is a load of the same magnitude, applied to the safety catches, which prevent the conveyance falling down the shaft if the rope breaks, or is detached from the conveyance, in the headgear.

In most situations these will be the most severe loads, and will determine the design of the main members of the conveyance. The code thus defines "safety critical" members which must be designed to withstand these very high loads. Other members are only required to carry the normal operating loads. The philosophy adopted in using this approach is that quite severe damage to a conveyance is acceptable in an emergency, provided it does not fall down the shaft.

The load factor defined for these emergency loads is 1.05, as it is recognised that they should not occur, but that if they do occur, some plastic deformation of the conveyance is acceptable. It was debated whether this load factor could perhaps be set at a lower value, of say 0.85, but once plastic deformation occurs member stability and joint performance cannot be properly assessed, so it was decided that any plastic deformation should be minimised.

An interesting philosophy underlies this load specification, in that the conveyance design load is set at a value which ensures that the wire rope will break prior to the conveyance itself. This ensures the maximum energy absorption capability, which minimises the risk of a disaster, should the tension in the rope be lost and the conveyance drop some distance in free-fall.

CONCLUSION

The dependence of motion generated loads on factors such as travelling velocity, operator skills, maintenance criteria, the structural stiffness and variations in the stiffness, has been demonstrated. These factors are frequently neglected by design procedures, and can lead to unexpected response of the structure. The empirical nature of much current quantification of the loads also lead to certain difficulties with the scope of situations for which load formulations are applicable. The possibility of accident or emergency conditions arising has also been discussed.

Code requirements which address these difficulties have been demonstrated by reference to a new code for the design of conveyances in mine shafts. Here it has been shown that many of the difficulties can be dealt with in a rational manner, but that there are necessarily assumptions which must be clearly recognised and their implications understood.

REFERENCES

1. SABS 0160, The General Procedures and Loadings to be adopted for the Design of Buildings, South African Bureau of Standards, Pretoria, 1990.
2. Devy S.D., Investigation of Dynamic Loads on Mine Shaft Steelwork with particular reference to Free State Geduld No. 5 Shaft. GDE assignment, University of the Witwatersrand, June 1982.
3. Calver A.H., Comments on the Shaft Steelwork Problem at the No. 4 Shaft at President Steyn Gold Mining Company Limited. AMRE Circular No 1/80, November 1980.
4. Krige G.J. (ed.), Commentary on SABS 0208, Design of Structures for the Mining Industry. Part 3: Conveyances, SAISC, Johannesburg, 1992.
5. Krige G.J., Lateral Loads on Overhead Travelling Cranes. In Trends in Steel Structures for Mining and Building, SAISC, Johannesburg, 1991.
6. COMRO, Design Guidelines for the Dynamic Performance of Shaft Steelwork and Conveyances. SDRC Project No 11964, Contract for Chamber of Mines research Organisation, Johannesburg, 1990.
7. SABS 0208, Design of Structures for the Mining Industry. Part 3: Conveyances, SAISC draft document, Johannesburg, 1992.

VIBRATION OF LOW-FREQUENCY FLOORS:
OFFICES AND SHOPPING CENTRES

Per-Erik Eriksson
Chalmers University of Technology
Department of Structural Engineering
412 96 Göteborg, Sweden

ABSTRACT

This paper describes a method for predictive calculations of the vibrational response of low-frequency floors, i.e. floors with the lowest natural frequency below 7-8 Hz, subjected to pedestrian traffic. In buildings used as offices and shopping centres this is the most usual internal cause of perceptible mechanical vibrations. The calculation method is based on a frequency domain model of the forces produced by walking people, which has been established through laboratory measurements, and on a modal model of the floor being examined. This representation facilitates a realistic consideration of the multi-mode response of an orthotropic floor subjected to broad band forces. The method further illustrates the limited applicability of dynamic floor models simplified as far as to take account of only the response in the fundamental mode of vibration. The dynamic behaviour of low-frequency floors is discussed and the issue of damping of normally designed and constructed floors is emphasized.

INTRODUCTION

Floor vibrations of small amplitudes may cause disturbance to occupants. The perception levels of vibrations are generally so low that they are not connected with any risk of damage to the structure itself. The vibrations are often caused by normal pedestrian traffic, although more severe vibrations may result if the floor is subjected to different types of rythmic activities.

The type of occupancy is an important factor when determining the acceptable level of vibrations. This paper concerns office and shopping center type buildings where possible annoyance will primarily be experienced by people sitting or standing still. Someone who is walking or physically active in some other sense will generally accept significantly higher levels of vibration.

The terms low-frequency and high-frequency floors were introduced by Wyatt [1] to distinguish between floors with the lowest natural frequency below and above 7-8 Hz respectively. Low-frequency floors have traditionally been called long-span floors, whereas high-frequency floors have been referred to as short-span or light-weight floors. The resonance frequency distinction is, however, more appropriate for describing the important differences between the two floor categories. The distinction is primarily due to the nature of the dynamic loads caused by pedestrians and rythmic activities. Both these activities cause a combination of impulsive and steady-state loads. The steady-state loads mainly have frequency components below 8 Hz and may thus cause a build-up of resonant vibrations in floors with resonance frequencies below this limit. For high-frequency floors, the impulsive part of the load is more important as it may result in transient resonant vibrations and a feeling of springiness, see [2].

For low-frequency floors, continuous resonant vibrations are the main problem . The reason is that their mass is generally sufficiently high to limit the vibration amplitudes caused by impulsive loads. Furthermore, it is commonly agreed that a criterion of constant acceleration is appropriate for assessment of continuous vibrations of frequencies between 4 and 8 Hz.

LOW-FREQUENCY FORCES CAUSED BY HUMANS

People in motion cause significant dynamic forces at low frequencies. These forces are of a broad-band nature with frequency components up to at least 40 Hz. Provided that any of the resonance frequencies of the floor falls within this frequency range, which is very likely, there is a risk that this will result in perceptible resonant vibrations.

The continuous dynamic forces from walking, running and rythmic activities have been extensively measured by Rainer and Pernica [3] and by the author [5]. The former utilise a load model based on dynamic load factors in the frequency domain, i.e. the Fourier components of the load divided by body weight. Major components are typically at the step or jumping frequency and at their higher harmonics. Measured forces due to handclapping, stamping etc. are presented in [4] in a similar manner.

For floors in office buildings and shopping centres the forces produced by *walking* are most important. A study of such forces is reported in [5], which is a reevaluation and extension of [6]. In a spectral representation, the largest contribution to these forces will be found at the step frequency f_s, see Fig. 1. There are, however, also sharp peaks in the spectrum at integer multiples of f_s. These are commonly called higher harmonics of the step frequency. The y-axis quantity in Fig. 1 is the power spectral density of the force S_F.

Considering now that normal step frequencies are found between 1.4 and 2.5 steps/second, or simply Hz, the peaks mentioned above will merge together when the source of excitation is not one but several persons. In [5] a design force spectrum is determined as an envelope function, E in Fig.1, and a "mean" function, M. The E functions connect the peaks of the force spectra obtained at different step frequencies. The M functions represent the force power around each harmonic, evenly distributed over a frequency band-width equal to the step frequency, i.e. the width between the different harmonics. The abrupt change in magnitude between the two parts of the E and M functions at 2.5 Hz illustrates the difference in force magnitude between the first and the higher harmonics of the step frequency f_s.

The influence on the force spectra due to several persons walking both in step as well as at different step frequencies is shown in [5]. For design purposes the E and M functions are multiplied by a magnification factor that takes account of the number of participating persons. For uncoordinated walking by groups of people, the magnification factors for the E and M functions have been found to be roughly $(n_p)^{0.8}$ and $(n_p)^{1.2}$ respectively above 2.5 Hz. The experimental data covers groups of up to 11 people.

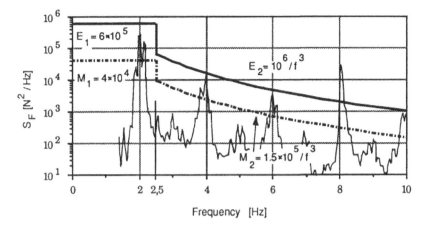

Figure 1. Envelope, E (–), and "mean", M (-··-), functions of the force spectral density design model. The spectral density at $f_s = 2$ Hz from one walking subject is also shown. Note that the vertical axis has a logarithmic scale.

LOW-FREQUENCY FLOORS

Low-frequency floors are floors with the lowest natural frequency f_1 below 7-8 Hz. As the resonance frequencies are proportional to the square root of the ratio of the modal (generalized) stiffness to the modal mass, it follows that a low natural frequency corresponds to a high mass to stiffness ratio. This feature is mainly found in floors constructed from precast floor elements of prestressed concrete and from steel beams acting compositely with a concrete slab for which relatively large spans can be achieved with limited structural depth.

Dynamic behaviour

For low-frequency floors the governing event for human perception of vibrations is usually continuous or steady-state vibrations. In a case when these vibrations are caused by pedestrians they will normally consist of several frequency components and are best quantified by the root mean square (rms) value of the displacement, velocity or acceleration. For the frequency range considered here, the acceleration is the most appropriate quantity with respect to human sensitivity, as mentioned above.

Composite floors and floors with T-shaped concrete elements are strongly orthotropic. This implies that they have a number of closely spaced resonance frequencies, each of which is connected to a specific mode of vibration. If the load is of a broad-band nature and has a spectral density $S_F(f)$, the rms value of the resonant acceleration in mode No. n is given approximately by Eq. 1, according to [6].

$$a_{n,rms} = \frac{\sqrt{S_F(f_n)\pi f_n}}{2M_n\sqrt{\zeta_n}}$$
(1)

f_n is the resonance frequency, M_n is modal mass (normalised such that the maximum mode shape value $\Phi_{n,max}$ equals unity) and ζ_n is the modal damping ratio of mode n. This means that

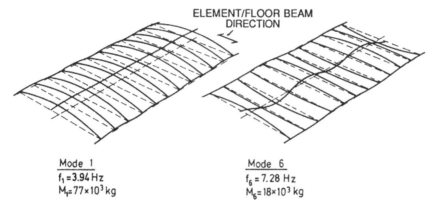

ELEMENT/FLOOR BEAM
DIRECTION

Mode 1
$f_1 = 3.94\,Hz$
$M_1 = 77 \times 10^3\,kg$

Mode 6
$f_6 = 7.28\,Hz$
$M_6 = 18 \times 10^3\,kg$

Figure 2. Calculated mode shapes, resonance frequencies and modal masses for two modes of a floor with two unrestrained edges and two edges simply supported by primary beams.

the resulting "modal" acceleration is therefore inversely proportional to M_n, which is given by the integral over the whole structure of the product of the mode shape squared and the mass distribution. M_n signifies how large a proportion of the floor is participating in the motion of this specific mode.

The most commonly used model for dynamic calculations for floors has been a single degree of freedom (SDOF) model, modelling only the fundamental mode of vibration of the floor, see e.g. [1,7]. The higher modes of many floors have only marginally higher resonance frequencies and may therefore be excited by almost the same dynamic forces due to walking as the fundamental mode. At the same time the modal mass is often lower for the higher modes, which means that the resulting acceleration may be higher in some of these modes than in the fundamental mode. This means that a calculation method based on a SDOF model will not reflect the behaviour of the floor properly. A lower modal mass, however, also means that a smaller proportion of the floor area will be affected by the maximum "modal" acceleration.

For a rectangular floor that is simply supported on vertically rigid supports all around, the modal masses of all modes are equal, provided that the mass may be approximated as being evenly distributed over the floor. For this case the fundamental mode approach is therefore more relevant. If, on the other hand, any of the edges are dynamically unrestrained, some of the higher modes may have significantly lower modal masses as shown in Fig. 2. This is often the case around an open space in the building and also in some cases along the walls. Indeed, many of the vibration problems in concrete element floors that have been reported in Sweden have been most strongly experienced near unrestrained edges.

If, furthermore, the forces applied to the floor are of a broad-band nature the floor will vibrate in all modes with natural frequencies within the frequency range of the excitation simultaneously. This means that the more strongly orthotropic the floor is, the higher the resulting acceleration response will be. This is another fact that is not reflected by a method based on a SDOF floor model. Calculation methods for floor response based on a SDOF model of the floor can therefore not be recommended.

Damping
This is an area of much confusion. The range of suggested modal damping ratios ζ_n for design purposes is as wide as from 0.5% to 12%, for virtually the same kind of floor structures. In a

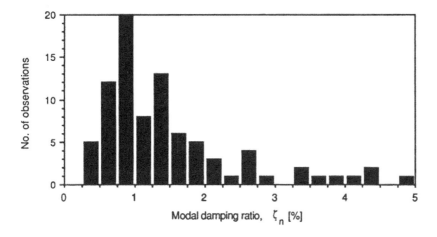

Figure 3. Experimentally determined modal damping ratios from FRFs or acceleration spectra for floors. Based on data from [5, 8, 9, 10, 11 & 12]. Note that several observations may refer to each floor as ζ_n is plotted also for higher modes.

series of experimental modal analyses of pre-stressed concrete element floors, performed by the author [5, 9], no modal damping ratios higher than 2.2% were found. Most modes had ζ_n values around 1% and in some cases as low as 0.5%. The floors that were tested had few or no partitions but were in half of the cases furnished and equipped with installations. Similar results have been achieved for composite floors by Osborne and Ellis [8].

Various reported damping ratios [5, 8 ,9 ,10, 11 & 12] for concrete element, steel joist and composite (steel-concrete) floors are presented in Fig. 3. From these graphs it is obvious that the damping ratios suggested in [1] and [7] (1.5 % for extremely bare floors, 3 % for normal, open-plan floors and 4.5 % for floors with partitions) are too high.

Dynamic modelling

The dynamic modelling of floors described here is aimed at establishing a *modal* model of the floor system. The modal model can be illustrated by a number of single degree of freedom systems each with a mass M_n, the modal mass, a stiffness K_n, the modal stiffness, and a damping ratio $\zeta_n = (C/C_{cr})_n$, the modal damping ratio. The resonance frequency f_n for mode n is then given by Eq. 2. Each mode is also associated with a mode shape $\Phi_{n,i}$. Once an appropriate physical model of the floor is established all the modal parameters except the modal damping ratio can be computed while ζ_n has, at present, to be guessed.

$$f_n = \frac{1}{2\pi} \sqrt{\frac{K_n}{M_n}} \tag{2}$$

When establishing the physical model of a floor, the limitation of the model is of great importance. This is a difficult issue due to the differences in behaviour at the serviceability and ultimate limit states (SLS and ULS) respectively. The ultimate limit state structural model is fairly straightforward with assumed simple supports working as such and "non-structural" components not contributing significantly to the load-carrying capacity. The serviceability limit states model is more complicated. For example, two longitudinally adjacent simply supported beams (ULS) may well behave as one continuous beam at lower load and deformation levels (SLS) and a nonstructural partition may provide a stiff support or a stiff link to a floor below or above or may appear as an increased appearent damping ratio. Exact models that would reflect the behaviour of a *constructed* floor can therefore not be made based on present knowledge. However, positive effects from for example partitions should not normally be taken into account as these can be moved and as their structural function is not normally properly established. Neither should the continuity of slabs or slab toppings be taken as a positive effect. If the underlying floor beams or elements are not designed for bending moment continuity, cracks will generally form over the supports after some time of use.

The bay width, orthogonal to the floor elements or floor beams, that is significantly set in motion, is another intricate matter. A rough proposal in [9], of a width not exceeding three times the span length, is based on comparison with experimental results. More detailed estimates of this parameter is given in [1] as a function of the slab flexibility relative to the flexibility of the floor beams. The maximum number of continuous spans that should be taken into account is limited to three in reference [9], while a number of two can be extracted from [1] although it is not explicitly stated.

For the calculation of the modal properties of the floor, a computer analysis is generally required, e.g. using a finite element program with an eigenmode solution facility. The accuracy and reliability of computed modal models of this type depends on several factors. The resonance frequencies are usually fairly accurate, although there is some uncertainty in the choice of a dynamic Young's modulus for concrete. The calculated mode shapes and modal masses depend on accurate modelling of the boundary conditions and the calculated modal masses also strongly depend on the limitation of the model.

A METHOD FOR VIBRATION PREDICTION

Modal model of the floor

The floor model that is used for prediction of the floor response to walking excitation according to the method set out below is a modal model established as described in the previous section. The modal model should describe all modes with a resonance frequency lower than 8 Hz. The formulae below are based on a mode shape scaling where $\Phi_{n,max}$ equals unity. The mode shapes can be scaled arbitrarily and some programs utilize other definitions than this. If that is the case the calculated modal mass should be multiplied by a factor $(1/\Phi_{n,max})^2$.

The modal damping ratio $\zeta_n = (C/C_{cr})_n$ should for normally constructed open plan floors of prestressed concrete element be chosen $\leq 1\%$. Higher damping may be expected under more severe vibrations but should not be used for the purpose of serviceability assessment of floors. For composite steel beam concrete slab floors the damping should not differ significantly, but this has not been experimentally verified by the author.

Estimation of vibration acceleration

The background to the calculation method presented below is shortly as follows. The dynamic loads applied to the floor are produced by walking people. If other sources of considerable dynamic excitation are acting on the floor, such as certain machines, a separate dynamic analysis should be carried out. One of the harmonics of the step frequency is assumed to coincide precisely with one of the resonance frequencies below 8 Hz. The corresponding mode is called the "governing mode" and is supposed to be excited by the force defined by the envelope function E in Fig. 1. The other modes are subjected to the "mean level" of the broad band force defined by the function M.

The calculation method is divided into two parts; one assuming that the load is a concentrated force (treading in place), which is especially applicable for offices, and another that treats the load as evenly distributed over parts of or over the whole floor. The latter is mainly intended for floors with more intensive pedestrian traffic.

Treading in place

The first stage of the analysis is to establish in which mode the acceleration rms value will be maximal, i.e. the "governing" mode. Using the envelope function E of the spectral density of the forces from single person excitation in Eq. 1 we have

$$a^I_{n,rms}=\frac{\sqrt{E(f_n)\pi f_n}}{2M_n\sqrt{\zeta_n}} \tag{3}$$

$Max(a^I_{n,rms})$ defines the governing mode for which henceforth the notation n = X is used. The load is assumed to be stationary in the room and to act at the weakest point for the governing mode, i.e. where $\Phi_X(x,y) = 1$. The modal accelerations for the other modes, calculated with the M instead of the E function, are therefore multiplied by the magnitude of the mode shape values at this point. The modal accelerations for all modes are then given by

$$a^{II}_{n,rms}(n_p)= \begin{cases} a^I_{n,rms} & ,n=X \\ |\Phi_n(...)|\sqrt{\dfrac{M(f_n)}{E(f_n)}}\,a^I_{n,rms} & ,n\neq X \end{cases} \tag{4}$$

$\Phi_n(...)$ is the mode shape value of mode n in the point where $\Phi_X(x,y) = 1$. The total estimated acceleration rms value (for N modes with $f_n \leq 8$ Hz) is then calculated as

$$a_{n,rms}=\sqrt{\sum_{n=1}^{N} (a^{II}_{n,rms})^2} \tag{5}$$

Distributed load

The load is, in this case, supposed to act uniformly distributed over a specific area of the floor A or over the whole floor ($A = A_{tot}$). It will therefore act in positions representing various mode shape values and its effect will have to be weighted with respect to the mode shapes. This is achieved by using the generalised (modal) spectral density of the force spectral density $S_{F,n}$. For a uniformly distributed load with the spectral density $S_q = S_F / A$, this quantity is given by Eq. 6, where $K_{\Phi,n}$ is called the mode shape factor and is determined from Eq. 7.

$$S_{F,n}=(K_{\Phi,n})^2 S_F \tag{6}$$

$$(K_{\Phi,n})^2=\frac{1}{A}\iint_A [\Phi_n(x,y)]^2 dxdy \tag{7}$$

If $A = A_{tot}$ and if the mass of the floor is approximately evenly distributed, then Eq. 7 equals the ratio between the modal mass and the total mass of the floor, i.e. $(K_{\Phi,n})^2 = M_n/M_{tot}$.

The governing mode (again $n = X$) is now found as the one with $\max(a^{III}_{n,rms})$ and may not be the same for this loading case as that above.

$$a^{III}_{n,rms} = \frac{\sqrt{(K_{\Phi,n})^2 E(f_n) \pi f_n}}{2M_n \sqrt{\zeta_n}} = K_{\Phi,n} a^{I}_{n,rms} \qquad (8)$$

The number of *walking* people present on the floor is taken into account by using the magnification factors previously given. The acceleration in each mode with $f_n \leq 8$ Hz is thus estimated using the following equation and is now a function of the number of *walking* people n_p (note that the magnification factors are the square roots of those in the earlier chapter);

$$a^{IV}_{n,rms}(n_p) = \begin{cases} (n_p)^{0.4} \, a^{III}_{n,rms} & ,n=X \\[2ex] (n_p)^{0.6} \sqrt{\dfrac{M(f_n)}{E(f_n)}} \, a^{III}_{n,rms} & ,n \neq X \end{cases} \qquad (9)$$

The summation of the modal responses are then carried out according to Eq. 5, substituting $a^{IV}_{n,rms}$ for $a^{II}_{n,rms}$

FLOOR ACCEPTABILITY

Limiting values in absolute terms for what levels of vibration are acceptable for a user of a building are very difficult to establish. There are several reasons for these difficulties. The sensitivity to vibrations differs substantially between different subjects. Knowledge about whether or not the vibrating structure is safe can lead to very different opinions of the serviceability of the structure. Another important parameter is the activity of the subject. The highest sensitivity to vertical vibrations at low frequencies is found for subjects lying horizontally, in which case the perception threshold for vertical vibrations is typically of the order 0.01 m/s². In comparison Wyatt [1] states that the head of a walking person is typically subjected to a vertical vibration acceleration of approximately 3 m/s². This does not, however, mean that a walking person will tolerate floor vibrations of this magnitude. A third parameter is the time of exposure to vibrations at a certain level. Both the time length of exceedance of a perceptible level of vibrations and the number of such occurrances during a certain time influence the user acceptance.

Most criteria that have been developed, have been fitted to certain calculation methods and to subjective ratings of existing floors to distinguish between acceptable and unacceptable floors. This is of course a highly relevant method and much valuable data has been collected in the process. However, as discussed above, none of the existing calculation methods known to the author seems to properly reflect the dynamic behaviour of an arbitrary floor structure. The main reasons for this are the single degree of freedom assumption and the assumed high damping values. Unfortunately, when introducing any more accurate concept for the calculation of the response of floors, the whole procedure of establishing criteria has to be repeated.

The International Standard ISO 2631-2 [13] represents a different approach. The base curves that are presented there are meant for evaluation of *measured* vibrations. This could be very good if better guidance was given on how to measure the vibrations. For example the rms acceleration limits that are suggested are not in any way related to exposure times although some guidance is given in an annex to the standard. The frequency dependence of the vibration limits (acceleration, velocity or displacement) presented in the standard is relatively widely used but has been questioned by e.g. Griffin [14]

For the evaluation of floor acceptability according to the method presented in the previous chapter it is, inter alia, suggested that comparisons to existing buildings are made using the same calculation assumptions. The owner of the building can then, at the design stage, more easily estimate whether or not the behaviour of the building will be expected to meet the user requirements. It should also be pointed out that the calculated acceleration levels represent "worst case" vibrations and are much higher than what would normally be measured on existing floors. The calculated levels should therefore not be assessed directly against sensitivity curves aimed at normally occurring acceleration levels.

CONCLUSIONS

Floor vibrations in office buildings and shopping centres are primarily caused by walking people. In some instances these vibrations can reach a magnitude that is unpleasant or even unacceptable to the occupants.

The dynamic forces from walking people can be divided into two parts; An impulsive part of higher frequencies originating from the heel strike and a continuous excitation of lower frequencies from the successive footsteps. For a single person walking with a constant step frequency the latter part will consist of rather distinctive frequency components at the step frequency (the largest component) and at the higher harmonics of the step frequency. For a group of people walking independently, the force spectrum of the continuous excitation will be flatter and it is more relevant to characterize it as a broad band process.

Low-frequency floors (here defined as floors with a lowest resonance frequency below 8 Hz) are mainly affected by the continuous part of the excitation from walking as this can cause a build-up of resonant vibrations.

Most modern floors are mainly designed for one-way action which means that they are usually strongly orthotropic. Such floors have a number of closely spaced resonance frequencies, any of which may coincide with a harmonic of the step frequency. This and other facts indicate that the commonly used "single-degree-of-freedom" simplification of the dynamic behaviour of a floor is too coarse an over-simplification when anticipating the behaviour of a variety of floor configurations.

Furthermore it has been proven in experimental studies (not reported in this document) that the damping of low-frequency floors is much overestimated in several published design methods. Modal damping ratios $(C/C_{cr})_n$ of $\leq 1\%$ for prestressed concrete floors are suggested by the author for the calculation of vibrations of small amplitudes.

The method for predictive calculations of floor vibrations presented here is believed to yield a good comparison between different floors. It is, however, fully appreciated that it may be somewhat too complex to be used by practising engineers in general as a design method and should therefore be simplified. In the absence of reliable data for quantifying the severity of occupancy vibrations, a comparative assessment of different floors is recommended.

REFERENCES

1. Wyatt, T.A., Design Guide on the Vibration of Floors, Publication 076, Steel Construction Institute, Berkshire, England, 1989.

2. Ohlsson, S.V., Floor vibrations and Human Discomfort, <u>PhD. thesis</u>, Div.of Steel and Timber Structures, Chalmers Univ. of Technology, Göteborg, Sweden, 1982.

3. Rainer, J.H., Pernica, G., Vertical Dynamic Forces from Footsteps, <u>Canadian Acoustics</u>, 1986, **2**, 12-21.

4. Vogt, R., Bachmann, H., Dynamische Kräfte beim Klatschen Fusstampen und Wippen, <u>Rep No.7501-4</u>, Inst. für Baustatik, Eidgenössige Techn. Hochschule, Zürich, 1987.

5. Eriksson, P-E., Vibration of Low-Frequency Floors – Dynamic Forces and Vibration Prediction, Unpublished report, Chalmers University of Technology, Göteborg, 1992.

6. Eriksson, P-E., Ohlsson, S., Dynamic Foofall Loading from Groups of Walking People, <u>Proc. of Symposium/Workshop on Serviceability of Buildings</u>, Nat. Res. Council Can., Ottawa, 1988, pp. 497-511.

7. Allen, D.E., Design Criterion for Walking Vibrations, <u>Proc. of Conf. on Serviceability of Steel and Composite Struct.</u>, Czechoslovak Scientific and Techn. Society, Committee for Steel Struct., Prague, 1990, pp. 98-105.

8. Osborne, K.P., Ellis, B.R., Vibration Design Criteria and Testing of a Long Span Lightweight Floor, <u>PD 61/89</u>, Dept. of the Environment, Building Research Establishment, Garston, England, 1989.

9. Eriksson, P-E., Vibration Serviceability of Pre-Cast Concrete Element Floors, <u>Report R95:1988</u>, Swedish Council for Building Research, Stockholm, 1988 (in Swedish).

10. Matthews, C.M., Montgomery, C.S., Murray, D.W., Designing Floor Systems for Dynamic Response. <u>Structural Engineering Report No. 106</u>, University of Alberta, Dept. of Civil Eng., Alberta, Canada, 1982.

11. Pernica, G., Effect of Architectural Components on the Dynamic Properties of a Long-span Floor. <u>Canadian Journ. of Civ. Eng.</u>, 1987, **4**, 461-467.

12. Rainer, J.H., Swallow, J.C., Dynamic Behaviour of a Gymnasium Floor. <u>Canadian Journ. of Civ. Eng.</u>, 1986, **3**, 270-277.

13. Evaluation of Human Exposure to Whole-body Vibration - Part 2: Human Exposure to Continuous and Shock-Induced Vibrations in Buildings (1 to 80 Hz), <u>ISO 2631-2</u>, The Int. Org. for Standardization (ISO), 1989.

14. Griffin M.J., <u>Handbook of Human Vibration</u>, Academic Press Ltd, London, 1990.

THE ULTIMATE STRENGTH OF A-SHAPED BRIDGE TOWERS UNDER CYCLIC LOADING

Yoshito ITOH
Department of Civil Engineering
Nagoya University, Japan

Arif KARTAWIDJAJA
Department of Civil Engineering
Nagoya University, Japan

Yuhshi FUKUMOTO
Department of Civil Engineering
Osaka University, Japan

ABSTRACT

This paper presents a basic knowledge of ultimate strength of A-shaped bridge towers under cyclic loading in the plane perpendicular to the bridge axis. The ultimate strength of the A-shaped bridge towers of Meiko-Nishi Cable-Stayed Bridge and Konohana Self-Anchored Suspension Bridge subjected to horizontal cyclic loading are studied. Experimental tests and numerical computations of model towers are carried out. The geometrical and material nonlinearities are considered in the finite element program and the numerical results are compared with the test data. Behavior of model towers with different type of steels is studied numerically. High tensile strength steels with both low yield ratio and high yield ratio are applied to the analytical model.

INTRODUCTION

In Japan, a large number of steel bridge towers have been constructed as parts of cable-stayed bridges or suspension bridges. In spite of the increasing popularity of steel bridge towers, only few experimental investigations have been conducted on their behavior under cyclic loading and even less investigation about A-shaped tower. Thus, the behavior of bridge towers under severe earthquakes has not been thoroughly understood.

This paper presents an experimental and numerical investigation of A-shaped bridge towers subjected to horizontal cyclic loading in the plane perpendicular to the bridge axis and presents the numerical analyses of model towers with application of a new type of steel.

OUTLINE OF THE TEST PROGRAM

The experimental tests of the A-shaped tower were carried out at Nagoya University by Fukumoto et al.[1]. SHAPE-1 was a prototype of *Meiko-Nishi Cable-Stayed Bridge,* which has center span length of 405 m. This tower has an A-shaped structure with straight legs until the base. For this type, four models of scale 1/30 with different member's sections were tested, two of them had bracing . Four models of the tower were subjected to a cyclic horizontal load and the constant vertical load.

The experimental tests of another type of the A-shaped tower were also carried out at Nagoya University by Fukumoto et al.[2]. SHAPE-2 was a prototype of *Konohana Self-Anchored Suspension Bridge,* which has center span length of 300 m. This tower has A-shaped structure with inward leaning legs below the stiffening girder. Four models of scale 1/20 were tested. Four models of the tower were subjected to the same load condition as described in the previous test.

For comparison with A-shaped towers, one model of *portal frame tower* (SHAPE-3) was also tested. All types of the model towers are shown in Figure 1. Sections of the members of the model towers are shown in Table 1.

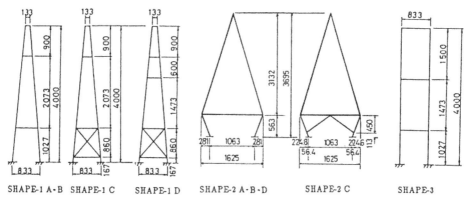

| SHAPE-1 A·B | SHAPE-1 C | SHAPE-1 D | SHAPE-2 A·B·D | SHAPE-2 C | SHAPE-3 |

Figure 1. Types of the model towers.

TABLE 1
Cross sections of the model towers

Tower Type	Column	Beam	Bracing
SHAPE-1 A	H100x100x6x8	H100x100x6x8	-
SHAPE-1 B	H150x75x5x7	H150x75x5x7	-
SHAPE-1 C	H100x100x6x8	H100x100x6x8	H100x50x6x8
SHAPE-1 D	H100x100x6x8	H100x100x6x8	H100x50x6x8
SHAPE-2 A	H125x125x6.5x9	H125x125x6.5x9	-
SHAPE-2 B	H125x125x6.5x9	H125x60x6x8	-
SHAPE-2 C	H125x125x6.5x9	H125x125x6.5x9	H100x50x5x7
SHAPE-2 D	H150x150x7x10	H150x150x7x10	-
SHAPE-3	H100x100x6x8	H100x100x6x8	-

Note : H height x flange width x web thickness x flange thickness in mm

Test Setup

In the test, cable loads which were attached at the top of the towers were assumed as constant vertical loads. A horizontal cyclic load, as simulation of earthquake load, was applied at the stiffening girder level. Constant vertical load P_{const} equal to 27.5% of the yield load was applied on the top of the model towers and an increasing horizontal cyclic loading was applied at the point A, as shown in Figure 2(a). The general arrangement of the test setup is shown in Figure 2(b). The model tower was bolted to the base. The horizontal lateral load was applied by a hydraulic jack (maximum capacity = 35 tonf) and the vertical load by a pair of gravity load simulators (total maximum capacity = 200 tonf). As the top of the model tower displaced laterally, the loading arms and load cell maintained the jack force always exerting a vertical load to the deformed tower. The horizontal jack was restrained against out-of-plane displacement and was connected to the tower by pin-connection. The out-of-plane displacement was restrained by four pairs of lateral bracings.

History of load-displacement and load- strain relationships, as well as maximum horizontal load, were experimentally investigated.

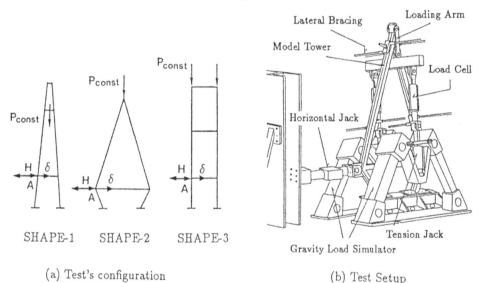

(a) Test's configuration   (b) Test Setup

Figure 2. Test's configuration.

Material Properties

Material properties of the model towers were measured with tensile coupons. The measured (average) values are shown in Table 2. Residual stress patterns of the column sections of the test towers were measured by the sectioning method. Measured residual stress patterns are considered in the following numerical calculations.

TABLE 2

Material properties $(1 \text{ kgf/cm}^2 = 0.0981 \text{ N/mm}^2)$

Tower Type	E $(\times 10^6 \text{ kgf/cm}^2)$	ν	σ_y (kgf/cm^2)	σ_u (kgf/cm^2)	ϵ_{ST} $(\times 10^{-6})$	E_{ST} $(\times 10^6 \text{ kgf/cm}^2)$
SHAPE-1 A	2.17	0.286	3285	-	-	-
SHAPE-1 B	2.18	0.270	2940	-	-	-
SHAPE-1 C	2.15	0.267	3285	-	-	-
SHAPE-1 D	2.15	0.267	3285	-	-	-
SHAPE-2 A	2.08	0.251	3549	5055	21360	0.040
SHAPE-2 B	2.10	0.259	3526	5057	22150	0.041
SHAPE-2 D	2.10	0.258	3304	5078	15620	0.039
SHAPE-3	2.10	0.286	3285	-	-	-

E = Young's Modulus, ν = Poisson's Ratio, σ_y = Yield Stress, σ_u = Ultimate Stress, ϵ_{ST} = Strain at onset of Strain-Hardening, E_{ST} = Strain-Hardening Modulus

NUMERICAL ANALYSES OF THE MODEL TOWERS

The ultimate strength and the load-displacement relationship of the model towers under the same loading conditions of the tests are computed numerically using the finite element program [3]. Geometrical and material nonlinearities are also considered in the calculation. The analysis is performed with the *displacement control strategy*. Constant vertical load P_{const} equal to 27.5% of the yield load is applied on the top of the model towers, an increasing horizontal cyclic displacement are applied at the point A as shown in Figure 2(a), and the horizontal load associated with these conditions is computed. The comparison of the numerical calculations with the experimental results are shown in Table 3. The *yield displacement* δ_{yo} and the *yield load* H_{yo} are defined as the displacement and load respectively at point A when the first yielding occurs in the structure. Figure 3 shows the comparison of numerical results with the experimental results of the model towers. In each figure, the experimental results are plotted as dashed line and the numerical results are plotted as solid line.

TABLE 3

Comparison of numerical calculation with experimental results

Tower Type	Calculation			Experimental	
	δ_{yo} (mm)	H_{yo} (tonf)	H_{max} cyclic (tonf) (1)	H_{max} cyclic (tonf) (2)	(1)/(2)
SHAPE-1 A	4.81	2.55	5.08	5.20	0.977
SHAPE-1 B	4.55	0.91	2.33	2.39	0.977
SHAPE-1 C	0.90	9.64	32.84	32.10	1.023
SHAPE-1 D	0.90	9.82	33.44	33.12	1.010
SHAPE-2 A	1.54	3.83	13.52	-	-
SHAPE-2 B	1.74	2.74	12.37	14.20	0.871
SHAPE-2 C	-	-	-	-	-
SHAPE-2 D	1.09	4.29	21.55	-	-
SHAPE-3	5.85	2.14	4.34	4.76	0.912

423

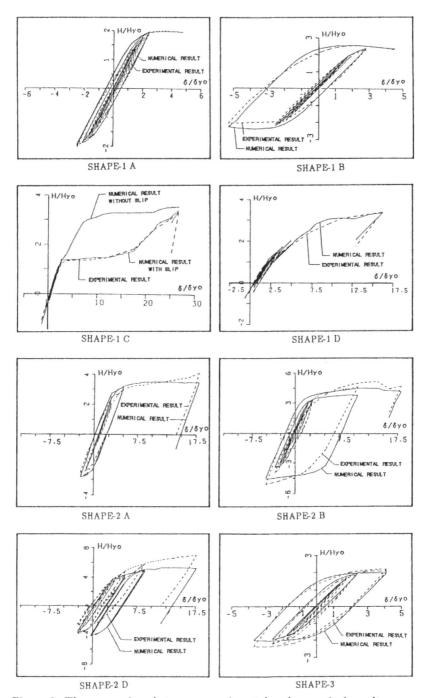

Figure 3. The comparison between experimental and numerical results.

As shown in Figure 3, the analytical model can simulate behavior of the model towers subjected to horizontal cyclic loading. In the test of SHAPE-1C, the base slipped and rigid body motion occurred. Considering the rigid body motion, the numerical results have good agreement with the test results. But these types of data with rigid body motion are meaningless. So the 'true' load-displacement curve is calculated numerically. Unfortunately, the rigid body motion occurred severely in SHAPE-2C and its test results are not considered.

Number of cycles in inelastic range has been too small, the ultimate strength and the deformation capacity of the model towers subjected to a cyclic load could not be determined. This problem is overcome when the model towers subjected to a cyclic load with new steel type are analyzed by increasing the number of cycles (as will be explained in the following section).

STUDY USING DIFFERENT TYPE OF STEELS

To prevent the structures from collapse against severe earthquake, seismic structural design with special consideration to safety has been developed which are based on up-dated research. Accordingly, steel for structural use is required to have high seismic properties. To this effect, new earthquake resisting steel which has *high tensile strength with low yield ratio* has been developed [4].

The deformation capacity of the model tower with the low yield ratio steel is studied numerically. For comparison purpose, the *high strength steel with high yield ratio* is also considered.

The stress-strain curves from measured data for both types of steel are simplified for the purpose of numerical modeling. For comparison purpose, it is assumed that both steels have the same ultimate stress. The simplified stress-strain relationship for both steels are shown in Figure 4. *Kinematic hardening rule* is employed in this study.

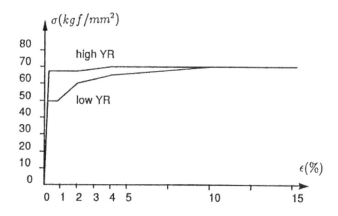

Figure 4. Stress-strain relationship for steels.

In this part, the ultimate strength and the load-displacement relationship of the real Meiko-Nishi Cable-Stayed Bridge tower with application of both type of steel is also computed numerically. The tower has 122 m height and consists of two types of stiffened box sections as shown in Figure 5. In the analytical model, thin-walled closed section beam elements are used. The rigidity of the stiffeners is uniformly smeared over the width of the plate. The equivalent plate thickness is used instead of the actual plate thickness.

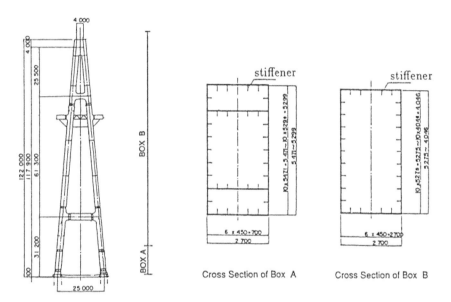

Figure 5. Real Meiko-Nishi Tower.

Deformation Capacity Factor

There are various ways to determine the deformation capacity of the structure. In this study, *displacement ratio method* is employed to determine the deformation capacity of the model towers. The deformation capacity ratios μ_{90} and μ_{max} are defined as follows:

$$\mu_{90} = \frac{\delta_{90}}{\delta_{yo}} \tag{1}$$

$$\mu_{max} = \frac{\delta_{max}}{\delta_{yo}} \tag{2}$$

Where δ_{90} corresponds to displacement after loading has decreased to a value of 90 % of H_{max} and δ_{max} corresponds to displacement when loading has reached to a maximum value H_{max}.

Cyclic Loading

Horizontal load-deflection relationship for the model towers subjected to constant vertical load and cyclic horizontal load is studied. The constant vertical loads equal to

27.5% of the yield load are applied on the top of the model towers and the *displacement control program*, as shown in Figure 6, is used in the calculation.

For a progressive sequence of loading and unloading, the line joining the peak points in the load-displacement curve of each loading sequence is termed the *skeleton curve*[5]. This skeleton curve and the displacement ratio method are used to determine the deformation capacity of the model towers.

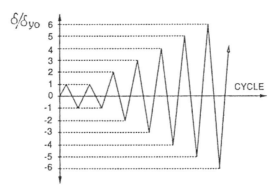

Figure 6. Displacement control program.

The comparison of results of both types of steels are shown in Table 4. Figures 7 and 8 show comparison of the results of different shapes of towers and different types of steels.

TABLE 4
Results of cyclic loading

Tower	Low YR $\sigma_y = 50$ kgf/mm^2					High YR $\sigma_y = 68$ kgf/mm^2				
Type	δ_{yo} (mm)	H_{yo} (tonf)	μ_{90}	μ_{max}	$\dfrac{H_{max}}{H_{yo}}$	δ_{yo} (mm)	H_{yo} (tonf)	μ_{90}	μ_{max}	$\dfrac{H_{max}}{H_{yo}}$
SHAPE-1 A	7.42	3.40	5.8	4.0	2.12	10.30	4.53	4.9	3.4	1.95
SHAPE-1 B	9.00	1.39	4.2	3.0	2.08	12.65	1.67	3.2	2.4	1.77
SHAPE-1 C	1.43	14.33	> 30	> 30	≈ 4.0	1.91	19.07	15.0	13.0	4.18
SHAPE-1 D	1.43	14.57	> 30	> 30	≈ 4.0	1.91	19.39	15.0	13.0	4.18
SHAPE-2 A	2.14	4.77	14.4	9.0	3.53	2.97	6.22	10.6	7.0	3.20
SHAPE-2 B	2.53	3.38	7.8	4.0	3.53	3.54	4.15	6.1	4.0	3.10
SHAPE-2 C	1.12	15.21	> 25	> 25	≈ 5.8	1.53	21.12	> 25	> 25	≈ 5.7
SHAPE-2 D	1.65	5.75	18.4	11.0	4.38	2.27	7.66	14.1	9.0	4.03
SHAPE-3	9.03	2.89	5.8	4.0	1.97	12.24	3.59	4.3	3.2	1.90
MEIKO	280	3370	7.9	4.0	1.57	383	4521	5.0	3.0	1.44

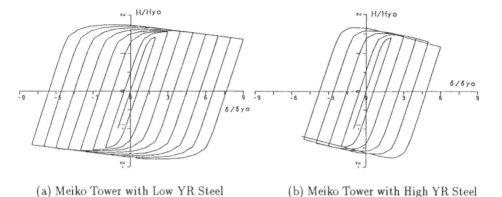

(a) Meiko Tower with Low YR Steel   (b) Meiko Tower with High YR Steel

Figure 7. Load-Displacement relationship of Meiko-Nishi.

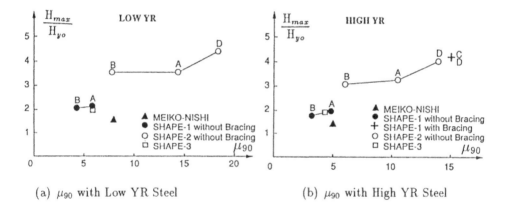

(a) μ_{90} with Low YR Steel   (b) μ_{90} with High YR Steel

Figure 8. Deformation capacities of towers subjected to cyclic loading.

The deformation capacities measured by the μ_{max} give too conservative values because they do not consider the post critical strength. The deformation capacities measured by the μ_{90} give more reasonable values. Table 4 shows that towers with low yield ratio steel have more reserve strength and deformation capacity in inelastic range in order to resist overload compared to towers with high yield ratio steel. SHAPE-2 has higher H_{max}/H_{yo} ratio than the others shape. SHAPE-1 and SHAPE-3 have almost the same H_{max}/H_{yo} ratio.

These deformation capacities are computed in the structural level and the real deformation capacities will be lower than these values, due to limitation of local buckling and joint rotation capacity. For the same reason, number of cycles in the calculation is limited.

For Meiko-Nishi tower subjected to horizontal monotonic loading, the local buckling has not occurred yet when maximum load is reached, so the deformation capacity is determined in the structural level. But for horizontal cyclic loading case, the local buckling will occur due to cumulative deterioration.

CONCLUSIONS

The following conclusions have been drawn from this study

1. Behavior of the model towers subjected to a horizontal cyclic load is significantly affected by the geometrical shape of their lower parts. The lower parts of SHAPE-1 and SHAPE-3 being geometrically similar have similar behavior. SHAPE-2 and the braced structures have slanted members so they have more capacity to withstand the horizontal load.

2. It is shown that SHAPE-2 model tower has higher H_{max}/H_{yo} ratio than the others shapes.

3. Towers with low yield ratio steel have more reserved strength and deformation capacity in inelastic range compared to towers with high yield ratio steel.

ACKNOWLEDGMENT

A part of this research was funded by Nippon Steel Corporation.

REFERENCES

1. Fukumoto, Y., Itoh, Y. and Katsuya, M., Theoretical and Experimental studies on In-plane Strength of Towers of the Meikonishi-Ohashi Cable-Stayed Bridge, NUCE Research Report No: 8101, 1981 (in Japanese).

2. Fukumoto, Y., Usami, T. and Itoh, Y., Experimental Studies on the Strength of Framed Tower of Hokko-Renraku Bridge, NUCE Research Report No: 8201, 1982 (in Japanese).

3. MARC Research Analysis Corporation, MARC Program Manual, Vol. A to Vol. E, 1990.

4. Simura, Y. and Kuwamura, H., Experiment on High-Strength Steel Beam with Different Yield Ratios, Transaction of The Architectural Institute of Japan, 1987, pp. 873 - 874 (in Japanese).

5. Wakabayashi, M., Design of Earthquake-Resistant Building , McGraw-Hill, 1986.

NEW KNOWLEDGE ON THE FATIGUE BEHAVIOUR OF STEEL STRUCTURES AND ESPECIALLY ON HOLLOW SECTION STRUCTURES UNDER DIFFERENT LOAD SPECTRA

Ö. BUCAK and F. MANG
Versuchsanstalt für Stahl, Holz und Steine, Universität Karlsruhe
Testing Center for Steel, Timber and Stones, University of Karlsruhe
Kaiserstr. 12, 7500 Karlsruhe 1, FRG

GENERAL

The object of the use of realistic load spectra is to make a more economical construction possible, and to allow an accurate safety assessment of such structures. Many constant amplitude (CA) tests with different notch cases and types of joints (T-, X-, N-, K, ...) made of hollow sections have been performed and the results were made available to the user as S-N-line catalogue [2] some years ago. Based on these investigations, design rules for the standard specification Eurocode 3 [1] have been done. In the following, the results of the fatigue investigations under variable amplitude on hollow section joints are given and the application of the Miner-rule for constructions under variable loads are shown, as it is required in Eurocode 3 [1] for this type of loading.

SYMBOLS AND DEFINITIONS

CA	Constant Amplitude	d_0	Diameter of the chord member CHS
VA	Variable Amplitude	t	Wall thickness
CHS	Circular Hollow Sections	t_0	Wall thickness of the chord member
RHS	Rectangular Hollow Sections	m	Slope of the S-N-line
R	Stress ratio $r = \min\sigma/\min\sigma$	σ	Stress [N/mm^2]
C	Constants	σ_m	Maximum stress of a load sequence [N/mm^2]
Pü	Probability of survival	S	Miner-sum
γ	Safety factor	S_R	Stress range [N/mm^2]; $S_R = (\sigma_{max} - \sigma_{min})$

d	Diameter	GL	Germanisch Lloyd
IABG	Industrie der Anlagenbetriebs-gesellschaft, Ottobrunn	LBF	Laboratorium für Betriebsfestigkeit, Darmstadt
WG3	ECCS-Working Group 3		

f_E Increase factor for the stress range $f_E = \dfrac{\Delta\sigma_{var. ampl.}}{\Delta\sigma_{const. ampl.}}$

f_A Load increment coefficient for the collective A (similar to f_E)

f_B load increment coefficient for the collective B (similar to f_E)

V_A Fullness of the load sequence A V_B Fullness of the load sequence B

TESTING PROGRAMME

Systematic tests on hollow section joints subjected to spectrum loading have been carried out at first with X- and K-type test specimens (fig. 1) under axial load applying load spectra known from literature [4, 7, 8, 11 etc.).

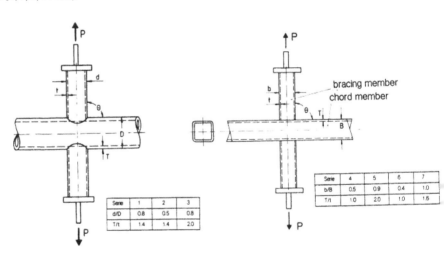

Serie	1	2	3
d/D	0.8	0.5	0.8
T/t	1.4	1.4	2.0

Serie	4	5	6	7
b/B	0.5	0.9	0.4	1.0
T/t	1.0	2.0	1.0	1.6

Fig. 1   X- and K-type joints made of hollow sections (circular and rectangular) for the tests carried out in Karlsruhe

TABLE 1
Data of the test specimens

Type of joints		chord member	bracing	material	R
(circular section)	X	Ø 101.6 x 5.6 Ø 101.6 x 5.6 Ø 101.6 x 8.0	Ø 51 x 4.0 Ø 82 x 4.0 Ø 82 x 4.0	Fe 360	+0.1 +0.1 -1.0
	K	Ø 101.6 x 5.6	Ø 51 x 4.0		+0.1
(square section)	X	100x100x4.0 100x100x6.0 100x100x4.0	50x50x4.0 90x90x3.0 40x40x4.0	Fe 360 Fe E 600	+0.1 +0.1 +0.1
	K	100x100x4.0	50x50x4,0	Fe 360	+0.1

The basis for this work are fatigue tests carried out under constant amplitude (S-N-lines). Based on the CA-tests, block tests were performed under the load sequence P = 1/3 and P = 2/3 (see fig. 2), which are known from literature [7, 8, 11 etc.), for cranes and crane railways.

By means of the load spectra "S-III Laplace" [7], North Sea" [7] and "North Sea Sequence" [12], a connection to the European Offshore Research was established. Two further load spectra, one structure subjected to wind (mast, 200 m high) the other one to wave loads (Gulf of Mexico) completed the testing program.

Fig. 2 shows a compilation of load spectra applied. The test values under load spectra have been compared with those of the Wöhler curves (S-N-lines) and the applicability of the Miner rule has been checked.

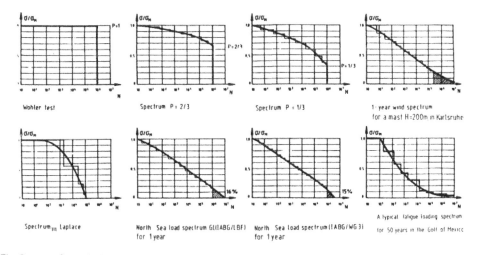

Fig. 2   Compilation of spectra applied

Endurance tests will be evaluated by means of the Miner-rule whose accuracy and applicability will be examined in the course of the present investigations with regard to the treated specimen form and the load spectra applied.

The investigations of CHS X-joints were carried out with test specimens of three different dimensions (2 diameter ratios - $d/d_0 = 0.5$ and 0.8; and two wall thickness ratios - $t_0/t = 1.4$ and 2.0). Using these specimens, the dependence of such hollow section joints on the existing geometric parameter ratios could be considered. The given parameter ratios lead to three different types of failure: a)   Joint failure in chord shear (see fig. 3)
b)   Crack starting from the weld toe of the web member (see fig. 4)
c)   Crack starting from the inside surface of the tube in the non-welded area of the chord (see fig. 5)

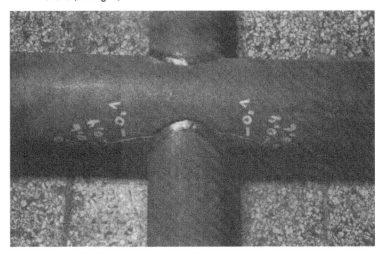

Fig. 3   Joint failure in chord shear ($d/d_0 = 0.5$ and $t_0/t = 1.4$)

Fig. 4   Joint failure from the weld toe of the web member (CHS; $d/d_0 = 0.8$ and $t_0/t = 1.4$).

Fig. 5   Joint failure from the inside surface of the tube in the non-welded area of the chord
(CHS; $d/d_0 = 0.8$ and $t_0/t = 2.0$)

A detailed report has been given in [4, 5 and 6]. In the following, further investigations will be presented. The failure mode for rectangular hollow section (RHS) joints is identical to that of circular hollow section (CHS) joints. Figure 6 shows the test specimen with the geometric parameter $b/b_0 = 0.5$ and $t/t_0 = 1.0$ after the failure.

Fig. 6   Joint failure in the chord shear ($b/b_0 = 0.5$ and $t_0/t = 1.0$)

From figure 5 to 7, the result of the investigations on X-type CHS-joints, and from figures 8 to 10, the results of X-type RHS-joints can be seen. The stress range ($S_R = \Delta\sigma$) is plotted in the vertical axis of these diagrams. The test results under load sequence have been plotted on the level of the maximum stress range in a collective. The tests were performed with the same testing machines, under the same test conditions without corrosion impact.

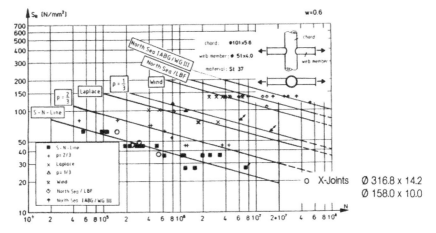

Fig. 7   Results of the fatigue investigations on CHS X-type joints under different load spectrum ($d/d_0 = 0.5$ and $t_0/t = 1.4$)

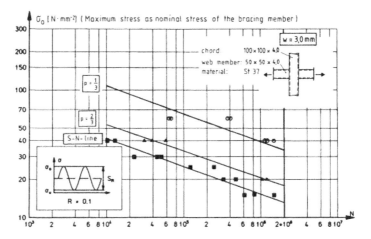

Fig. 8   Results of the fatigue investigations on RHS X-type joints under load spectrum ($b/b_0 = 0.5$ and $t_0/t = 1.0$)

The determination of the S-N-lines for different load collectives has the same slope as the S-N-lines for constant amplitude tests. After completion of the first tests under the collectives P = 2/3, Laplace S_{III} and P = 1/3, we found out that the slope of each S-N-line is the same. Therefore, we decided to perform the tests under load collectives on one stress range level, and to draw a fatigue life line with the same slope as used under constant amplitude by means of the mean value of the load cycle numbers of the test points. From tests on small and L-type specimen tests it is known that the slope can become smaller, e.g. bigger values for m. In favor of considering the scatters, this positive phenomenon has been ignored.

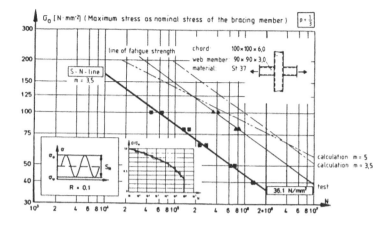

Fig. 9   Results of the fatigue investigations on RHS X-type joints under load spectrum
P = 1/3 (b/b$_0$ = 0.9 and t$_0$/t = 2.0)

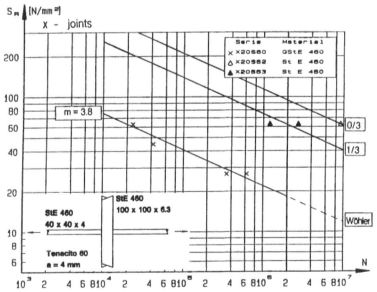

Fig. 10   Results of the fatigue investigations on RHS X-type joints loaded under different load
spectrum; material Fe E 600

An important question to be answered concerns the size of the crack, when a joint type loses its serviceability. Since the definition of the bearing capacity of the applied testing load proved to be only a bulk value, measurements of the crack growth were conducted. As a criterion for discon-necting the testing machine, an elongation of the test pieces from 0.1 up to 3 mm, divided into various stages, was fixed. The limitation of the maximum elongation of the test pieces to 3 mm proved to be sufficient, since the crack reached over half of the width of the web chord when using this value. Nevertheless, even in this failure condition, the test piece could still bear the maximum testing force.

The EGKS-standards [7] indicate four different load cycles or failure criteria for such investigations.

N1   Load cycle by a strain reduction of 15% (ϵ - 15% = 0.85 ϵ) close to the first crack.

N2   Load cycle when the first discernible crack occurs.

N3   Load cycle through wall cracking (crack goes through the whole wall thickness).

N4   Load cycle at the end of the test.

where the load cycles N1 and N2 only slightly differ from each other.

Before defining the cut-off criteria, the relation between the load cycle numbers have been determined by means of single tests. One of the test results is given in figure 11.

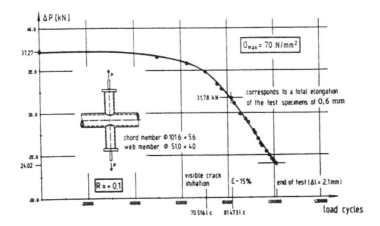

Fig. 11   Reduction of the upper loads (simultaneous increase of the lower load) when testing a hollow section joint upon a constant oil transporting amount with a testing machine without internal control system in dependence on the crack or load cycle number.

In the figure, each stage corresponds to an elongation of the test specimen by 0.1 mm
In order to clarify the effect of the shape influence on the service fatigue strengths, besides X-type joints, K-type joints made of circular and rectangular hollow sections were also tested under identical loading collectives up to now according to fig. 2. A test rig, which has been used for previous investigations of K-joints, was applied in order to gain the truss forces as they occur for lattice structures.

The results of the investigations on the fatigue strength of K-type truss joints are presented in figure 12. Figure 13 shows one of the specimens after failure. The type of failures is identical to that of X-type joints according to figures 3 and 6.

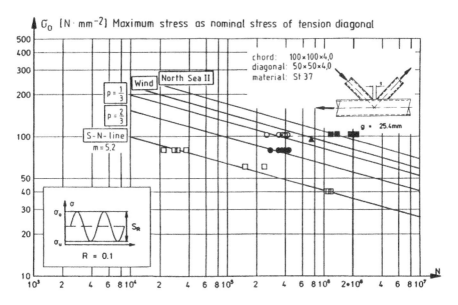

Fig. 12   Results of the fatigue investigations (RHS K-type joints) under different load spectra
(b/b$_0$ = 0.5; T/t = 1.0)

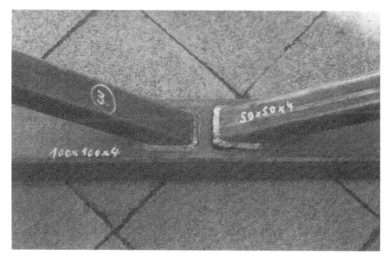

Fig. 13   Test specimen after failure; K-type joint (b/b$_0$ = 0.5)

PROPOSAL FOR THE UTILIZATION OF TEST RESULTS

Through the comparison of stress ranges for a given load cycle, increase factors are obtained from the test values for the stress range in depend-ence on the collectives.

$$f_E \;=\; \frac{\Delta\sigma_{Collective}}{\Delta\sigma_{W\ddot{o}hler}}$$

They are for the collectives

2/3	=>	$f_E = 1.3$
1/3	=>	$f_E = 1.8$
North Sea LBF	=>	$f_E = 3.0$
North Sea WG3	=>	$f_E = 3.4$
Wind	=>	$f_E = 3.0$

APPLICATION OF THE PALMGREN-MINER-RULE

Most of the new design rules recommend the application of the Miner-rule for the consideration of varying load (load sequence), e.g. for the proof of fatigue strength. For this reason, the meaningfulness of the Miner-rule has been checked in the scope of this work. The results of the Miner evaluation are presented graphically by means of fig. 14.

The application of the Palmgren-Miner-Rule <u>did not</u> result in a good accordance in the field investigated (fig. 14).

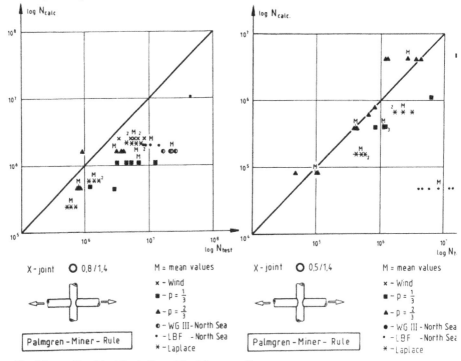

Fig. 14   Graphical illustration of the Miner results

This statement is applicable to all possibilities of modified S-N-lines (different cut-off limits; different slope etc.) [5] and fig. 15.

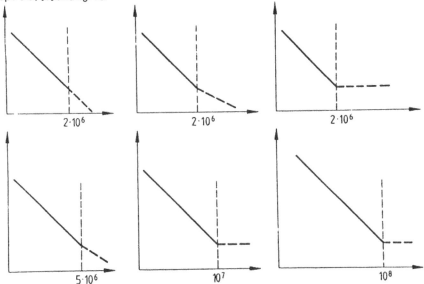

Fig. 15   Different cut-off limits of S-N-lines for the application of the Palmgren-Miner-rule

From figure 16, the frequency of the Miner sum determined from the tests can be seen. The results of the investigations made in Karlsruhe are recorded comparatively into the diagram according to Schütz/Zenner [3].

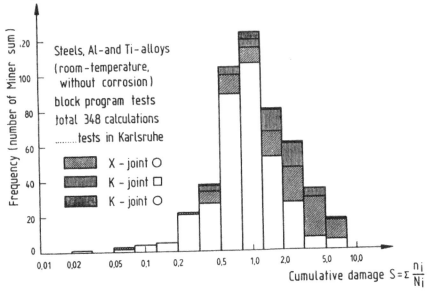

Fig. 16   Failure frequency; presentation according to [3] supplemented by tests carried out in Karlsruhe

It can be concluded from available tests that the cumulative damage results in values below 1.0 for load collectives with a higher fullness, as for example 2/3, and in values between 3.0 and 4.0 for a smaller fullness.

For the practical application of the Miner-rule, the cumulative damage indicated in table 2 is recommended for different types of collectives.

TABLE 2
Cumulative damages in dependence on the type of collective
S_{min} is the lowest value determined during the tests in Karlsruhe
$S_{proposed}$ is the sum of damages to be used for the calculation.
The poor single results due to weld defects and roughnesses of the specimens
were not considered.

Form of load spectrum Collective	Damage accumulation All test results (All types of failure) for all R-values		Damage accumulation non considering the series with the crack on the inside of the tube and $R \geq 0.1$	
	S_{min}	$S_{proposed}$	S_{min}	$S_{proposed}$
2/3	0.27	0.3	0.27	0.5
1/3	0.34	0.5	2.11	3.0
Laplace	1.87	----	1.87	2.5
North Sea (LBF-IABG)	0.48	1.0	3.18	4.0
North Sea (LBFWG III)	0.56	1.0	1.59	4.0
Wind	0.33	1.0	1.47	2.8

These indications are valid for structures with small tube dimensions as they occur for crane systems, masts etc. Presently, experimental investigations are being carried out with hollow sections of a bigger dimension and bigger wall thickness. The results of big-size joint tests (X-joints) with the dimensions chord member Ø 316.8 x 14.2 and web member Ø 158 x 10.0 are plotted in the diagram as shown in fig. 7.

PROPOSAL FOR THE DESIGN OF HOLLOW SECTION COMPONENT PARTS UNDER LOAD SPECTRUM

As the design calculation of the structural component parts by using the Miner-rule or the relative Miner-rule did not give good results in the past, new design methods were sought for practical application.

While considering the load spectrum with the area under the corresponding staircase curves, and the endurance lines for these collectives determined by tests according to figures 7, 10 and 12 as well as the test result given in [5], it occurs to one that the maximum stresses or alternatively the endurance (load cycles) for certain stress levels increase with the decreasing fullness coefficient.

The fullness of a load spectrum is an integration of each level over the number of cycles (N).

The following hypothesis is formulated using this reverse effect:

"The ratio of the fullness (area) of two load spectra behaves inversely proportional to the load increment coefficients of these load spectra".

According to the hypothesis, the following is valid for the load spectra A and B:

$$\frac{V_A}{V_B} = \frac{f_B}{f_A}$$

V_A fullness of the load spectrum (collective) A
V_B fullness of the load spectrum (collective) B
f_A load increment coefficient for the load spectrum (collective) A
f_B load increment coefficient for the load spectrum (collective) B

The load increment coefficient is defined as follows:

$$f_i = \frac{\text{maximum stress for a joint under load spectrum collective i (VA)}}{\text{maximum stress for the same according to the Wöhler tests (CA)}}$$

Fig. 17 shows the determination of fi-values (schematically).

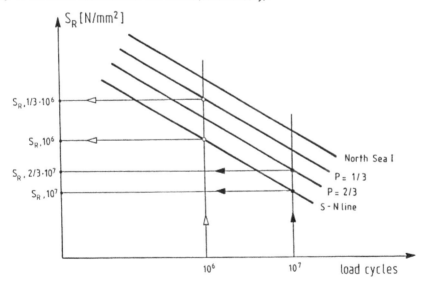

Fig. 17   Determination of the load factors (schematically)

In general, the above mentioned hypothesis can be expressed for any load spectrum i as follows:

$$V_i \cdot f_i = C \quad \text{where C is a constant, which has to be determined empirically by tests.}$$

The argument for the above hypothesis is that the fullness coefficient or alternatively the product of the fullness coefficient and the corresponding load increment coefficient is considered as physical value and defined as follows:

- The fullness coefficient represents the quantity of energy applied on the structural component part in the form of deformation energy.

- The product of the fullness coefficient and the load increment coefficient represents a physical limit, a measuring number, which shows the load bearing capacity of a structural component part for any load spectrum (collective), where the constant amplitude (Wöhler) test is taken for reference collectives.

The advantage by using this method is that only one reference S-N-line (Wöhler = CA) is required in total, in this case the Wöhler-line for the structural component part is to be designed. The proof for the fatigue strength for any other load spectrum can be done using this recalculation factor.

In order to obtain the life endurance line for any other load spectrum (collective), it is only necessary to carry out constant amplitude tests for a certain structure or to use constant amplitude tests already available [2].

The disadvantage of this method is that at the moment the lowest value is obtained by using this method.

This hypothesis was checked for various types of load spectra (collectives) by means of extensive investigations.

Table 3 contains the maximum and minimum values of the load increment coefficient calculated using the test data and the measuring number for the design according to the figures 5, 6, 7, 9 and other test values according to [5].

TABLE 3
$V_{colletive}$, $f_{collective}$-design for different load spectra

Definition of the load spectrum (Block collective)	Fullness[1] $V_{collective}$	Load increment coefficient		Proposed ratio value[2] for the design
		Min. $f_{coll.min}$	Max. $f_{coll.max}$	$f_{coll.design}$
2/3	$6.586 \cdot 10^7$	1.31	1.68	1.3
1/3	$4.395 \cdot 10^7$	1.82	2.98	1.8
North Sea LBF	$2.330 \cdot 10^7$	3.15	5.60	3.0
North Sea WGIII	$2.164 \cdot 10^7$	3.40	6.80	3.4
Wind	$2.201 \cdot 10^7$	2.87	4.80	2.8

[1] determined for a maximum stress range of $S_r = 100$ N/mm^2 and a collective range of 10^6 cycles

[2] due to the small number of test data, one starts from the lowest value, while the minimum value is rounded off to the lower side (presently, $f_{coll.design}$ is similar to the increase factor f_E).

A constant close to $7.5 \cdot 10^7$ for a collective range of 10^6 loading cycles is obtained by multiplying the corresponding fullness coefficients with the measuring values on the right hand column of the table. This value is, however, only an approximative value, that represents an acceptable calculation value for all described load collectives.

If the fullness coefficient for a new collective be now obtained by integrating the staircase curve, the load increment coefficient can be calculated by determining the ratio of the limiting value $7.5 \cdot 10^7$ to the given fullness coefficient.

EXAMPLE FOR THE CALCULATION

The proof of the fatigue strength is to be given for a portal crane made of circular hollow sections. The calculation is to be made for a load sequence (collective) of $P = 0/3$. The load cycle number is $2 \cdot 10^6$.

For the given joint geometry, a load cycle number of $2 \cdot 10^6$, a probability of survival of 50% ($P_{\ddot{u}} = 50\%$, and a bearable stress range of 42.0 N/mm^2 are determined from the Wöhler test (CA-test)

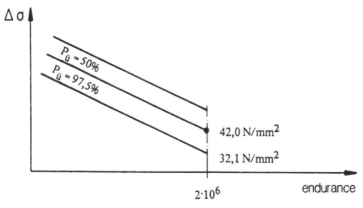

Conversion of the bearable stress range of $P_{\ddot{u}} = 50\%$ to $P_{\ddot{u}} = 97.5\%$ can be done by means of the factors from the table in [5].

$$\Delta\sigma\ 50\% \ / \ \Delta\sigma\ 97.5\% \ = \ 1.31$$

Thus, this results in $\Delta\sigma_{P_{\ddot{u}} = 97.5\%} = 42.0 / 1.31 = 32.06$ N/mm^2

(This value can be also taken from the standards).

For a collective $P = 0/3$, the fullness $V_{0/3}$ is obtained (for a maximum stress range of $S_R = 100$ N/mm^2 and for a collective range of 10^6 cycles.

$$V_{0/3} = 2.115 \cdot 10^7$$

The load increment coefficient $f_{0/3}$ is calculated through

$$f_{0/3} \quad = \quad \frac{C}{V_{0/3}} \quad = \quad \frac{7.5 \cdot 10^7}{2.115 \cdot 10^7} \quad = \quad 3.54$$

The stress range $\Delta\sigma_{P\ddot{u} = 97.5\%}$ = 32.06 N/mm^2 calculated for $P_{\ddot{u}}$ = 97.% is multiplied with this load increment coefficient.

$$\Delta\sigma_{P\ddot{u} = 97.5\%} \cdot f_{0/3} = 32.06 \cdot 3.54 = = 113.5 \text{ N/mm}^2$$

γm = 1.25 (from the chapter "Proposal for a safety concept") [5]

$$\frac{\Delta\sigma 97.5; 0/3}{\gamma m} = \frac{113.5}{1.25} = 90.8 \text{ N/mm}^2$$

Thus, the allowable maximum stress range for this joint geometry under spectrum loading $P = 0/3$; for $2 \cdot 10^6$ cycles, is 90.8 N/mm^2.

REFERENCES

1. N.N.: Eurocode 3, Design of Steel Structures, August 88

2. Mang, F., Bucak, Ö. and Klingler, I.: Wöhlerlinien-Katalog für Hohlprofilverbindungen Studiengesellschaft für Anwendungstechnik von Eisen und Stahl e.V., Düsseldorf, Juni 1987, Januar 1988

3. Schütz and Zenner: Schadensakkumulationshypothesen zur Lebensdauer-vorhersage bei schwingender Beanspruchung
    Zeitschrift für Werkstofftechnik 1973
    Part 1, No. 1 p. 25-33
    Part 2, No. 2 p. 97-102

4. Mang, F. and Ö. Bucak: Investigations into the Fatigue Behaviour of Hollow Section Joints Subjected to Spectrum Loading
    ISOPE '91, Edinburgh 11.-15.08. 1991

5. Bucak, Ö.: Bewertung des Ermüdungsverhaltens von Hohlprofilkonstruk-tionen (Evaluation of the Fatigue Behaviour of Hollow Section Joints) PhD Thesis, University of Karlsruhe, May 1990

6. Bucak, Ö. and Mang, F.: Investigations into the Fatigue Behaviour of CHS-Joints Subjected to Spectrum Loading, Third International Symposium on Tubular Structures, Lappeenranta, September 1 and 2, 1989

7. N.N.: Stahl in Meeresbauwerken
    Proceedings of the International Conference in Paris, Oct 15-18, 1981
    EUR-Bericht Nr. 7347 (1981), p. 439-483 and Plenary Session No. 5

8. Bierett, G.: Einige wichtige Gesetze der Betriebsfestigkeit geschweißter Bauteile aus Stahl. Schweißen und Schneiden, Heft 11, 1072, p. 429-434

9. Haibach, E.: Modifizierte lineare Schadensakkumulationshypothese zur Berücksichtigung des Dauerfestigkeitsabfalls mit Fortschreiten der Schädigung, LBF Darmstadt, TM 59/70

10. Lipp, W.: Zur Lebensdauerabschätzung mit dem Blockprogramm und Zufallslastenversuch LBF Darmstadt, TM 76/80

11. DIN 15018: Krane, Grundsätze für Stahltragwerke; Berechnungen, Ausgabe April 1974

12. Sonsino, C.M.; Klätschke, H.; Schütz, W. and Hück, M.: Standardized Load Sequence for Offshore Structures
WASH-1
LBF Report No. FB 181 (1988)
IABG-Report No. TF 2347 (1988)

13. Sonsino, C.M. and Lipp, K.: Übertragbarkeit des an Winkelproben ermittelten Betriebsfestigkeitsverhaltens auf große Rohrknoten für die Offshore Technik
5. LBF Kolloquium in Darmstadt, March 8-9, 1988
Report No. TB-180 (1988)

ESTIMATING WELDMENT LIFE UNDER VARIABLE LOADS

FARREL J. ZWERNEMAN
Associate Professor
Civil Engineering, Oklahoma State University
Stillwater, Oklahoma 74078-0327
USA

KURT D. SWENSSON
Project Engineer
Stanley D. Lindsey and Associates Ltd.
300 Galleria Parkway, Suite 450
Atlanta, Georgia 30339-3149
USA

KARL H. FRANK
Professor
Civil Engineering, The University of Texas
Austin, Texas 78712
USA

ABSTRACT

Results of fatigue tests on steel weldments loaded under variable amplitude stress-time histories are presented. The data from these tests are used to assess the advisability of neglecting damage produced by small cycles in the stress-time history when estimating the fatigue life of a structure.

INTRODUCTION

A highway bridge in service is required to support stress cycles which vary widely in range and mean stress level. Laboratory tests have generally been conducted using stress cycles with constant range and mean. To use available constant amplitude data to estimate the fatigue life of a bridge in service it is necessary to convert the variable amplitude service loading to an equivalent constant amplitude loading. This conversion is generally accomplished using the rainflow method [1] to count cycles and Miner's rule [2] to assess damage.

When Miner's rule is used to assess damage, the question arises as to whether cycles below the infinite life fatigue limit should be included in the damage summation. Several researchers [3,4,5] have proposed analytically attenuating the damage produced by small cycles in a variable amplitude history. The effect of this attenuation is to change the slope of the fatigue life curve when the stress-time history contains cycles below the constant amplitude fatigue limit. Other researchers contend that this attenuation does not always occur [6,7]. In fact, for some stress-time histories there can be an acceleration in damage [7,8].

To investigate the cause of these differences in experimentally observed behavior, fatigue tests were conducted using several different variable amplitude stress-time histories. These histories vary from a relatively simple superposition of sine waves to more complex histories recorded from a bridge in service. The results of these tests will be used to assess the advisability of analytically attenuating damage produced by small cycles in a stress-time history.

BACKGROUND

Damage models may be categorized as linear, nonlinear, or interaction. In a linear model, damage assigned to individual cycles in a variable-amplitude history is computed directly from the fatigue-life equation, $N = A S_r^{-m}$. In this equation, N is number of cycles to failure, A is an empirical constant equal to the N-axis intercept, S_r is stress range, and m is the slope of the S_r-N curve. To determine total damage produced by the history, damage produced by individual cycles making up that history is added linearly. It is assumed that all cycles produce damage and that cycles of different sizes do not interact, i.e., retard or accelerate growth. The best-known and most widely used linear fatigue-damage model is Miner's rule. Miner's rule can be combined with the standard fatigue-life equation to produce an expression for effective stress range [9].

$$S_{reff} = [\Sigma(n_i/N_t)S_{ri}^m]^{1/m}$$

where S_{reff} is the effective stress range, n_i is the number of cycles applied at stress range S_{ri}, and N_t is the total number of cycles to failure. The effective stress range is the stress range of a constant-amplitude stress-time history which will produce the same fatigue damage as a variable amplitude history, for the same number of cycles.

The linear damage model has been used by numerous researchers to estimate fatigue lives of test specimens. The linear model provided a conservative estimate of fatigue lives for specimens tested by Barsom and Schilling et al. [9,10], but this work was predominately directed at the effect of cycles above the constant amplitude fatigue limit. Fisher et al. [6], Schijve et al. [11], Heuler and Seeger [12], and Dowling [13] conducted tests with large percentages of cycles below the fatigue limit and concluded that cycles below the fatigue limit can be damaging when they are part of a variable-amplitude history containing some cycles above the fatigue limit. Fisher and Dowling have recommended that all cycles in a variable-amplitude history be linearly added in the damage summation.

Nonlinear damage models are based on the assumption that damage produced by individual cycles in a variable-amplitude history is affected by the presence of other stress cycles in the history. Damage produced by a cycle is assumed to depend on such things as the relative magnitude of the cycle within the history, the mean level of cycle within the history, or the crack length at the time that cycle is applied. This type of model is still not a strict interaction model, because damage is not treated as a function of prior load history on a cycle-by-cycle basis.

Nonlinear models have been proposed by Albrecht and Friedland [3], Haibach [4], Tilly and Nunn [5], Gurney [8], and Swensson [14]. The Albrecht/Friedland, Haibach, and Tilly/Nunn models make use of Eq. 1, but either ignore or attenuate damage produced by cycles below the constant-amplitude fatigue limit. The Gurney model is less dependent on Eq. 1; damage is taken to be a function of the ratio of minor cycle stress range to major cycle stress range. The Swensson model also employs Eq. 1, but applies a correction factor which is dependent on the maximum and mean stresses in a cycle. Gurney's and Swensson's models differ from the other nonlinear models described above in that these models can predict an increase in the damage produced by a variable amplitude stress-time history while the other models predict a decrease. Experimental results have been published to support all of these nonlinear models.

Interaction models have been developed which estimate fatigue crack growth on a cycle-by-cycle basis [15,16]. Interaction models generally include some means of considering the retarding effect of overloads and may include the accelerating effect of underloads. Available interaction models were developed for the aerospace industry, and require a much more detailed knowledge of the stress-time history than would be available for bridges.

EXPERIMENTAL PROGRAM

Fatigue tests were conducted with tee-shaped weldments made from A572 Grade 50 steel plate. Specimen dimensions are shown in Fig. 1 and a photograph of the specimen bolted to the load frame is shown in Fig. 2. An upward force is applied to the stem of the tee through the knife-edged attachment. Loads were applied using a closed loop servo-hydraulic system. Variable amplitude load-time histories were produced with a microcomputer substituted for the standard function generator. Load was continuously monitored with a peak detector and intermittently monitored with a digital oscilloscope. Fatigue cracks initiate on the tension side of the tee stem at the fillet weld toe and grow up through the stem.

Stress-time histories used in this research include constant amplitude (CA), superimposed sine (SS), test truck (TT), and traffic (TR). The CA history was used to establish a base for estimating damage due to variable amplitude loads. The SS, TT, and TR histories are variable amplitude stress-time histories representing the passage of single or multiple trucks over a bridge.

The SS, and TT histories are shown in Figs. 3 and 4. The SS history is a simple approximation of the variable amplitude loading produced by a truck passing over a bridge. The SS loading is simpler to handle, both analytically and experimentally, than a true bridge loading

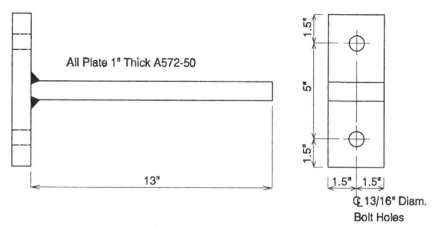

All Plate 1" Thick A572-50

13"

1.5"

5"

1.5"

1.5" 1.5"

¢ 13/16" Diam.
Bolt Holes

Figure 1. Specimen Dimensions

Figure 2. Specimen in Load Frame

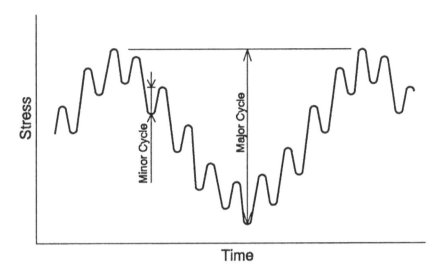

Figure 3. Superimposed Sine Stress-Time History

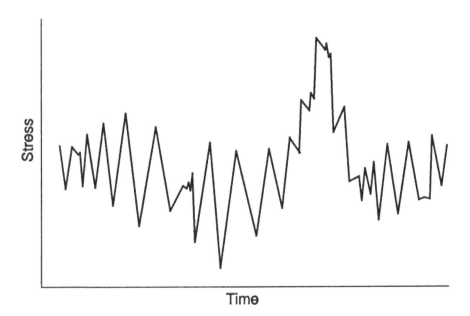

Figure 4. Test Truck Stress-Time History

because of the regular way in which stress varies with time. The TT history was recorded on a cantilever bridge as a truck passed over the bridge at a speed of 50 mph (80.5 km/hr) [17].

The components of the TR histories were obtained at the same time and place as the TT history. The TR histories represent portions of ten minutes of strain data recorded while the bridge was in normal service. During this ten minute period, nine significant loadings took place (corresponding to the presence of single or multiple trucks on the bridge) separated by periods of light vehicle traffic. TR histories used in the tests reported here were developed from portions of this recorded data, and include combinations of loadings representing both significant events and light vehicle traffic [14].

EXPERIMENTAL RESULTS

Results of constant amplitude fatigue tests are shown in Fig. 5. A total of 30 constant amplitude tests were performed at stress ranges from 12.0 ksi (82.8 MPa) to 33.5 ksi (231 MPa). A least squares regression line and 95% confidence limits are shown on the figure. Only specimens tested to failure are included in the statistical analyses. For comparison, a line representing an American Association of State Highway and Transportation Officials (AASHTO) [18] category B detail for redundant members is shown with the data. The sloped portion of the category B line matches the lower confidence limit fairly well, but the category B fatigue limit is considerably above whatever fatigue limit exists for the tee-shaped weldment.

Results of variable amplitude tests are plotted in Fig. 6. Data are plotted as effective stress range versus total number of cycles to failure. Cycles are counted using the rainflow method and damage is summed using the linear model. Along with the data are plotted the regression line and confidence limits from the constant amplitude tests. For comparison to data with low values of effective stress range, the regression line is extrapolated to the most extreme variable amplitude data point; confidence limits are stopped at the boundary of the constant amplitude data.

In Fig. 6, it can be seen that the linear model routinely overestimates fatigue life for all variable amplitude load-time histories tested here other than the traffic histories. The obvious consistent difference between the traffic histories and the other histories is the magnitude of the effective stress range. Fatigue lives for stress-time histories with effective stress ranges below approximately 12 ksi (82.8 MPa) are accurately estimated.

Regarding the effect of cycles below the constant amplitude fatigue limit, it is instructive to examine the SS history with the smallest minor cycles. This SS history is made up of one cycle at 33.3 ksi (230 MPa) and 9 cycles at 5.1 ksi (35.2 MPa). The effective stress range for this history is 18.1 ksi (125 MPa), assuming all cycles are damaging. The estimated fatigue life for this history is 3.93×10^6 cycles; the measured fatigue life was 1.83×10^6 for one of the two specimens tested under this history and 3.39×10^6 for the other. To incorporate the minor cycles into the damage sum, it was necessary to extrapolate the regression line for the constant

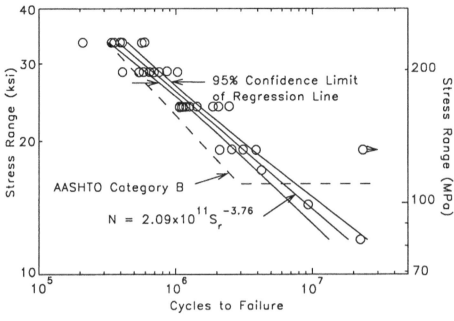

Figure 5. Constant Amplitude Data

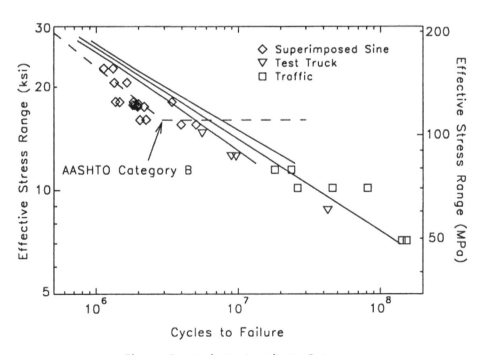

Figure 6. Variable Amplitude Data

amplitude data to a stress range of 5.1 ksi (35.2 MPa). At this constant amplitude stress range, 457×10^6 cycles would be required to fail the specimen. Using cycle count as an indicator since fatigue limit was not experimentally determined, it would normally be assumed that 5.1 ksi (35.2 MPa) is below the fatigue limit. If the 5.1 ksi stress range is accepted as being below the fatigue limit, 90 percent of the cycles in the considered SS history are below the limit and the linear model is unconservative.

CONCLUSIONS

It is recognized that a linear model will not provide an accurate estimate of fatigue life for all stress-time histories. Some researchers have concluded that the linear model is conservative while others have concluded that it is unconservative. Work reported here has shown that a linear model can produce highly unconservative results, even when a large percentage of the cycles are below the fatigue limit and all cycles are included in the damage sum.

The linear model has been shown to be accurate for stress-time histories measured from a bridge in service. It is likely that this accuracy is more the result of a fortunate combination of cycles causing acceleration, cycles causing retardation, and cycles causing no damage, than it is a result of accurate behavioral modeling. However, until a more sophisticated model can demonstrate improved accuracy for the stress-time history of primary concern (that produced by a truck on a bridge) the simplicity inherent in the linear model should result in its continued use.

REFERENCES

1. Endo, T., Mitsunaga, K., Takahashi, K., Kobayishi, K., and Matsuishi, M., "Damage Evaluation of Metals for Random or Varying Load—Three Aspects of Rainflow Method," Mechanical Behavior of Materials, Proceedings of the 1974 Symposium on the Mechanical Behavior of Materials, The Society of Materials Science, Japan, 1974.

2. Miner, M. A., "Cumulative Damage in Fatigue," J. Appl. Mech., Trans., ASME, 67, pp. A159-A164, 1945.

3. Albrecht, P., and Friedland, I. M., "Fatigue-limit Effect on Variable-Amplitude Fatigue of Stiffeners," J. Structural Div., 105 (ST12), 2657-2675, 1979.

4. Haibach, E., (Contribution to Discussion) In Fatigue of Welded Structures, Proceedings of the Conference, The Welding Institute, Brighton, England, Vol. 2, pp. xx-xxii, 1970.

5. Tilly, G. P., and Nunn, D. E., "Variable Amplitude Fatigue in Relation to Highway Bridges," Proc. Inst. Mech. Eng. Appl. Mech. Group, 194(27), 259-267, 1980.

6. Fisher, J. W., Mertz, D. R., and Zhong, A., "Steel Bridge Members Under Variable Amplitude Long Life Fatigue Loading," National Cooperative Highway Research Program Report 267, Transportation Research Board, National Research Council, Washington, D. C., 1983.

7. Zwerneman, F. J., and Frank, K. H., "Fatigue Damage Under Variable Amplitude Loads," Journal of the Structural Division, ASCE, Vol. 114, No. 1, 67-83, January 1988.

8.   Gurney, T. R., "Fatigue Tests on Fillet Welded Joints to Assess the Validity of Miner's Cumulative Damage Rule," Proc. Roy. Soc. London, **A386**, 393-408, 1983.

9.   Schilling, C. G., Klippstein, K. H., Barsom, J. M., and Blake, G. T., "Fatigue of Welded Steel Bridge Members Under Variable-Amplitude Loadings," National Cooperative Highway Research Program Report 188, Transportation Research Board, National research Council, Washington, D. C., 1978.

10.   Barsom, J. M., "Fatigue-Crack Growth Under Variable-Amplitude Loading in ASTM A514-B Steel," Progress in Flaw Growth and Fracture Toughness Testing, ASTM STP 536, American Society for Testing and Materials, 147-167, 1973.

11.   Schijve, J., Vlutters, A. M., Ichsan, and Provo Kluit, J. C., "Crack Growth in Aluminum Alloy Sheet Material Under Flight-Simulation Loading," Int. J. Fatigue, **7**(3), 127-136, 1985.

12.   Heuler, P. and Seeger, T., "A Criterion for Omission of Variable Amplitude Loading Histories," Int. J. Fatigue, **8**(4), 225-230.

13.   Dowling, N. E., "Estimation and Correlation of Fatigue Lives for Random Loading," Int. J. Fatigue, **10**(13), 179-185, 1988.

14.   Swensson, K. D., "The Application of Cumulative Damage Theory to Highway Bridge Fatigue Design," Master's Thesis, Department of Civil Engineering, The University of Texas at Austin, 1984.

15.   Engle, R. M., and Rudd, J. L., "Analysis of Crack Propagation Under Variable Amplitude Loading Using the Willenborg Retardation Model," AIAA Paper No. 74-369, AIAA/ ASME/SAE 15th Structures, Structural Dynamics and Materials Conference, Las Vegas, Nevada, April, 1974.

16.   Newman, J. C., "A Crack Closure Model for Predicting Fatigue Crack Growth Under Aircraft Spectrum Loading," Methods and Models for Predicting Fatigue Crack Growth Under Random Loading, ASTM STP 748, Edited by J. B. Chang and C. M. Hudson, American Society for Testing and Materials, 53-84, 1981.

17.   Hoadley, P. W., "Estimation of the Fatigue Life of a Welded Steel Highway Bridge from Traffic Data," Master's Thesis, Department of Civil Engineering, The University of Texas at Austin, 1982.

18.   American Association of State Highway and Transportation Officials, Standard Specifications for Highway Bridges, 14th Ed., 1989.

HYSTERETIC BEHAVIOR OF FRAMES WITH THREE DIFFERENT COLUMN-BASES

SHIGETOSHI NAKASHIMA AND TOSHIAKI SUZUKI
Associate Professors/Department of Architecture
Osaka Institute of Technology
5-16-1 Ohmiya, Asahi-ku, Osaka 535, JAPAN

ABSTRACT

The column bases in a steel structure play a very important role in
connecting the columns to the foundation. This report describes the
mechanical behaviors of 1-story 1-span portal rigid frames each connected
to different types of conventional column bases.

INTRODUCTION

There are three types of column bases, which have been used conventionally
in steel structures, namely exposed type, encased type and embedded type.
The Japanese aseismatic design standards regulate that the story drift
shall not exceed 1/200 of each story height. Thus, the column bases must
have required degrees of fixity. To ensure sufficient rotational rigidity
and also to facilitate repair of the column bases when they are damaged,
the column bases are often encased in reinforced concrete. However, the
mechanical properties of the column bases after such reinforcement/repair
greatly differ from those of the column bases before such reinforcement/
repair. Therefore, the hysteretic performance, stress distribution,
height of the column inflection point, etc. of the 1-story 1-span rigid
portal frames under lateral loading will be investigated as the supporting
conditions of these columns are changed.

EXPERIMENT PLANNING

Table 1 shows the experiment plan. As shown in Fig. 1, the supporting
conditions of the columns for the 1-story 1-span rigid portal frames as
reduced to an approximately 1/4 scale model were varied. Namely, these
steel frames were each connected to one of five different types (i.e.,
pinned type, fixed type and three conventional types) of column bases. In
all cases of encased type column base, the length of each side of the
reinforced concrete section was 3 times the outside dimension of the steel
square tube. In all cases of embedded type column base, the width of the

foundation beam was taken as 3 times the outside dimension of the steel square tube. As shown in Fig. 1, lateral load was applied repeatedly to the centroid of steel beam section with a constant axial compression on the columns.

Table 1
Experiment plan

Column base type	No.	Specimen name	N/N_u	Remark	Material No.
Fixed type	1	K-11	0%		
	2	K-12	15%		I
	3	K-13	30%		
Pinned type	4	P-11	0%		II
Exposed type	5	R-11	0%	A.Bt 4-M12	
	6	R-12	15%	B.Pl 190x190x16	
	7	R-21	0%		I
	8	R-22	15%	A.Bt 8-M12	
	9	R-23	30%	B.Pl 190x190x16	
Encased type	10	N-21	0%	Height h=200 (2D)*	
	11	N-22	15%	Section bxd=300x300	
	12	N-31	0%	Height h=300 (3D)*	I
	13	N-32	15%	Section bxd=300x300	
Embedded type	14	U-11	0%		II
	15	U-12	15%	Depth l=150 (1.50)*	I
	16	U-13	30%		I
	17	U-21	0%		II
	18	U-22	15%	Depth l=250 (2.50)*	I

※ D: Outside dimension of tubular column

Table 2
Mechanical properties of materials

Steel	Size	σ_y (N/mm²)	σ_u (N/mm²)	E x10⁵(N/mm²)	ε %	Material No.
Beam	H-150·75	362	479	1.94	27.3	I
Column	□-100·100	362	491	1.95	32.4	II
		407	482	1.80	24.7	I
Anchor bolt	M-12	402	442	2.20	32.3	II
		327	465	2.14	35.3	I
Base plate	Pl-12	300	438	2.11	39.6	II
		284	446	2.11	51.3	I
		285	446	2.20	41.6	II
	Pl-16	273	416	2.08	65.5	I
Chord bar	D13	362	520	1.81	21.3	
Top hoop	D10	332	474	1.88	23.2	I
Hoop	D 6	421	536	1.88	25.6	

Concrete	σ_c (N/mm²)	σ_t (N/mm²)	Material No.		σ_c (N/mm²)	σ_t (N/mm²)	Material No.
Beam	23.2	3.04	I	Encasement	25.2	3.94	I
	22.1	3.87	II				

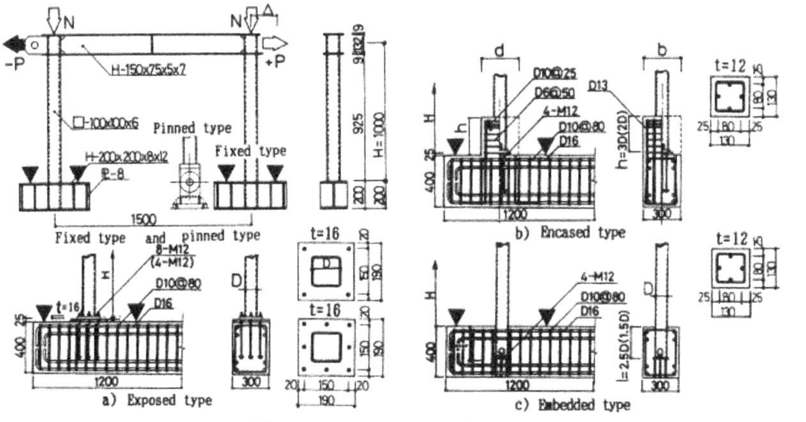

Fig. 1 Test specimen

RESULTS OF THE EXPERIMENTS AND DISCUSSION

Hysteresis loops

Fig. 2 shows the relations between the lateral force (P) and the story deflection angle (R). The story deflection angle is the value obtained by dividing the displacement of the column top (Δ) by the height from the base plate lower surface to the centroid of the beam (H). In case of the frames having exposed type column bases, they showed relatively small loop areas if the axial compression was absent; however, their loop areas would become larger and the gradient of the hysteretic curves between the time of removal of lateral force and the time of reverse loading would become steeper if the axial compression was involved. The energy-absorbing capacity of the frame with this type column bases is larger than that of the frame with pinned column bases but is considerably smaller than that of the frame with fixed type column bases. However, even when the base plates are not of the moment resisting type as in this test specimen and furthermore no axial force act on the column, the loop will have a certain bulge and the strength and rigidity of the frame will increase if the column bases are encased in reinforced concrete. If the axial compression is involved, the loop takes a more stabilized spindle shape. If the column base is embedded in reinforced concrete with an embedment depth of 2.5 times the outside dimension of the tubular column, a stabilized spindle shaped loop is obtained. If embedment depth is taken as 1.5 times the outside dimension of the tubular column, an inverted S-shaped loop indicating a slip phenomenon is obtained in case the axial force on the column is absent; however, if the axial force is present, a stabilized spindle-shaped loop will result.

Moment distributions and inflection point heights of steel column
Fig. 3 and Fig. 4 show the bending moment distribution of steel column and the heights of inflection points as the load is varied, respectively. It could be understood from these figures that the bending moments on the column head and the steel beam end of the frames connected to encased or embedded column bases are smaller than those of the frame connected to exposed type column bases and this indicates that the reinforcement/repair with column base encasement is effective on the frame to receive lateral force. At an initial loading stage, the height of the inflection point of the tubular column having the exposed column base when the axial compression is present is nearly identical with that of the tubular column whose column base is fully fixed, but this height lowers as the lateral load increases. It can also be understood in case of the encased type that the height of inflection point decreases as the lateral force increases. The apparent point of fixation is located at a mid-height of the encasement at the initial loading stage. The shifting of the inflection point due to the load variation is relatively small in the frame having the embedded column bases. In this case, the apparent point of fixation at the initial loading stage is found at a location 1.3 times the tubular dimension below the top of the foundation beam. For the clarification of the behavior of the frames connected to the different types of column bases, the mechanical properties of the column base itself should be studied in greater detail.

Fig. 2 P-R relations

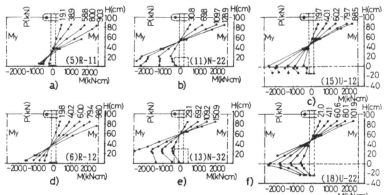

Fig. 3 Moment distributions of steel columns

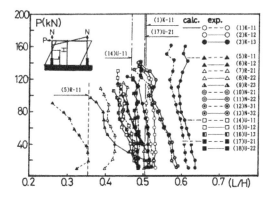

Fig. 4 Inflection point height ratios

CONCLUSIONS

From this experiment, the following facts were clarified.
1) In the frames connected to all types of column bases, the strength and
   rigidity of the frame become large and the area of the hysteresis loops
   becomes large, as the axial compression increases.
2) The inflection points of the frames having exposed type or encased type
   column bases lower as the lateral force acting on them increases;
   however, such lowering does not occur in the frames whose column bases
   are sufficiently embedded in foundation beams.
3) The apparent points of end fixation of the encased type column bases
   and the embedded type column bases are located at the mid-height of the
   encasement and at a point 1.3 times the tube dimension below the top of
   the foundation beam respectively.

REFERENCE

1. Nakashima, S. and Igarashi, S., Influence of details of column bases on
   structural behaviour, Prereport 13th congress, IABSE, Seminar sessions,
   Helsinki, 1988, pp. 705-710.

INFLUENCE OF STRENGTH AND STIFFNESS ON THE NATURAL FREQUENCY OF UNBRACED STEEL FRAMES

LOK TAT SENG & TAN KANG HAI
School of Civil and Structural Engineering
Nanyang Technological University, Nanyang Avenue
Singapore 2263

ABSTRACT

Vibration studies have been conducted on a range of fixed-base medium rise plane unbraced rigid-jointed steel frames. The results indicate the influence of strength and stiffness on the fundamental natural frequency of vibration. Results of forty-three unbraced frameworks are presented and these are compared with existing methods for determining the fundamental natural frequency. Expressions for the natural frequency which incorporates the lowest elastic critical load and the rigid-plastic collapse load are proposed for 4, 7 and 10 storey frameworks. It is shown that the ratio of these parameters have an influence on the accuracy of the dynamic response of unbraced steel frames.

INTRODUCTION

Over the years, advances in construction techniques, structural design philosophy, economic choice of high strength materials and better understanding of sophisticated structural forms have led to significant reductions in the overall weight of buildings. The traditional heavy cladding and static block partition walls have been replaced by lightweight materials and movable screens. The unobstructed interiors which provide for the freedom of layouts in buildings are increasingly utilised in the construction of tall buildings, particularly in land-scarce environments. Consequently, modern tall buildings are much more susceptible to dynamic forces. Oscillatory motion of the building caused by dynamic forces such as those arising from wind loading, service machinery and heavy traffic causes unpleasant sensations to occupants and exarcebates the propagation of cracking of finishes. Dynamic stresses resulting from such effects could also exceed static values. Therefore, a knowledge of the natural frequency of vibration is necessary to determine the design loading and response of the building.

The determination of the dynamic characteristics of structures is a relatively complex task and recourse to the computer is the general practice of engineers. However, empirical expressions for approximating the fundamental natural frequency have been proposed. The formula recommended by the Structural Engineers Association of California, and other approximations[1] are widely used for structures such as reinforced concrete shear wall buildings and braced/unbraced steel frames. The Canadian Building Code[2] recommended a simple relationship based on the height of the building alone to estimate the natural frequency. However, care should be taken to interpret the conditions under which the formulas may be considered reasonably accurate for preliminary design. Ellis[3] suggested simple relationships between natural frequencies and building height on the basis of extensive tests[4] and evaluation undertaken over a number of years. The simple relationships proposed by him are also dependent on the height of the building alone. It was suggested that such approximations could provide good estimates of the natural frequency of vibration of tall buildings.

Roberts and Wood[5] adopted the Grinter substitute frame as the model for evaluating the natural frequency of frameworks. By making use of the dynamic stability functions developed by Armstrong[6], they were able to evaluate the natural frequencies and the corresponding modal shapes of the structure. A natural frequency occurs when the determinant of the dynamic joint stiffness at any selected joint becomes singular. The technique is suitable for manual computation because the maximum size of the necessary matrices is limited to two. However, a small computer is required to facilitate the calculation. In contrast, Ko et al.[7] used the established finite element technique in their detailed study of the dynamic characteristics of tall buildings. The sophisticated modelling entails significant effort but meaningful results could be obtained for more complicated structures.

This paper reports on studies conducted on forty-three medium-rise plane unbraced steel frames. The aims are to determine the fundamental natural frequency of these frames and to compare the results with established empirical expressions. One consequence of this study is to establish the influence of strength and stiffness on the natural frequency and to propose empirical relationships based on these parameters for the groups of steel frames examined.

VIBRATION STUDIES ON UNBRACED STEEL FRAMES

The frames examined here have been reported[8]. The forty-three frames were rectangular in elevation, comprising three groups of four, seven and ten storeys in height and varied from two to four or five bays. The storey height was constant at 3.75 metres but two bay widths, 7.5 metres and 5 metres were considered. The preliminary sections were increased as appropriate to satisfy the limit on sway deflection at the serviceability limit of 1/300 of each storey height. However, it should be noted that while the sway limit is well within acceptable values, it does not necessarily lead to satisfactory dynamic comfort.

The frames were subjected to second-order elasto-plastic computer analysis under the appropriately factored combined loading. As the factored loads were taken as the reference loads for the analysis, a value of $\alpha_f > 1$ indicates that the factored load level for the ultimate limit state has been achieved. Two other important parameters were also derived from computer analyses; the lowest elastic critical load (α_c) and the rigid-plastic collapse load (α_p) factor. In practice, these parameters may be computed from approximate methods that are available in standard text.

The lowest elastic critical load (α_c) and the rigid-plastic collapse load (α_p) parameters derived from the previous study[8] for the frames were used to examine the influence such parameters have on the natural frequency. A graph of natural frequency versus the stiffness parameter $\ln(\alpha_c/\alpha_p)$ for the frames considered in that study is shown in Figure 1. The fundamental natural frequency was obtained from a standard plane frame vibration analysis program.

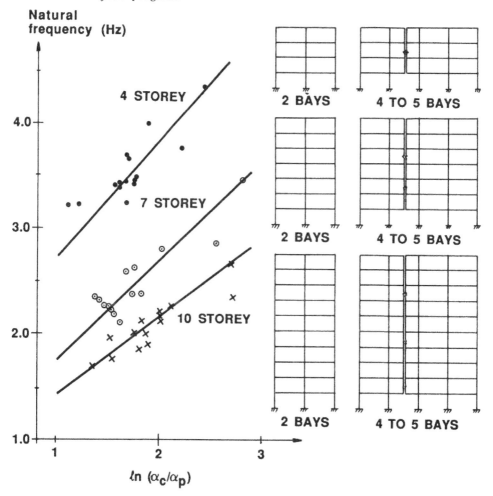

Figure 1. Relationship between $\ln(\alpha_c/\alpha_p)$ and natural frequency

A best-fit straight line has been drawn for each group of frames to deduce the relationship between the parameters. The results could be interpreted as a measure of the effect of overall lateral stiffness and strength on the natural frequency of the frames. Expressions relating the stiffness parameter and the natural frequency for each group may be derived as follows:

A. Four storeys

$$f = 1.12 \ln(\alpha_c/\alpha_p) + 1.6 \quad (Hz) \tag{1}$$

B. Seven storeys

$$f = 0.93 \ln(\alpha_c/\alpha_p) + 0.8 \quad (Hz) \tag{2}$$

C. Ten storeys

$$f = 0.70 \ln(\alpha_c/\alpha_p) + 0.7 \quad (Hz) \tag{3}$$

The lower constants proposed for the taller frames in the expressions reflect the relative flexibility of taller and slender structures. It should be noted that the expressions are applicable only to the range of stiffness parameters considered here.

COMPARISON AND DISCUSSION OF RESULTS

The natural frequency calculated from equations 1, 2 or 3 have been compared with established empirical expressions. Table 1 lists the comparisons of the three groups of frames, together with those calculated from several established empirical expressions. Values shown in column 4 are calculated from any one of the three equations that correspond to the number of storeys in the structure examined while values in columns 5 to 8 are those derived from widely-used formulas[1].

The expression used in column 5 has been suggested to be particularly applicable to reinforced concrete shear wall buildings and braced steel frames. This expression was included in Table 1 for comparison purpose only. The other formulas were appropriate for moment/unbraced steel frames.

The proposed expressions used to calculate the fundamental natural frequency of the 4, 7 and 10 storey skeletal frameworks provided consistently better approximations than the more established methods. The expressions are dependent on the inherent design frame properties rather than on the physical dimensions of the structure. The results of the fundamental natural frequency derived from the proposed expressions indicated a maximum error of approximately 10% and they do not fluctuate in the manner calculated from other expressions. The ratio of α_c/α_p varies from 3.2 to 11.4, 4.0 to 16.6 and 3.9 to 15.3 for the four, seven and ten storey frames respectively[8]. It should be noted that although the first frame in each group had values of α_c/α_p greater than 9, it does not necessarily imply that all the frames which overestimated the 'exact' frequency had high values of this stiffness parameter.

TABLE 1

Comparison of results for various established empirical expressions

	Frame	'Exact' frequency (Hz)	Ratio of calculated to computer values for:				
			Eqn (1),(2) or (3)	$\dfrac{\sqrt{D}}{0.091H}$ (ref1)	$\dfrac{10}{N}$ (ref1)	$\dfrac{46}{H}$ (ref3)	$\dfrac{1}{c_t L^{0.75}}$ (ref1)
	1	3.74	1.09	0.76	0.67	0.82	0.41
	2	3.22	1.08	1.39	0.78	0.95	0.48
	3	4.33	1.00	0.54	0.58	0.71	0.36
	4	3.65	0.95	1.00	0.68	0.84	0.42
4	5	3.45	1.03	0.82	0.72	0.89	0.45
	6	3.42	1.01	0.83	0.73	0.90	0.45
S	7	3.41	0.99	1.18	0.73	0.90	0.45
T	8	3.39	0.99	1.32	0.74	0.90	0.45
O	9	3.99	0.93	0.58	0.63	0.77	0.39
R	10	3.67	0.95	0.77	0.68	0.84	0.42
E	11	3.21	0.92	1.02	0.78	0.96	0.48
Y	12	3.20	0.91	1.14	0.78	0.96	0.48
S	13	3.45	1.04	0.82	0.72	0.89	0.45
	14	3.42	1.04	1.02	0.73	0.90	0.45
	15	3.39	1.00	1.32	0.74	0.90	0.45
	1	2.85	1.11	0.57	0.50	0.61	0.35
	2	2.10	1.10	1.22	0.68	0.83	0.48
	3	3.45	0.99	0.38	0.41	0.51	0.29
7	4	2.62	0.93	0.80	0.55	0.67	0.39
	5	2.37	1.02	0.68	0.60	0.74	0.43
S	6	2.18	1.03	0.91	0.66	0.80	0.46
T	7	2.26	0.97	1.01	0.63	0.78	0.45
O	8	2.26	0.96	1.13	0.63	0.78	0.45
R	9	2.80	0.98	0.47	0.51	0.63	0.36
E	10	2.58	0.91	0.63	0.55	0.68	0.39
Y	11	2.32	0.91	0.81	0.62	0.76	0.44
S	12	2.34	0.89	0.89	0.61	0.75	0.43
	13	2.37	1.05	0.68	0.60	0.74	0.43
	14	2.23	1.00	1.15	0.64	0.79	0.45
	1	2.34	1.11	0.49	0.43	0.52	0.33
	2	1.91	1.06	0.84	0.52	0.64	0.40
1	3	2.64	0.98	0.35	0.38	0.46	0.29
0	4	2.26	0.96	0.58	0.44	0.54	0.34
	5	1.87	1.05	0.61	0.53	0.66	0.41
S	6	1.77	1.00	0.79	0.56	0.69	0.44
T	7	1.70	0.97	0.94	0.59	0.72	0.45
O	8	2.21	0.98	0.42	0.45	0.56	0.35
R	9	2.14	0.92	0.53	0.47	0.57	0.36
E	10	1.96	0.90	0.67	0.51	0.63	0.39
Y	11	2.12	1.02	0.54	0.47	0.58	0.36
S	12	2.14	0.99	0.53	0.47	0.57	0.36
	13	2.01	1.00	0.69	0.50	0.61	0.38
	14	2.02	0.95	0.46	0.50	0.61	0.38

D = base dimension in the direction of motion (metres)
H = total height of the building (metres)
N = Number of storeys in the building
C_t = constant (taken as 0.035 for steel structures)
L = building height (feet)

An interesting feature of the comparison tabulated in column 5 of Table 1, which is strictly for braced frames, shows that this widely-used expression overestimated the natural frequency by as much as 39% for the lower-rise frames and underestimated in all cases for the tallest frames. All the frames which were overestimated in column 5 were either four or five bays. This tendency appears to suggest the influence of relative stiffness associated with the corresponding relative dimensions of multi-bay frames.

Predicted results provided by the expressions shown in column 6 consistently underestimated the natural frequency for all the frames. The best results were restricted to the two groups of low-rise frames derived from the expression by Ellis[3] shown in column 7. In a limited number of cases for the four storey frames, his approximation was better than the proposed estimate. Results derived in column 8 indicates poor estimates of natural frequencies of unbraced steel frames.

CONCLUSIONS

Studies on the natural frequency of forty-three medium-rise plane skeletal steel frames have been conducted. The results from this study were compared with established expressions. Alternative expressions were proposed for 4, 7 and 10 storey plane unbraced steel frameworks. The expressions depend on the inherent design parameters rather than the physical dimensions of the frame. It is suggested that such parameters better characterise the response of structures. The effects of vertical loads, cladding and composite action on the natural frequency were not considered in the study. The stiffening effect of the latter actions could compensate for the reduction of natural frequency caused by vertical loading. It is recommended that care should be exercised in the interpretation and application of simplified expressions for the evaluation of natural frequencies of unbraced frames.

REFERENCES

[1]   Newmark, N.M. and Hall W.J., Earthquake spectra and design, Earthquake Engineering Research Institute, Berkeley, California, 1982.

[2]   National Building Code of Canada, National Research Council of Canada, Ottawa, Canada, 1990.

[3]   Ellis, B.R., An assessment of the accuracy of predicting the fundamental natural frequencies of buildings and the implications concerning the dynamic analysis of structures, <u>Proc. ICE</u>, London, Pt.2, Sept 1980, 69, 763-776.

[4]   Jeary, A.P. and Ellis B.R., The accuracy of mathematical models of structural dynamics, <u>Building Research Establishment</u>, Aug 1981.

[5]   Roberts, E.H. and Wood, R.H., A simplified method for evaluating the natural frequencies and modal shapes of multistorey steel frames, <u>The Structural Engineer</u>, No.1, Mar 1981, 59B.

[6]   Armstrong, I.D., Dynamic stability functions for continuous structures, <u>Herriot-Watt University</u>, Edinburgh, 1969.

[7]   Ko, J.M., Lau, S.L. and Wong, C.W., Finite element modelling of the dynamic characteristics of tall buildings, <u>The Structural Engineer</u>, No.24, Dec 1989, 67, and Correspondence in No.7, 68, Apr 1990.

[8]   Anderson, D. and Lok, T.S., Design studies on unbraced, multistorey steel frames, <u>The Structural Engineer</u>, No.2, June 1983, 61B, 29-34.

FATIGUE TESTS OF CROSS-BEAM CONNECTIONS IN PLATE GIRDER HIGHWAY BRIDGES

ICHIRO OKURA and YUHSHI FUKUMOTO
Department of Civil Engineering
Osaka University
2-1 Yamadaoka, Suita, Osaka 565, JAPAN

ABSTRACT

To investigate the effects of connection details between main girder flange
and concrete slab on the cracking at cross-beam connections, fatigue tests
were conducted for specimens consisting of a concrete slab and cross-beam
connections. The cracks always occurred at the cross-beam connections with
stud shear connectors, regardless of their number and arrangement. They
also occurred at the cross-beam connections with slab anchors. Reduction
of the local stresses which govern the cracks can be secured by removing
the stud shear connectors and slab anchors from the cross-beam connections.

INTRODUCTION

Fatigue cracks as shown in Figure 1 occur at the connections of cross beams
to main girders in many plate girder highway bridges in the urban areas of
Japan. Three types of cracks are initiated in the connection plate
installed between the top flanges of a cross beam and main girder. Type 4
crack is produced at the toe on the web side of the fillet weld to connect
the web to the top flange of a main girder. These cracks are caused mainly
by the rotation of a concrete slab due to slab-deformation caused by wheel
loads[1]. Therefore, fatigue tests were carried out to examine the effect
of the concrete-slab rotation on the cracking, with special attention to
the connection details between the main girder flange and the concrete
slab.

FATIGUE TESTS

Figure 2 shows the fatigue test specimens. They consist of cross-beam
connections and a concrete slab in a stripped form. Series A and B
correspond to the cross-beam connections at exterior and interior main
girders, respectively. In Series B, negative moment is introduced in the
concrete slab on the cross-beam connection.

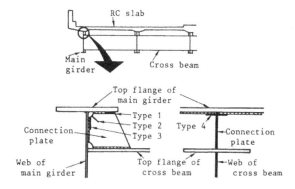

Figure 1. Fatigue cracks at cross-beam connections.

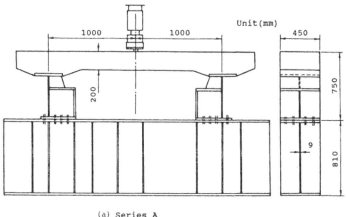

(a) Series A

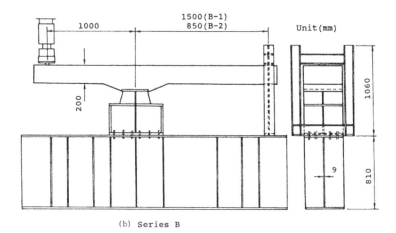

(b) Series B

Figure 2. Fatigue test specimens.

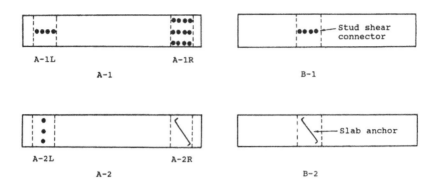

Figure 3. Connection details between girder flange and concrete slab.

To investigate the effects of connection details between the girder flange and the concrete slab on the cracking, the number of stud shear connectors and their arragement were changed at each cross-beam connection, as shown in Figure 3. In Connection A-2R and Specimen B-2, a slab anchor was installed.

The following were observed in the fatigue tests:
-Similar cracks as shown in Figure 1 occurred at all the cross-beam connections, regardless of the connection details between the girder flange and the concrete slab.
-In Series A, there existed no order in initiation of Types 1 and 4 cracks, while in Series B, Type 4 crack always followed Type 1 crack.

CHARACTERISTICS OF LOCAL STRESSES

Referring to Figure 4, the membrane stress σ_{my} in the vertical direction in the connection plate and the plate-bending stress σ_{by} in the girder web are the local stresses which govern the initiation of Types 1 and 4 cracks, respectively[2].

Using a finite element analysis, the membrane stress σ_{my} and plate-bending stress σ_{by} were decomposed into the stress components corresponding to the vertical force Q and the horizontal force S shown in Figure 4. The values of these stress components are listed in TABLE 1. Here, σ_{myQ} and σ_{byQ} are the stress components of σ_{my} and σ_{by} produced by Q, respectively. σ_{myS} and σ_{byS} are the stress components of σ_{my} and σ_{by} produced by S, respectively. Connections A-0 and B-0 correspond to the cross-beam connections in which neither stud shear connectors nor slab anchors are installed between concrete slab and girder flange. In the finite element analysis of both models, a concentrated load was applied vertically to the edge of the girder flange just above the connection plate.

The following are drawn from TABLE 1:
-The stress σ_{my} is mostly produced by the vertical force Q. The stress σ_{by} is initiated equally by both Q and S.
-The stress values of Connections A-0 and B-0 are much smaller than those of any other connection model. Thus, removing stud shear connectors and slab anchors from the cross-beam connections makes the membrane stress σ_{my} as well as the plate-bending stress σ_{by} smaller.

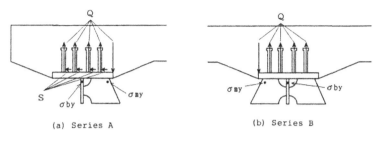

(a) Series A   (b) Series B

Figure 4. Forces acting on girder flange.

TABLE 1
Comparison of stress components corresponding to Q and S

Connection	Q (kN)	S (kN)	σ_{my} (MPa)	σ_{myQ} (MPa)	σ_{myS} (MPa)	σ_{by} (MPa)	σ_{byQ} (MPa)	σ_{byS} (MPa)
A-1L	49.0	41.6	−238.2	−241.2	3.0	43.5	8.7	34.8
A-1R	49.0	56.3	−291.3	−295.2	3.9	96.2	49.0	47.2
A-2L	49.0	57.8	−240.8	−244.7	3.9	67.6	18.6	49.0
A-2R	49.0	51.7	−345.2	−348.6	3.4	31.1	−13.9	45.0
A-0	49.0	0.0	−183.9	−183.9	0.0	− 9.8	− 9.8	0.0
B-1	49.0	0.0	−276.4	−276.4	0.0	−10.2	−10.2	0.0
B-2	49.0	0.0	−249.7	−249.7	0.0	−10.3	−10.3	0.0
B-0	49.0	0.0	−180.0	−180.0	0.0	− 7.4	− 7.4	0.0

CONCLUSIONS

The main results are as follows:
-Cracks are initiated at the cross-beam connections with stud shear connectors regardless of their number and arrangement. and these cracks are also initiated at the cross-beam connections with slab anchors.
-To reduce the stresses causing Types 1 and 4 cracks, it is recommended that neither stud shear connectors nor slab anchors should be placed at the cross-beam connections.

REFERENCES

1. Okura, I., Takigawa, H. and Fukumoto, Y., Structural parameters governing fatigue cracking in highway bridges. Proc. of JSCE, Struct. Eng./Earthq. Eng., 1989, pp. 423s-426s.

2. Okura, I. and Fukumoto, Y., Fatigue of cross beam connections in steel bridges. IABSE, 13th Congress, Helsinki, 1988, pp. 741-746.

FATIGUE BEHAVIOUR OF CONSTRUCTIONS MADE OF HIGH-STRENGTH STEELS

F. MANG, Ö. BUCAK, E. KOCH[*] and H. STAUF[**]

Testing Center for Steel, Timber and Stone, University of Karlsruhe, Kaiserstr. 12, 7500 Karlsruhe[*]
Mannstaedt Werke GmbH & Co, Postfach 14 62, 5210 Troisdorf[**]

ABSTRACT

The advantages of high-strength steels such as small sectional area and resulting lower dead weight and their disadvantages (such as notch sensitivity) have been known for a long time. High-strength steels under fatigue loading are advantageous in the area of low cycle fatigue or for constructions with a higher level of tensile mean stress, when the maximum stress might exceed the allowable stress for conventional steels.

In this paper, the results of fatigue investigations on hollow section connections (fillet weld, butt weld, longitudinal rib, transverse rib) and hollow section joints (X-joints and K-joints) are presented.

Investigations on hollow section structures under variable amplitude (VA) are presented, and the application possibilities of Miner's rule are checked. Presently, investigations are carried out on 2 different grades of high-strength steels StE 460 (Fe E 460) and StE 600 (Fe E 600).

The second part of the paper presents results on full-scale fatigue tests (HE 400 A and HE 300 AA profile; StE 460) on specimens from H-profiles with transverse attachments. A comparison of the test results on high-strength steels to those of mild steels (StE 235 and StE 355) is given.

As a result of the small number of full-scale test results for welded StE 460 (Fe E 460) and StE 690 (Fe E 690) and inconsistencies with small-scale test specimen data in the high-cycle region, these steels have so far been excluded from standardisation. For this reason, new experimental investigations are necessary, which are presently performed in Karlsruhe on structural members (full-scale tests). First results presented in this paper, show a better fatigue life for structures made of high-strength steels.

LIST OF SYMBOLS

σ_{max}	maximum nominal stress in a constant amplitude loading cycle $(= \sigma_o)$
σ_{min}	minimum nominal stress in a constant amplitude loading cycle $(= \sigma_u)$
$\Delta\sigma_m$	maximum stress range in the spectrum
S_R	stress range $(= \Delta\sigma \quad \sigma_{max} - \sigma_{min})$
S_{Ra}	stress range on the outer surface of the tension flange for H-profiles
S_{Ri}	stress range on the inner surface of the tension flange for H-profiles
R	stress ratio $(\sigma_{min}/\sigma_{max})$
CHS	circular hollow sections

RHS   rectangular hollow sections
CA   constant amplitude
VA   variable amplitude
OPB   out-of-plane bending
N   number of cycles to failure
m   slope of the S-N-line
p   type of load spectra
ü   overlap [%]

INTRODUCTION

The reason for the application of high-strength rolled products is that effective dimensions and sectional areas can be economically produced through a prefabrication in the rolling mill and thus, the favourable properties of high-strength steels can be utilized for structures. In the last years, the structural steel market increasingly tends to products with profiles of bigger wall thicknesses and to those of higher strength steels. Hi-tech areas such as the offshore industry and high building construction additionally require a high toughness as well as improved welding properties of these steels, e.g. welding without preheating of rolled products with bigger wall thicknesses.

The publication of the German standard DASt-Ri 011 (Feb 1979) "Application of Weldable Fine Grain High-Strength Structural Steels StE 460 and StE 690 for Steel Structures" [2], increased the application of high-strength structural steels.

In the meantime, high-strength structural steels have found a broad field of application such as in the construction of tanks, pipelines, vehicles, cranes and steel framed structures. This development is essentially connected with the progress in steel production as well as in welding technology. The previously mentioned DASt-standard [2] is presently revised.

Its economical application matches with the utilization of higher allowable stresses for predominantly statically loaded constructions compared to those of "conventional" structural steels (Fe 360 and Fe 510). In this case, those structural members come off favorably that are subjected to tensile stress or structures that are loaded under fatigue load spectra with a lower fullness ratio of the spec-tra. The fullness ratio describes a spectrum shape, which is the area below the curve (fig. 11) divided by the area, which has been determined by CA-tests ($\Delta\sigma$ · endurance = area determined by CA-tests) [13].

The latest editions of Eurocode 3 [1] on steel structures do not indicate any regulations for structures made of high-strength steels. Therefore, the first results of experimental investigations on high-strength steels are provided in this paper. As far as possible, they have been compared with the results of conventional steels.

In [4] and [5], structural shapes are indicated for which the same reduction factors are valid for the dimension as they have been previously worked out for the mild steels Fe 360 and Fe 510. These structural shapes are bending stiffened knees of frames (L-joints) for which the failure occurred on these positions through by-passing the forces to the more stiffened corners and plastic buckling resulting from this, as well as through T-joints under moment loading. With this, the torsional stiffness of the joints has been compared with the stiffness of joints of conventional steels. In this context, they resulted in a good accordance with each other. In the near future it is planned to continue these investigations with high-strength steels under static load in Karlsruhe and in Delft.

An ECCS [7] and a CIDECT [9] research program on the fatigue behaviour of the most important notch cases as well as on X- and K-type joints made of the high-strength steels StE 460 TM and StE 600 TM is presently performed in Karlsruhe. In the following, the first results achieved by CIDECT are provided.

PRINCIPLE INVESTIGATIONS ON HIGH-STRENGTH STEELS

Fatigue test results obtained from K-joints in rectangular hollow sections in three steels are shown in fig. 1. They indicate a parallel shift of the S-N-lines corresponding to an increase in fatigue strength with increase in steel strength. This is the influence of stress concentration. Tests on small specimens (for ex. fillet welded cruciform specimens) show no influence of the steel grade for cycles about 10^6. The reason for this might be the relatively low stress concentration, besides this, the mutual support of the non cracked structural members is important with regard to fatigue life.

However, as a result of the small number of test results for welded StE 460 and StE 690, and inconsistencies (parallel shift of S-N-lines in full-scale tests) with small scale test specimen data in the high cycle regime (small scale test specimens: the S-N-line is in the same stress range for $2 \cdot 10^6$ load cycles), these steels have so far been excluded from standardisation work.

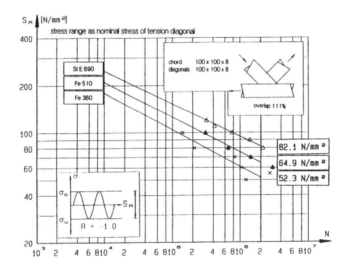

Fig. 1   Test series with the same joint geometry but with different materials [4,6]

Based on the draft for Eurocode 3, the stress range ($S_R = \Delta\sigma$) is taken as a basis for future diagrams. From fig. 2, the justification for the application of the $\Delta\sigma$-concept for welded structures made of high-strength steels can be derived.

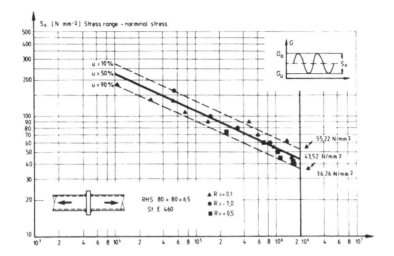

Fig. 2   Justification of the application of the $\Delta\sigma$-concept of welded structures made of high-strength steels [5]

DESIGN OF LONGITUDINAL ATTACHMENTS (RIBS) ON HOLLOW SECTIONS MADE OF HIGH-STRENGTH STEELS.

The advantages of high-strength steels such as small sectional area and resulting lower dead weight as well as their behaviour for structures with a higher level of mean stress (σ_M) have been known for a long time. A disadvantage of high-strength steels is their notch sensitivity for welded constructions in addition to high material cost. For some time it has been recommended to design the structure in accordance with the material. One of these recommendations is the design of longitudinal attachments. The simplest way is to shear off or saw off the attachments from the flat material, and to weld them to the required position by means of fillet welds. In Eurocode 3 [1], this notch case has been classified into category 50, 71 or 80, depending on the length of the attachment. With an attachment length of 200 mm (corresponding to detail category 50 according to Eurocode 3) for high-strength steels, the experimental investigations carried out in Karlsruhe resulted in a value of about 75.6 N/mm^2 for a probability of survival of 97.5% (fig. 3).

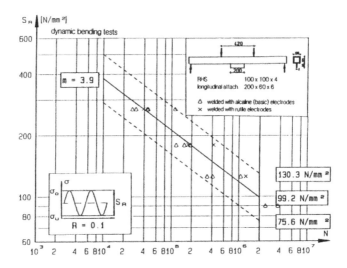

Fig. 3   Results of the experimental investigations on hollow sections with longitudinal attachments, material StE 600 TM; probability of survival p_s = 2.5%; 50% and 97.5%

If the abrupt stiffness increase on the same structural details is avoided through a continuous transition with a radius r = 40 mm, and if the transition has been worked off by grinding after welding of the attachment by means of a butt weld, better values of the stress ranges are obtained for $2 \cdot 10^6$ load cycles. The results of these investigations can be taken from figure 4. In the same figure, they are plotted comparatively with the results of the rectangular attachments.

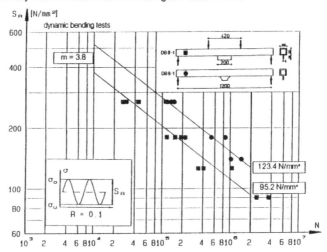

Fig. 4   Comparison of the results of fatigue tests (probability of survival = 50%) on test specimens with longitudinal attachments; different design of the longitudinal attachments; material StE 600 TM

Through the comparison of the values for a survival probability of 50%, an increase of the values S_R

at $2 \cdot 10^6$ load cycles resulted in a factor around 1.3; where for the endurances, an increase factor of about 3.0 was obtained. These different factors are to be put down to the relatively low slope of the S-N-line. The detail categories for longitudinal stiffeners of 45, 71 or 90 (r $\geq$ 150 mm) as well as the slope of the S-N-line with 3.0 within the load cycle range smaller than $2 \cdot 10^6$ indicated in Eurocode 3, do not correspond to the test results gained at our institute.

From previous experimental investigations on full-scale tests [10, 11,15 etc.] it can be derived that, in our opinion, a slope of 4.0 or 5.0 [10, 11, 15] seems to be better compared to the slope of 3.0 which has been mainly ascertained for small-scale test specimens.

Since, through working off of the stiffener end in the joint area, only a small stiffener length remained, further experimental investigations have been carried out with a rectangular stiffener l = 120 mm for clarifying the influence of the length of the attachments. In figure 5, the results for a survival probability of 50% are plotted comparatively. The fatigue life of a construction is influenced by the stiffener length as it can be seen in fig. 5. The influence, however, is not so high.

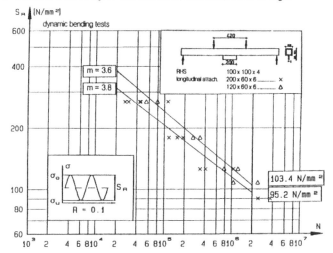

Fig. 5   Comparison of the results of tests on specimens with longitudinal attachment in dependence on the stiffener length; material StE 600 TM

Figures 6 and 7 show the test specimens after failure with the smooth transition radius r initially formed by machine or gas cutting of the gusset plate before welding, and subsequently grinding of the weld area parallel to the longitudinal direction of the attachments.

Fig 6   Test specimen with rectangular stiffener L = 200 mm after failure; material StE 600 TM

Fig. 7   Test specimen with longitudinal attachment (grinded) after failure
The smooth transition radius r is initially formed by machine or gas cutting of the gusset
plate before welding, and subsequently grinding of the weld area parallel to the longitudinal
direction of the attachment

Figures 8 and 9 show the hardness distributions of these test specimens. The values for HV
correspond to those determined for designed structures. This points to the fact that the tested specimens have been welded in practical conditions.

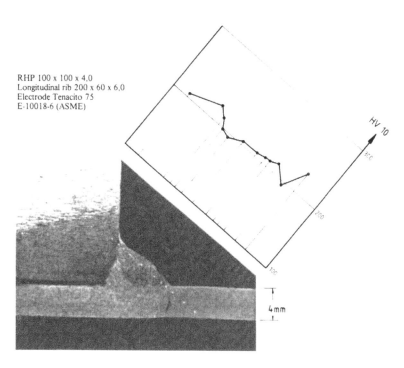

RHP 100 x 100 x 4,0
Longitudinal rib 200 x 60 x 6,0
Electrode Tenacito 75
E·10018-6 (ASME)

Fig. 8   Macro section with hardness distribution, longitudinal attachments; material StE 600 TM

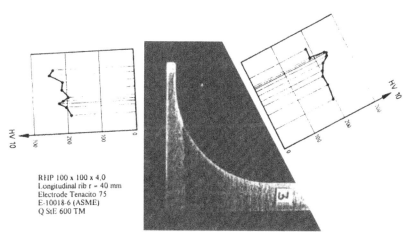

RHP 100 x 100 x 4,0
Longitudinal rib r = 40 mm
Electrode Tenacito 75
E·10018-6 (ASME)
Q StE 600 TM

Fig. 9   Macro section with hardness distribution, longitudinal attachments with radius r = 40 mm;
         material StE 600 TM

First tests carried out on specimens with longitudinal stiffener under variable amplitude have been
finished. The results are presented graphically in fig. 10. The spectra used are taken from the German
standards for crane construction DIN 15018 [8]; these are the spectra p = 0/3; p = 1/3 and p = 2/3.
These spectra are presented graphically in fig. 11. In fig. 10 a parallel shift of the fatigue life line for
the spectra can be seen. Through the application of the Miner-rule on test specimens with

dinal attachments, we obtained Miner sums between 0.56 up to 0.88 for a load sequence of p = 2/3, and between 2.58 up to 2.65 for a load sequence of p = 1/3. With the calculation according to Bucak [13] there was a good agreement of the new test data. The indications according to [13] are valid for lowest fatigue data. A report on this will be given in another paper.

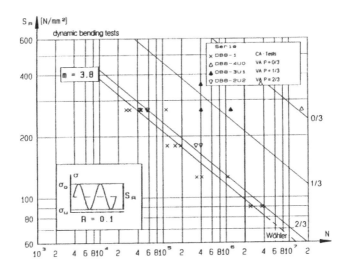

Fig. 10 Results of experimental investigation on specimens with longitudinal attachments under different load spectra (R = + 0,1); material StE 600 TM

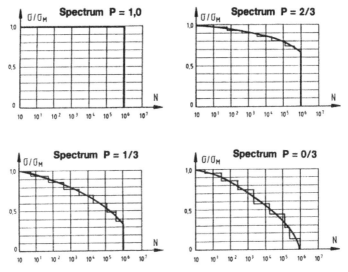

Fig. 11 Load spectra used by the tests such as in fig. 10
$\Delta\sigma$ = stress range in each level of the spectrum
$\Delta\sigma_M$ = maximum stress range in the spectrum

A further variant for welding longitudinal attachments is as follows. The rectangular attachments are fixed, connected by means of fillet welds, and at the end of the attachments the welds are welded longer to approximately 25 up to 30 mm. Thus, it could be achieved that the ends of the weldments do not coincide with the location of the change in stiffness. This very simple and economic variant with regard to production has been subjected to a fatigue test with constant amplitude (CA). In fig. 12, the results are plotted comparatively with those of the standard variant. An increase of the stress range of about 25 to 30% at $2 \cdot 10^6$ load cycles can be seen. Fig. 13 shows these test specimens after failure. The end of the weld is not the end of the stiffeners. The slope of the S-N-line is m = 3, however, only 7 tests have been performed with test specimens different to those used in fig. 4.

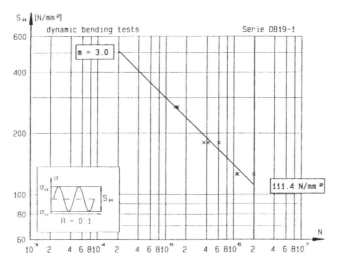

Fig. 12 Results of the experimental investigations on specimens with longitudinal attachments for which the weld have been extended above the end of the stiffener

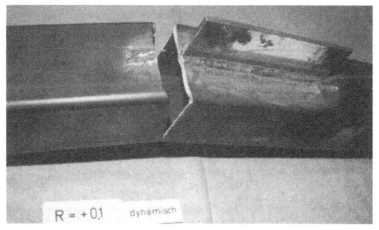

Fig. 13 Test specimen with longitudinal attachments for which the weld has been extended above the end of the stiffeners after failure

BUTT WELDED JOINTS ON HOLLOW SECTIONS MADE OF HIGH-STRENGTH STEELS

The next group of notch cases are butt welded hollow sections. Tenacito 75 (E-10018-6 according to ASTM) for the material StE 600 TM and Tenacito 60 (E-8018-6 according to ASTM) for the material StE 460 TM were chosen as electrodes. The edge preparation was V-shaped with an included angle of $\alpha = 60^0$, and a gap of approximately 1.5 to 2.0 mm. The test specimens have been welded without preheating. The results of the first investigations can be taken from fig. 14. As expected, the bearable stresses are much higher compared to the categories of Eurocode 3 (category 56 for rectangular hollow sections). Unacceptable lacks of fusion were stated for 6 test specimens. They have been subjected to a fatigue test in order to determine the decrease of load cycles. These 6 test results are also plotted in the diagram of fig. 14. The evaluation results in a decrease of the stress range of 72% at $2 \cdot 10^6$ load cycles where the bearable cycles decreased to about one fifth. For butt welded specimens under variable amplitude loading we obtain a Miner sum between 0.14 up to 0.57.

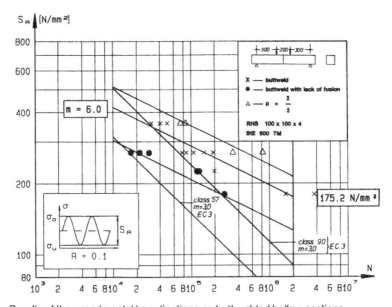

Fig. 14   Results of the experimental investigations on butt welded hollow sections

Tests are being continued.

HOLLOW SECTIONS WITH TRANSVERSE ATTACHMENTS

Some hollow sections were provided with a transverse attachment with the dimensions 100 x 60 x 6.0. The weldings (fillet weld) were carried out with rutile electrodes as well as with Tenacito 75 (basic electrodes).

A common evaluation of all test data available can be taken from fig. 15. It becomes evident that there is no big difference between specimens welded with basic electrodes and those welded with rutile electrodes. In contrast, the test specimens made of the material StE 460 TM are in the lower part of the scatter area. The slope of the S-N-lines is flatter with a value of 4.6 compared to the indications made in Eurocode 3. The fatigue strength at $2 \cdot 10^6$ cycles for a probability of survival $p_s = 97.5\%$ of 123 N/mm^2 is far higher than that given in Eurocode 3 for mild steels St 37 (Fe 360) and St 52 (Fe 510) (i.e. category 80).

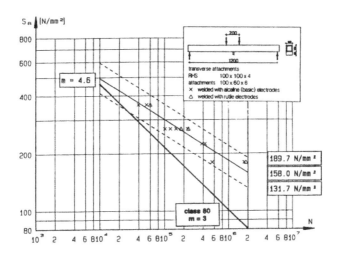

Fig. 15 Results of the experimental investigations on hollow sections with transverse attachments; material StE 600 TM

Fig. 16 shows a test specimen after failure and fig. 17 the macro section with hardness distribution.

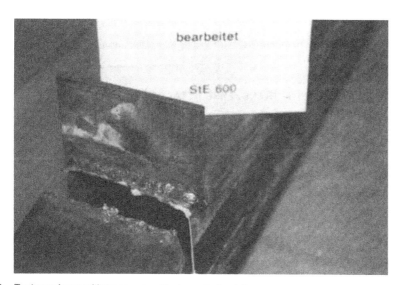

Fig. 16 Test specimen with transverse attachment after failure

In the test under variable amplitude the Miner sum was between 0.18 up to 0.26 for a load sequence of p = 2/3.

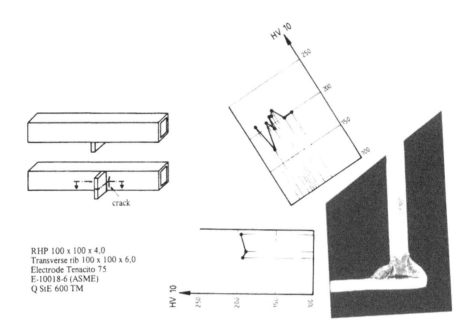

RHP 100 x 100 x 4,0
Transverse rib 100 x 100 x 6,0
Electrode Tenacito 75
E-10018-6 (ASME)
Q StE 600 TM

crack

Fig. 17   Macro section with hardness distribution, specimen with transverse attachment

ROLLED BEAMS WITH TRANSVERSE ATTACHMENTS (WEB STIFFENINGS)

Since weldings with fillet welds can provide serious notches through fatigue loading, a good design of stiffenings and similar connections is important for the utilizability of the beams' fatigue life. Such notch cases can be designed in different ways according to fig. 18. With this it is to be distinguished between

- stiffenings ending outside the tension area (form I)
- stiffenings with hole at the corner (form II)
- and completely welded stiffenings (forms IIIa and IIIb)

The difference of the forms IIIa and IIIb is that for form IIIa the stiffener plates are adapted to the contour of the rolled section, whereas for the specimens of form IIIb the stiffener plates have been diagonally cut at the neck area of the sections and finally have been filled with weld material.

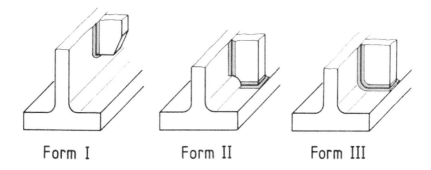

Form I   Form II   Form III

Fig. 18   Survey about the fatigue tests on rolled sections with web stiffening (schematically)

In the scope of a research program on thick-walled sections, experimental investigations and extensive finite element calculations have been carried out in Karlsruhe. The experimental investigations
were carried out as 3-point and 4-point bending tests on the sections HE 400 A and HE 300 AA. The
fabrication of the specimens was realized in a steel construction firm by means of arc welding. The
fabrication of the specimens was done step by step and several welders were engaged so that a production scatter was guaranteed. First results in an S-N-diagram (figs. 19 and 20) show that

- there are no serious differences between the results of various designs and thus, no degradations
   for the "classical sharper" notches (form II) are necessary
- extensive safety reserves are existing for the classifications to be done according to Eurocode 3
   (categories 80 (t < 12 mm) and 71 (t > 12 mm))
- post weld heat treatment of the wall thicknesses could result in additional advantages.

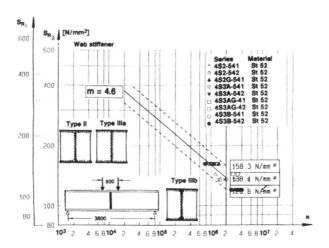

Fig. 19   Results of the 4-point bending tests on I-sections with web stiffening
          S_{Ra} = outer surface of the tension flange
          S_{Ri} = inner surface of the tension flange

90% of the specimens made of St 52-3 (Fe 510 DD) showed also cracks in the compression flange of the specimen after fatigue tests. Figures 21 and 22 illustrate the results of the dye penetration method in place of the specimens investigated. It can be derived that in the case presented, the residual stresses were very high, which means that industrial fabrication was done for the test specimens. In contrast to this, no cracks in the compression area could be found in specimens made of high-strength steels StE 460. An explanation for this is that the residual stresses on the structural member might be lower compared to the yield limit of the material, and therefore a more favourable fatigue behaviour could be derived from this.

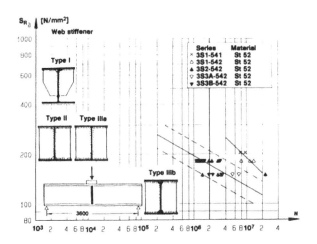

Fig. 20   Results of the 3-point bending tests on I-sections with web stiffening

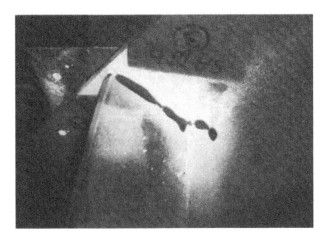

Fig. 21   Results of the dye penetration method in the compression area of the specimens SA III b 3, Type II, test specimen no. 3

485

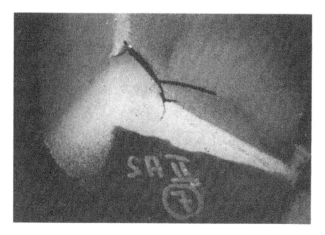

Fig. 22   Results of the dye penetration method in the compression area of the specimens SA II 7,
Type II, test specimen no. 7

In 4-point bending tests, the specimens of the type I reached load cycles of $20 \cdot 10^6$ and more. It has become critical for this type, if the load has eccentricities (OPB). This was the reason for various damages on steel bridge structures in Japan and Europe. Investigations for clarifying the influence of out-of-plane bending moments are being made at present.

Parallel to this, some series made of the high-strength material FE 460 Kg N have been checked. The results of these investigations are plotted comparatively in fig. 23. It can be seen a better fatigue life for specimens made of high-strength steel Fe 460 Kg N.

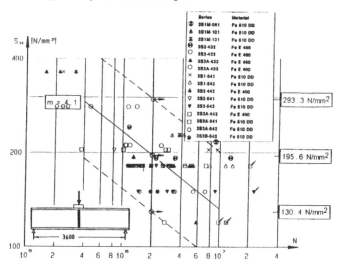

Fig. 23   Results of the 3-point bending fatigue tests on rolled sections made of high-strength steel
Fe 460 KG N with web stiffening in comparison with data from mild steel St52-3

The results recommend a new classification of this type of notch in Eurocode 3, part 9 and similar specifications (category 100 up to 114 instead of 71 and 80). In addition it could be shown, that the negative influencing parameters could be eliminated or rather reduced to a minimum by a modified technology of steel fabrication and rolling.

CONCLUSIONS

In this paper, the first results of experimental investigations made on the high-strength steels StE 460 TM (Fe E 460) and StE 600 TM (Fe E 600) are presented and compared with those of Eurocode 3. Although, only a few experimental investigations have been performed, it appears that we obtain another slope compared to that indicated in Eurocode 3. In order to be able to make reliable statements on this, further investigations will be necessary. In addition, it has been shown that an increase of fatigue strength values can be gained by an additional treatment of the attachments relevant for the material involved.

REFERENCES

[1]   Eurocode 3. "Design of Steel Structures, Part 1, General Rules and Rules for Buildings"
       Nov 1989
       Nov 1990, Editorial Group
       Sep 1989 Eurocode 3, Annex K

[2]   DASt-Ri 011 (1979)."Hochfeste, schweißgeeignete Feinkornbaustähle StE 460 und StE 690,
       Anwendung für Stahlbauten" (High-Strength Weldable Fine Grain Structural Steels StE 460
       and StE 690, Application for Steel Structures). Stahlbau Verlag Köln

[3]   DIN 18 808 (1984). "Stahlbauten, Tragwerke aus Hohlprofilen unter vorwiegend ruhender
       Beanspruchung" (Steel Structures Consisting of Hollow Sections Predominantly Static
       Loaded). Beuth Verlag Köln

[4]   Mang, F. and Ö. Bucak (1990). "Behaviour of Structures of High-Strength Steels", RILEM Work-
       shop on Needs in Testing Metals, Naples

[5]   Mang, F.; Ö. Bucak and I. Klingler (1987/88). "Wöhlerlinien Katalog für
       Hohlprofilverbindungen" (S-N-Line Catalogue for Hollow Section Joints). Studiengesellschaft
       für Anwendungstechnik von Eisen und Stahl e.V., Düsseldorf

[6]   Mang, F. and Ö. Bucak (1977/1981). "Untersuchungen an Verbindungen von offenen und
       geschlossenen Profilen aus hochfesten Stählen" (Behaviour of Joints made of Rolled Sections;
       I, H and ..), P. 11 and P. 71. Studiengesellschaft für Anwendungstechnik von Eisen und Stahl
       e.V., Düsseldorf

[7]   N.N. "Fatigue Behaviour of Hollow Section Joints made of High-Strength Steels", ECSC
       research programme No. 7210-SA 116; Contractor. Fa. Kloeckner-Mannstaedt-Werke,
       Troisdorf

[8]   EN 10025 (1991). "Hot Rolled Products on Non Alloy Structural Steels; Technical Delivery
       Conditions"

[9]   EN 10113, part 1 (1990). "Hot Rolled Products in Weldable Fine Grain Structural Steels"

[10] Becker, F.; R. Henrion; L. Sammet; F. Mang and Ö. Bucak. "Ermüdungsverhalten stumpfgeschweißter Trägerverbindungen" (Fatigue Behaviour of Butt Welded Rolled I-Sections), Report EUR 13077 DE

[11] Mang, F. and Ö. Bucak (1984). "Fatigue Behaviour of Welded Tubular Joints, Design Proposal and Background Information", Proceedings of the Conference "Welding of Tubular Structures", Boston

[12] N.N. "Fatigue Behaviour of Hollow Section Joints made of High-Strength Steels", CIDECT Project 7 L

[13] Bucak, Ö. (1990)."Bewertung des Ermüdungsverhaltens von Hohlprofilkonstruktionen" (Evaluation of the Fatigue Behaviour of Hollow Section Structures), thesis at the University of Karlsruhe, Versuchsanstalt für Stahl, Holz und Steine

[14] DIN 15018 (1974). "Krane, Grundsätze für Stahltragwerke, Berechnungen" (Cranes), Beuth Verlag, Köln

[15] Brozetti; Wardenier; Mang, F.; Dutta, D. and Grothmann . "Background Documentation", Chapter 9, Document 9.03. "Background Information on Fatigue Design Rules for Hollow Sections", part 1 "Classification Method - Statistical Evaluation", Eurocode Editorial Group. First draft Jan 1991, CIDECT No. 7 M-1/91

[16] Zirn, R. (1975) "Schwingfestigkeitsverhalten geschweißter Rohrknotenpunkte und Rohrlaschenverbindungen" (Dynamic Stress Behaviour of Welded Tubular Joints and Tubular Butt Joints). Techn. Wiss. Ber. MPA Stuttgart, Heft 75-01

[17] Maeda, T; K. Uchino and H. Sakurai (1969). "Experimental Study on the Fatigue Strength of Welded Tubular K-Joints". IIW-Doc. XV-269-69,

[18] Kurobane, Y. and M. Konomi (1973). "Fatigue Strength of Tubular K-Joints, S-N-Relationships Proposed as Tentative". IIW-Doc. XV-340-73

[19] Kurobane, Y.; Y. Makino and M. Sagawa (1970). "Low-Cycle Fatigue Research on Tubular K-Joints". IIW-Doc. XV-291-70

[20] Uchino, K.; H. Sakurai and S. Sugiyama (1973). "Experimental Study on the Fatigue Strengh of Welded Tubular K-Joints". IIW-Doc. XV-690-73, IIW-Doc. XV-344-73

[21] Toprac, A.A. and B.G. Louis (1970). "The Fatigue Behaviour of Tubular Constructions". IIW-Doc. XV-293-70

Seismic Design

A NEW APPROACH TO THE SEISMIC DESIGN OF STEEL STRUCTURES

CARLO CASTIGLIONI & PIERLUIGI LOSA
Structural Engineering Department
Politecnico di Milano, ITALY

ABSTRACT

An extensive experimental research has been performed at Politecnico di Milano, Italy, imposing constant amplitude cycles on full scale steel members. By varying from test to test the amplitude of the displacement cycles, collapse envelopes were defined relating the displacement cycles's amplitude Δv to the number of cycles to failure N for the different profiles analysed. By means of testing and extensive numerical simulation of the seismic behavior of beams and beam-columns, it has been shown that a linear cumulative damage model (Miner's rule) is applicable in the case of members subjected to large plastic strains. It is shown how the structural behavior factor (q-factor) can be related to the damage index computed by Miner's rule.

INTRODUCTION

Structures in seismic areas are usually designed so that part of the energy input during a severe earthquake is dissipated through inelastic deformations. In order to prevent failure, the value of these inelastic deformations must be limited, depending on the available local and global ductility of the structure. Design forces are usually derived from inelastic response spectra which provide, as a function of the period T, the normalized pseudoacceleration required for a specified level of inelastic response. In the most recent Seismic Codes (e.g. [1,2]) these inelastic spectra are obtained modifying a linear elastic design spectrum by means of a factor (q-factor) which takes globally into account the dissipative capacity of the structure.

The q-factor can be defined as the minimum ratio between the value of peak ground acceleration leading to structural failure and the one corresponding to first yielding. Its correct evaluation generally requires several dynamic analyses for different ground motions. In case of multi-degree of freedom systems (MDOF) this type of analysis results very cumbersome, and therefore justifies the development of simplified methods which have been proposed by many authors [3].

The proposed methods for the assessment of the q-factor for steel structures can be grouped into three main categories [3]:
- methods based on the theory of the ductility factor;
- methods based on the extension to MDOF systems of the results of dynamic inelastic response of single degree of freedom (SDOF) systems;
- methods based on the energy approach.

The theory of the ductility factor (based on the hypothesis of equivalence between the maximum displacement of a SDOF elastic system and that of an elastic plastic one) leads to a definition of the q-factor substantially coincident with the global ductility, unless limited by local ductility requirements. In the methods of the first group, the theory of the ductility factor is used for the evaluation of the q-factor through some parameters characterizing the post-elastic behavior of steel structures which are easier to determine (e.g.[4]). Some authors propose to use it to interpret the results of inelastic dynamic analyses (e.g.[5]).

Limitations to the use of the ductility factor theory for assessing the q-factor lay in that it requires the hypothesis of structural regularity and of global collapse mechanism. In fact, the method requires that the plastic deformations are uniformly distributed within the structure.

Also the methods of the second group (e.g. [6-8]) require hypothesis of structural regularity and global collapse mechanism. The critical point in the extension to MDOF systems of the results obtained for SDOF ones is the definition of the number of parameters which are necessary to characterize the pattern of yielding events in multy-story frames. In fact, in MDOF systems, different damage patterns could correspond to the same maximum inelastic displacement.

Furthermore, the theories based exclusively on the concept of ductility complitely disregard the effect of the loading history: the number of excursions in the plastic range as well as the number of reversals; hence they can estimate the real behavior of the structure only approximately, because they cannot quantify the damage actually accumulated by the structure.

The methods of the third group (e.g.[9-11]) are the most general ones, and are based on the concept that a structure collapses when it cannot anymore dissipate the seismic energy input. They do not require any hypothesis of regularity or of global collapse mechanism, and can interpret rather accurately the damage accumulation mechanism in the structure. However their application is rather lengthy and difficult, involving the evaluation of the energy input absorbable by the structure without collapsing. Furthermore, these methods requires the availability of a large number of experimental full-scale test data and/or the availability of models for both the constitutive law of the material and damage accumulation.

It is evident that the non-linear structural response to a strong earthquake is not univocally determined, but can be characterized by means of some damage parameters which allow a global evaluation of the actual structural damage, in terms of both dissipated energy and ductility demand. Actually, the less the number of cycles with a large ductility demand, the less important the dissipated energy aspect in the damage characterization, while a large number of cycles with a small ductility demand is usually associated with a damage accumulation mechanism that is strictly dependent on the absorbed energy.

It is the authors' opinion that the "ideal" method for the assessment of the q-factor should be of easy and rapid applicability (as those methods of the first group), but should also mantain the completeness of informations about the damage accumulation process in the structure, characteristic of the methods of the third group. Hence, in this paper, a method is proposed for the assessment of the q-factor, based on the comparison between elastic and inelastic response of a SDOF system, with reference to a cumulative damage parameter. In this way, the proposed method allows an evaluation of the q-factor based on both the global ductility demand and the damage accumulation.

EXPERIMENTAL RESEARCH

Testing program

At the Structural Engineering Department of Politecnico di Milano, Italy, an experimental testing program is being carried out that encompasses cyclic quasi-static tests to be performed on 45 full scale rolled steel beams (fig. 1) of the commercial shapes HE220A, HE220B and IPE300 (15 tests for each type).

Two different testing procedures are followed:

1. quasi-static tests with imposed displacements cycles of constant amplitude Δv;
2. quasi-static tests with imposed displacements following a "random" path, previously obtained by dynamic numerical simulation [5,12] of the seismic response of similar elements.

Research in the field of seismic and cyclic behavior of steel structures has been widely performed, both experimentally and numerically by the Steel Construction Group of the Structural Engineering Dept. of Politecnico di Milano [12]. During some of these previous research works, experimental data were obtained and numerical simulation models developed and calibrated. In particular, a model was developed by Castiglioni & DiPalma [13] enabling a fair good correlation between experimental results and numerical simulation, which was later introduced in a dynamic simulation code [5]. Model [13], accounting for structural damage due to both local buckling and low-cycle fatigue, was calibrated on a small number of tests carried out following the ECCS Recommended Testing Procedures [14]. According to these Recommendations, groups of three displacement cycles are imposed to the specimen, each group having an increased cycle amplitude. This loading procedure, although allowing a correct assessment of the global behavior of the specimen under testing, is not suitable for the assessment of the single aspects governing such behavior. In fact, by varying the amplitude of the displacement cycles, strain hardening effects are superimposed to reductions in load carrying capacity due to local buckling and fatigue crack propagation. For this reason the present research project focused on constant amplitude testing, as other authors previously suggested [15]. Objectives of the research are:

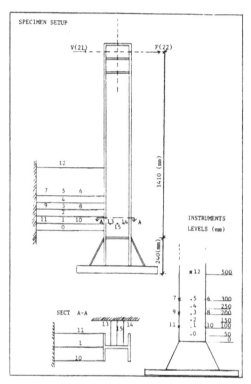

FIGURE 1

1. collection of a larger number of informations than previously available about the development of damage in the profile by means of 15 displacement transducers applied to the specimens in the plastic hinge zone (fig. 1). These instruments, connected to a computer, allowed continuous monitoring of local buckling and damage accumulation during the tests;

2. improvement of the calibration of the numerical model [13] on the new experimental results;

3. experimental assessment of the collapse criteria and of the cumulative damage model proposed in [16] where, based on numerical simulations using model [13], Castiglioni & Goss proposed to adopt a linear cumulative damage model to interpret the experimental results by Castiglioni & DiPalma [13].

Presently 25 tests were already performed (11 HE220A, 13 HE220B and 1 IPE300). Of these, 8 HE220A and 9 HE220B were performed imposing to the specimens displacement cycles with a constant amplitude (Δv) and zero mean value (v_m). The available experimental results are extensively presented and discussed elsewhere [17,18]. Hereafter, the main conclusions drawn in [17,18] with regard to constant amplitude tests on HEA and HEB specimens are summarized:

a. Strain hardening effects are much greater in HE220B rather than in HE220A profiles;

b. Deterioration effects, causing a reduction in the load carrying capacity, hysteresis loops area and stiffness, begin earlier in HEA beams (having larger width to thickness ratios of both the web and flange plates) rather than in HEB ones;

c. Once local buckling took place, the hysteresis loops stabilize, and the rate of reduction in load carrying capacity decreases with increasing the number of cycles imposed to the specimen; this effect is common to both HEA and HEB profile types;

d. Corresponding to the last load reversal before failure, the load carrying capacity of the specimen is nearly 70% of the yield strength for HEA beams, while nearly 100% for HEB beams. This is due to the large strain hardening which is typical of HEB beams that in the first cycles show an increase of the load carrying capacity of more than 40% of the yield strength;

e. The two profiles showed different failure modes, the HEA by steady crack propagation due to low-cycle fatigue effects, the HEB by some kind of brittle fracture, of both flange and web, either at specimen-to-base welds or at plastic hinge, where, due to large localized distorsions, surface cracks usually develop a few cycles after local buckling of the flange plates;

f. In both HEA and HEB beams, plastic deformations developed in a well restricted area of the beam, while the remaining portion of the specimen remain elastic up to collapse. This confirmed the correctness of the adoption in numerical simulation models of the hypothesis of concentrated plastic hinge;

g. From the carried out tests, it has been possible to obtain for both HEA and HEB beams the relationships between cycle amplitude Δv (normalized on the yield displacement v_y) and number of cycles to failure N, which are presented in the following figs. 2 and 3, plotted in a log-log scale.

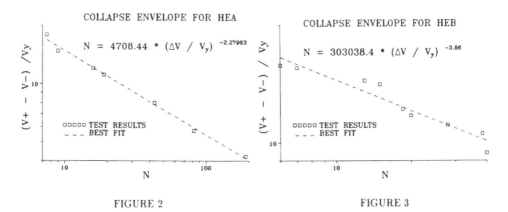

FIGURE 2   FIGURE 3

It can be noticed that for both profiles, the $\Delta v/v_y$ - N relationship can be fitted by means of an exponential function of the type $N=a*(\Delta v/v_y)^b$, with a and b constant parameters to be defined and calibrated on the experimental test results, for the different types of profiles. In a log-log scale, such an equation plots as a straigt line. the equation of such line is given in the same figures. This is in good agreement with previous experimental results [19,20], and with the results obtained numerically in [16].

These $\Delta v/v_y$-N diagrams are very similar to Woehler diagrams, commonly adopted in high-cycles fatigue. This fact induced to check the possibility to extend the applicability of a linear damage cumulation model based on Miner's rule [23] (usually adopted in high-cycle fatigue) to members subjected to large cyclic strains in the plastic range (as those experienced by a steel member during an earthquake).

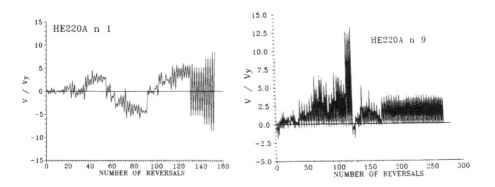

FIGURE 4   FIGURE 5

Random loading tests

In order to assess the seismic behavior of the different shapes, as well as to check the validity of the damage accumulation model [16], some random displacement histories were numerically obtained (figs.4-9) by means of the dynamic numerical simulation code presented in [5].

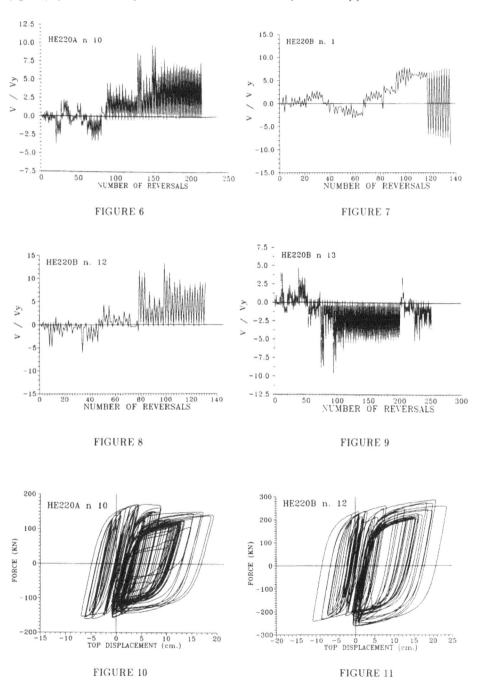

FIGURE 6

FIGURE 7

FIGURE 8

FIGURE 9

FIGURE 10

FIGURE 11

Model [5], set up to simulate the dynamic behavior of steel columns under compression and bending, consists of a rigid bar connected to the ground by a "cell" where all the deformability of the member is concentrated, and a structural lumped mass applied on the top. The behavior of the "cell" follows a constitutive law for the material and accounts for damage according to the rules and the model proposed in [13]. Some artificial accelerograms were generated starting from EC-8 response spectra [21]. These accelerograms were used as seismic input for steel members similar to the test specimens, whose dynamic behavior was simulated by means of model [5]. The oscillograms numerically obtained as output from the simulation were imposed in a quasi-static way to the specimens under testing. To date 4 HE220A and 3 HE220B beams were tested under random cycles. As an example, figs. 10 and 11 show the histeresis loops for HE220A n.10 and HE220B n.12 specimens.

PROPOSED CUMULATIVE DAMAGE MODEL

In general, the probability of failure P_f may be defined as the probability that an index D of the damage cumulated in the member (or the structure) exceeds a limit value γ, which in reality is a random variable but, under simplyfing assumptions, can be assumed deterministically:

$$P_f = P\ [D > \gamma] \tag{1}$$

In the literature, many damage model were presented, leading to different definitions of the damage parameter D (e.g. [22]). Among the different proposals, the one based on the hypothesis of linear proportionality between the cumulated damage and the number of cycles undergone by the member (proposed by Miner [23]) is very simple and is commonly adopted in high cycle fatigue. Under this hypothesis, the damage index can be considered as the ratio of the number n_i of cycles (having a constant amplitude Δv_i) imposed to the member to the number of cycles to failure N_{fi} having the same amplitude. In the case of variable amplitude loading, Miner's rule leads to define the damage index D as

$$D = \sum_{i=1}^{L} \frac{n_i}{N_{fi}} \tag{2}$$

where L is the number of groups of n_i cycles having the same stress (or strain) level.

The value D=1 indicates failure, although the variability associated with the problem under examination cannot but lead to a rather large scatter of the results. This is obviously to be expected whenever it is tried to interpret deterministically a stochastic process. The validity of an extension of these concepts to the case of members under seismic loading is not immediate, but must be experimentally verified. In fact method [23] was originally proposed for members under cycles in the elastic range (i.e. having a strain amplitude $\Delta\epsilon < 2\epsilon_y$, ϵ_y being the yield strain), while seismic loading is characterized by large plastic strains (i.e. by strain cycles with $\Delta\epsilon >> 2\epsilon_y$) with a collapse mechanism involving local buckling effects.

For assessing the applicability of Miner's rule to members under seismic loading it was hence applied the following procedure:
1. Define the $\Delta v/v_y$-N curves, already described (figs. 2 and 3).
2. By some cycles counting method [24], obtain from the seismic oscillograms (fig. 4-9) the numbers of cycles n_i for each cycles amplitude $\Delta v_i/v_y$.
3. From the $\Delta v/v_y$-N curves obtained at point 1., define the number of cycles to failure N_{fi} corresponding to each cycles amplitude $\Delta v_i/v_y$.
4. Compute the damage index D according to equation (2).

Two different counting methods were considered, the "Rainflow" and the "Simple Range Counting" (SRC) method [24]. Furthermore, as some HE220B specimens failed prematurely, due to fracture in the specimen-to-base weldments, for HEB specimens two $\Delta v/v_y$-N curves were considered, one comprising all test data (curve I), the other comprising only data of tests which did not show premature weld failure (curve II).

The results of the random tests were processed according to the proposed procedure, applying both

Rainflow and SRC counting methods to the random loading histories imposed to the specimens up to collapse.

The following TABLES 1 and 2 summarise the damage indexes associated to collapse of the specimens for the random tests.

	$D = \displaystyle\sum_{i=1}^{L} \dfrac{n_i}{N_{fi}}$	
TEST	RAINFLOW	SRC
HE220A n.1	1.002	0.999
HE220A n.9	0.986	1.190
HE220A n.10	0.986	1.320
HE220A n.11	1.083	1.270

TABLE 1

		$D = \displaystyle\sum_{i=1}^{L} \dfrac{n_i}{N_{fi}}$	
TEST	CURVE	RAINFLOW	SRC
HE220B n.1	I	1.061	0.924
HE220B n.1	II	0.890	0.790
HE220B n.12	I	0.696	0.598
HE220B n.12	II	0.633	0.617
HE220B n.13	I	0.363	0.297
HE220B n.13	II	0.394	0.370

TABLE 2

It can be noticed that, for HEA specimens, a unit value for Miner's damage index is in fair good agreement with the experimental reality, while for HEB specimens, there is a good agreement only for specimen HEB n.1, while for specimens n.12 and n.13, which failed at weldments, Miner's index at collapse results smaller than 1. Furthermore, it can be noticed that Rainflow counting method always leads to a definition of Miner's index rather close to unity, while SRC method (at least for HEA specimens) leads to higher values of the damage index. This means that it leads to an underestimation of the specimen strength, being on the safe side.

It should be added that a $\Delta v / v_y$-N curve was also obtained fitting all data of tests failed at weldments. An evaluation of Miner's index based on this curve lead to higher values of the damage index, although still lower than 1.

PROPOSED PROCEDURE FOR ASSESSING q-FACTOR

The previous results confirm those obtained by Castiglioni and Goss [16], and lead to the applicability of a linear damage accumulation model based on Miner's rule in the case of steel members under seismic loading. This fact leads to some interesting design considerations related to both the assessment of the q-factor and, eventually, to some new design approach overcoming those based on the q-factor itself.

In what follows, reference is made to the assessment method proposed by Ballio et al. [5], who defined, for SDOF systems, the "optimal" design q-factor as the intersection between the curve representing, in terms of ductility demand, the non-linear dynamic response with the line representing the response of the same member, obtainable by means of a linear elastic dynamic analysis (fig. 12). Under the hypothesis of validity of the ductility factor, in [5] the optimal q-factor for cantilever members and cross-bracings of single story buildings, was obtained comparing the maximum value of the top displacement obtainable by a dynamic non-linear analysis, accounting for structural damage, with that obtained by a conventional dynamic linear analysis.

The "optimal" q-factor is defined in [5] as that value beyond which a linear dynamic analysis is no more on the safe side because the global ductility demand estimated by a non-linear dynamic analysis is larger than that resulting from a linear elastic analysis.

Although on the safe side, method [5] leads to an evaluation of the q-factor which doesn t allow an evaluation of the actual damage cumulated in the member. This fact doesn't allow a clear evaluation of the actual safety factor associated with the adopted design value of the q-factor.

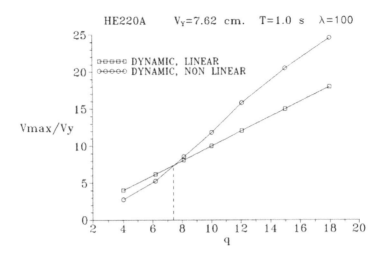

FIGURE 12

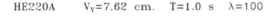

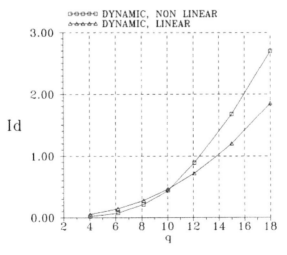

FIGURE 13

To overcome these limitations, with reference to simple cantilever members with a structural mass lumped on the top, having a period T and a slenderness ratio λ, the following procedure for the assessment of the q-factor is proposed. The structural typology, although very simple, can be referred to columns of single story frames as well as to columns of shear type buildings.

Between the damage index I_d computed for the same member by a dynamic linear and a dynamic non-linear analysis and the q-factor, a relationship holds of the same kind of that existing between global ductility demand and the q-factor (fig. 13).

In fact, it can be noticed that initially the damage index computed by means of a dynamic linear analysis is greater than that computed by means of a non-linear dynamic analysis. The two curves intersect at a point beyond which an evaluation of the cumulated damage based on the linear dynamic analysis is no more on the safe side. The intersection point can be regarded as the "optimal" q-factor, based on the equivalence between the cumulative damage evaluated by means of linear and non-linear dynamic analysis.

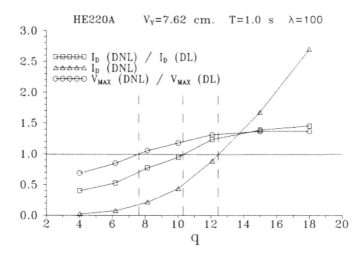

FIGURE 14

Fig. 14 shows that the proposed damage index I_d is strictly connected to the q-factor assessments based on both ductility demand and damage equivalence. In particular it is evident that basing the assessment of the q-factor on equivalence of global ductility leads to a more conservative evaluation than the equivalence of damage. The latter results however still very conservative with regard to failure, with a safety factor, for the presented case, of nearly 2 (I_d = 0.5, when $I_d(DNL)/I_d(DL)$ =1).

CONCLUSIONS

In this paper it is shown that it is posible to apply Miner's rule for the evaluation of cumulative damage in steel members subjected to seismic loadings. Furthermore a procedure for the assessment of the q-factor based on the equivalence of Miner's damage index computed by means of a linear and a non-linear dynamic analysis is presented. The proposed method leads to an evaluation of the q-factor that, although as simple as those methods based on global ductility considerations (and the ductility factor theory), allows to keep into account the actual damage cumulation process within the structural member. Further research work is needed and presently underway to validate and extend the proposed model; still, these early results, relative to SDOF systems, seem very promising, for the possible development of a new approach of general validity for the seismic design of steel structures.

REFERENCES

1. Eurocode-8: "European Code for Seismic Regions", draft, 1989
2. Structural Engineering Assocition of California - Seismology Committee: "Tentative Lateral Force Requirements", 1985.
3. Guerra C.A., Mazzolani F.M., Piluso V., "Evaluation of the q-factor in steel framed structures: state-of-art", Ingegneria Sismica, Vol. 7, n.2, 1990, pag. 42-62.
4. Cosenza E., De Luca A., Faella C., Mazzolani F.M., "On a simple evaluation of structural coefficients in steel structures", 8th ECEE, Lisboa, Portugal, september 1986
5. Ballio G., Castiglioni C.A., Perotti F., "On the assessment of structural behavior factor for steel structures", 9th WCEE, vol.5, Tokyo 1988.
6. Giuffre' A., Giannini R., "La duttilita' delle strutture in cemento armato", ANCE-AIDIS, Roma, 1982.
7. Palazzo B., Fraternali F., "L'uso degli spettri di collasso nell'analisi sismica: proposta per una diversa formulazione del coefficiente di struttura", Proc. 3rd National Conference on Seismic Engineering in Italy, Roma, 1987.
8. Cosenza E., De Luca A., Faella C., Piluso V., "A rational formulation for the q-factor in steel structures", 9th WCEE, Tokyo, 1988.
9. Como M., Lanni G., "Aseismic toughness of structures", Mecanica, vol. 18, 1983, pag. 107-114.
10. Kato B., Akiyama H., "Seismic design of steel buildings", ASCE, Journal of the Structural Division, Vol. 108, n.8, August 1982.
11. Akiyama H., "Earthquake resistant design based on energy concept", 9th WCEE, Vol. 5, Tokyo, 1988.
12. Seismic design of steel structures, Selected Papers 1985-1989, by Researchers of the Steel Construction Group, Structural Engineering Dept., Politecnico di Milano, 1990.
13. Castiglioni C.A., DiPalma N., "Steel members under cyclic loads: numerical modelling and experimental verifications", Costruzioni Metalliche, n.6, 1988.
14. ECCS (European Convention for Constructional Steelwork), "Recommended testing procedure for assessing the behavior of structural steel elements under cyclic loads", Techn. Publ. n. 45, 1986.
15. Krawinkler H. et al., "Recommendations for expeerimental studies on the seismic behavior of steel components and materials", The John Blume Earthquake Eng. Research Center, Stanford, CA, Rept. 61, 1983.
16. Castiglioni C.A., Goss G., "Adopting Miner's Rule in low cycle fatigue", Techn. Rept. n. 8/89, Struct. Eng. Dept., Politecnico di Milano, 1989.
17. Castiglioni C.A., Losa P.L., "Local buckling and structural damage in steel members under cyclic loading", Proc. X WCEE, Madrid, 1992
18. Losa P.L., "Elementi strutturali in acciaio soggetti ad azioni sismiche: modellazione numerica e validazione sperimentale dell'applicabilita' di un modello di accumulazione lineare del danno", Laurea Thesis, Struct. Eng. Dept., Politecnico di Milano (in italian).
19. Coffin L.F., "A study of the effects of cyclic thermal stresses on a ductile metal" Trans. ASME, Vol. 76, 1954
20. Manson S.S., "Behavior of materials under conditions of thermal stress", Heat transfer Symp., NACA TN 2933, 1953
21. Ballio G. et al. "Generation of artificial accelerograms for assessing q factors", Internal Rept. T.C. "Safety and Loadings", W.G.1.3, "Seismic Design", E.C.C.S.
22. Yao J.T.P, et al. "Stochastic fatigue, fracture and damage analysis", Structural Safety, n.3, 1986, pag. 231-267.
23. Miner M.A., "Cumulative damage in fatigue", Journal of Applied Mechanics, Sept. 1945.
24. Bannantine J.A., Comer J.J., Handrock J.L, "Fundamentals of metal fatigue analysis", Prentice Hall, Englewood Cliffs, NJ.

BEHAVIOUR OF STEEL FRAMES WITH SEMI-RIGID CONNECTIONS UNDER EARTHQUAKE LOADING

K. TAKANASHI[†], A. S. ELNASHAI[*], A. Y. ELGHAZOULI[*] and K. OHI[†],
[†] Institute of Industrial Science, University of Tokyo, Japan
[*] Imperial College of Science, Technology and Medicine, London, UK

ABSTRACT

This paper presents a preliminary account of the results from a series of tests on two-storey steel frames with fully-welded (rigid) and semi-rigid beam-to-column connections. The experiments were conducted using monotonic, cyclic and on-line computer-controlled earthquake testing. The rapid assessment of comparative results show that semi-rigid frames exhibit ductile and stable hysteretic behaviour and may be used effectively in earthquake-resistant design.

INTRODUCTION

For static design, modern codes (EC3 [1], AISC [2]) have opened the door to the use of semi-rigidly connected steel frames by giving broad classifications of connection stiffness and by allowing their use in design, provided all other code requirements are met.

On the other hand, reliance on the rigidity of fully-welded connections under earthquake loading has recently come under question in Japan, as a consequence of difficulties associated with quality control of welding processes. This cast doubts over the integrity, hence reliability, of this type of construction when subjected to severe inelastic cyclic straining.

Conventionally, semi-rigidly connected frames are considered to be inappropriate for seismic design purposes mainly due to their excessive flexibility. Their advantages, in terms of lower construction costs and simple fabrication, are hitherto not utilised in seismic design.

Recent analytical and experimental work has indicated that semi-rigid frames may be used to advantage in low and medium seismicity zones. The comparative experiments conducted by Astaneh and his co-workers [3] indicated that fully-welded frames may not be the optimum solution, and that semi-rigid designs may perform adequately under simulated earthquake loading. An analytical study by Parra Rosales [4], using nonlinear dynamic analysis, showed that semi-rigid frames have the advantage of a longer effective period hence attract lower inertial loads. This may offset the effect of the increased flexibility, hence resulting in a most satisfactory earthquake-resistant design solution.

To further progress this subject and to provide detailed information of the seismic response parameters of semi-rigid frames in comparison with fully-welded alternatives, experimental work was undertaken on two storey single bay steel frames. This formed part of the ongoing joint earthquake engineering research programme (JEERP) between Imperial College, London, and the Institute of Industrial Science, Tokyo. {Previous work under JEERP focussed on the seismic performance of composite frames [5]}

The tests were conducted using monotonic, cyclic and on-line earthquake loading, with the aim of providing well-controlled tests results to verify the feasibility of semi-rigid frames and for the calibration of analytical models [6,7]. A preliminary account of selected results from these recently-completed tests is given below.

TEST SET-UP

A general layout of the testing arrangement is given in Figure 1. All frames were tested horizontally. One half of the two-storey single bay frames was used, with symmetry accounted for in the boundary conditions. Consequently, vertical displacements were restricted at the beam ends, thus representing the points of contraflexure. The frames were fixed at the base, whilst hydraulic actuators were used to apply displacements and/or loads at the two floor levels.

Figure 1. General view of testing assembly

In all tests, at least 26 strain gauges were provided in addition to displacement and force readings Hence, sufficient information on the detailed response including various aspects such as capacity, local and overall buckling, joint rotation, beam-to-column strength ratio and moment distribution was obtained. Considerable attention was given to measuring the change in angle at the beam-column connection. For this purpose, a simple system was chosen after verification through predictive analysis. An online computer was employed for all control and data acquisition operations.

SPECIMEN DETAILS

Member and cross-sectional dimensions are given in Figure 2 showing the general layout of a typical frame. Rigid connections were fully welded, whilst the semi-rigid connection consisted of top and seat angles, and two side angles as shown in Figure 2. In all frames, hot rolled steel H profiles were used for the columns, and the beams were built-up from welded steel plates.

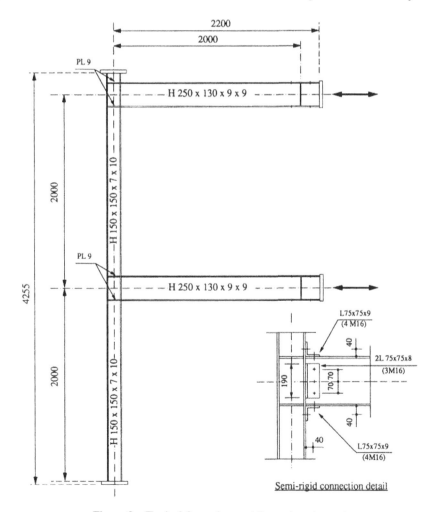

Figure 2.   Typical frame layout (dimensions in mm)

The rigid frames were appropriately designed. Codes such as AISC [1] and EC3 [2] give average stiffness coefficients and classify connections accordingly. Consequently, design of a semi-rigid frame will make use of assumptions that have not been verified. Therefore, the semi-rigidly connected frames were not designed; the rigid frame design was retained whilst replacing the rigid connection with a semi-rigid counterpart. The shortcoming of this approach was compensated for in the procedure for load amplitude calculation, as discussed below.

TABLE 1
Material Properties

	σ_y (t/cm^2)	σ_u (t/cm^2)	ε_{ult} (%)
Beam Plates	2.98	4.31	32
Column Flange	3.06	4.32	29
Angle 75x75x9	3.01	4.50	29
Base Plate	2.67	4.22	36

Average material properties obtained from test samples are given in Table 1, where σ_y, σ_u and ε_{ult} are the yield stress, ultimate stress and ultimate strain, respectively.

LOADING PROCEDURES

Table 2 gives a summary of the testing procedures for the frame specimens, referred to as either semi-rigid (SRB) or rigid (RGD); A_{max} is the maximum acceleration amplitude used.

TABLE 2
Testing Sequence and Procedure

Model Reference	Testing Method	m_1 (Kg)	m_2 (Kg)	T_1 (sec)	T_2 (sec)	A_{max} (g)
SRB01	Monotonic	--	--	--	--	--
RGD02	Cyclic	--	--	--	--	--
SRB02	Cyclic	--	--	--	--	--
RGD03	Pseudo-dynamic	8,400	8,400	0.62	0.19	0.54
SRB03	Pseudo-dynamic	8,000	8,000	0.60	0.20	0.45

The first test on the semi-rigid frame SRB01 was conducted under monotonic loading to obtain the static capacity necessary for the calibration of accompanying predictive analysis and the choice of the earthquake scaling factor for the dynamic on-line test. The second floor actuator was used in a displacement-control mode, where the displacement was incremented in one direction. The second floor actuator force was then measured and used to drive the first floor actuator, which was operating on a load-control basis. The ratio between the second and first floor loads was maintained at a 2:1 ratio.

The cyclic tests on models RGD02 and SRB02 were conducted under the same hybrid displacement/load control procedure described for monotonic loading. Based on the second storey yield displacement (δ_y) estimated from predictive analysis, three cycles were applied at each displacement of δ_y, $2\delta_y$, $4\delta_y$, $6\delta_y$, etc. In some cases close to failure, only one cycle per amplitude was applied.

RGD03 and SRB03 were tested using the pseudo-dynamic testing technique. Two approaches were considered regarding the earthquake tests on the rigid and semi-rigid frames RGD03 and SRB03. One option was to design each system to the same peak ground acceleration. In which case, testing under the same ground acceleration and its multiples would provide an objective yard-stick for comparison. An alternative approach would be to select the section dimensions and other design details to suit other testing criteria. An estimate has to be obtained of the yield and ultimate capacities of the structures, then testing proceeds with input motions that are different; each reflecting a fixed ratio of design-to-testing peak ground acceleration. Although the former approach is recommended, it was not adopted due to lack of appropriate design criteria, as mentioned earlier. Inspite of several inherent uncertainties such as the definition of yield and ultimate capacities, the latter approach was more feasible for use in this test series.

Based on predictive analysis [6,7] and following the approach described above, part of the 1940 Imperial Valley (El Centro) NS component acceleration record, of 15 second duration, was chosen to conduct the tests. The masses and peak ground acceleration were appropriately selected in order to satisfy a fixed capacity-to-testing ratio in both frames, i.e. the same scaling factor to the frame capacity, while maintaining a similar fundamental period as shown in Table 2. Equal concentrated masses at both storeys were assumed in the pseudo-dynamic algorithm; 8400 Kg for RGD03 and 8,000 Kg for SRB03. Peak ground acceleration of 0.54g was chosen for RGD03, whereas the corresponding value for SRB03 was 0.45g.

EXPERIMENTAL RESULTS

Model SRB01 tested under monotonic loading behaved in a ductile manner. Figure 3 depicts the load versus deflection curve at the two floor levels of the frame. Yield was observed at a second storey force of approximately 2.7 tons. As loading proceeded, yielding and large distortion of the top and seat angles, under tension and compression respectively, was clearly observed. Near the end of the test, at a second floor displacement of about 350 mm, large local buckling was observed near the base of the first floor column. At the termination of the test, the load supported by the frame was just over 5 t.

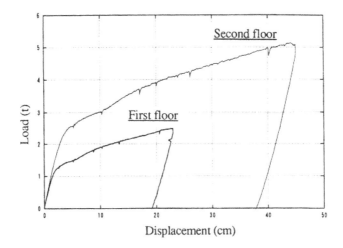

Figure 3   Load-displacement relationships for SRB01

Figure 4 gives the load versus displacement relationships for model RGD02 at the two floor levels. The yield load was approximately 3.9 t. The main purpose of this test was to verify the analytical prediction of the stiffness and capacity of the rigid frames under cyclic loading. The load-displacement curves for SRB02 at the two floors are given in Figure 5. Yield was achieved at a load value of about 2.3 t, and the envelope curve was similar in shape to the monotonic response. Again, under cyclic loading, the semi-rigid frame exhibited adequately stable hysteretic behaviour with large ductility levels. The test was terminated at a displacement of 200 mm at the second floor with the ultimate load gradually increasing.

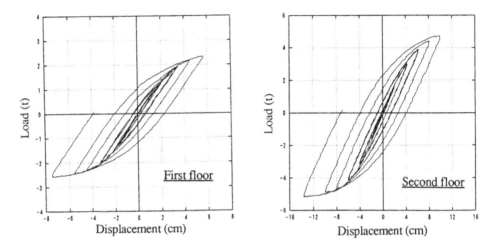

Figure 4   Load-displacement relationships for RGD02

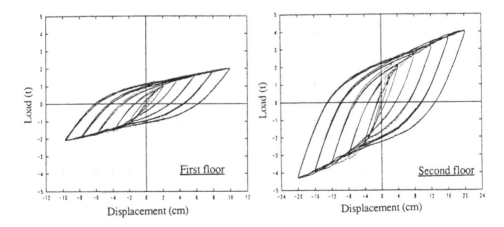

Figure 5   Load-displacement relationships for SRB02

The displacement response histories at both floors from the two pseudo-dynamic tests on models RGD03 and SRB03 are given in Figures 6 and 7, respectively. Both frames exhibited very stable hysteretic behaviour with no drop in capacity. In terms of maximum displacements, RGD02 gives a maximum displacement of approximately 70 mm and 120 mm at the first and second storey levels, respectively. The corresponding values for SRB03 are 40 mm and 85 mm, respectively, which are considerably lower than the rigid frame.

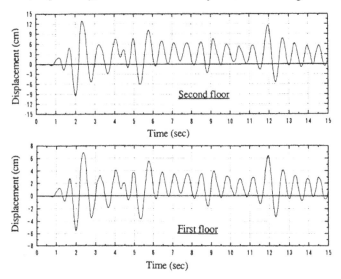

Figure 6   Displacement response histories for RGD03

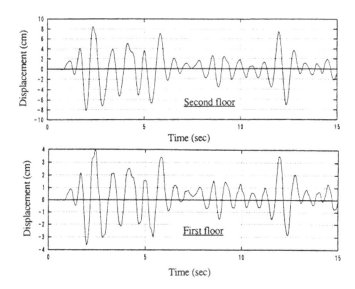

Figure 7   Displacement response histories for SRB03

CONCLUDING REMARKS

The preliminary results presented above indicate that semi-rigidly connected steel frames provide adequate and, in some cases, favourable earthquake-resistant qualities. It was shown that semi-rigid frames do exhibit a ductile and stable hysteretic behaviour. The stiffness as well as yield and ultimate capacities of semi-rigid frames were lower than similar rigid frames under monotonic or cyclic loading. However, tested under a consistent capacity-to-testing peak and ground acceleration ratio, the semi-rigid frame response displacements were lower than the fully-welded case. It is to be noted that there are several other testing options that have not been adequately considered in this study; final confirmation of the feasibility of steel frames with semi-rigid connections for seismic design is still some way away.

Full processing of the experimental data is underway to provide more detailed information on the response and to accurately quantify the seismic response parameters of all frames.

ACKNOWLEDGEMENTS

The tests described above were funded by the Institute of Industrial Science. The British Council in Japan (Dr. R. Soweden; Science Officer) kindly provided full funding for the Imperial College group's (Drs. Elnashai, Elghazouli and Izzuddin) travel and subsistence expenses throughout the duration of the testing programme. The test set-up and control software was the work of Mr. Y. Shimawaki and Mr. H. Kondo, both Technical Officers at IIS. Mr. P. Madas, IC, provided the analytical prediction of the connection behaviour.

REFERENCES

1. EC3, Eurocode No 3, "Common Unified Rules for Steel Structures", Commission of the European Communities, April 1990.

2. AISC-LRFD, Specifications for the Design, Fabrication and Erection of Structural Steel for Buildings, American Institute of Steel Construction, Chicago, IL, 1986.

3. Nander, M.N. and Astaneh, A., Experimental Studies of a Single Storey Steel Structure with Fixed, Semi-rigid and Flexible Connections, Report No. UCB/EERC-89/15, Earthquake Engineering Research Center, University of California, Berkeley, California, August 1989.

4. Rosales, J.G.P., Seismic resistance of steel frames with semi-rigid connections, MSc Dissertation, Engineering Seismology and Earthquake Engineering Section, Imperial College, University of London, August 1991.

5. Elnashai, A.S., Takanashi, K., Elghazouli, A.Y. and Dowling P.J., Experimental Behaviour of Partially-encased Composite Beam-Columns under Cyclic and Dynamic Loads, Proc. of the Institution of Civil Engineers, Part 2, June 1991, pp. 259-272.

6. Izzuddin, B.A. and Elnashai, A.S., ADAPTIC, A Program for Adaptive Large Displacement Elastoplastic Dynamic Analysis of Steel, Concrete and Composite Frames, Engineering Seismology and Earthquake Engineering, Imperial College, Research Report No. ESEE 89/7, 1989.

7. Madas, P.J. and Elnashai, A.S., A Component-Based Model for the Response of Beam-Column Connections, Tenth World Conference of Earthquake Engineering, Madrid, Spain, July 1992.

RESPONSE OF A BRACED STEEL FRAME TO SEISMIC ACTIONS: NUMERICAL SIMULATION AND EXPERIMENTAL VERIFICATION

ORESTE S. BURSI[1], KURT H. GERSTLE[2] and PUI-SHUM B. SHING[2]

[1]Department of Structural Mechanics and Design Automation,
University of Trento, Mesiano, 38050 Trento, ITALY
[2]Department of Civil, Environmental and Architectural Engineering,
University of Colorado, Boulder, CO 80309-0428, USA

ABSTRACT

This paper compares analytical simulations with the results of cyclic and pseudodynamic tests conducted on a $\frac{1}{2}$-scale concentrically braced frame specimen. The study is intended to validate a finite element model and an equivalent truss model for simulating the strength and stiffness of bolted bracing connections in the context of frame analysis. The specimen represents the lower story of a three-story concentrically braced steel frame and the pseudodynamic tests were conducted with a newly developed substructuring approach. From the comparison of numerical and experimental results, in terms of the global load-displacement hystereses, energy dissipation, and displacement time histories, it is concluded that both the global and local behavior of the test specimen can be well captured by means of the proposed bracing connection models. Furthermore, properly detailed bolted bracing connections exhibited a good ductile behavior, thus providing a good energy dissipation mechanism. Finally, it has been shown that the results of conventional frame analysis, in which the flexibility of moderately bolted bracing connections is neglected, can underestimate the story drift by 28.3% and overestimate the base shear by 57.8% under a severe seismic load.

INTRODUCTION

The seismic performance of concentrically braced steel frames (CBF), such as the ones shown in Fig. 1, has been widely studied. The results of these studies have led to design recommendations recently adopted in the Seismic Provisions for Structural Steel Buildings - LRFD [1] and Eurocode 8 [2]. In particular, in order to avoid undesired premature failure of CBF under severe seismic load conditions, stringent strength requirements are specified [1,2] for designing bracing connections of the type shown in Fig. 2. Nevertheless, no specific guidelines are available to evaluate the strength of such connections. In frame analysis, bracing connections are often modeled as ideal pins or rigid joints, thus neglecting the flexibility contributed by gusset plate-to-member fasteners and the gusset plates themselves. In reality, bracing connections are expected to affect the sway of braced frames and the local stress distribution in beams and columns around the joints. To evaluate the strength and stiffness of bracing connections through detailed stress analysis, a plasticity-based interface element has been developed [3] for modeling bolted clip-angle and welded gusset plate-to-member fasteners. The finite element model has been validated with tests of isolated bolted and bolted-welded bracing connections [4]. Furthermore, to allow a realistic assessment of the behavior of braced frames, a phenomenological model based on an equivalent truss concept (Fig. 3) has been proposed [5]. The latter has to be calibrated with the finite element model.

The main objectives of this study are to examine the accuracy of the aforementioned models in the context of frame analysis, to investigate the influence of connection flexibility on the response of a braced frame, and to evaluate the performance of a CBF, including that of bolted bracing connections, designed in accordance with the new LRFD provisions [1], under severe seismic load conditions. To this end, a $\frac{1}{2}$-scale plane frame specimen representing the bottom story of a three-story CBF was tested under cyclic and pseudodynamic loads. The latter tests were conducted with a newly developed substructuring approach [6], in which only the bottom story of a multistory frame needs to be tested, while the upper stories are modeled in an on-line computer. The experimental results are compared to frame analyses conducted with an equivalent truss model using the computer program ANSR-III [7]. In this paper, the analysis method, experimental procedure, and numerical and experimental results, in terms of the global response, plastic energy dissipation, and failure mechanism of the frame specimen at various seismic load levels, are presented. Furthermore, the performance of the frame specimen under severe seismic load conditions is investigated. Finally, the influence of the flexibility of bracing connections on the overall frame performance is commented upon.

TEST STRUCTURE AND ANALYTICAL MODELING

In this study, a diagonal bracing system representing a single bay of a three-bay, three-story office building, which was designed with the "Equivalent Lateral Force Procedure" specified in the 1988 Uniform Building Code [8], is considered. Because of the limited capacity of the loading apparatus, the frame was designed for UBC Seismic Zone 3. The geometric configuration and member sizes of the prototype frame are shown in Fig. 4. The design complies with the AISC-LRFD Manual [9] and the recently proposed seismic provisions [1]. The lateral resistance of the frame was provided by a single diagonal brace at each story. While this may not be a most desirable lateral load resistance system, in view of the limited energy-dissipation capability associated with the buckling of braces, this bracing system, nevertheless, simplified the test procedure and specimen fabrication, and allows for a better comparison of experimental and numerical results. It also induced severe tensile forces on the bracing connections.

The tests were carried out on a $\frac{1}{2}$-scale frame specimen, representing the bottom story of the prototype frame, as shown in Fig. 5. Similitude rules were applied to the frame members as well as to the bracing and beam-to-column connections. To allow the use of standard member sizes, the similitude rules were slightly violated, but it was shown by numerical simulations that a reasonable dynamic similitude can be attained between the prototype frame and the scale model [6]. A typical bracing connection used in the scale model is shown in Fig. 6. It consisted of a $\frac{1}{4}''$ (6 mm) thick gusset plate fastened to the beam and column members with 3 x $2\frac{1}{2}$ x $\frac{3}{16}''$ (76 x 63 x 4.8 mm) double clip angles which have a yield stress F_y equal to 53 ksi (365 MPa). To attain the axial strength of the bracing [1], three bolts were used for each clip angle connection [9]. The gusset plate buckling strength was computed with a conservative procedure based on the Whitmore section [10]. This type of connections is designated as 3x3 bracing connections in the following. Furthermore, to provide additional data for validating the analytical models, 2x2 bolted fasteners were used in a few tests. As shown in Fig. 6, two types of beam-to-column connections were used. One is Web Angle (WA) connections (Fig. 6a) and the other is Extended End-Plate (EP) connections of $\frac{1}{2}''$ (12.7 mm) thick (Fig. 6b). The ratio of the rotational connection flexibility F to the beam flexibility $\frac{L}{EI}$ computed for the WA and the EP beam-to-column

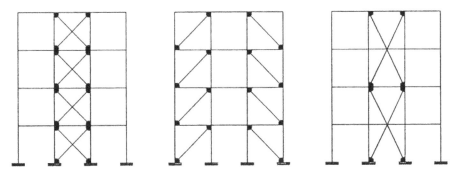

Figure 1. Concentrically braced frames.

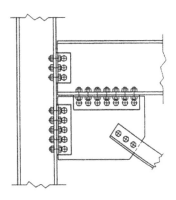

Figure 2. Bolted beam web-to-column
and bracing connections.

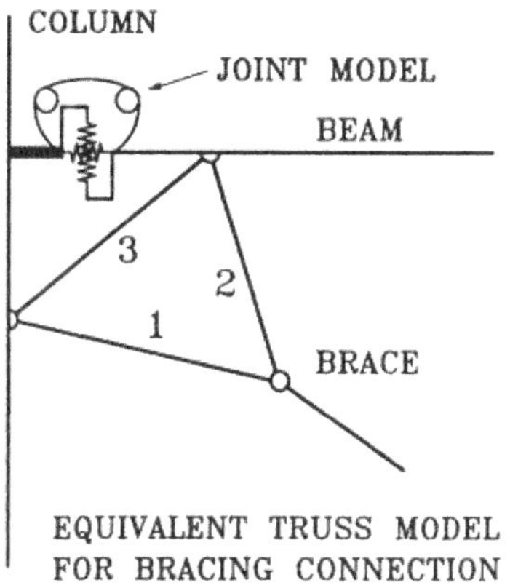

Figure 3. Phenomenological model for beam-
to-column joint and bracing connection.

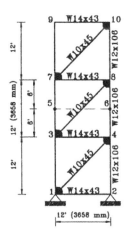

Figure 4. Prototype structure.

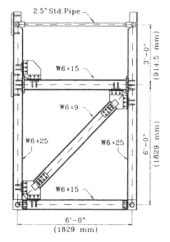

Figure 5. Test structure.

connections are 5.7 and 0.039, respectively. These values are well within the range for which a pinned or a rigid connection can be assumed for the beam-to-column joint [11]. A-36 steel has been used for all members while the actual average yield stress F_y of the diagonal braces used is 47.7 ksi (329 MPa).

The bracing connection used has been analyzed with a refined finite element model developed within the framework of hardening plasticity, in which the bolted clip-angle fasteners are simulated with interface elements [3]. This analysis has been used to simulate the behavior of isolated bolted bracing connections [4] and to provide data for calibrating the phenomenological model used for the subsequent frame analysis. As shown in Fig. 3, the phenomenological model for bracing connections is based on an equivalent truss [5], which can capture both the rotational constraint and axial flexibility introduced into beam-to-column joints and diagonal bracings, respectively. A bilinear axial force-axial deformation relation with a kinematic hardening rule is used for each truss member to simulate the nonlinear behavior of bracing connections. The model, however, cannot capture the buckling of the gusset plate.

To check the validity of the phenomenological model in the context of frame analysis, the bracing connection model is incorporated into the static and dynamic analyses of the frame specimen using the computer program ANSR-III [7]. While only a single story was tested by means of the pseudodynamic method (Fig. 7) using a substructuring approach, the dynamic analyses are conducted with the entire three-story, single-bay scale model. Elastic-plastic beam-column elements based on the plastic hinge concept are used to model the beam and column members. Four beam-column elements with distributed plasticity and geometric nonlinearity [12] are used to model the buckling of the diagonal brace by introducing an initial imperfection. The depth of a column is modeled by a rigid beam extending from the center line to the flange of the column. The masses are lumped at nodes 3, 7 and 9 of the frame model which are numbered in the same fashion as shown in Fig. 4. The masses at the first and second stories are assumed to be 0.204 kip-sec^2/in (35.7 kN-sec^2/m) and that at the third story is 0.08 kip-sec^2/in (14 kN-sec^2/m). Rayleigh damping is used and the damping ratio adopted for the fundamental mode is 2.3% that was measured in pseudodynamic free-vibration tests.

TEST PROGRAM AND PROCEDURE

Quasi-static Cyclic Tests

To identify the lateral load-vs.-lateral displacement response characteristics of the braced frame, it was subjected to a series of quasi-static cyclic load reversals. The specimen was loaded with a servo-controlled electrohydraulic actuator at the top of the first story. Tests were conducted with both 2x2 and 3x3 bracing connections at small deformation levels. Four tests were performed under load control on the frame with and without a diagonal brace to characterize the elastic response of both the frame and the connections. Two tests were performed under displacement control, with gradually increasing fully reversed lateral displacement cycles. Three equal amplitude cycles were imposed at each displacement level and the amplitude was progressively increased in steps of 0.1 inch.

Pseudodynamic Tests

A multilevel substructuring algorithm has been developed and implemented in a finite element-based pseudodynamic test program, in which part of a structure can be modeled in a computer with beam-column and truss elements while the remaining portion is tested experimentally. An unconditionally stable implicit direct time integration scheme is used to evaluate the dynamic response. A modified Newton-Raphson iterative approach based on the initial structural stiffness is adopted in the solution process. The initial stiffness can be estimated by means of an analytical model. The detailed test methodology and theory have been covered in [6].

In this study, only the lowest story of the braced frame, in which damage was expected to be most severe, was tested, while the two upper stories were modeled in a computer. The schematic of the test setup is shown in Fig. 7. The experimental and analytical substructures were partitioned at the mid-height of the second story, where the inflection points were assumed to be located. This assumption is not exactly true but has no significant impact on the response of the frame. The partition line is shown in Fig. 4, where nodes 3, 5, and 6 are the interface nodes. It should be noted that the diagonal brace at the second story is considered to be part of the analytical substructure and is, therefore, not shown in Fig. 7. The mass is assumed to be distributed in the same fashion as in the frame analyses using the ANSR-III program.

Two actuators were used to control the horizontal displacements at nodes 3, 5, and 6 (i.e. DOF 1, DOF 2, and DOF 3 shown in Fig. 7) of the experimental substructure. By means of a rigid link, the displacements at DOF 2 and DOF 3 (i.e., d_2^M and d_3^M) were slaved. Even though all degrees of freedom at the nodal points were retained, those corresponding to zero mass were not controlled except for the interface degrees of freedom. This is identical to static condensation. The restoring forces corresponding to DOF 1, 2 and 3 were measured by load transducers in the actuators and the rigid link. However, since the diagonal brace in the second story was considered as part of the analytical substructure, its axial force had to be computed and added to the force measured at DOF 1 to obtain the total restoring force.

For this test series, the NS component of the 1940 El Centro ground motion record was used. The peak ground acceleration was scaled progressively from 0.04 g (minor level) to 0.30 g (severe level). Prior to these tests, pseudodynamic free-vibration tests were conducted to measure the inherent damping in the structure in a presumably linearly elastic range.

NUMERICAL SIMULATIONS AND CYCLIC TEST RESULTS

For conciseness, only the main results will be presented in the following. The hysteresis loops of story shear vs. story displacement obtained from the quasi-static cyclic test with 3x3 bracing connections and extended end-plate joints are shown in Fig. 8. The inelastic hysteretic behavior observed is mainly due to the yielding of the gusset plate-to-member fasteners, thus indicating that a substantial amount of energy can be dissipated in the bolted clip angles. The result of a numerical simulation using the phenomenological model for bracing connections is also plotted in Fig. 8. It can be observed that the hysteretic behavior of the test frame is well captured even though a simple bilinear law and kinematic hardening rule have been adopted for the truss members in the phenomenological model (Fig. 3). The analytical model overestimates the energy dissipation by about 11.3%.

To assess the influence of beam-to-column connections on the response of the braced frame, the two limiting cases of pinned and rigid (Fig. 9) joints are considered. The former represents the web-angle (WA) connection (Fig. 6a), whereas the latter approximates the extended end-plate (EP) connection (Fig. 6b). Analyses with 3x3 bracing connections have been carried out for both conditions, and the load-displacement envelopes obtained at the maximum displacement of each cycle in the pull and push directions are shown in Fig. 9. The brace is subjected to tension in the push direction and compression in the pull direction. As expected, the rigid-jointed frame has a stiffer response and higher resistance than the pin-jointed frame. It must be noted that the equivalent truss elements are specified to be stronger in compression than in tension to account for the yielding of the bolted fasteners. Hence, the frame appears to be stronger in the pull direction than in the push direction. Finally, the bottom story displacement of the test frame is computed in a conventional way, i.e., the bracing connection flexibilities are neglected and the brace is assumed pin-ended, while the beams are assumed to be rigidly connected to the columns. The comparison of the conventional frame analysis and the analyses with flexible bracing connections shows that the conventional analysis tends to overestimate the stiffness and the strength of the test frame. However, the diagonal brace buckles at an earlier stage in the conventional analysis because of using a longer brace.

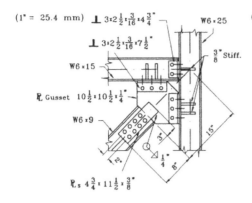

Figure 6a. Details of bolted bracing and web angle connections.

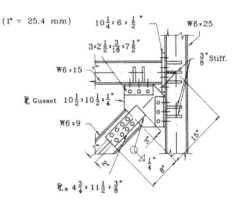

Figure 6b. Details of bolted bracing and extended end-plate connections.

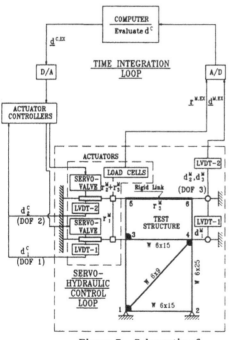

Figure 7. Schematic of pseudodynamic test setup [6].

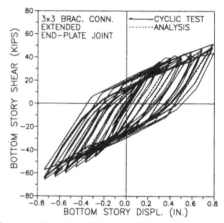

Figure 8. Hysteresis loops of story shear vs. story displ. of a single-story braced frame.

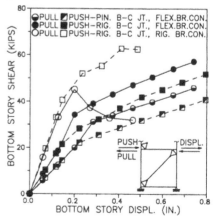

Figure 9. Load-displ. envelopes from analyses with rigid and flexible bracing connections.

NUMERICAL SIMULATIONS AND PSEUDODYNAMIC TEST RESULTS

Global Frame Response

The displacement time history plotted in the compressed time frame and the bottom story shear vs. the bottom story displacement obtained from the pseudodynamic test conducted with a 0.15-g peak ground acceleration and with 2x2 bracing connections are shown in Figures 10 and 11, respectively. In the same figures, the analysis results obtained with ANSR-III are shown as well. An equivalent viscous damping of 2.3% measured from a prior pseudodynamic free-vibration test is used in the numerical simulation. As indicated in the figures, the experimental results correlate well with the numerical simulation. The small nonlinearity is due to the yielding of the gusset plate-to-member fasteners.

Figure 12 shows the comparison of the story shear-vs.-story displacement response of a braced frame with 3x3 bracing connections and the response with 2x2 connections. It can be seen that the maximum base shear developed in the structure with 2x2 connections is 43.9% less than that of the frame with 3x3 connections. Furthermore, the 2x2 bracing connection was able to dissipate a substantial amount of energy through the slippage and yielding of the bolted fasteners. This is shown in Fig. 13, where the deformation of a bracing connection along the bracing direction is plotted against the brace axial force.

The pseudodynamic test results with 0.2-g and 0.3-g peak ground acceleration and 3x3 bracing connections are shown in Figures 14 to 17. In both tests, there were severe yielding of the gusset plate-to-member fasteners and out-of-plane buckling of the diagonal brace. Therefore, the displacement time histories show a significant period elongation when compared to the test result shown in Fig. 10. Nevertheless, the frame exhibited a relatively ductile behavior and both the bracing connections and the diagonal brace were able to dissipate a substantial amount of energy, although the buckling of the diagonal brace led to a substantial drop of lateral resistance.

Response of Diagonal Brace

To examine the accuracy of the frame model, the hysteretic behavior of the diagonal braces which includes the axial deformation of the bracing connections, obtained from the pseudodynamic tests and numerical simulations with 0.2-g and 0.3-g peak ground accelerations are shown in Figures 18 and 19, respectively. The numerical results show a good agreement with the experimental results. The observed differences are largely due to the fact that the analytical brace model cannot capture the stiffness degradation caused by the local buckling of the brace. In fact, the brace sections exhibited severe local flange buckling. The effective length factor K was computed for the diagonal brace, based on the brace buckling strength of 79.6 kips (354 kN) and 79.9 kips (355.4 kN), which were measured in the above two tests, and the average yield stress of 47.7 ksi (329 MPa). Using the clear span length of the brace, i.e. the clear distance between the ends of the bracing connections (70.2 inches (1783 mm)), and the AISC-LRFD [9] buckling strength formula, a K factor of about 1 is obtained, which corresponds to a pinned-pinned end condition. This indicates that the gusset plate connections provided little restraint to the out-of-plane bending.

EFFECT OF BRACING CONNECTION ON SEISMIC RESPONSE

To explore the effect of the bracing connection on the seismic response of a braced frame, the test frame is modeled with and without the bracing connection flexibility taken into consideration. The beam-to-column joints are assumed to be rigid. The envelopes of the maximum story displacement and story shear force are compared in Figures 20 and 21, respectively. It can be seen that for the worst case, the displacement response of the analysis without bracing connections modeled appears to be 28.3% smaller than the response with flexible bracing

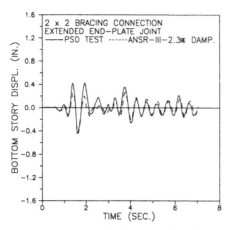

Figure 10. Story displacement histories for El Centro with 0.15-g peak acceleration.

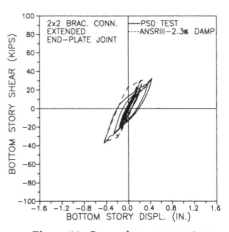

Figure 11. Story shear versus story displacement for El Centro with 0.15-g.

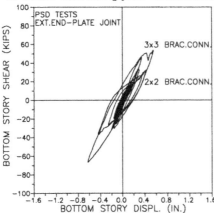

Figure 12. Story shear versus story displacement for El Centro with 0.15-g.

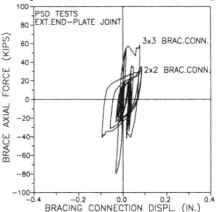

Figure 13. Brace axial force vs. bracing connection deformation for El Centro with 0.15-g.

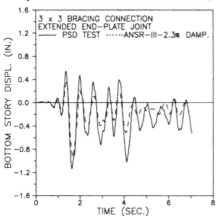

Figure 14. Story displacement histories for El Centro with 0.20-g peak acceleration.

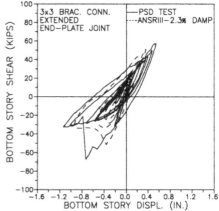

Figure 15. Story shear versus story displacement for El Centro with 0.20-g.

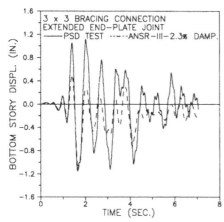

Figure 16. Story displacement histories for El Centro with 0.30-g peak acceleration.

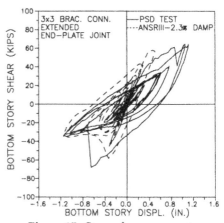

Figure 17. Story shear versus story displacement for El Centro with 0.30-g.

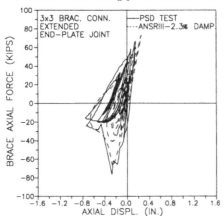

Figure 18. Brace axial force vs. axial deformation for El Centro with 0.20-g.

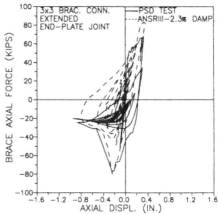

Figure 19. Brace axial force vs. axial deformation for El Centro with 0.30-g.

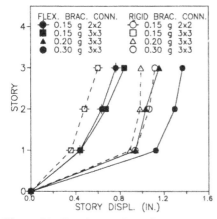

Figure 20. Envelopes of maximum story displacements from analyses with rigid and flexible bracing connections.

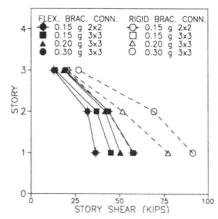

Figure 21. Envelopes of maximum story shears from analyses with rigid and flexible bracing connections.

connections (Fig. 20), while the base shear is 57.8% larger. These results show that for moderately bolted bracing connections, the appropriate modeling of connection flexibility is important.

CONCLUSIONS

In this paper, the seismic response of a flexibly-connected steel braced frame is studied by means of pseudodynamic testing and numerical simulations. The experimental study was carried out with a newly developed substructuring approach and the analyses have been conducted with a phenomenological model that is capable of capturing the linear and nonlinear response of bracing connections. The conclusions reached are as follows:

1. The frame designed in accordance with the proposed AISC Seismic Provisions [1] exhibited a relatively ductile behavior. The moderately bolted (2x2) bracing connections were capable to dissipate a substantial amount of energy by the slippage and yielding of the bolted clip angles.
2. The K factor for the out-of-plane buckling of the diagonal brace connected with bolted gusset plates has been found to be close to unity.
3. Satisfactory correlations have been obtained between the test results and numerical simulations using the bracing connection model.
4. The conventional frame analysis without considering the flexibility of bracing connections could underestimate the load capacity of a frame by assuming a longer effective length for the diagonal braces. In the case of seismic response, the conventional analysis could overestimate the story shear but underestimate the lateral displacement. As a result, for sway-critical frames, it is important to consider the flexibility of bracing connections.

ACKNOWLEDGMENTS

The first author would like to acknowledge the support of C.N.R. while visiting at the University of Colorado. The test specimen was contributed by several Colorado steel fabricators. The development of the pseudodynamic test technique and the experimental study are financially supported by the National Science Foundation under Grant No. BCS-8658100 and MTS Systems Corporation. The writers are also grateful to Dr. M.T. Vannan for his assistance in the experimental work. However, opinions expressed in this paper are those of the writers, and do not necessarily reflect those of the sponsors.

REFERENCES

1. AISC, Seismic Provisions for structural steel buildings - LRFD, AISC, Chicago, 1990.
2. Commission of the European Communities, Eurocode 8, Luxembourg, 1988.
3. Bursi, O.S., Elastic-plastic modeling of structural fasteners for steel bracing connections. J. Construct. Steel Research, 1992, (submitted).
4. Bursi, O.S., Gerstle, K.G., Sigfusdottir, A. and Zitur, J., Behavior and analysis of bracing connections for steel frames. J. Construct. Steel Research, 1992, (submitted).
5. Bursi, O.S. and Gerstle, K.H., Analysis of flexibly-connected braced steel frames. J. Construct. Steel Research, 1992, (submitted).
6. Vannan, M.T., The pseudodynamic test method with substructuring applications. Ph. D. Thesis, CEAE Department, University of Colorado at Boulder, CO., 1991.
7. Oughourlian, C.V. and Powell, G.H., ANSR-III - General purpose computer program for nonlinear structural analysis. Report No. UCB/EERC-82/21, Berkeley, CA., 1982.
8. UBC, Uniform building code, International Conference of Building Officials, CA. 1988.
9. AISC, Manual of Steel Construction - LRFD, AISC, Chicago, 1986.
10. Gross, J.L., Experimental Study of Gussetted Connections. In Proc. National Steel Construction Conference, ed. AISC, 1989, pp. 11.1-22.
11. Gerstle, K.H., Effect of connections on frames. J. Construct. Steel Research, 1988, **10**.
12. Chen, P.F. and Powell, G.H., Generalized plastic hinge concepts for 3D beam-column elements. Report No. UCB/EERC-82/20, Berkeley, CA., 1982.

SEISMIC BEHAVIOUR OF LOW-RISE STEEL BUILDINGS

LUIS CALADO[*] and ANTÓNIO LAMAS[**]
Instituto Superior Técnico
Dep. of Civil Engineering
Av. Rovisco Pais, 1096 Lisboa Codex, Portugal

ABSTRACT

The main goal of the research presented in this paper was to obtain numerical data for the evaluation of the behaviour factor (q) of low-rise steel buildings. Two structural systems have been analyzed: simple frames and diagonally braced frames. These systems were studied for different natural frequency, slenderness and axial load level. Regarding the seismic action real and artificial accelerograms with two levels of duration, 15 and 25 seconds, were employed. To characterize failure a numerical index based on damage accumulation was developed. It can be obtained from static tests and used to assess failure in dynamic studies. A parametric analysis on models of one-floor buildings was then conducted and results indicating important influences on the q factors are reported.

INTRODUCTION

Low-rise steel buildings with various structural systems, in particular portal frames, represent a substantial part of the investments in steel structures, especially in Portugal where they are used for such purposes as light industries, shops, supermarkets and entertainment buildings.

Being low-rise, recommendations to secure seismic resistance are often inadequately considered in design, with the consequence that during recent earthquakes [1] some of the most affected buildings are of this type. On the other hand, earthquake regulations, like the Eurocode 8 under preparation for structures in seismic regions [2], do not give explicit attention to low-rise steel buildings. In fact, the current draft of this code proposes the same behaviour factor (q) for all the frame steel structures with or without cross-bracings. As it will be shown, this can be conservative for low-rise frames and may contribute to maintain the insufficient attention given to their design against earthquake action.

---

[*] Assistant Professor and [**] Professor of Steel Structures

The aim of the research developed at the Department of Civil Engineering of the Technical University of Lisbon was the better understanding of the seismic behaviour of low-rise buildings and the simultaneous identification of the main parameters influencing the values of the behaviour factor.

THE BEHAVIOUR FACTOR

The method of the behaviour factor enables the use of simple elastic analysis to estimate the nonlinear response of a structure under earthquake loading. It considers the capacity of the structure to dissipate energy through ductile behaviour and thus acts as a "corrective" factor.

The draft of Eurocode 8 states that " the behaviour factor can be obtained as the average ratio between the seismic intensity inducing an ultimate limit state in the structure, taking into account its nonlinear behaviour, and the design seismic intensity used with a conventional linear model ".

From a practical point of view, the behaviour factor corresponds to the ratio between the seismic intensity a_u (in the sense of the peak value of the accelerations) which causes an ultimate limit state (failure) and the seismic intensity a_y associated to the elastic limit state of the structure:

$$q = a_u / a_y \qquad (1)$$

FAILURE

The assessment of failure in a dynamic loading process is difficult, requiring a definition, in particular for any numerical simulation of the structure under an earthquake or an equivalent static cyclic loading involving a step-by-step analysis on time.

In the context of seismic behaviour failure can be associated to a significant reduction of the global capacity of the structure to dissipate the energy transmitted by the earthquake, and this occurs in the ductile elements through hysteretic dissipation of energy. Most of this energy is dissipated by alternate plasticity but buckling, local buckling, fracture and low cyclic fatigue, which are phenomena causing deterioration of resistance, must be considered in the evaluation of the dissipated energy.

If an elastic-perfectly plastic behaviour of the ductile element is considered as a basis, the reduction of its capacity to hystereticaly dissipate energy may be expressed by:

$$\eta_i = A_i / A_{yi} \leq \gamma \qquad (2)$$

where η_i is the *normalized hysteretic energy* in the ith cycle, γ its value at failure, A_i is the hysteretic energy dissipated and A_{yi} is the energy dissipated if the element had an elastic-perfectly plastic behaviour. A_i and A_{yi} may be obtained from force-displacement diagrams as shown in Figure 1.

In the present study only beam-columns and diagonal braces were considered as dissipative structural elements. The hysteretic behaviour of these elements can be numerically simulated down to very low values of the *normalized hysteretic energy*. However, for numerical economy and structural consistency it is not advisable to assume that failure occurs for a very low η_i (Figure 2).

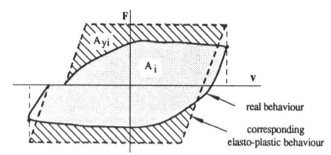

Figure 1. Typical force-displacement cycle for a beam-column

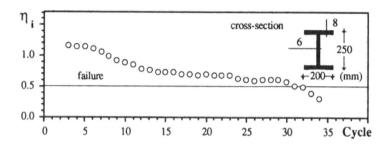

Figure 2. Variation of the *normalized hysteretic energy* and failure criterion for a beam-column under the loading history proposed by the testing recommendations of the ECCS [3]

For a *normalized hysteretic energy* of 50% the structural elements show clearly large plastic deformations due to buckling, local buckling and a large propagation of cracking over all the cross-section [3].

A MODEL FOR DAMAGE ACCUMULATION

Under the variable nature of the seismic loading the described assessment of failure is difficult to apply and led to the identification of the *accumulated damage (D)* as a better estimation of the structural deterioration. It is based on the classical low-cycle fatigue model for steel and uses the assumption of a linear damage accumulation [4-5]. After n cycles of different amplitudes it is expressed by:

$$D = a \sum_{i=1}^{n} (\Delta\zeta_{pi})^{c} \le \partial \qquad (3)$$

where, in a generic cycle i, $\Delta\zeta_{pi}$ is the plastic part of a "normalized" deformation quantity such as strain, rotation or deflection. The value of D at failure is represented by ∂ and defines the *damage index*.

The *damage index* ∂ is then a measure of the ductile hysteretic capacity of the structural element. It may be used to indicate failure under cyclic loading: a limit to the recorded *accumulated damage*.

NUMERICAL MODELS

The behaviour of beam-column associations and diagonal bracings was numerically modelled as described in earlier works [3,6,7].

The model of the beam-column consists of a rigid bar with an elasto-plastic "cell" at the lower end, where all the geometrical and mechanical properties are "concentrated", subjected to vertical load and horizontal displacements at the top (Figure 3). The bar is assumed to be initially straight with the applied horizontal displacement causing bending about a principal axis of the cross-section. Shear deformation is disregarded. The equivalence between the model and the beam column is imposed in the elastic range by equating the moment of the horizontal force associated to the displacement of the upper end to the maximum elastic bending moment at the lower "cell".

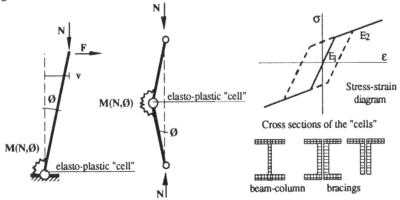

Figure 3. Numerical model of a beam-column and a bracing element

The model of the diagonal bracing has two elements connected by an elasto-plastic "cell" (Figure 3). The equivalence between the model and the diagonal bracing is similarly imposed in the elastic range by equating the critical load to the bending moment in the "cell" at the elastic limit. At the end connections of the diagonal bracing the effects due to joint slippage and ovalization of holes can also be considered.

The constitutive stress-strain relation of the material is assumed to be elasto-plastic with kinematic strain-hardening. When compared with other more complex stress-strain relations this was shown by Castiglioni [8], for the cyclic behaviour of steel bent sections, to give good results without excessive computing time.

The models were adapted to reproduce at the elasto-plastic "cells" the effects of local buckling, fracture and low-cycle fatigue. Local buckling was modelled using the concept of variable effective widths from plate theory and the strain energy density criterion proposed by Sih and Madenci [9] was used to simulate fracture. The low-cycle fatigue was modelled using the Miner's rule [4].

The comparison between numerical results and experimental tests indicates a good agreement over most of the loading ranges (errors less than 10%) either in terms of the force-displacement diagram or in terms of resistance, ductility and energy dissipation (Figure 4).

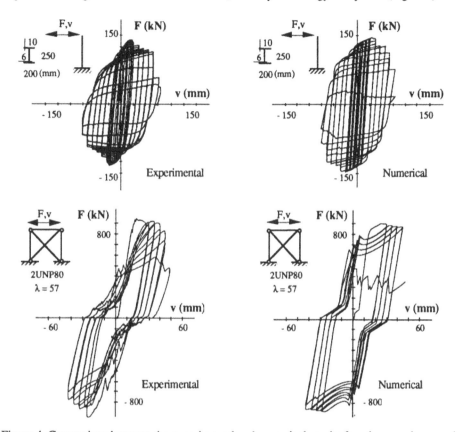

Figure 4. Comparison between the experimental and numerical results for a beam-column and a bracing system

The deformation quantity considered for the assessment of the *accumulated damage* ($\Delta \zeta_{pi}$ in expression 3) was the average plastic longitudinal strain calculated over the critical cross-sections or "plastic cells", normalized in respect to the elastic limit value.

PARAMETERS INFLUENCING THE ACCUMULATION OF DAMAGE

After several experimental tests and numerical simulations [3,5], the parameters a and c (expression 3) were set equal to unity for beam-columns and diagonal bracings.

Different cyclic loading histories were stochasticaly generated and applied statically to dissipative elements of both types. The numerical simulations were carried out until the normalized hysteretic energy decreased to 20%, and the values of ∂ were calculated from the corresponding diagram of the *accumulated damage* for $\gamma \approx 50\%$. Table 1 summarizes the values thus obtained. The standard deviation coefficient achieved in all cases is between 5 and 9%, justifying the assumption [5] that the *damage index* is almost independent from the loading history (Figure 5). In these figures the superposition of the values of $D(\eta_i)$ for different loading histories is presented, each point corresponding to the *accumulated damage* up to a given cycle in a loading history. Therefore, just a few (or even just one) static numerical simulations, or experimental tests, are sufficient to obtain an adequate average value for ∂.

TABLE 1. Average damage indexes ∂ for beam-columns and bracing elements (Fe360)

		N/Ncr=0.00	N/Ncr=0.10	N/Ncr=0.20	N/Ncr=0.40
	$\lambda = 50$	91	40	32	21
	$\lambda = 75$	92	36	28	20
IPE300	$\lambda = 100$	94	31	20	18
		$\lambda = 75$	$\lambda = 100$	$\lambda = 125$	$\lambda = 150$
	2 UNP 80	54	51	48	46
	2L100x65x7	46	43	42	42

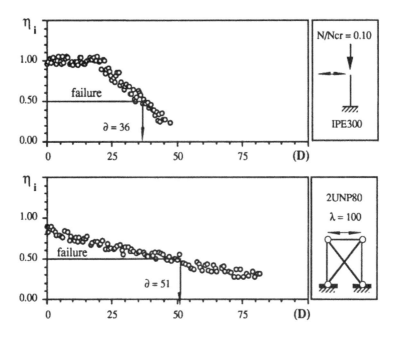

Figure 5. Superposition of values of $D(\eta_i)$ for different loading histories applied to a beam-column and a bracing element without dissipation of energy at the ends (Fe360)

For beam-columns the most important parameter influencing ∂ is the level of axial load since it reduces the bending strength of the element.

For bracing elements the *damage index* decreases when the most important parameter, the slenderness, increases. However, this effect is not very pronounced in cross-bracings since their cyclic behaviour is mostly governed by the element in tension.

q FACTORS FOR ONE-FLOOR BUILDINGS

One-floor buildings were modelled as cantilever structures assembled from beam-columns, with or without symmetric cross bracings.

The dissipative structural elements which were considered are characterized in Table 1. In the parametric study (Table 2), structures with different natural frequencies (f), assumed independent from the axial loading on the columns, were studied.

TABLE 2. Variation of parameters for the study of one-floor buildings

A		$\lambda = 75$	N/Ncr=0.00	N/Ncr=0.10	N/Ncr=0.20	N/Ncr=0.40
			f=0.50 Hz	f=0.50 Hz	f=0.50 Hz	f=0.50 Hz
			f=0.67 Hz	f=0.67 Hz	f=0.67 Hz	f=0.67 Hz
			f=1.00 Hz	f=1.00 Hz	f=1.00 Hz	f=1.00 Hz
			f=1.50 Hz	f=1.50 Hz	f=1.50 Hz	f=1.50 Hz
B			column $\lambda = 75$	brace $\lambda = 100$	column $\lambda = 100$	brace $\lambda = 150$
		N/Ncr=0.10	f=0.67 Hz	f=0.67 Hz	f=0.67 Hz	f=0.67 Hz
		N/Ncr=0.40	f=1.50 Hz	f=1.50 Hz	f=1.50 Hz	f=1.50 Hz

Real accelerograms (from the Array SMART[*]) and artificial ones (generated from the power spectrum proposed in the Portuguese Code for Actions - RSA[**]) were considered for dynamic loading in a step-by-step integration in time (0.10 sec) of the equation of motion using the Wilson - θ method. The accelerograms were normalized with respect to the gravity acceleration (g) to obtain a unit value for the peak acceleration (Figure 6). Earthquakes of different intensities were "produced" amplifying these "normalized" accelerograms.

---

[*] Earthquakes recorded between 1980 and 1985 in special seismographic stations located Northeast of Taiwan, near the city of Lotung, and commonly referred as Array SMART (Strong Motion Array of Taiwan) [10]. They are different in duration (15 and 25 sec), sequence of peaks and with respect to epicentral distance, hypocentral depth, magnitude and hypocentral azimuth.

[**] The RSA [11] proposes spectra with two different durations (15 and 25 sec) for three types of site conditions: hard rock or soft rocky formations; coarse-grained granular materials of medium relative density; and loose coarse-grained granular materials with low relative density values.

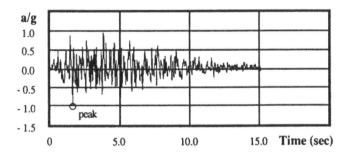

Figure 6. Example of a normalized accelerogram generated from a 15 sec spectrum
proposed for a coarse-grained foundation in the RSA [11]

As referred to previously, the behaviour factor q may be obtained from expression 1. For an accelerogram with a peak value a_y the *accumulated damage* in the dissipative structural elements is null because no plastic deformations occur. For consecutive amplifications of the accelerogram, the *accumulated damage D* reaches ∂ (obtained from previous static analysis) at the end of the seismic signal. The corresponding peak value is a_u. For the studied structures failure occurs when both dissipative elements fail.

Influence of the seismic action and of the natural frequency of the structure
The calculated behaviour factors were different for each seismic action considered (Figure7), even for those corresponding to the same intensity, magnitude and epicentral distance. At the same time, for each earthquake q is smaller for structures with higher frequency as predicted in [2]. An attempt to interpret the causes of these differences was not conclusive and further research is required.

Influence of the axial load level on the columns
This parameter has the same influence on the values of q as it has on the values of the *damage index* (Figure 7). Although predictable, this is a significant conclusion for code recommendations which should, therefore place a limit on the axial load of cantilever columns to enable their consideration as dissipative elements.

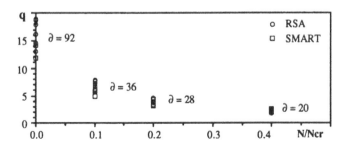

Figure 7. Influence of the axial load level
(Structure of Type A: IPE300, f=1.50 Hz and λ=75)

Influence of the cross bracings

The association of cross bracings to cantilever columns leads to structures with a better seismic behaviour and, consequently, with a higher behaviour factor (Table 3). The relative influence of the bracings is proportional to the *damage indexes* of both dissipative elements. For all the cases analyzed, failure in beam-columns and bracing elements occurred approximately at the same time.

TABLE 3. Behaviour factors for one-floor buildings (reliability of 90%)

structure type **A** Table 2	$\lambda = 75$	N/Ncr=0.00	N/Ncr=0.10	N/Ncr=0.20	N/Ncr=0.40
	f=0.50 Hz	20.1	6.2	3.7	1.9
	f=0.67 Hz	17.1	5.2	3.9	1.7
	f=1.00 Hz	15.6	5.7	3.5	1.7
	f=1.50 Hz	11.9	4.9	2.8	1.6
structure type **B** Table 2		column $\lambda = 75$ brace $\lambda = 100$		column $\lambda = 100$ brace $\lambda = 150$	
		N/Ncr=0.10	N/Ncr=0.40	N/Ncr=0.10	N/Ncr=0.40
	f=0.67Hz	8.0	8.3	8.0	6.0
	f=1.50Hz	8.3	7.0	7.5	5.3

It is interesting to compare the q values summarized in Table 3 with those proposed in the draft of Eurocode 8 [2] for simple cantilever structures, which are between 1.5 and 2.0. If these values assume safety factors of the order of 2, they are only of the same magnitude as those of Table 3 for medium levels of the axial load on the columns.

CONCLUSIONS

The defined *damage index* ∂ is a good measure of the ductile hysteretic capacity of the structural elements. Since it is "independent" of the loading history it is a promising indicator of failure, which can be easily calculated from statical numerical or experimental tests, and is thus suitable to structural classification under the application of standard cyclic tests.

The knowledge of the *damage index* of each structural element seems the correct procedure to predict the global structural failure. Further research to substantiate this statement is being conducted.

Although heavy in computer time, the assessment of the behaviour factor (q) used in the reported study seems the most appropriate one. The influence of the seismic action and the natural frequency on q is an open field for research.

The influence of the axial load on the columns was shown to be important and justifying consideration in the codes, but it diminishes when they are associated to cross-bracings. Further research on the relative importance of the two structural schemes on the values of q is also being studied.

The simulated type of low-rise buildings may have very large behaviour factors, in particular when cantilever columns are associated to bracings. The values proposed in the codes may be unsafe for simple cantilevers very highly loaded and too conservative in some other practical cases.

REFERENCES

1. Bockemohle, L. W., Earthquake behaviour of commercial-industrial buildings in the San Fernando valley, Proc. 5th World Conference on Earthquake Engineering, Rome, 1, pp. 76-81, 1973.

2. Eurocode 8 - Common Unified Rules for Structures in Seismic Regions, Commission of the European Communities, May, 1989.

3. Calado, L., Caracterização do comportamento de estruturas metálicas sob acções sísmicas. Phd thesis, Technical University of Lisbon, Portugal, 1989 (in Portuguese)

4. Miner, M., Cumulative damage in fatigue, Journal of Applied Mechanics, ASCE, vol. 67, September, A159-A164, 1945.

5. Calado, L. and Azevedo, J., A model for predicting the failure of structural steel elements. Journal of Constructional Steel Research, vol.14, Nº1, pp. 41 - 64, 1989.

6. Ballio, G. and Calado, L., Steel bent sections under cyclic loads - Experimental and numerical approaches, Costruzioni Metalliche, Nº 1, pp, 1 - 23, 1986.

7. Ballio, G., Calado, L., Leoni, F. and Perotti, F., Numerical simulation of the cyclic behaviour of steel subassemblages, Costruzioni Metalliche, Nº 5, pp. 269 - 294, 1986.

8. Castiglioni, C., Modellazione numerica di sezioni inflesse in acciaio soggette a carichi ciclici: influenza del legame costitutivo. Costruzioni Metalliche, Nº 3, 1987.

9. Sih, G. and Madenci, E., Crack growth resistance characterized by the strain energy density function. Engineering Fracture Mechanics, vol.18, pp. 1159-1171, 1983.

10. Abrahamson, N., Bolt, B., Darragh, R., and Penzien, J., The SMART 1 Accelerograph Array (1980-1987): A review. Earthquake Spectra, Vol. 3, Nº 2, 1987.

11. Regulamento de Segurança e Acções para Estruturas de Edifícios e Pontes. Imprensa Nacional - Casa da Moeda, Lisboa, 1986 (in Portuguese).

Steel Triangular Plate Energy Absorber for Earthquake-Resistant Buildings

Keh-Chyuan Tsai and Ching-Ping Hong
Department of Civil Engineering
National Taiwan University
Taipei, Taiwan, Republic of China

ABSTRACT

Under cyclically applied load, the mechanical properties and the ductility capacity of thirteen steel triangular plate energy absorbing device are studied. Special welding details are developed in order to minimize the welding distortions of the triangular steel plate during the fabrication of the device. The degree of the effectiveness of the proposed energy absorbing devices for buildings in severe seismic environment is investigated using psuedodynamic testing procedures for a two story steel frame. The test results showed that the proposed device is a promising alternative for buildings to resist earthquake forces. The paper concludes that the seismic responses of the frame equipped with the proposed energy absorbing device can be adequately predicted using conventional frame response analysis procedures.

INTRODUCTION

In order to achieve economical earthquake resistant construction, a building must be able to absorb and dissipate large amounts of seismic energy. In recent years, a number of researchers have investigated techniques of increasing the building energy absorbing capacity through the use of a steel plate added damping and stiffness (ADAS) device [2,5,8]. These studies have confirmed that ADAS elements using X-plates are suitable for buildings in regions of high seismic risk. However, as the ADAS elements used in these studies were made of X-shaped steel plates bolted together through two ends of each plate, the stiffness of the device have been found very sensitive to the degree of the end restraints, i.e. the tightness of the bolt [8].

Since the bending curvature for a transverse load applied at the end of a triangular plate is uniform over the full height of the triangular plate, the plate can deform well into the inelastic range without large curvature concentration thereby dissipating energy efficiently. Recent research results have shown that casted steel triangular added damping and stiffness (TADAS) device possesses adequate stiffness and strength [6], however, the ductility and the energy dissipation capacities of the casted TADAS device may not be adequate to survive severe earthquakes. In this paper, results of a combined analytical and experimental research program [7] conducted recently at the National Taiwan University to study the behavior of the welded steel TADAS elements are discussed. The proposed device consists of several triangular plates welded in parallel to a common base plate. If the elastic stiffness, ultimate strength and ductility capacity of the device can be accurately predicted, economical proportions of beam, column and brace in the earthquake resistant structural system can be achieved. The objectives of this investigation are to assess the mechanical properties, ductility capacity as well as the degree of the effectiveness of the proposed TADAS devices for buildings in severe seismic environment.

BASIC CHARACTERISTICS OF THE TADAS DEVICE

As shown in Fig. 1, when a finite displacement of the top of the triangular plate with respect to the bottom of the plate is imposed, the bending curvature is uniformly distributed over the full height. Therefore, the yielding can occur simultaneously over the full height. Assuming the base of the triangular plate is fully restrained and considering the flexural deformation only, the theoretical elastic lateral stiffness of a steel TADAS element can be obtained:

$$K_d = \frac{N E B t^3}{6 h^3} \tag{1}$$

where E is the young's modulus, N is the number of the triangular plates, t is the plate thickness, B and h are the base width and the height of the triangular plate, respectively. The yield displacement, Δ_y, and the yield strength, P_y, of the element are:

$$\Delta_y = \frac{F_y h^2}{E t} \tag{2}$$

and

$$P_y = \frac{F_y N B t^2}{6 h} \tag{3}$$

respectively, where F_y is the tensile yield stress. If the yield displacement, Δ_y, is divided by the height of the triangular plate, h, the yield rotational angle, γ_y, is then:

$$\gamma_y = \frac{F_y h}{E t} \tag{4}$$

From Eqs. 1 to 4, it can be seen that the height to thick ratio, h/t, of the plate is an important parameter to the mechanical properties of TADAS device. The effects of the h/t ratio of the triangular plate on the stiffness, K_d, and yield displacement, Δ_y, are depicted in Fig. 2 and Fig. 3. It is clear that the stiffness of the device increases rapidly as the height of plate decreases or the thickness increases. Further more, it is illustrated in Figs. 2 and 3 that it is possible to chose a small h/t ratio for the plate to achieve a large device stiffness while keeping the same h/t ratio but changing

531

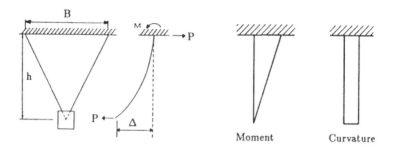

Fig.1 Basic behavior of triangular plate under load

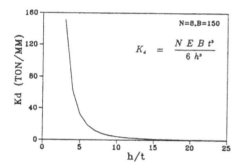

Fig.2 effect of plate heigh to thickness
ratio on TADAS stiffness

Fig.3 effect of plate heigh to thickness
ratio on TADAS yield displacement

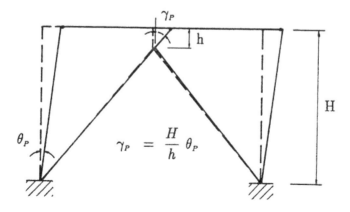

Fig.4 Energy dissipation mechanism for the TADAS frame

the height of the triangular plate to let the device yield at specific displacement. When properly combined with the braces in a frame, the TADAS element can be proportioned to provide not only additional frame lateral stiffness but also hysteretic damping when specific frame deformation is reached [7]. Nevertheless, it is important to estimate the plastic rotational demand imposed on the TADAS device. This can be conveniently accomplished through the use of energy dissipation mechanism constructed by assuming rigid plastic behavior of the members. Mechanism for a TADAS frame is shown in Fig. 4. From the geometry of the mechanism, the plastic rotational demand of the TADAS device can be determined as follows:

$$\gamma_p = \frac{H}{h}\theta_p \tag{5}$$

where H is the story height and θ_p is the plastic frame drift. From Eq. 5, it is obvious that the TADAS rotational demand grows as plate height decreases. When a frame drift of 0.02 radian is reached under a sever earthquake, a TADAS rotational demand of about 0.20 radian is likely to occur for a typical H/h ratio of 10.

CYCLIC TESTS OF STEEL TADAS DEVICE

Based on these mechanical characteristics, a total of thirteen TADAS specimens, about two for each device shown in Table 1, were fabricated for this experimental study. Fig. 5 shows the experimental set up where the end of each triangular plate is pin connected to a horizontal channel with slotted holes. The channel is connected to a servo-controlled hydraulic actuator. As shown in Fig. 6, Specimen 0B2 for the pilot test is made from eight triangular plates connected to a 35 mm thick base plate using full penetration welds. Under the cyclically applied load as shown in Fig. 6, Specimen 0B2 exhibited stable hysteretic behavior with no signs of stiffness or strength degradation for a rotation amplitudes of up to 0.16 radian before fracture of plate occurred.

Although this test, the pilot test in this experimental program, have shown that the proposed TADAS device using the aforementioned welding details can sustain a large number of yielding reversals and dissipate a significant amount of energy, the plate were found severely distorted during the fabrication due to uneven heat distribution from the welding of the triangular plate. Subsequent heat treatments of the specimen had to be followed before the testing. Moreover, after examining the cause of the fracture of the plate, the welding details as shown in Fig. 6 were proved to be impractical to assure the quality of the full penetration welds as the spacing between the plates is too small. As a result, an improved welding detail as shown in Fig. 7 is developed. In this detail, each triangular plate is inserted into the slotted base plate before the fillet welding and the plug welding are applied to connect the triangular plate to the front and the back, respectively, of the base plate. As shown in Fig. 7, under the cyclically increasing load, Specimen 2B2 exhibited extremely stable hysteretic behavior with no signs of stiffness or strength degradation for a rotation amplitudes of more than 0.29 radian.

From Fig. 6 and Fig. 7, it is evident that the theoretical elastic stiffness of the proposed TADAS device agrees very well with the experimental response. The effect of the strain hardening is pronounced as shown in Fig. 7. A resistance of 1.5 times the plastic

ADAS Element	N	B (mm)	t (mm)	h (mm)	Kd (t/mm)	Δ_y (mm)	Py (Ton)
A1	8	150	20	305	1.2	5.48	6.6
A2	8	150	20	220	3.1	2.88	9.1
A3	8	150	20	135	14.0	1.06	14.9
B1	8	150	35	415	2.6	5.80	14.8
B2	8	150	35	305	6.4	3.11	20.2
B3	8	150	35	190	26.7	1.20	32.4

All material is of ASTM A36

Table 1 Schedule of triangular plate energy absorbing devices

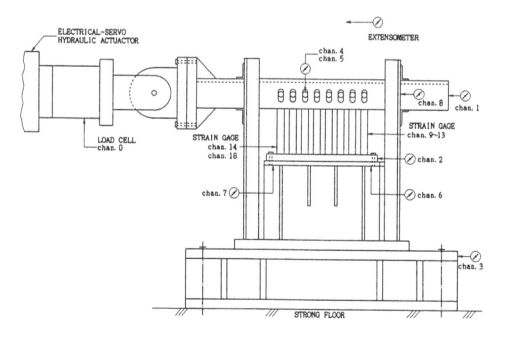

Fig.5 Experimental set-up for cyclic test of TADAS devices

534

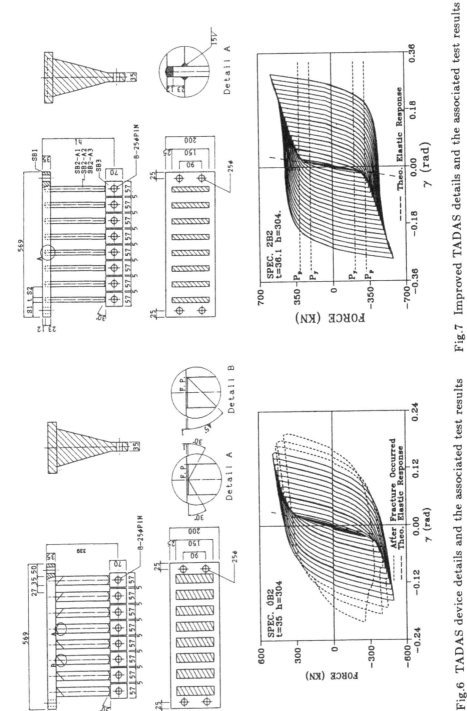

Fig.7 Improved TADAS details and the associated test results

Fig.6 TADAS device details and the associated test results

capacity of Specimen 2B2 had been developed when a rotation angle of 0.20 radian was reached. The improved welding details have been found very successful for the remaining specimens. With proper welding procedure, there was no significant welding distortion occurred during the fabrication of the specimens when the improved details were used. The tests have shown that the proposed TADAS device using the improved welding details can sustain an extremely large number of yielding reversals and dissipate a great amount of energy. In general, the ultimate cyclic rotational capacities attained from the tests of the remaining specimens are larger than 0.25 radian.

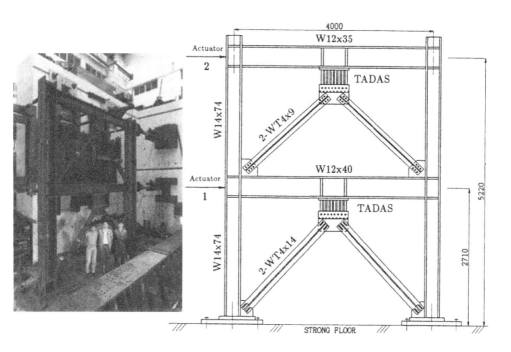

Fig.8 Experimental set-up for pseudodynamic testing of a 2-story TADAS frame

PSEUDODYNAMIC TESTS OF A 2-STORY TADAS FRAME

The effectiveness of the TADAS devices for buildings in high seismic risk were further investigated using pseudodynamic testing procedure for a two-story steel frame. Fig. 8 shows the member sizes of the test frame. All steel is of A36 material. The TADAS device used in the second floor consists of five triangular plates of 325 mm high, 177 mm wide and 35 mm thick each, while eight such plates are used in the TADAS device in the first floor. The 50 mm thick base plate under each column was bolted to a 90 mm thick plate, through which tied down was used to anchor the test frame to the strong floor. A mass of 0.019 $ton/sec^2/mm$ was assumed for each floor. The central-difference integration scheme was implemented in this two degree-of-freedom pseudodynamic test program [4,7]. A series of free vibration test was conducted first to compute the vibration periods and the friction force within the system. During the

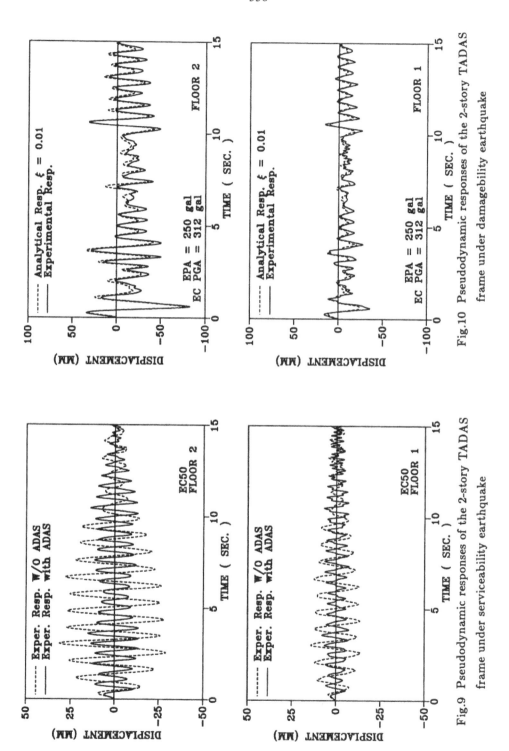

Fig.10 Pseudodynamic responses of the 2-story TADAS frame under damagebility earthquake

Fig.9 Pseudodynamic responses of the 2-story TADAS frame under serviceability earthquake

course of the study, the effects of the TADAS devices on the seismic responses of the frame without the device were also investigated by removing the connecting pins between the device and the braces. The first and the second mode period of the frame without the TADAS devices activated are 0.881 and 0.237 second, respectively. The first and the second mode period of the frame with the TADAS devices are 0.573 and 0.201 second, respectively. The coulomb damping effects were found essentially constant over a wide range of displacement amplitude [7]. The coulomb damping effects in the system were subsequently compensated by introducing two constant forces in the direction opposite the velocity vector similar to what have been done in other single degree of freedom pseudodynamic testings [1]. A viscous damping of 0.01 of the critical damping was assumed for all pseudodynamic tests of the frame.

In order to study the elastic and inelastic responses of the TADAS frame, several ground acceleration records with different intensities were used in this study. Here, only some important observations from the tests of the TADAS frame under the 1940 El Centro Earthquakes (EC), scaled to a PAG of 50 gal and an EPA of 250 gal, are presented. As shown in Fig. 9, the frame responses were greatly reduced when the TADAS devices were activated under the El Centro Earthquake scaled to a intensity of PGA = 50 gal. As shown in Fig. 10 for the EPA = 250 gal event, the maximum roof drift response of the frame is 0.016 radian and the TADAS rotation demand is 0.14 radian, much less than the rotational capacity of a typical TADAS device tested previously.

In Fig. 10, the experimental results were also compared with the analytically predicted responses. Since the frame had gone through several tests where inelastic deformations and strain hardening of TADAS device had occurred, the analytical predicted responses do not exactly agree with the experimental results. However, considering the much better agreements in other earlier tests not discussed in this paper and that the TADAS device was modeled by an equivalent prismatic beam member using simple bilinear moment-rotation relationships [3], the overall agreements between these analytical and the experimental responses are entirely satisfactory.

CONCLUSIONS

Based on this series of test, following conclusions can be drawn:

- Distortion of the triangular plates in welded TADAS can be resolved by proper welding details.

- Elastic stiffness and ultimate strength of the proposed steel TADAS device can be accurately predicted.

- Deformation capacity of the proposed TADAS device is larger than ±0.25 radian.

- Seismic responses of building structure can be effectively controlled by using the proposed steel TADAS device.

- Seismic responses of building frames equipped with the proposed TADAS devices can be accurately predicted by traditional nonlinear frame response analysis computer program. The hysteretic behavior of the TADAS device can be adequately characterized by an equivalent prismatic beam member using the elastic strain-hardened bilinear moment-rotation relationships.

ACKNOWLEDGEMENTS

The authors gratefully acknowledge the supports of the Sinotech Foundation for Research and Development of Engineering Science and Technology and the National Science Council of Republic of China. The valuable suggestions provided by Prof. Yang-Feng Su of National Central University are very much appreciated.

REFERENCES

[1] Balendra, T, Lam, K.Y., Liaw, C.Y. and Lee, S.L., "Behavior of Eccentrically Braced Frame by Pseudodynamic Test", Journal of Structural Engineering, Vol. 113, No. 4, 1987.

[2] Hanson, R.D.,"Basic Concepts and Potential Applications of Supplemental Mechanical Damping for Improving Earthquake Resistance," Proceedings, ATC Seminar on Base Isolation and Passive Energy Dissipation, San Francisco, California, March 1986.

[3] Kannan, A.E. and Powell, G.H., "DRAIN-2D, A General Purpose Computer Program for Dynamic Analysis of Inelastic Plane Structures", Report No. UCB/EERC-73/22, Earthquake Engineering Research Center, University of California, Berkeley, 1973.

[4] Shing, P.B. and Mahin, "Pseudodynamic Test Method for Seismic Performance Evaluation: Theory and Implementation", UCB/EERC-84/01, Earthquake Engineering Research Center, University of California, Berkeley, 1984.

[5] Su, Y.F. and Hanson, R.D., "Seismic Response of Building Structures with Mechanical Damping Devices," Report UMCE 90-2, Department of Civil Engineering, University of Michigan. Ann Arbor, Michigan, 1990.

[6] Su, Y.F., "Aseismic Design of Building Structures with ADAS Devices", Report to the Sinotech Engineering Consultants, Inc. Su and Structural Engineers Corporation, October 1990.

[7] Tsai, K.C. and Hong, C.P., "Experimental Studies of Steel Triangular Plate Energy Absorbing Device for Seismic Resistant Structures", Center for Earthquake Engineering Research, National Taiwan University, August, 1992.

[8] Whittaker, A., Bertero, V.V. and Alonso, J., "Earthquake Simulator Testing of Steel Plate Added Damping and Stiffness Elements," Report No. 89/02, Earthquake Engineering Research Center, University of California, Berkeley, CA. 1989.

Buildings

ECONOMIC DESIGN AND THE
IMPORTANCE OF STANDARDISED CONNECTIONS

S.M.C. FEWSTER
Allott Bros & Leigh Limited
Ickles Forge
Fullerton Road
Rotherham S60 1DJ

E.V. GIRARDIER
Steel Construction Industry Federation
820 Birchwood Boulevard, Warrington, Cheshire WA3 7QZ

G.W. OWENS
Steel Construction Institute
Silwood Park, Ascot, Berkshire SL5 7QN

ABSTRACT

A simple analysis of the constructional steelwork industry is made, together with the relevance of the connections content. The range of variants to be taken into account when specifying and designing connections are identified. It is suggested how these can be rationalised in order to achieve higher productivity and how, for example, the number of different bolts utilised can be dramatically reduced. Finally the timing and the role of the design of the connection is discussed and a summary of good practice given.

INTRODUCTION

The value and costs of a typical steel frame can be approximated as shown in Table 1.

TABLE 1
The Industry makeup

Item	Value
Basic materials	40
Added Value	60
Total	100

The basic materials are, in the main, composed of the steel, bolts and welding electrodes. Added value is the work carried out by the steelwork contractors in the form of detailed engineering, fabrication and erection. Virtually all of this added value is taken up in the definition, organisation and execution of the joining and handling of the material, whether this be in the offices, fabrication shops, or on the construction site.

In simple and brief terms, the industry's activity i.e. its added value is spent on either joining material or handling it. The remaining 'office' activities e.g. management, planning, drafting, are a service to these two functions. An analysis of the industry's organisation and employees shows that, when taking into account both fabrication and erection, approximately 50% of the added value is spent on making connections with the other 50% spent on handling. The handling can be further analysed as shown in table 2.

TABLE 2
Typical Handling Analysis per Finished Component

Work Activity	Number of handling occasions
Rolling Mills to Store	1
Handle and sort	2
Store to Workshops	1
Progress in Workshops:	
Fittings	4*
Main Components	5*
Sort and to Construction Site	2
Off-load and Sort	2
Erect and Position	4*†
Total	21

† This includes an allowance for 'time spent on the hook' as well as for the number of times handled ie 5 minutes hook time = 1 handling, 15 minutes hook time = 3 handlings.

The items marked with * amount to two thirds of the total and are those which are directly affected by the type of connection, its standardisation and repetition or their lack. In addition it should never be overlooked that the direct time spent making connections is never 100% but is always diluted due to the handling function. In practice the workshop handling can be easily doubled or halved, dependent on the connections and their repetition. In other words at least half of the time and cost of the handling is connection dependent.

The value of a typical steel frame can thus be simply re-stated as given in Table 3.

TABLE 3
The Connections Content

Item	Value
Basic materials	40
Connections	30
Handling:	
Basic	10
Connections Dependent	20
Total	100

The conclusion is that **the connections account for a minimum of 50% of the total value** of constructional steelwork. As a general principle, this applies to virtually all structures, both buildings and bridges. In designing and specifying structures it is difficult to alter the weight of materials used by more than ± 10%. It is, however, very easy to revise the connection work content by amounts well in excess of ± 50% e.g. by the necessity to stiffen locally the 'inadequate' main section in the connection zone, or by the time spent searching for that 'special' bolt with the whole of the erection front waiting.

THE DIVERSITY OF CONNECTIONS

There is a wide diversity of connection types which can be used even in simple straightforward structures:

Double angle web cleats
Top and bottom cleats
End plates
Fin plates

All of these can be utilised in making simple connections. Full moment and semi-rigid connections increase the range.

In carrying out these connections there are at least available:

60 different angles, thicknesses and grades
40 different plate widths, thicknesses and grades

Furthermore, the range of bolts which can be used can very easily reach 100 in number arrived at by:

3 different common grades
3 different diameters
10 variations in length

these can be used in:

40 different configurations

The range of welds in type, preparation, procedure and consumable can be limitless but, even with a great deal of discretion and limited only to fillet welds, it is reasonable to state that there are 20 variations.

The result of all of this is that even the conservative and discriminating designer has a choice of 3,000 different variants, see Table 4. The indiscriminate designer could well extend this range to some 10,000 in number. Even more, these numbers can be readily extended by the designer should his selected main members prove to be inadequate to carry the forces in the connection zone, we then enter into a whole new arena called local stiffening.

TABLE 4
A Limited List of Connection Variants

Bolted Cleats		Welded Plates		
Connection Type	2		2	Connection Type
	x		x	
Different Angles	10		10	Different Plates
	x		x	
		Bolts	Weld	
Bolt Numbers	4	4	1	Weld Type
	x	x	x	
Bolt Grade	2	2	1	Weld Preparation
	x	x	x	
Bolt Diameter	2	2	4	Weld Procedure
	x	x	x	
Bolt Length	10	6	3	Weld Consumable
	___	___	___	
	3200	1920	240	

It should not be overlooked that these figures and options are only for **simple connections in simple structures**. Furthermore, in the detailing of the connection a whole new set of parameters come into being such as the cross-centres, backmarks and pitches of holes and the size and type of notch, but these are more to do with the detailer and are not of primary concern to the designer and of this dissertation.

However, if all these detailing aspects are taken into account it is easy to visualise tens of thousands of different configurations. Such a range of options can only lead to inefficient construction and production and indeed it does so. It is no coincidence to note that those countries which have achieved nationally applied rationalised connections are those with the highest productivity and lowest man hour utilisation.

It is clearly possible and desirable for any country to achieve a range for beam end connections limited to:

2 connection types - from bolted cleats, welded end plates, and welded fin plates
2 different angles, thicknesses and grades
4 different flats, thicknesses and grades
2 bolts - 1 diameter and grade, only 2 different lengths
4 welds - 1 type, 1 preparation, 2 positions and 2 consumables

and for these to apply to a minimum of 80% of its simple connections.

TOWARDS HIGHER PRODUCTIVITY

Three nations lead the world in the terms of structural steel. Table 5 indicates the size of their markets and their percentage share as at 1989/90.

TABLE 5
Constructional Steel Market Share 1988/89

Country	Population (millions)	Steel home market (k tonnes)	% share of construction market
Japan	122	10,400	64
USA	242	5,200	51
UK	56	1,250	61

During the 1980's the UK competitiveness and market share have shown a significant increase. This competitiveness is demonstrated by the graphs in Figure 1 and is mainly as a result of much greater flexibility of labour, investment in high-tech capital equipment, and the adoption of new design and construct techniques such as composite flooring and through deck stud welding, as well as a more realistic and competitive approach to corrosion and fire protection.

FIGURE 1
Price of steel -v- concrete in the last 10 years

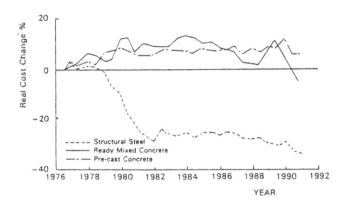

Cost Indices Related to Inflation

However, the two leading nations in competitiveness, particularly in low man hour terms, are Japan and USA where, besides having the previous competitive features, the industry in these nations is much more construction led and rationalised in its approach to the specification and design of connections.

The USA and Japan place much greater emphasis on site performance, mainly due to their attitude to speed of construction and the role of the general contractor respectively.

In the rest of the world structural steel industry needs to recognise that the construction site is not only its shop window - or where it comes face to face with its customer - but it is where the performance of its product is measured against that of its competitor and where it has most effect with regard to the performance of the majority of the other building trades.

It is possible to bring to bear a great breadth and depth of resource to the production process for both steel production and steelwork fabrication but site construction is strictly limited by the site, its capacity, and most particularly by the crane hook - unlimited cranes and 'sky hooks' are just not available. The Americans and Japanese have developed, on a national basis, a rationalisation and simplification of their connections which has led to a significant degree of repetitiveness and consistency. This, combined with their attitude to construction, has led to greatly increased productivity. The Australians have followed suit. It is the next major step which the UK must and is undertaking.

An ideal example of what leads to low productivity, and yet which can be revised to lead to high productivity, is the approach and attitude to bolts. As demonstrated in section 2 the range of bolts utilised can easily be 100 in number and this can even be applied to an individual project. This is amply demonstrated in Table 6 which is part of an analysis carried out on a recent London fast build, construction influenced, project. It is for a building of 19,000 sq.m. involving 1,520 tonnes of structural steelwork which was erected and completed, including floor deck, in 15 weeks using one full-time crane hook with the back-up of a second. This is certainly fast and efficient erection. Yet, as Table 6 shows, the structure used 70 different bolts. The erection contractor considered that with rationalised and construction led connections he could have increased his productivity by a further 10-20%.

TABLE 6

19,000 sq.m. building, 1,520 tonnes of Steelwork

Beam and Column - long span - Office Building

Bolts	Type	Lengths	Quantity
Grade 8.8	M12	3	640
Grade 8.8	M16	2	150
Grade 8.8	M20	19	25,769
Grade 8.8	M24	11	1,340
Grade 8.8	M30	6	740
H.S.F.G.	M20	13	815
H.S.F.G.	M24	2	30
H.S.F.G.	M30	14	3,414
2	8	70	32,898
Grades	Types	Lengths	

The single largest significant factor in the high number of different bolts is the UK practice of part threading in order to achieve a full shank in the shear plane, and ordering in 5mm increments. This is in spite of the fact that the latest British Standard, BS5950, allows for the threaded portion being in the shear and bearing and by far the majority of bolts are designed with this in mind.

The exact location of each different bolt has to be communicated to the steel erector who has to identify, store, and have to hand each of these 70 bolts even when 30 metres in the air. If the UK were to adopt a one grade, one diameter, fully threaded bolt concept then the range used on this contract could have been reduced from 70 to 4 and the erection contractor would have achieved his improved productivity in one simple step.

The benefits of adopting fully threaded bolts and reducing the range to a total of 4 for a minimum of 80% of all usage are, of course, much wider than just the steel erector (1-4).

- Larger numbers off - bulk purchases
- 'Just-in-time' purchasing
- Smaller stock range and totals

- Less handling - no sorting
- Faster assembly
- Eliminates site bolt location lists
- Much faster erection
- Reduction in errors
- Significantly reduced wastage

A recent study (5) has demonstrated that these bolts will perform satisfactorily in general structural steelwork.

TIMING AND THE ROLE OF DESIGN

There are, in principle, two different approaches to the design of the frame and its connections. Firstly where a consultant designs the frame and the contractor usually the connections and, secondly, design and construct where the same engineer often designs both frame and connections.

In the first case the effect of the joint behaviour and design is too often ignored and even in the second case this sometimes occurs. The trend, however, is towards the design of the frame and its joints at one and the same time. The demand and efficiency of this approach will further increase due to a combination of:

Interactive designs e.g. semi rigid design
Composite construction using the structural benefits of composite connections
Integrated design and detail computer programs
Growth in design and construct contracts

BUT - Unless there is a rationalisation of connection design and components on a national basis then a realistic achievement of these benefits will be very difficult if not impossible to achieve.

Where the design of the connection is at a later and separate time then it is incumbent upon the frame designer to provide at the least the following information:

Design Code to be used
A statement of the overall design concept and of the members and connections
Design drawings showing sizes, grades, and positions of all members
The forces and their combinations to be transmitted by each connection
Whether the force shown is factored or unfactored

There then needs to be agreement between the connection designer and the frame designer as to the principles of the connections and their design before any work takes place. Even before considering the connections themselves it needs to be recognised that the layout of the framing can have a profound impact on their execution and cost. Most structural designers will advocate against placing columns and building setbacks off the common grid. But how many architects and engineers fully consider the cost implications of moving columns a few millimetres off centre, beams eccentric, or skew connections at random positions? Yet high cost, and often horrendous, connections result.

It is not possible to itemise all the economies of connection design in one short presentation but there are some points of which the designer can and should be aware, always bearing in mind that the connections account for 50% of the installed cost; that handling is the single largest factor, and this significantly increases if bolting and welding are combined in the same member or it is welded to two or more sides; that bolts should be one material grade for one diameter and never more than two grades/diameters on one project; that connection materials should be from one grade and from a very limited range of section and plate sizes; that workmanship, eg assembly and/or welding, should be concentrated on selected components and not spread equally over every piece.

Perhaps the common greatest sin of any designer which still prevails is the goal of minimum weight. When taken in isolation this leads, virtually without fail, to uneconomic designs with either the customer or the contractor, or both, paying too much for the building or its construction. It can be shown that the cost of one man hour of fabrication time is equal to the cost of approximately 75kg of material* and that the fabrication costs of fitting one stiffener to a typical 400 deep beam is equivalent to 50kg of material. In other words, taking into account detailing, fabricating, and inspecting, each and every stiffener must save 70kg of main material before it can even be considered to be a worthwhile proposition.

*Based on a fabrication labour rate of £25 per hour and plain steel at £350 per tonne.

Similarly if the current erection crane hook productivity is 4 pieces erected per working hour, and it often is, then consider the benefits to everyone if this were to be 12 pieces erected per hour, and it easily can be.

The connection design and the rationalisation of its components is the single biggest item affecting this productivity of both construction and production.

CONCLUSIONS

A summary of the points from this presentation then is:

1. Minimum weight does not usually result in minimum cost - avoid local stiffening.

2. Make every effort to ensure that all main components, beams and columns, are on grid centre lines.

3. The connection design should be carried out at the same time as the member design.

4.	If the design is separated then the same full design information has to be given to the connections designer as used by the frame designer.

5.	Make full use of national standards for design guides and components.

6.	Do not combine bolting and welding in the same member and concentrate any necessary fabrication in selected members.

7.	Design for minimum erection hook time.

8.	Rationalise on the number of different sections and plates used for making fittings.

9.	Use fully threaded bolts of only 1 grade per diameter - and only 2 lengths in total on a project.

Connections are where the majority of hours are either spent or mis-spent

Make sure your design effort reflects this

REFERENCES

1.	GIRARDIER, E.V. Customer Led, Construction Led, Steel Construction, BCSA, Vol 7 No 1 February 1991.

2.	MACPHERSON, I.G. The right attitude to erection, Ibid.

3.	ROGAN, B. Efficient erection, Ibid.

4.	MILLER, P.W. The construction led steelwork contractor, Ibid.

5.	OWENS, G.W. The use of fully threaded bolts for connections in structural steelwork for buildings. (To be published. The Structural Engineer)

QST SHAPES, A NEW GENERATION OF ROLLED SECTIONS FOR ECONOMICAL STEEL CONSTRUCTION

FRANK VAN REST, M.Sc.
TradeARBED, Inc.,
825 Third Avenue
New York, N.Y. 10022, USA

ABSTRACT

With the development of the Quenching and Self-Tempering (QST) process by the Luxembourg steel manufacturer ARBED S.A., a new generation of High Strength Low Alloy (HSLA) steel shapes with yield strengths up to 70,000 psi or 500 MPa has become available. The disadvantages of classical HSLA shapes: poor weldability, low toughness and a high price have been overcome. Use of HISTAR (HIgh STrength ARbed) QST shapes in building applications can lead to considerable cost savings. Those savings will consist of material cost savings, obtained by a reduction of the weight of the structures, fabrication cost savings, foundation cost savings and savings in handling and transportation cost. Since the implementation of the QST process on an industrial scale at the ARBED Differdange works in Luxembourg in the summer of 1990, a large number of projects using high strength QST shapes has been realized throughout the world.

INTRODUCTION

The use of HSLA structural shapes (F_y > 50,000 psi or 345 MPa) in steel buildings has always been limited. This has been caused by a variety of reasons: stiffness considerations and code restrictions have been important factors, but a limited availability, a poor weldability, a low toughness and a high price have been equally important in limiting the acceptance of HSLA structural shapes.

With the development of its QST process, the Luxembourg steel manufacturer ARBED S.A. has succeeded in overcoming most of the limitations common to HSLA steel shapes.

QST PROCESS

Traditionally, HSLA structural shapes have been produced by adding alloying elements to the ladle and by rolling at controlled temperatures. However, the alloying content needs to remain limited as it influences the weldability and the toughness of the steel in a negative way. Furthermore thermomechanical (TM) or controlled rolling is restricted by the mechanical power of the rolling mills. Subsequently the production of HSLA steels has been limited to the lighter shapes.

These limitations have been overcome by the QST process. In this process the desired yield strength is obtained by an in-line heat treatment after the last rolling pass. An intense watercooling is applied to the whole surface of the shape so that the skin is quenched. Cooling is interrupted before the core of the shape is affected by quenching and the outer layers are tempered by the flow of heat from the core to the surface during the temperature homogenization phase. A schematic representation of the QST process is given in Figure 1 and in Figure 2 the quenching of a W 14 x 730 Jumbo section is shown.

The QST process limits the necessity to add expensive alloying elements and is therefore a cost-effective process. The manufacturer still has the option to roll at controlled temperatures, which allows for the production of very heavy sections (flange thickness up to 5 inches) while using a mild steel.

PROPERTIES AND WELDABILITY

Due to its metallurgical principle, the QST process results in steel properties that were previously considered incompatible:

. a high yield strength (up to 70,000 psi or 500 MPa for the entire section range);
. an outstanding low temperature toughness;
. an excellent weldability due to a low carbon equivalent value[1].

In Figure 3 the stress-strain curves for QST steels of the Grades 50 and 65 are shown. Note that the F_u/F_y ratio and the elongation are at a high level for both grades.

Figure 4 shows the Transition Temperature for different yield strengths for both the classical TM process and for the QST process for a material thickness of 2 and 5 inches. In Figure 5 the relationship between carbon equivalent value (CE) and yield strength is shown for both the classical TM and the QST process.

1. The carbon equivalent value (CE) is a figure which is calculated from the chemical composition of a steel. The following formula is the most commonly used: CE (%) = % C + % Mn/6 + % (Cr+Mo+V)/5 + % (Cu+Ni)/15 (C = carbon, Mn = manganese, Cr = chromium, Mo = molybdenum, V = vanadium, Cu = copper, Ni = nickel).

Extensive weldability evaluations performed at ARBED's research facility in Esch-sur-Alzette, Luxembourg [1] have shown the excellent weldability of the QST shapes. Using different welding processes and heat inputs, QST shapes with a flange thickness of up to 5 inches have been butt-spliced with full penetration single bevel groove welds. All welding has been performed <u>without preheating</u>. The tensile tests and Charpy V-Notch impact tests have shown that for a variety of heat inputs the tensile properties of the joined material are not influenced and the toughness of the base material (BM), heat affected zone (HAZ) and fusion line (FL) remains on a high level. In <u>Figure 6</u> the toughness of the BM, HAZ and FL of a QST section in Grade 65 with a thickness of 2 inches are compared to that of a classical TM section in Grade 65.

<u>Figure 7</u> shows the results of Charpy V-Notch impact tests performed on the core area (web-flange intersection) of a 5 inch thick QST section in Grade 65, which has been spliced with full penetration groove welds. Again the welding has been performed <u>without preheating</u>. Note that the Charpy V-Notch impact toughness requirement of 20 ft-lbs. at 70 °F, as required by the Specifications for Structural Steel Buildings [2, 3] of the American Institute of Steel Construction, Inc. (AISC) to guarantee the reliability of splices using full penetration welds in heavy shapes subject to primary tensile stresses, is easily met in the as rolled <u>and</u> the as welded condition.

The American Welding Institute, Inc. (AWI), a scientific institute specialized in welding research has conducted an independent assessment of the weldability of QST Jumbo sections in Grade 65 in the United States [4, 5]. 5 inch thick sections have been spliced with full penetration groove welds using Shielded Metal Arc Welding (SMAW), Flux Core Arc Welding (FCAW) and Submerged Arc Welding (SAW) and a variety of heat inputs. Commercially available welding consumables have been used and <u>no preheat</u> has been applied.

The results of the AWI weldability tests confirm the excellent weldability of QST steels.

<u>Figure 8</u> shows the guaranteed transition temperatures and maximum carbon equivalent values for HISTAR QST sections in the Grades 50, 60, 65 and 70 for different material thicknesses.

ECONOMY

An economical use of QST HSLA shapes is limited to the members that can be designed to higher allowable stresses. As higher stresses result in larger deformations (the E-modulus of QST HSLA steel is virtually equal to that of its lower strength equivalents), the shapes whose sizes are determined by deformation criteria are not suited to be designed in HSLA steel.

In multi-storey buildings these members will be found in the floor framing and to some extent in the lateral force resisting frame.

Elements most economically designed in HSLA steel are tension and compression members, for which the deformation consists of axial lengthening and shortening only.

In figure 9 the economy of ASTM A 36, A 572 Grade 50 and HISTAR Grade 65 is compared for a column section. Figure 10 illustrates the weight savings and fabrication cost savings that can be obtained by substituting a heavy built-up compression member with a rolled section in HISTAR Grade 65.

In figure 11 the economy of ASTM A 36, A 572 Grade 50 and HISTAR Grade 65 is compared for a tension chord in a truss. An important part of the cost savings in this example originates from a decrease in weld volume, caused by a reduction in cross sectional area of the tension member.

The economy of HSLA steels in lateral force resisting frames has not been thoroughly analyzed. Naturally lateral displacement (drift) is an important design criterium. However, the fact that not all moment frames are currently being constructed with the lowest strength steel available is a strong indication that there might be use for HSLA steels with F_y > 50,000 psi.

An interesting concept in moment frame design is the omitting of web doublers and continuity plates or stiffeners in the columns (figure 12) by using HSLA steels. The column will be kept the same size, so the lateral stiffness will remain equal; the high yield strength is used only in the connection design.

The Uniform Building Code [6] specifically limits the yield strength to F_y = 50,000 psi for Special Moment Resisting Frames in seismic regions. Certainly this is due to concerns regarding the ductility of HSLA steels. All mechanical tests performed on HISTAR Grades 65 and 70 have indicated that these materials have a ductility behavior that is similar to that of its lower strength equivalents. The tests have included tensile tests, toughness tests, full scale buckling tests and full scale bend tests.

CONCLUSION

The development of the QST process has led to the availability of a new series of steels that have overcome most of the disadvantages of the classical HSLA steels. Excellent weldability, outstanding low temperature toughness and a competitive price are guaranteed. QST shapes will be highly competitive for compression and tension members, as deformation is not a decisive criterium. Weight savings will vary from 15 - 25 %. Cost savings are potentially higher because of fabrication cost savings.

The economy of QST shapes in lateral force resisting frames needs to be analyzed further, but there are certainly indications that HSLA steel shapes will lead to cost savings for that type of application as well. The code limitation of F_y = 50,000 psi for Special Moment Resisting Frames needs to be addressed based on available data and possible future research.

REFERENCES

1. de la Hamette, J., Panunzi, C., Becker, F. and Dengler, J.M., QST-Beams, A New Generation of High Performance Products. ARBED Research, Esch-sur-Alzette, Luxembourg, 1989
2. AISC, Inc., Load and Resistance Factor Design Specification for Structural Steel Buildings, Chicago, IL, 1986
3. AISC, Inc., Specification for Structural Steel Buildings - Allowable Stress Design and Plastic Design, Chicago, IL, 1989
4. Sutter, V., Weldability of QST Beams, AWI, Inc., Knoxville, TN, 1992
5. Kaufmann, E.J. and Fisher, J.W., Evaluation of Mechanical properties of Welded TMCP Jumbo Sections, Lehigh University - ATLSS, Bethlehem, PA, 1992
6. ICBO, Inc., Uniform Building Code, Whittier, CA, 1988

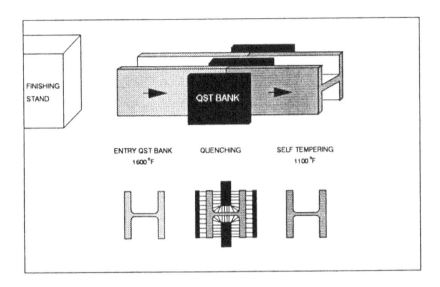

Figure 1. Quenching and Self-Tempering (QST) Treatment of Beams
         in the Rolling Heat

Figure 2. Quenching of a W 14 x 730 Jumbo Section

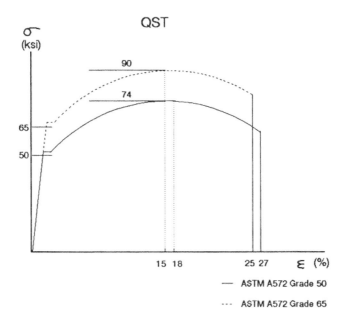

Figure 3. Stress-Strain Curves of QST Grade 50 and Grade 65 Steels

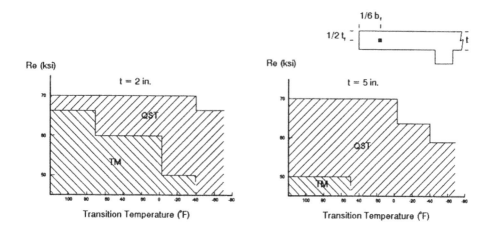

Figure 4. Transition Temperature and Yield Strength for QST Treatment and TM Rolling for Different Product Thicknesses

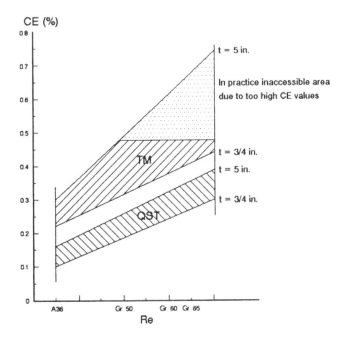

Figure 5. Yield Strength and Carbon Equivalent Value for QST
Treatment and TM Rolling for Different Product
Thicknesses

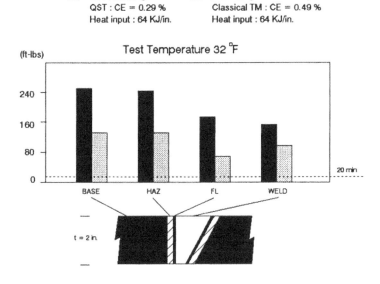

Figure 6. Charpy V-Notch Impact Testing for QST Treatment and
TM Rolling of 2 Inch Thick Welded Section in Grade 65

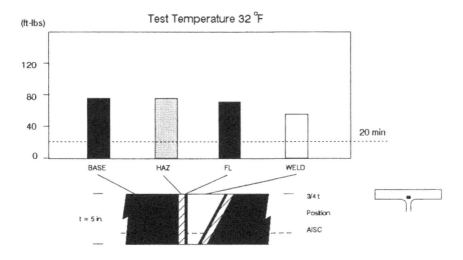

Figure 7. Charpy V-Notch Impact Testing of 5 Inch Thick Welded
           Section in Grade 65 with QST Treatment

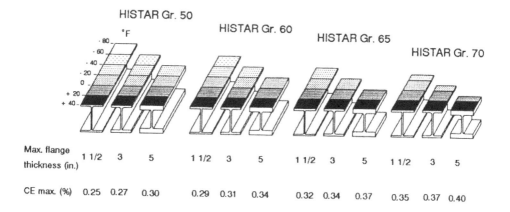

Figure 8. Guaranteed Transition Temperatures for Charpy V-Notch
           Impact Testing (30 ft-lbs.) and Maximum Carbon
           Equivalent Values

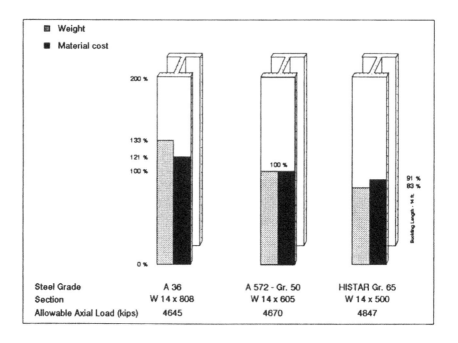

Figure 9. Weight and Material Cost for Different Yield Strengths - Column Application

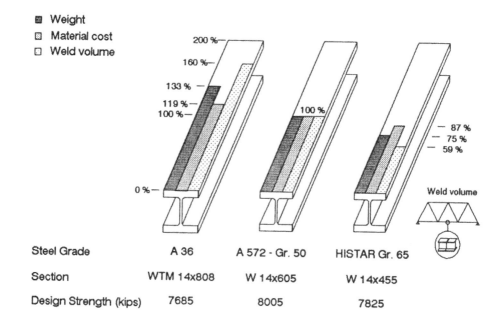

Figure 10. Weight, Material Cost and Weld Volume for Different Yield Strengths - Tension Chord Application

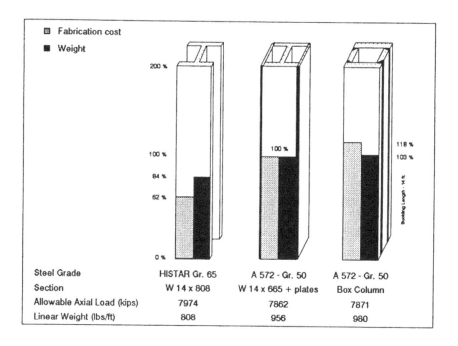

Figure 11. Weight and Fabrication Cost for Different Yield Strengths - Built-Up Column Application

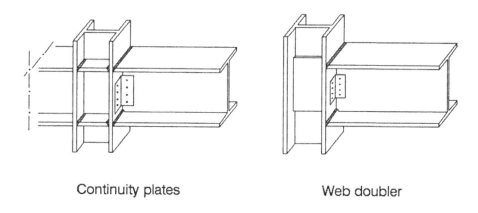

Continuity plates   Web doubler

Figure 12. Continuity Plates and Web Doublers - Moment Connection

STAINLESS STEEL IN STRUCTURAL APPLICATIONS

GJ VAN DEN BERG, P VAN DER MERWE
Materials Laboratory, Faculty of Engineering,
Rand Afrikaans University.
PO Box 524, Auckland Park, Johannesburg, 2004.
Republic of South Africa.

SYNOPSIS

In recent years a need has developed for a more comprehensive design specification for cold-formed and hot-rolled stainless steel structural members. Because of the limited life of carbon steel in bridges and structures that are exposed to moderate aggressive atmospheres, the low chromium containing ferritic stainless steels can be an alternative to carbon steels.

Research work over the past six years jointly undertaken by the Rand Afrikaans University and the University of Missouri Rolla, resulted during 1991 in the publication by the American Society of Civil Engineers (ASCE) of a specification for the design of cold-formed stainless steel structural members and connections.[1] This specification contains both the Allowable Stress Design Method and the Load and Resistance Design Method.

This paper will discuss the reasons for the need of stainless steel design specifications, the characteristics of certain types of stainless steels suitable for structural applications and will give a brief overview of the new design specification.

GENERAL REMARKS

A high proportion of the total raw steel in the world is used merely to replace steel which through corrosion reverts as rust to something very much like the one originally mined. Coating processes such as painting, epoxy coating, galvanizing and electroplating assist to stem this wasteful process of attrition, but often provide only partial success.

The engineer as part of his responsibility towards society is expected to be energy and cost conscious and therefore exercise care in the selection of materials. His choice is, however, restricted due to the lack of suitable information, to the relatively inexpensive but corrosion-prone carbon and low alloy steels and the relatively expensive stainless steels, of which the ability to resist corrosion often by far, exceeds the requirements when structures are to be designed to function in moderately aggressive atmospheres.

According to statistics released in 1989 by the Federal Highway Administration in the United States of America[2] it was revealed that 238 357 (41%) of the nations 577 710 bridges are either structurally deficient or functionally obsolete. The definition of structurally deficient does not necessarily mean that the bridges are unsafe. The bridges are either closed or restricted to lighter vehicles only because of deteriorated structural components. The average age of all United States rural bridges was found to be 36,6 years. When taking into account the life cycle cost of a structure, stainless steel can be a cost effective material for structural applications.

The basic price structure of the austenitic and ferritic type stainless steels and their ability to resist corrosion are such that the development of design criteria for the use of these materials in structural applications, will contribute towards filling the gap that currently exists in the range of materials available for structural purposes. The specific type of austenitic stainless steels under consideration in this discussion are Type 304 and 316 and the ferritic stainless steels are Type 409, 430 and a modified Type 409 steel, designated 3CR12, a corrosion resisting steel, which has been developed by the South African based specialty steel producing company, Middelburg Steel and Alloys. The specification for the design of structural members cold-formed from austenitic and ferritic stainless steels[1] needs to be updated every five years. In order to revise a design specification, the design criteria have to be updated. The development of the design criteria constitutes the primary objective of this investigation. Additional to the above research, heavy hot-rolled sections are increasingly becoming available from Japanese mills. This opens a whole new field for structural steel research aimed at generating information required for safe and durable stainless steel structures.

CLASSIFICATION OF STAINLESS STEELS

Stainless steel is the name given to a large group of special iron based alloys containing at least 11 percent chromium, the upper limit of chromium content being set by practical considerations at about 30 percent. By the addition of other elements such as nickel, molybdenum, copper, titanium and niobium, all of which exert a beneficial effect in increasing the resistance to corrosion or to provide resistance to specific chemical attack, there is a large group of alloys produced today, providing a broad base of usefulness. All grades of stainless steels are suitable for exposure to atmospheres that are free from contamination even when the humidity approaches 100 percent. However, as corrosive conditions of the atmosphere or environment become more severe, due to contamination, discriminating selection of grade becomes important. The stainless steels can generally be classified into the following groups:

- Austenitic Stainless Steels
- Ferritic Stainless Steels
- Martensitic Stainless Steels
- Precipitation Hardening Stainless Steels

The primary difference among the different groups is their response to heat treatment rather than composition or ability to resist specific chemical attack or corrosion. Since iron is an allotropic element, that is, it can exist in more than one crystal form, carbon and low-alloy steels are allotropic. At room temperature, carbon and low-alloy steels normally are a mixture of ferritic (body centred cubic iron) and iron carbide. When they are heated to a temperature above 732° C (the temperature depends on the carbon content) their crystal structure of ferrite changes to austenite (face centred cubic iron) and the iron carbide dissolves in the austenite. When the austenite structures are cooled to various temperatures below 732° C the austenite transforms into various structures depending on the precise composition and rate of cooling. The austenite normally reverts to a form of ferrite or a mixture of ferrite and iron carbide. Allotropy is thus the phenomenon that permits quench hardening of carbon and low-alloy steels. Reference is often made in this document to the austenitic stainless steels and the ferritic stainless steels. The fundamental differences between these two groups will be highlighted in the following paragraphs.

Austenitic Stainless Steels

These steels have chromium (18 to 30 percent) and nickel (6 to 20 percent) as major alloying elements. The carbon content is kept low. For some grades a portion of the nickel is replaced by manganese in a ratio of approximately 2 parts of manganese for each part of nickel replaced. The austenitic phase is stabilized by the presence of sufficient nickel. These steels are predominantly austenitic at all temperatures and do not appreciably change their crystal form during heating and cooling and cannot appreciably be quench hardened, because cooling and hence the cooling rate, does not affect the crystal structure. The principal characteristics, are the ductile austenitic condition, rapid hardenability by cold working and excellent corrosion resistance. Besides being referred to as austenitic stainless steels, they are sometimes called chromium-nickel, work-hardening or 18-8 stainless steels.

Ferritic Stainless Steels

Most of the ferritic stainless steels have chromium in the range of 11,5 to 18 percent as the sole major alloying element. The absence of nickel (austenite stabilizer) in all types but Type 409 steel, provides the ferritic structure which characterizes these steels, since chromium is a ferrite stabilizer. Type 409 steel contains 0,5 percent nickel as well as a small amount of titanium. The titanium is added to reduce the harmful effects of grain boundary carbide precipitation, which is also a common occurrence in ferritic stainless steels, and to assist in stabilizing the ferritic structure. In this respect titanium is more powerful than chromium. The limited carbon content and other minor alloying additions provide for a micro structure which is predominantly ferritic at room temperature as well as at elevated temperatures and an inability to harden by quenching. In this respect they bear resemblance to the austenitic grades, because they are also not allotropic.

STAINLESS STEELS UNDER CONSIDERATION

The majority of structures in civil engineering, when corrosion is a problem, will make use of either the austenitic or ferritic types of stainless steels. Thus only these two type of steels will be discussed. The chemical compositions of the different steels are given in Table 1.[3]

Austenitic Stainless Steels

Stainless Steel Type 304: Stainless steel Type 304 is commonly available and is used in a wide range of applications such as, architectural, brewing industry, cook ware, cryogenic plants, food and diary processing equipment, heat exchanger tubes and supports, pressure vessels and process plant parts to name a few.

Although stainless steel Type 304 is less corrosion resistant than Type 316 stainless steel, it has a wider field of application as it is more price competitive.

Stainless steel Type 304 has a corrosion resistance in industrial areas where there is a combination of moisture, carbonaceous and other pollutants. Besides good corrosion properties, this steel offers excellent formability and weldability properties.

TABLE 1
Chemical Composition of Selected Stainless Steels

ALLOYING ELEMENT	304	316	409	430	3CR12
Carbon	0,080 max	0,080 max	0,080 max	0,120 max	0,02
Manganese	2,000 max	2,000 max	1,000 max	1,000 max	0,90
Phosphorous	0,045 max	0,045 max	0,045 max	0,045 max	0,04
Sulphur	0,030 max	0,030 max	0,045 max	0,030 max	0,04
Silicon	0,750 max	1,000 max	1,000 max	1,000 max	0,40
Nickel	8,00-10,50	10,0-14,0	0,500 max	0,750 max	0,60
Nitrogen	0,100 max	-	-	-	0,02
Titanium	-	-	6xCmin 0,750 max	-	0,25
Chromium	18,0-20,0	16,0-18,0	10,5-11,75	16,0-18,0	11,5

Stainless Steel Type 316: This steel contains 2 to 3% molybdenum which enhances the corrosion resistance in reducing media, and the resistance to pitting and crevice corrosion in chloride solutions. Since molybdenum is a potent ferritizer, the nickel content has to be increased in order to prevent delta ferrite formation. An added benefit of molybdenum addition is the improved high-temperature creep and stress-rupture strength. This steel is more corrosion resistant than Type 304 stainless steel.

Ferritic Stainless Steels

Stainless Steel Type 409: Stainless Steel Type 409 is one of the least expensive stainless steels because of its relatively low chromium and nickel content. Unlike the distinctive characteristics of the ferritic stainless steels, modifications have been made to Type 409 stainless steel, which have greatly improved its formability. Stainless steel Type 409 only belongs to the family of stainless steels due to its chromium content. In the strict sense it is only a corrosion resisting steel. Generally this steel has poor welding properties and is susceptible to embrittlement at low temperatures. One of the major applications for stainless steel Type 409 is in the automotive industry as exhaust pipes, vehicle bodies, mufflers and silencers.

Stainless Steel Type 430: This type of steel is a low-carbon plain chromium steel. It has good corrosion resistance in mildly corrosive environments. In the annealed condition, Type 430 stainless steel is ductile, does not harden excessively during cold work, and can be formed using a large variety of roll forming or mild stretch-bending operations, as well as the more common drawing and bending processes. This steel has limited weldability and should not be used in the as-welded conditions for dynamic or impact loaded structures. It is liable to brittle fracture at sub-zero temperatures. The chemical composition is listed in Table 1.[3] This steel is normally produced in accordance with the requirements of ASTM Standard Specification A176-85a.[4] Traditional applications of Type 430 stainless steel are in areas where mildly corrosive conditions occur, or where scaling resistance at moderate temperatures are required, for example: mining applications such as ore storage and materials handling equipment, process and materials handling equipment in the sugar industry and heat resisting applications up to 750° C.

Type 3CR12 Corrosion Resisting Steel: Type 3CR12 corrosion resisting steel was developed by the specialty steel producing company Middelburg Steel and Alloys from AISI Type 409 stainless steel. The aim with the development of this steel was to create a low chromium steel of which the mechanical properties and the weldability would be superior to that of Type 409 stainless steel. The chemical composition of this steel falls typically within the limits of Type 409 stainless steel, except for nickel, manganese and titanium. The carbon and nitrogen levels are kept low and it has therefore improved toughness over AISI Type 409 and 430 stainless steels in both the annealed and welded conditions.

Melvill et al[5] reported that it was shown in tests, that an improved performance could be obtained over AISI Type 304 and 430 stainless steels, in applications such as ore hoppers and shutes. Type 3CR12 steel is less expensive than AISI Type 304 stainless steel. It has advantages in terms of toughness and weldability over AISI Type 430 stainless steel and the corrosion and corrosion abrasion resistance should be sufficient to make it suitable as a substitute for certain applications. Further applications for Type 3CR12 steel are headgear shute liners, surface ore car liners, skip liners, transfer shute liners, municipal refuse containers, autoclave brick trolleys, structures, and railway coal wagons.

Ball and Hoffman[6] studied the micro structure of Type 3CR12 steel. A comprehensive discussion on chemical composition and micro structure, corrosion resistance, forming properties, toughness, weldability and physical properties was done by Van der Merwe,[7] and can be recommended as reference.

DESIGN SPECIFICATIONS FOR STAINLESS STEEL

Due to the difference in mechanical behaviour of the stainless steels compared to the carbon and low alloy steels, the AISI specification[8] for carbon and low alloy steels does not apply to the design of stainless steel members.

In 1968 the first edition of the Specification for the Design of Light Gage Cold-Formed Stainless Steel Structural Members[9] was published by the American Iron and Steel Institute. This specification was mainly based on the extensive research conducted by Johnson and Winter at Cornell University[10,11] as well as experience and knowledge gained from the carbon and low alloy steels. The scope of that specification was limited to the use of annealed and strain flattened austenitic stainless steel Types 201, 202, 301, 302, 304 and 316.

The findings of the additional research work conducted at Cornell University by Wang and Errera[12,13], who investigated the performance of structural members cold-formed from cold-rolled austenitic stainless steels and the study by Errera, Tang and Popowich[14] on the strength of bolted and welded connections in stainless steels, were incorporated in the 1974 edition of the AISI design specification for stainless steels.[15]

The stainless steel design specification[15] at that stage lacked a considerable amount of design provisions in comparison with the AISI design specification for carbon and low alloy steels.[8] Research activities are centred around the carbon and low-alloy steels, simply because of the much higher usage of those materials.[7]

A modern approach is to base design specifications on the probabilistic approach. In 1984 the Canadian Standard Association published their cold-formed design specification for carbon and low alloy steels, entitled Steel Structural Members.[16] In 1986 the American Institute of Steel Construction published their specification on the probabilistic approach for hot-rolled and built-up members, fabricated from steel plates.[17] In addition to the above, the University of Missouri-Rolla conducted extensive research since 1979 on the development of load and resistance factor design criteria (LRFD) for cold-formed carbon and low alloy steels.

In order to develop a new design specification for cold-formed stainless steel members based on the allowable stress design and probabilistic approach, a research project was initiated in July 1986 at the University of Missouri-Rolla under the sponsorship of the American Society of Civil Engineers initiated by the International Chromium Development Association.

The first aim of the study by Lin and Yu, supported by Galambos as consultant, was to revise the 1974 edition[15] of the AISI allowable stress design specification for cold-formed stainless steel structural members. The second aim of their investigation was to develop load and resistance factor design criteria for cold-formed stainless steel structural members. This new design specification[1], issued during 1991, provides design rules for the structural application of AISI Types 201, 301, 304 and 316 stainless steels, annealed and cold-rolled in 1/16-, 1/4- and 1/2-hard tempers and AISI Types 409, 430 and 439 stainless steels in the annealed condition.

It should be noted that there is a commonality of many aspects on the design specifications for the carbon and low-alloy steels and the stainless steels. The feasibility and desirability of merging the two design specifications is not totally impossible, but is not currently under consideration.

DIFFERENCES IN DESIGNING

The designing of cold-formed stainless steel is similar to that of cold-formed carbon steel. However, since the mechanical properties of stainless steels are more complex than those of carbon steels, the design procedures for the former are occasionally more involved.

In order to account for the different response to load between stainless steels and carbon and low-alloy steels, certain modifications to the design equations are needed.

Non-Linear Stress-Strain Relationships

For carbon and low-alloy steels, a single stress-strain curve of the sharp-yielding type (for virgin material) is assumed to be valid for tension and compression. In contrast to this, stainless steels are categorized as having gradually yielding stress-strain behaviour. Aspects that should be considered include the proportional limit, F_p, which could be considerably lower than the yield-strength, F_y. Moduli such as the initial modulus, E_o, defined as the slope of the initial part of the stress-strain curve, the tangent modulus, E_t, defined as the slope of the tangent to the stress-strain curve at each value of stress, and the secant modulus, E_s, defined as the ratio of the stress to the strain at each value of stress. Other moduli, such as the tangent shear modulus, G_t, defined as the tangent to the shear stress-shear strain curve and the secant modulus, G_s, defined as the ratio of the shear stress to the shear strain at each value of stress are also important. Figure 1 identifies a number of these properties.

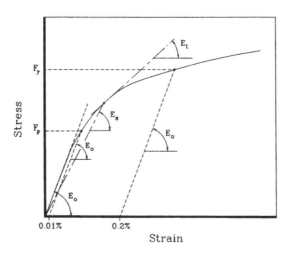

Figure 1. Typical stress-strain behaviour of a stainless steel

Inelastic Buckling
Owing to the relatively low values of the proportional limit of stainless steels, flat elements and members subjected to compression, shear, or bending may buckle at stresses that exceed the proportional limit, hence inelastic buckling. Plasticity reduction factors are being used to modify the design equations that have been derived for elastic buckling. These are listed in Table 2.

Local Distortion
When local distortions in flexural members under nominal service loads must be limited, the design flexural strength is determined at a stress equal to the critical local buckling stress, multiplied by a factor that depends on the amount of distortion that is allowed. Normally, this factor will vary between 0,75 and 1,2. The plasticity reduction factor for inelastic buckling for compression given in Table 2 is used to determine the critical local buckling strength.

Safety and Resistance Factors
Owing to the lack of design experience and the lack of sufficient test data for statistical analysis, relatively large safety factors and resistance factors are found in design specifications for stainless steels.

TABLE 2
Plasticity Reduction Factors for Inelastic Buckling

Type of Buckling Stress	Plasticity Reduction Factor
Local Buckling	
Compression	
Stiffened Elements	$(E_t/E_o)^{1/2}$
Unstiffened Elements	E_s/E_o
Bending	E_s/E_o
Shear	G_s/G_o
Column Buckling	
Flexural Buckling	E_t/E_o
Torsional Flexural Buckling	E_t/E_o and G_s/G_o
Beam Buckling	
Lateral Torsional Buckling	E_t/E_o and G_s/G_o

E_o = initial elastic modulus
E_t = tangent modulus
E_s = secant modulus
G_o = initial shear modulus
G_s = secant shear modulus

Determination of Deflections
A reduced modulus $E_r = (E_{ts}+E_{cs})/2$ is stipulated for the calculation of deflections. In this equation, E_{ts} is the secant modulus corresponding to the stress in the tension flange, and E_{cs} is the secant modulus corresponding to the stress in the compression flange. The different response of the material to tension and compression is accounted for by this

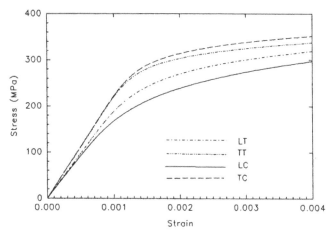

FIGURE 2   TYPICAL STRESS—STRAIN CURVES FOR STAINLESS STEEL

reduced modulus, as well as by the likelihood that the stress under service load in the extreme fibre may be higher than the proportional limit.

Anisotropy

Since four stress-strain curves are needed to describe the stress-strain behaviour of a stainless steel, care should be taken in the selection of the values of mechanical properties and plasticity reduction factors for design purposes. Figure 2 shows a typical set of four stress-strain curves for a stainless steel. Note the lower curve for longitudinal compression. The longitudinal axes of structural members will normally coincide with the longitudinal direction as defined above.

Flexural and Torsional Flexural Buckling of Columns

For the same reasons as mentioned above, the tangent modulus theory for column buckling is used to predict the failure of axially loaded compact compression members. Figure 3 shows the difference between the design approaches for a carbon steel and a stainless steel with identical yield-strength values against the Euler buckling curve.

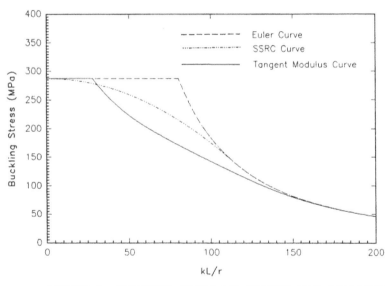

FIGURE 3 FLEXURAL BUCKLING STRESS FOR COLUMNS

Lateral Torsional Buckling of Beams

For inelastic lateral torsional buckling, the stainless-steel design specification[1] requires the use of a plasticity reduction factor, given in Table 2, based on the tangent modulus approach in conjunction with the equation that would otherwise be used for elastic behaviour in carbon and low-alloy steels. The parabolic equation used in the design specification[8] for inelastic buckling in carbon and low-alloy steels cannot be used for stainless-steel design. This effect is illustrated in Figure 4.

Limitations of Width-to-Thickness Ratios

Where pleasing appearance is of importance, the width-to-thickness ratio of flat elements has to be reduced to minimize local distortion of the elements. These ratios for stainless steel are different from those for carbon and low-alloy steels.

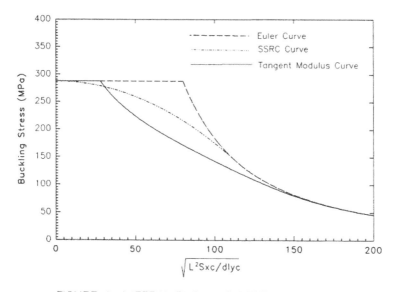

FIGURE 4 LATERAL BUCKLING STRESS FOR BEAMS

CONCLUSIONS

In this paper the need for stainless steel design specifications was outlifted. The characteristics and applications of stainless steels suitable for structural applications were discussed. Finally an overview was given of the differences between designing in carbon and low alloy steels and stainless steels.

ACKNOWLEDGEMENTS

The authors would like to acknowledge the financial assistance received from Chromium Centre.

REFERENCES

1.   American Society for Civil Engineers. Stainless Steel Design Specification. 1991.

2.   Farricone, P., Bridges Under Surveillance, Civil Engineering, May 1990.

3.   Lula, R.A., Stainless Steel, American Society for Metals.

4. American Society for Testing and Materials, <u>Standard Specification for Stainless and Heat-Resisting Chromium Steel Plate, Sheet and Strip</u>, Designation A176-85a.

5. Melvill, M.L., Mahony, C.S., Hoffman, J.P., Dewar, K., The Development of a Chromium Containing Corrosion Resisting Steel, Paper Presented at the <u>Third South African Corrosion Conference</u>, 1980.

6. Ball, A., Hoffman, J.P., A Microstructure and Properties of a Steel Containing 12% Chromium and Designated 3CR12 (Chromweld), <u>Metal Technology</u>, Vol. 8, Sept. 1981.

7. Van der Merwe, P., Development of Design Criteria for Ferritic Stainless Steel Cold-Formed Members and Connections, Ph.D. Thesis, University of Missouri-Rolla, 1987.

8. American Iron and Steel Institute, <u>Cold-Formed Steel Design Manual</u>, 1986.

9. American Iron and Steel Institute, <u>Specification for the Design of Light Gage Cold-Formed Stainless Steel Structural Members</u>, 1968.

10. Johnson, A.L., The Structural Performance of Austenitic Stainless Steel Members, Ph.D Thesis, Cornell University Report No. 327, 1966.

11. Johnson, A.L., Winter, G., Behaviour of Stainless Steel Columns and Beams, <u>Proceedings of ASCE</u>, Vol. 92, No. ST1, February 1967.

12. Wang, S.T., Errera, S.J., Behaviour of Cold-Rolled Stainless Steel Members, <u>Proceedings of the First International Specialty Conference on Cold-Formed Steel Structures</u>, University of Missouri-Rolla, 1971.

13. Wang, S.T., Cold-Rolled Stainless Steel, Material Properties and Structural Performance, Report No. 334, Cornell University, 1969.

14. Errera, S.J., Tang, B.N., Popowich, D.W., Strength of Bolted Connections in Stainless Steel, Report No. 355, Cornell University, 1970.

15. American Iron and Steel Institute, <u>Stainless Steel Cold-Formed Structural Design Manual</u>, 1974 Edition.

16. Canadian Standard Association, <u>Cold-Formed Steel Structural Members</u>, CSA Standard Can 3-S-136, 1984.

17. American Institute of Steel Construction, <u>Load and Resistance Factor Design Specification for Structural Steel Buildings</u>, Chicago Illinois, 1986.

THE USE OF STEEL IN REFURBISHMENT

FEDERICO M. MAZZOLANI
Professor of Structural Engineering
Engineering Faculty, University of Naples
80125 Naples, Italy

ABSTRACT

This paper is devoted to emphasize the main problems which arise when one uses steelworks to consolidate existing structures in refurbishment operations.

The flexibility of these technological systems is shown by means of the examination of the constructional details which can be designed to upgrading the load bearing capacity of structural members made of the commonly used materials (masonry, wood, reinforced concrete and steel too).

INTRODUCTION

The refurbishment of existing buildings is today an emerging activity, which recalls more and more the interest of the building industry.

The old masonry buildings are very often damaged by the age and by the ravages of time and, therefore, they require structural consolidation and functional rehabilitation [1].

But also more recent buildings made of reinforced concrete sometimes need refurbishment operations due to their bad state of preservation.

The development of these activities acknowledges steel as a suitable material both from structural and architectural point of view. In fact, refurbishment requires constructional systems which guarantee flexibility of execution and simplicity of erection [2].

For this purpose, the constructional steel offers "prefabricated" types of technology, which allow the designers to find "ad hoc" solutions and to achieve optimum results tailored to specific requirements [3].

From the structural point of view, the analysis of several practical examples collected from all over the world shows that steel construction is widely and successfully used at all levels of consolidation of already existing structures. These levels are differentiated in order of importance according to the main constructional phases, which are commonly defined in order of importance as safeguard, repairing, reinforcing, restructuring [4].

574

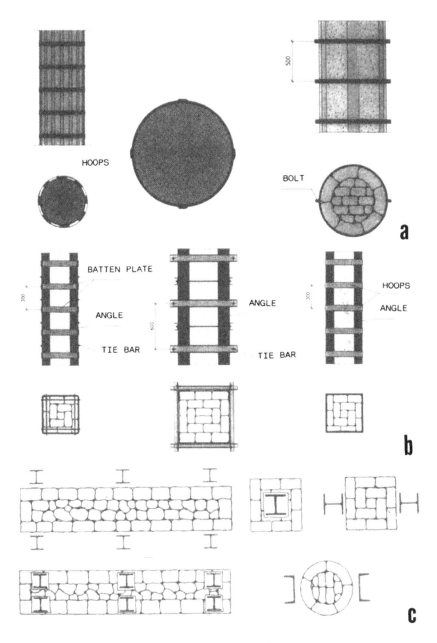

Figure 1. Consolidation of masonry columns

WHEN USING STEEL

It is interesting to observe that steelworks can be conveniently used for the consolidation of all kind of structures, both old and new, made of all common constructional materials, i.e. masonry, timber, reinforced concrete and also steel itself.

In addition, the use of steel is very effective in strengthening structures against earthquakes, when upgrading operation are requested by law [5].

When the building to be consolidated is a historial monument, its restoration is the most delicate operation for the designer and the use of steel gives him further advantages [6,7].

In fact, in accordance with the basic criteria of the various International Restoration Charts, the existing building must be conserved and integrated by new works necessary to ensure both adeguate safety level and suitable functionality [8,9].

Such new works must have an unmistakable modern and reversible feature, by means of methods and materials which can be removed without damaging the existing structure. A logical application of these principles undoubtedly shows that steel has the necessary pre-requisites of being a new material with "reversible" characteristics, particularly suitable to take its place alonside the materials of the past, in order to form together integrated structural systems [10].

CONSOLIDATION OF MASONRY ELEMENTS

The load bearing capacity of masonry elements must be improved whether when they are cracked due to the damaging of external unexpected actions (i.e. earthquake) or also when the structure as a whole must be upgraded in order to resist new loading conditions which are imposed in case of re-use of the building.

Masonry columns, when damaged, are usually repaired by means of steel hoopings. The lateral restraining of the material produces a sensible increasing of the vertical load bearing capacity. In case of circular columns the hooping can be made by means of vertical bar with rectangular cross-section, which are forced by horizontal steel rings (fig.1 a). In the past, this forcing operation was made by heating these rings at high temperature and using the shortening due to cooling for introducing the lateral prestressing of the column. Nowaday, two half rings can be forced by means of bolts.

In case of square or rectangular cross-sections, angle shapes can be used as vertical elements in the corners (fig.1 b). They can be connected in different ways: by means of internal ties integrated by batten plates, by means of channels connected by external ties or by means of horizontal rings.

When it is necessary to transfer an important part of the total vertical load acting on the masonry panel to a new steel structure, the new steel columns can be inserted in proper grooves or simply placed in adherence to the masonry (fig.1 c). Both hot rolled shapes and cold-formed sections can be used to this purpose, according to statical and esthetical requirements.

CONSOLIDATION OF WOODEN FLOORS

The masonry buildings are usually integrated by floor structures made of wood. It is very often necessary to strengthening the wooden parts (beams and deck) because they are usually in a bad state of conservation.

Many systems have been proposed to improve the bending capacity of beams (fig.2). Two main ways can be followed, according to whether it is convenient to introduce from the bottom or from the top of the beams the additional steel elements.

In cases a, b, c, d and e of fig. 2 the steel reinforcements are added in different forms, from the

576

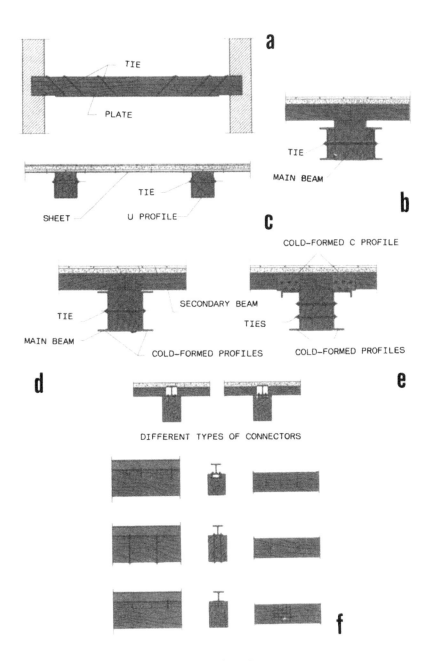

Figure 2. Consolidation of timber beams

577

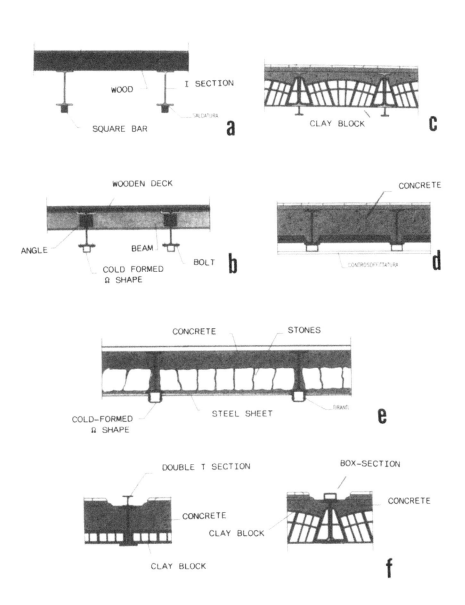

Figure 3. Consolidation of steel beams

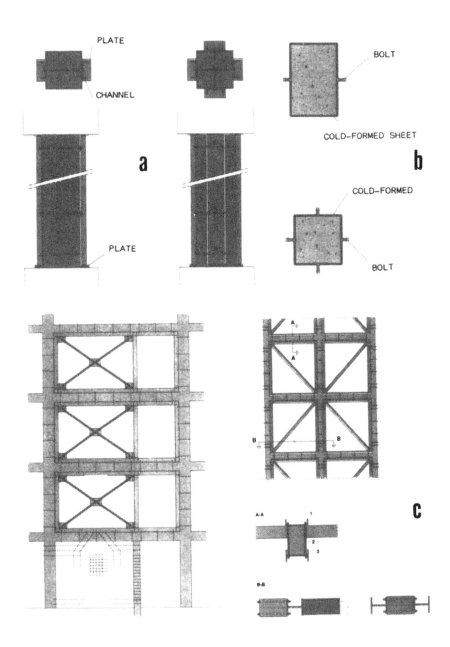

Figure 4. Consolidation of r.c. columns and frames

simple plate at the bottom (case a) to the hot-rolled double T sections (case b) or to the more advanced use of cold-formed profiles, which can be designed case by case according to the feature of the structure to be consolidated (cases c, d and e).

When the original shape of the beam must be conserved because it has particular interest from the historical point of view, it is necessary to follow the second way, namely by operating on the top of the beam (fig.2 f).

The final result corresponds to a composite wood-steel system, which considerably increases strength and rigidity of the original structure. In all cases, such cooperation between the new and the old material must be guaranteed by using appropriate connecting systems from simple ties to different types of studs.

CONSOLIDATION OF STEEL FLOORS

At the beginning of this century, the use of wooden beams for floor structures has been gradually substituted by the ancient I sections (fig.3). The steel beams were integrated firstly by a wooden deck (fig.3 a and b) and, later on, by clay blocks (fig.3 c), concrete (fig.3 d) or stones (fig.3 e). In all this cases the necessity to increase the section modulus can be very easily fulfilled by adding appropriate steel shapes to the lower flange, under form of square bars, T sections, cold-formed profiles with Ω shape or box sections.

When it is not allowed to operate from the bottom, the additional steel element can be connected to the top flange (fig.3 f).

The connection between the old and the new steel requires to pay particular attention to its state of conservation. In many cases welding is not allowed due to the umpure composition of the old material and the use of bolting is therefore advisable.

CONSOLIDATION OF R.C. STRUCTURES

The increase of the load carrying capacity of r.c. columns can be obtained by adding, in one or in two directions, a couple of hot-rolled steel sections (fig.4 a), which are connected together by means of appropriate ties. The use of cold-formed shapes (channels, angles and plates) allows to obtain a continuous resisting perimeter (fig.4 b), where the forcing effect is given by means of bolts. To improve the adherence between the concrete surface and the external steel sheet, the injection under pressure of gluing materials is recommended.

When the seismic upgrading of a r.c. skeleton is requested, we can use steelworks in order to provide shear resistant elements, which increase the capacity of the structure to resist new horizontal loading conditions. A reticular shear-wall is obtained as a composite structure, where the r.c. frame is integrated by cross-bracings made of steel profiles (fig.4 c). Each steel bracing is inserted in the mesh of the r.c. frame and the connection between both materials must be guaranteed by means of bolts or ties alongside the perimetral frame of the steel diagonals. Together with the advantage of an easy erection, this system gives the possibility to have openings for doors or windows, by using - if necessary - appropriate shapes for the diagonals or introducing only one diagonal per mesh.

The strengthening as well as the reparation of r.c. beam-to-column joints is usually fulfilled by means of angles and batten plates, which are located around the r.c. members (fig.5 a). The steelwork is welded and, possibly, glued to the concrete surface. The size of the additional elements depends upon the requested amount of increase in shear and bending capacity.

The increase of the inertia of r.c. beams can be obtained by integrating the r.c. section with steel

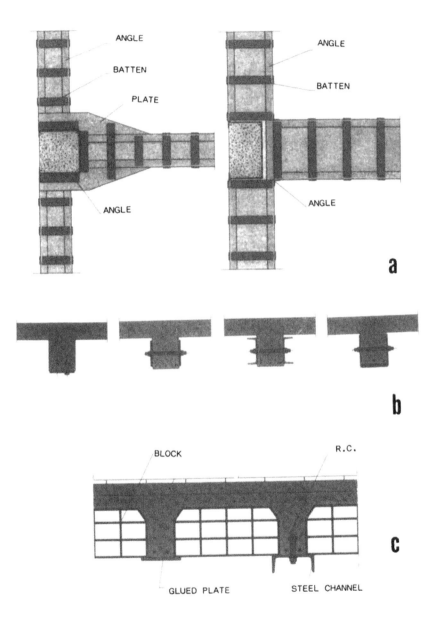

Figure 5. Consolidation of r.c. beams

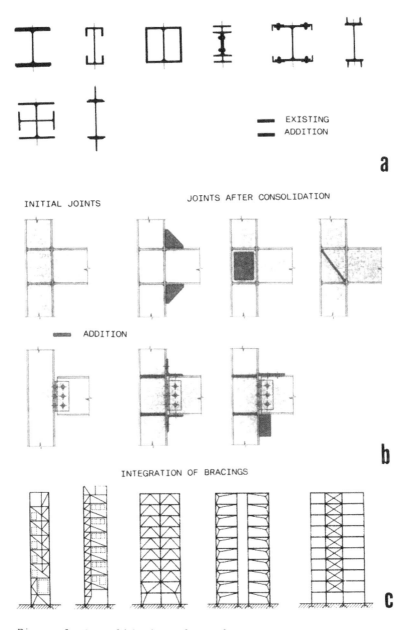

Figure 6. Consolidation of steel components

plates or profiles, which are connected to the concrete by means of bolts or ties and glue (fig.5 b). The same system can be used for strengthening the floor structures composed by r.c. and clay blocks (fig.5 c)

CONSOLIDATION OF NEW STEELWORKS

The use of steel in strengthening new steelworks is the simplest case. In fact, it is very easy to add integrative elements to the existing structure by means of the same connecting techniques (fig.6). The section modulus of double T profiles of beams or columns can be increased in different way by welding or bolting plates or/and shapes (fig.6 a), which transform the original section according to the new requested capacity.

It happens, for instance, when a structure must be upgraded to resist seismic actions, because of the recent inclusion of the building in a new seismic area. In this case, not only strength but also ductility must be improved, particularly in the joints.

Fig.6 b shows different systems for strengthening the two classical types of joints (rigid joint and pin-ended joint), by means of the introduction of stiffening elements. In the first case, the moment capacity is improved. In the second joint the integration is designed to introduce a given capacity to resist bending actions, which is pratically unexisting in the original joint.

The improvement in resisting horizontal actions can be easily obtained by increasing the cross-section of diagonal bracings in case of braced structures or introducing new bracings in case of moment resistant structures (fig.6 c).

REFERENCES

1. MAZZOLANI, F.M., L'acciaio e il consolidamento degli edifici. Acciaio 12/1985 & 1/1986.

2. MAZZOLANI, F.M., Refurbishment, Arbet-Tecom, 1990.

3. MAZZOLANI, F.M. and MANDARA, A., L'acciaio nel consolidamento. ASSA, 1991.

4. MAZZOLANI, F.M., Refurbishment and Extensions: The case for steel. Proceedings of the International Symposium of I.C.S.C., Luxembourg, may 1990.

5. MAZZOLANI, F.M. and MELE, G., Criteri di progetto per l'adeguamento antisismico dell'Ospedale Cardarelli di Napoli. Quaderni di Teoria e Tecnica delle Strutture, 1988.

6. MATACENA, G. and MAZZOLANI, F.M., Interventi di restauro con strutture di acciaio. Proceedings of the "Giornate Italiane della Costruzioni in Acciaio C.T.A." Palermo, october 1981.

7. MATACENA, G. and MAZZOLANI, F.M., Aspetti evolutivi della prevenzione sismica in Calabria: due esempi di restauro. L'industria delle Costruzioni, july/august 1986.

8. CANDELA, M., MANDARA, A., MAZZOLANI, F.M., L'uso dell'acciaio nel restauro degli edifici di culto in Campania. Proceedings of the "Giornate Italiane sulla Costruzione in Acciaio C.T.A." Capri, october 1989.

9. MAZZOLANI, F.M. and MANDARA, A., L'uso dell'acciaio negli interventi di restauro nell'edilizia monumentale nell'Italia Meridionale. Acciaio, 10/1989.

10. MAZZOLANI, F.M., and MANDARA, A., L'acciaio nel restauro. ASSA, 1992

DESIGN ASPECTS OF STEEL BUILDINGS WITH "NORAIL" BRIDGE CRANES

DORON GOUSSINSKY

Goussinsky Engineering & Manufacturing Co. Ltd,
P.O.Box 10033, 26110 Haifa, Israel

JOACHIM SCHEER & HARTMUT PASTERNAK

Institut für Stahlbau, Technische Universität Carolo-Wihelmina,
PF 3329, W-3300 Braunschweig, Germany

ABSTRACT

The NoRail crane concept inverts the overhead crane principle. Whereas conventional cranes have wheels travelling on rails, the NoRail crane has short rails attached to the bridge travelling on wheels mounted on the cantilevers of columns or suspension rods fixed to the roof truss. Benefits of this innovative design are both in cost savings on the steel structure of the building as well in material handling.

INTRODUCTION

Since 1873, overhead travelling cranes have been constructed in such a way that the travelling wheels, which are mounted on the endtrucks roll along craneways that are supported or suspended on the framework of the building.

Using the new "NoRail" design, short rails are mounted in the endtrucks of the crane. These rails run along a series of stationary wheels. The rails are designed to be somewhat longer than the maximum distance between three adjacent support points, so that the crane is always supported by at least two wheels on each side (Fig.1). Due to this design, the long conventional crane truck becomes superfluous. This concept influences also the design, construction and use of the building.

Figure 1. NoRail crane

CONSTRUCTION AND DRIVE SYSTEM OF NORAIL CRANES

The NoRail cranes (Fig.2) resemble the conventional bridge cranes in the construction of the main girder (the same rules and cata-logues can be used). The difference lies in the design and the arrangement of the endtrucks, the craneway, the support (travel-ling) wheels, as well as in the driving units. The endtrucks are rigidly fastened to the main girder. Rail elements, which trans-mit the load to the supporting wheels, are arranged at the bottom side of the endtrucks.

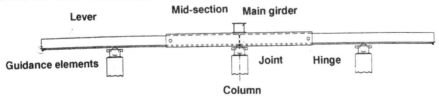

Figure 2. Construction of the NoRail crane

The horizontal guidance of the crane is accomplished by the guidance elements. With the direct arrangement of the rail elements below the endtrucks, the long crane track of the conventional design becomes unnecessary.

The endtruck consists of a mid-section and two levers. Both levers are connected to the mid section by means of hinges, and to each other, by means of a joint. This solution provides even distribution of vertical forces and assists in overcoming unequal settlement of the building foundations or support wheels.
Lever lengths (e + l in Fig. 3) of up to 1.4 times the distance between the supports is acceptable [1].

The mid section serves as support for the main girder(s) a well as for the transmission of the resulting force R (Fig. 3).

The hinges and levers carry the load G (G/R ≤ 0.5). At various positions of the crane on the supports, the mid-section can be loaded by an additional force D, pushing up from below onto the pressure pad at the joint. The hinges transmit the force from the mid-section onto the levers.

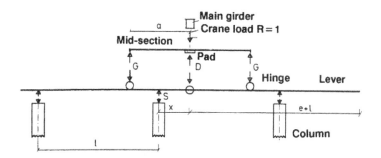

Figure 3. Static system of the endtruck

The joint ensures that the levers are vertically displaceable relative to the mid section. This motion is limited by the pressure pad mounted near the joint. The clearance between the mid section and the lever is about 20 to 30 mm.

The load case "Crane in the vincinity of the gable" (Fig. 4) may also be decisive for the dimensioning of the endtruck. Here the joint lies close to the last support point and the endtruck is supported on two wheels only. This situation is accounted for in

the calculations by a bigger force D. The load S on the support
point at the gable is bigger than that emerging from conventional
cranes mainly because the NoRail crane's hook comes nearer to the
gable.

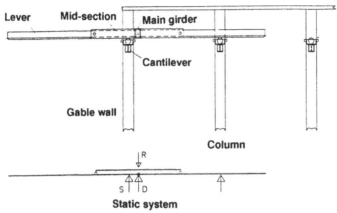

Figure 4. Crane in the vinicity of the gable

Various types of NoRail crane drives are available. The wire rope
drive (Fig. 5a) is the solution most frequently used. For cranes
with spans of up to 12 m, a wire rope drive system with a central
drive (mounted on the main girder) may be adopted. For cranes
with larger spans, a system with two seperate wire rope drives
(on the mid-section of each of the endtrucks) is used. The wire
ropes, usually of 8 to 12 mm diameter are fixed along the wall.
The sag of the wire ropes is taken up by hooks, situated along
the wire ropes at intervals of 6 to 8 m. Alternative solutions
are caterpillar drives (Fig. 5b), friction wheel drives and
infinite wire rope drives.

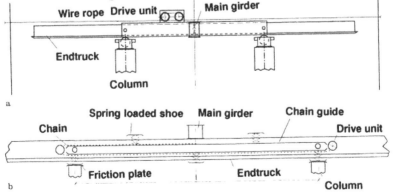

Figure 5. Types of drives. (a) Wire rope drive, (b) Caterpillar
         drive

LOADS

The vertical forces S and D as well as the distance a as a function of e (Fig.3), may be obtained (regardless of the type of crane drive) from Table 1:

Table 1. Vertical Force Values

e/l	S/R	D/R	a/l
0.10	0.90	0.64	0.50
0.20	0.80	0.36	0.50
0.30	0.70	0.16	0.50
0.40	0.60	0.04	0.50
0.42	0.58	0.03	0.50
0.47	0.54	0.08	0.54

The crane load R includes the lifting load, as well as the forces arising from the main girder, the weights of the crab, hoist and the lifting gear. The weight of the endtrucks has not yet been considered. The given values are valid only when the endtruck rests at least two supports. For the position when the endtrucks protrude from the gable wall, special considerations are called for. The factors for the dynamic effects of the load and the hoisting class and for the loads caused by travelling or self weight factors according to ISO 8686, Annex F [2] and DIN 15018 [3] may be regarded as an upper bound for NoRail cranes.

Vertical loads on the supports of NoRail cranes are much smaller than those of conventional cranes, even after considering the weight of the endtrucks and the track beams.

For the determination of the lateral forces, the approximation according to [2,3] for conventional cranes may be used. The ratio between the crane bridge span and the distance between the guide elements of the NoRail crane is significantly smaller than the same ratio on the conventional crane. For the NoRail crane this results in a reduced skew angle and lower lateral forces. Due to the larger mass of the NoRail crane its center of gravity and its center of rotation are located nearer to the center of the bridge, thus resulting in an additional decrease of the lateral forces. Although, the higher wheel loads of the NoRail crane contribute to some increase of the lateral forces, the total amounts of lateral forces in the NoRail design are significantly smaller than those of conventional cranes.

According to DIN 15018, the longitudinal forces resulting from starting and braking of the crane may be obtained from the smallest transmissable friction forces, which arise at the unloaded crane, adding 50% to account for dynamic influences. For NoRail cranes with wire rope drives, these forces are very small, since the crane is supported on wheels without brakes.

DESIGN AND CONSTRUCTION OF BUILDINGS

The design and construction of single-storey buildings with NoRail cranes doesn't basically differ from those with conventional cranes. These buildings may be designed on the basis of existing codes. Additional serviceability criteria are given in [1].

Due to the elimination of the craneway girders in buildings with NoRail cranes, a special solution for the transfer of buffer forces is necessary. Fig. 6 shows one of several possible solutions. The buffer load is transferred by a diagonal stiffener to a strong sheeting rail and to the side bracing and from there to the two adjacent column foundations.

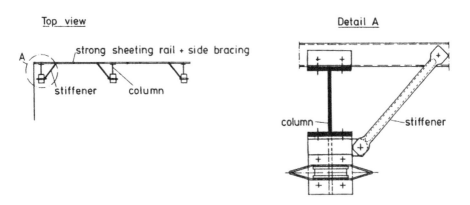

Figure 6. Absorbing buffer forces (suggested solution)

A similar problem exists for the free-standing support columns in an open yard. Fig. 7 presents two suggestions by which the buffer forces can be transferred to two adjacent foundations.

The support wheel and its connection to the column is shown in Fig. 8.

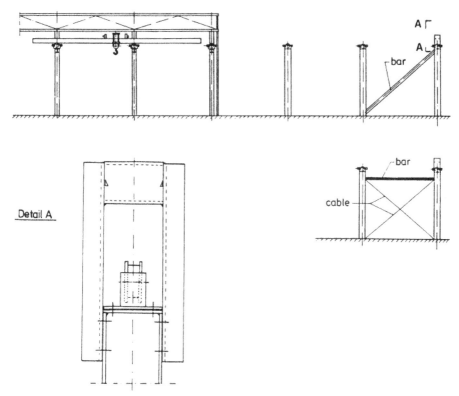

Figure 7. Absorbing buffer forces by free-standing columns

Figure 8. Support wheel

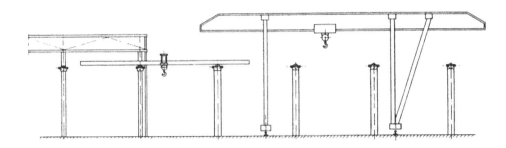

Figure 9. Combination of a NoRail bridge crane and a gantry crane

SPECIAL FEATURES

The mode of operation of NoRail cranes allows for several special features in the configuration of the building: Crane travel "tracks" that cross each other are feasible. This provides better flexibility of material flow. For instance, due to the absence of continuous crane rails the transfer of loads between neighbouring bays or buildings arranged at right angles to each other is simplified. Crossings between a bridge crane and a gantry crane is also possible (Fig. 9).

Better utilization of the heigth of the building is achieved, by providing gate openings in the longitudinal walls up to the heigth of the eaves.

Fig.10 shows an example of a hangar with a NoRail bridge crane. The support wheels are mounted on suspension rods fixed to the roof trusses.

By cantilevering the endtrucks out of the gable wall of the hall, the attained distance between the wall and the crane hook can be reduced as compared with conventional bridge cranes. Therefore, the length of the building can be utilized.

Until now, it has been customary to separate the hall from the service installations. The latter were usually located in side aisles or were separately arranged in other buildings. For buildings with NoRail cranes it is practical to integrate these service installations into the gable area of the hall and at the same time, utilize this area for the crane levers. The distance from the hook to the gable wall may amount to about 1.4 to 1.5 times the distance between the support wheels, if the endtrucks are not allowed at all to protrude from the gable wall.

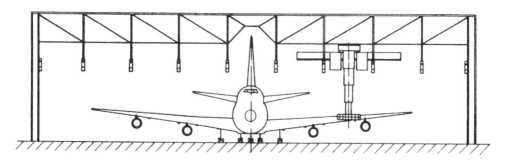

Figure 10. Hangar with NoRail crane

ECONOMY

Buildings of 21 x 90 m or 21 x 60 m with eave struts at a height of 10 m were chosen for comparative calculations. The required lifting height of the hook amounted to 7.5 m. A 20 t two-girder bridge crane (with lifting class H2 and loading group B3 according to DIN 15018) and alternatively a 6.3 t single-girder bridge crane (H2/B3), both of conventional design, were chosen for comparison with NoRail construction of equal lifting capacity. The frame distance was set to 5.0 m and 7.5 m. Hot rolled shapes with a yield stress of 240 N/mm² were used. Sheeting elements, intermediate supports of the gable walls and purlins were not considered.

Results are:
The building with the length of 90 m, with a frame distance of 7.5 m and the 20 t NoRail crane is 18% lighter and 17% less expensive than the conventional solution. For the 6.3 t crane, the respective values are 14% and 13%.

The building with the length of 60 m, the frame distance of 5.0 m and the 20 t NoRail crane is by 11% lighter and 10% less expensive than the conventional solution. For the 6.3 t crane, the respective values are 10% and 7%.

For very short halls, the application of NoRail cranes is not advisable if the steel consumption is decisive. In general, the economy of NoRail cranes increase
- with increasing load capacity of the crane and
- with increasing length of the building.

CONCLUSIONS

In many cases, NoRail cranes offer a suitable alternative for conventional cranes, if the cost of the building, the crane system and the simplified handling of materials are considered.

In certain cases it might also be possible to replace conventional cranes by NoRail cranes with higher loading capacity, without the need to reinforce the framework of the building.

Eight NoRail cranes have been operating since 1986 with complete reliability in Israel and since 1990 in Australia and Germany.

The NoRail crane design has been patented in Europe, Japan and U.S.A. and other countries.

REFERENCES

1. Goussinsky, D., Pasternak H. and Baars H., Hallen mit schienenlosen Brückenkranen - Berechnung, Konstruktion und Wirtschaftlichkeit (Single-storey buildings with NoRail bridge cranes). Bauingenieur 65(1990)247-253
2. ISO 8686-1 (draft), Cranes - Design principles for loads and load combinations. General, 1989
3. DIN 15018/1: Krane; Grundsätze für Stahltragwerke, Berechnung (Cranes; Steel Structures, Design). Berlin: Beuth 1984

On the Development of a Computer Simulator for Tests of Steel Structures

R. Spangemacher, G. Sedlacek
Institute of Steel Construction - RWTH Aachen
Mies-van-der-Rohe Str. 1; 5100 Aachen; Germany

1.   Starting point - Laboratory tests

To engineers in practice and students the loading and deformation behaviour of steel structures can be best demonstrated by laboratory tests. In these tests loading is increased incrementally in a load or deformation controlled way to obtain characteristic load deformation curves which are used to determine serviceability limits or ultimate strength, [1]. The test results may also be used for the development or justification of design rules for steel structures. Full scale member tests, however, are expensive and need a lot of time for the planning, execution and documentation of the tests. Therefore a supplementary alternative procedure in form of numerical simulations is of great concern.

2. Features of numerical simulations

2.1   What are the advantages of numerical simulations ?

The efficiency of member tests can be increased by numerical simulations during the preparation phase of the testing, the execution phase and in the evaluation and documentation phase of the testing.

By numerical simulations during the preparation phase of the testing the expected behaviour of members can be predicted and analysed in view of assumptions for the boundary conditions of the test member, the influence of the geometries and material properties. This helps to design optimum test members and to identify apriori the weak points of the members. This helps to find the best positioning of strain gages and transducers. Such investigations can also be used to find an optimum test matrix and to reduce the number of member tests to a minimum.

During the execution phase the predicted theoretical results can be compared with test results. This allows to verify whether the testing procedure is working well or whether there are any surprising unexpected effects. In consequence either the testing procedure can be modified or the simulation model can be improved.

In the evaluation and documentation phase of testing the influence of parameter variation can be investigated by the simulation technique without carrying out further tests. That helps to develop simplified engineering formulae. Any questions that may arise from test results (e.g. amount of scatter etc.) can also lead to further studies by simulations.

2.2   Tasks and Aims of a Simulator

In the following the software tool PROFILE-FEM-3D, an electronic numerical simulator for steel member tests, is presented. This software allows to perform realistic member tests for research and education and to substitute partially physical member tests, except pilot tests. By the use of graphical presentation methods this software can give the same or even better insight into the deformation- , straining- or stress-behaviour of the test pieces than physical tests can do.

In the development of PROFILE-FEM-3D attention was paid to the following features:
-       visual presentation of all investigated reactions of complex space structures under loading, which are calculated by a background finite element program.
-       if wanted a comfortable presentation of the time or load dependant behaviour by visual animation
-       applicability to students that wish to learn the background of the design rules governing the serviceability and ultimate limit state design.

2.3 How to work with the simulator ?

PROFILE-FEM-3D represents an integrated package of the following components:

PROFILE-PRE	-	for FEM - Preprocessing
PROFILE-CALC	-	for FEM - Calculations
PROFILE-POST	-	for FEM - Postprocessing

By PROFILE-PRE the full testing situation can be put into the computer. It comprises comfortable interactive and menu-controlled procedures to establish the input file for the net-discretisation, the geometrical and material description of the elements, the determination of the boundary conditions and the description of the numerical simulation procedure (load or deformation controlled) for the finite element background program.

The preprocessor takes advantage of the fact that steel members are mainly build up by plates, that can be identified by their surface. On this surface the program offers first proposals for the net-discretisation, which are sufficient for most of the applications.

In an additional step the net-discretisation may be improved by using additional net-refinements, by condensation, adjustments of some net-points and by providing cutouts or other details. The program also checks plausibility conditions for the mesh at each step of the working procedure.

The geometrical and material properties of each element are then defined using a databank containing a number of typical stress-strain curves for different steel grades. Finally the description of the numerical simulation procedure, i.e. load-controlled or deformation controlled procedure, is specified.

As result of the preprocessing a full input-data file is generated which can be used directly for the analysis. Fig. 1 illustrated how the different aforementioned steps are carried out.

By PROFILE-CALC The input file from the PROFILE-PRE program can be directly used by any qualified finite element program which the user chooses as a background calculator.

By PROFILE-POST the evaluation and visual presentation of the results is possible. The "test specimen" can be moved, rotated and zoomed in any position on the monitor. This is reached by using translation and rotation options which are controlled by slide buttoms for each degree of movement which are controlled by a computer mouse. The movements are such that they look like real time movements.

The results of the mechanical behaviour, as stress distributions (σ_x, σ_y, τ_{xy}), strain distributions (ϵ_x, ϵ_y, γ_{xy}), filtered or manipulated values like plastic strain distribution ($\epsilon_{pl,x}$, $\epsilon_{pl,y}$, $\gamma_{pl,xy}$), distribution of main stresses (σ_I, σ_{II}) or equivalent stresses ($\sigma_{equiv.}$) are presented as coloured patterns on the surface of the member. The full range of the rainbow's colors can be used, e.g. deep red for maximum tension stresses and deep blue for the maximum compression stresses.

As an alternative it is also possible to get the numerical results for particular finite elements. The values will be presented in a window, which on request gives the data for each time increment.

By applying the animation technique the time history of the loading and the reaction can be demonstrated similarly to a real time test.

3. Demonstration of the use of the numerical simulator PROFILE-FEM-3D for the case of 3- point bending tests

3.1 Introduction - Ultimate limit state design

In the following the use of this program is demonstrated for the case of 3-point-bending tests and for the derivation of a formula for the rotation capacity of rolled or welded beams form these tests, [2].

The rotation capacity is a property of plastic hinges that are used as a model for plastic zones in steel structures in the plastic hinge theory.

The values of the rotation capacity may be used for a safety check in which the rotation requirement (e.g. the plastic angle in the hinge at the ultimate limit state) is compared with the rotation capacity (i.g. the maximum rotation angle for which the plastic resistance $M_{pl.Rd}$ is maintained by the steel member).

The rotation requirements may be determined with the plastic hinge model [3,4], fig 2.

The rotation capacity is derived from a comparison of the actual moment rotation curve of a beam with a plastic hinge obtained in tests with the theoretical moment rotation curve which is assumed in the plastic hinge theory, fig 2.

Mostly the actual rotation angle ϑ is substituted by the related value R [5,6,7], fig 3.

$$R_{req} = \frac{\vartheta_{req}}{\varphi_{pl}}$$

$$R_{cap} = \frac{\varphi_{rot} - \varphi_{pl}}{\varphi_{pl}} = \frac{\vartheta_{cap}}{\varphi_{pl}}$$

where φ_{pl} is related to a "substituted beam", see fig 3.

To cope for the scatter in the determination of the rotation capacity, the characteristic values ϑ_{cap} and R_{cap} may be corrected by the partial safety factor γ_M^*, see part 3.5.

$$\vartheta_{req} \leq \frac{\vartheta_{cap}}{\gamma_M^*} \qquad\qquad R_{req} \leq \frac{R_{cap}}{\gamma_M^*}$$

3.2 Determination of the rotation capacity by tests and numerical simulations

The rotation capacity of I-profiles determined by 3-point-bending tests is mainly influenced by the following parameters:

- geometrical properties of the section (b; t; h; w; b/t; h/w)
- the length of the span governing the moment gradient (L=½·l; L/b)
- the steel grade of the material (f_u; f_u/f_y)

In analyzing test documentations on 3- point bending tests from literature the following failure modes can be identified that influence the shape of the moment rotation curve:

- shear failure in particular for short beams
- flange buckling due to bending with symmetrical as well as antimetrical flange buckling modes.
- web buckling

A comparison of the results of two numerical simulations with test results of 3-point-bending tests is shown in fig. 4 and 5. Figure 4 shows the typical deformation of a beam where flange buckling due to bending moments is governing the rotation behaviour, and figure 5 shows a comparison where instability of the web due to extreme shear forces determines the failure mechanism. The figures demonstrate that it is possible to get good results in simulating rotation tests by using FEM-methods.

3.3 Parameterstudy

In the evaluation and documentation phase of testing a parameter-study was carried out with a reference beam called TBC350, for which each single parameter could be varied, the other parameters being kept constant, fig. 6.

The variations comprised the geometrical properties:

- width and thickness of flange and web
- span of the beam

and the material properties:

- stress-strain curve according to different steel grades.

For material properties typical true stress-strain curves of Steel St 37, StE 460 and Steel StE 690, according to [8], were used.

In the following some main results of this parameter study are presented. Fig. 7a shows the moment-rotation-curves for different steel grades. With higher steel grades the load carrying capacity increases however the rotation capacity is decreased.

This becomes clearer when the results are plotted in the form of the non-dimensional moment-rotation curve, fig. 7b. For the reference beam TBC350 the value of the rotation capacity decreases from R = 8,6 for a beam of steel grade St 37 to a value of R = 1,6 for a beam of high strength steel grade StE 690.

The effect of the variation of the b/t-ratio are given in, fig. 7a and b. We can see, that the rotation capacity decreases when the b/t-ratio increases.

3.4 Formula for the calculation of the rotation capacity of I-profiles

According to the results of the parameter studies the influence of the material grade can be described by a hyperbolic behaviour, whereas all the other parameters have approximately a linear influence .

The formula for the rotation capacity was therefore structured such, that starting from the reference beam these effects were taken into account, fig. 9.
It reads:

$$R_i = R_0 \left(\frac{f_u}{f_y} \right) + \Delta R \left(\frac{b}{t}, \frac{L}{b}, t, K_\vartheta \right)$$

where - f_u/f_y represents the material grade
    - t   the thickness of the flange
    - b/t the slenderness ratio of the flange
    - L/b the moment gradient of the beam
    - K_ϑ and the stiffness of the web

The best fit formula to calculate the rotation capacity of I-profiles is shown in fig. 10.

A graphical presentation of the rotation capacity of a profile HEB 220 is given in fig. 11 a and b. The values of the rotation capacity R and ϑ, which can be reached by profiles with different steel grade, are plotted against the span length of the beam.

The accuracy of the formula was checked by FEM-calculations the results of which are also plotted in, fig 11a and b .

3.5 Calibration of the formula with test results

The calibration was carried out according to Annex Z of Eurocode 3 [1]. The comparison of the experimental results R_{test} with the results from the formula $R_{formula}$ is given in fig. 12, [5,6,9,10,11,12].

From the statistical evaluation according to Annex Z a mean value correction $b_m = 1,06$ and an error term $S_\delta = 0,185$ could be found. This leads to a value $\gamma_M^* = 1.50$ to be applied to get design values.

4. Summary

Design rules in steel structures are mainly based on the results of member tests. Such full scale member tests are expensive and there is a need to have alternatives.

Education in the design of steel structures should include an insight into the physical behaviour of steel members at the limit states to understand the background of the rules. It would be ideal to involve students in the performance of member tests to give them this insight.

That is the background on which a computer based simulation program for the simulation of tests of members made in steel was developed. This program uses modern preprocessing, qualified background FEM- programs and postprocessing tools to enable a student to define the geometric and material related properties of the member to be tested, to calculate the behaviour of the member along the loading path and to document the stress, strain and deformation states in each loading phase. This program can also used for research purposes.

The use of this program is demonstrated for the case of 3- point bending tests for the determination of a formula for the rotation capacities of rolled and welded sections.

5. Literature

[1]   Eurocode 3; Design of Steel Structures; Part 1: General Rules and Rules for Buildings, 1991; Vol. 1 Chapter 1 to 9; Vol. 2 Annexes

[2]   Spangemacher,R.: Zum Rotationsnachweis von Stahlkonstruktionen, die nach dem Traglastverfahren berechnet werden; Dissertation RWTH Aachen 1992.

[3]   Baker J.F.; Horne, M.R.: New methods in the analysis and design of structures in the plastic range. British Welding Journal 1 (1954), page 307.

[4]   Roik, K.-H.; Lindner J.: Einführung in die Berechnung nach dem Traglastverfahren, Stahlbau Verlags-GmbH Köln 1972

[5]   Lukey, A.F.; Adams P.F.: Rotation Capacity of Wide-Flange-Beams under moment Gradient; ASCE Journal of the Structural Devision, Vol. 95, No. ST 6, pp. 1173-1188, Paper 6599; June 1969

[6]   Roik K.-H.; Kuhlmann U.: Teil 1; Rechnerische Ermittlung der Rotationskapazität biegebeanspruchter I-Profile; Der Stahlbau 56 (1987), S. 312-327.
        Teil 2: Experimentelle Ermittlung der Rotationskapazität biegebeanspruchter I-Profile; Der Stahlbau 56 (1987), S. 353-358.

[7]   Kemp,A. R.: Slenderness limits normal to the plane bending for beam- coloumns in plastic design
        Journal of constructional steel research, vol 4 (1984), p.135-150.

[8]   Dahl,W.; Hesse,W.; Krabiell,A.; Zur Verfestigung von Stahl und dessen Einfluß auf die Kennwerte des Zugversuchs; Stahl und Eisen 103 (1983), Heft 2, S. 87-90

[9]   Sedlacek,G.; Spangemacher,R.; Dahl, W.; Hubo,R.; Langenberg,P.: Projekt P 169 Untersuchung der Auswirkungen unterschiedlicher Streckgrenzenverhältnisse auf das Rotations- und Bruchverhalten von I-Trägern; Studiengesellschaft Stahlanwendung e.V. - Forschung für die Praxis; 1992

[10]   Sedlacek,G.; Spangemacher,R.; Dahl,W.; Langenberg,P.: Elastisch plastisches Verhalten von Stahlkonstruktionen; Anforderungen und Werkstoffkennwerte; ECSC-F6 Projekt 7210-SA/113; Final report 1992.

[11]   Petersen Chr.; Rotationskapazität in Fließgelenken als Grenzzustandsgröße; DAST - Deutscher Ausschuß für Stahlbau; Bericht aus Forschung, Entwicklung und Normung 17/1990; S. 43-54

[12]   Defourny,J.; D' Haeyer,R.; Elasto-Plastic behaviour of steel structures; ECSC-F6 Project 7210-SA/204 Final report autumn 1992

Description of
the Surface

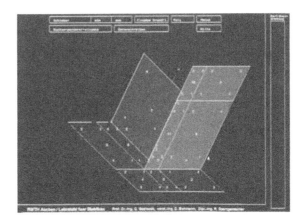

Automatic net
Generation

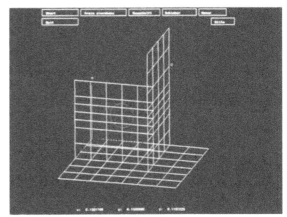

Description of
the Elements
- Material
- thickness
- etc.

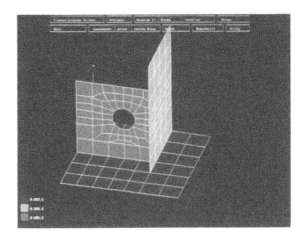

Fig. 1: FEM - Input - Generation by PROFILE - PRE

a) **Requirements by plastic hinge theory**

b) **Rotation capacity from 3- point bending tests**

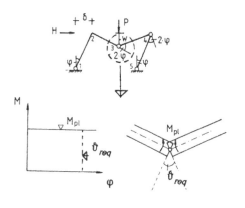
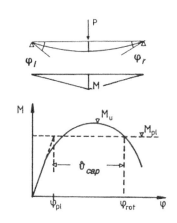

Fig. 2: Rotation requirement and rotation capacity in terms of actual plastic angles

a) **φ from a substituted Single Span Beam**

b) **Resistance from 3- point bending**

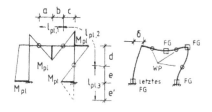
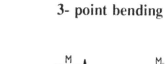
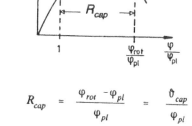

$$\varphi_{pl} = \frac{1}{2} \cdot \frac{M_{pl}}{EI} \cdot l_{pl}$$

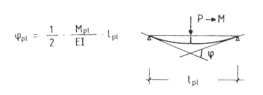

$$R_{cap} = \frac{\varphi_{rot} - \varphi_{pl}}{\varphi_{pl}} = \frac{\vartheta_{cap}}{\varphi_{pl}}$$

Fig. 3: Related Rotation capacity R of "plastic hinges"

Test

FEM

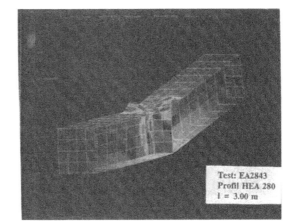

Comparison

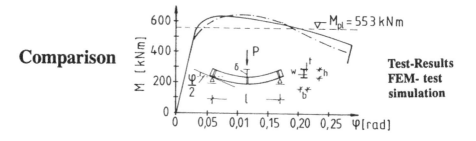

Fig. 4: Test and FEM- results: antimetric local flange buckling bending on the strong axis

Test

FEM

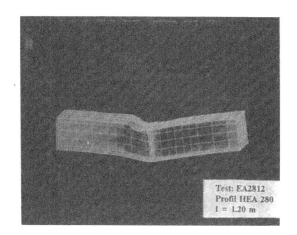

Comparison

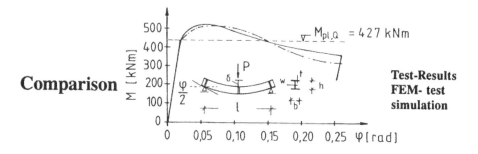

Fig. 5: Test and FEM- results: shear failure

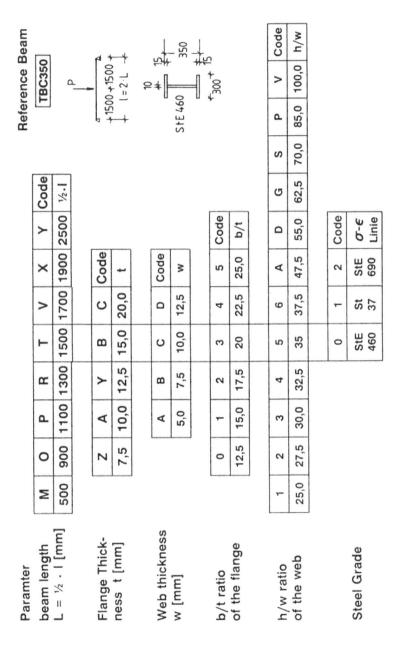

Fig. 6: Parametermatrix with the reference beam TBC350

a) **Rotation Capacity defined by ϑ**

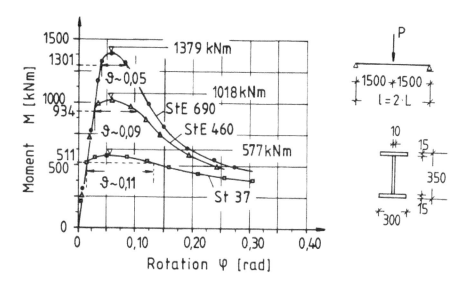

b) **Rotation Capacity defined by R**

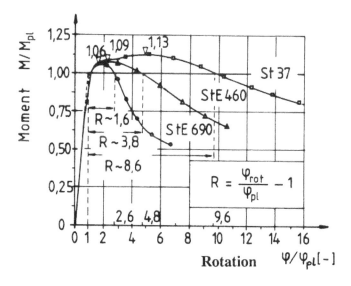

Fig. 7: FEM - Results with the reference beam TBC 35[ ]
Same geometrie, but different steel grades

a) Moment-Rotation-curves

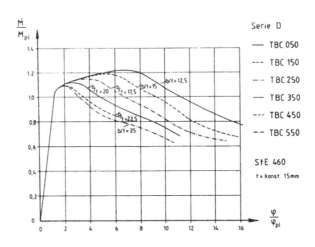

b) Rotation Capacity

expressed by ϑ

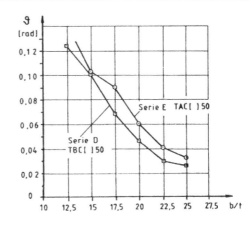

expressed by R

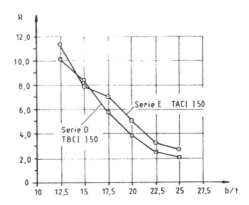

Fig. 8: FEM - Results: Variation of the b/t- ratios
Serie D: TAC□50 t = 15 mm
Serie E: TAC□50 t = 10 mm

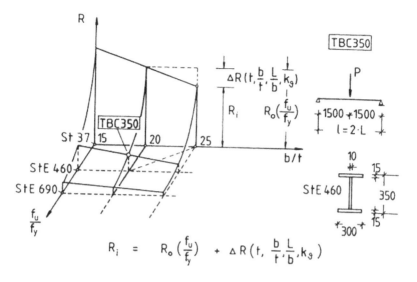

$$R_i = R_o\left(\frac{f_u}{f_y}\right) + \Delta R\left(t, \frac{b}{t}, \frac{L}{b}, k_\vartheta\right)$$

Fig. 9: Idea of a Rotation Capacity Formula

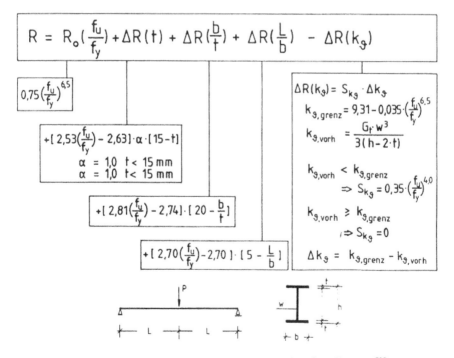

Fig. 10: Formula for the rotation capacity for I- profiles due to bending on the strong axis

a)   Rotation Capacity defined by R

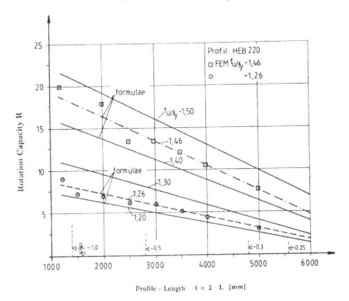

b)   Rotation Capacity defined by $\vartheta \quad = R \cdot \varphi_{pl}$

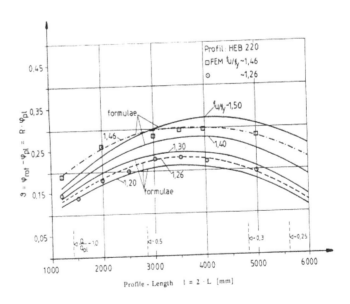

Fig. 11:   Rotation Capacity for any I- profile
for example profile HE 220 B

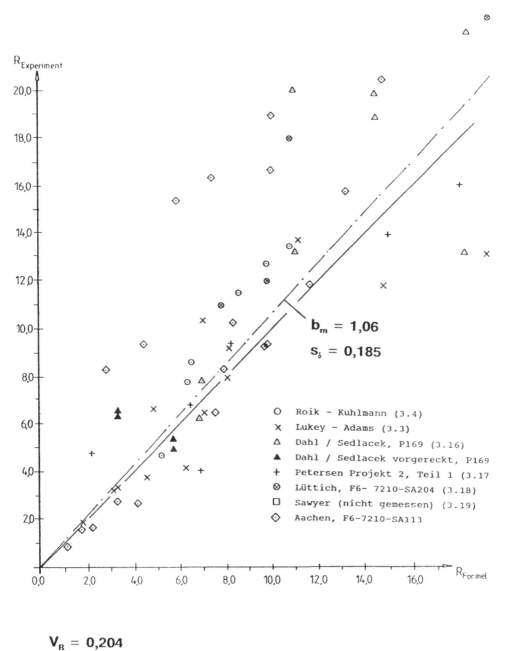

V_R = 0,204 → $V_R = 0,204$

$\gamma_M^{\cdot} = 1,50$

Fig. 12: **Determination of design values**

IN-PLANE STABILITY INVESTIGATIONS FOR PRACTICAL PITCHED-ROOF STEEL FRAMES: ARE THEY REALLY NECESSARY?

HANS SCHOLZ
Department of Civil Engineering
University of the Witwatersrand
Johannesburg, South Africa

ABSTRACT

The results of a study are presented comparing in-plane stability aspects, ductility demand and sway limitations of pitched-roof steel frames designed by simple plastic analysis. The object of the study has been to identify domains governed by stability, ductility or sway. It has been concluded that for practical frameworks in-plane stability is never critical. Sway limits, ductility requirements and practical slenderness limits govern the design.

INTRODUCTION

Pitched-roof frames are a common solution for single-storey industrial structures. It has been demonstrated that plastic analysis and design of pitched-roof frames are most economical. When adopting simple plastic analysis premature failure due to various other aspects must be avoided. Control of such failure aspects should be simple, otherwise the benefit of the simple plastic analysis is lost-one might just as well adopt a more rigorous frame analysis. The object of the presented paper has been to focuss on overall frame and member stability, frame ductility and sway displacement requirements. The depth requirements from Ref 1, the geometric requirements from Ref 2 and the allowable sway deflection limits of design specifications are compared. Domains where stability aspects, ductility or deflection requirements dominate are identified.

FRAME AND MEMBER STABILITY

With reference to previous work by the Author, the simple depth requirements of Fig 1 have been evolved, controlling ultimate

613

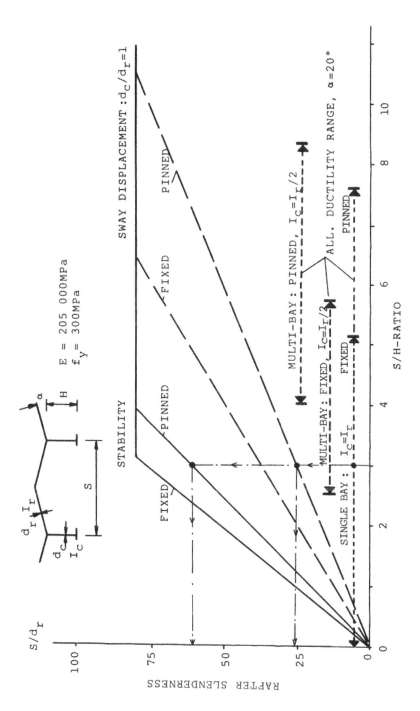

FIGURE 1. STABILITY, SWAY AND DUCTILITY DOMAINS OF PITCHED-ROOF FRAMES

limit state losses due to overall frame and member buckling.
Fig 1 appears at the end of this paper. A distiction is made
between pinned and fixed bases. The expression given in Fig 1
identifies domains of acceptable and unacceptable depth parameters for stability reasons. One can meet a stability problem by
increasing the depth of members. This effectively translates
into a higher design load level. A relevant proposal appears in
Ref 1.

FRAME DUCTILITY

First plastic hinges in pitched-roof sway frames undergo plastic rotations, which must not be excessive to avoid premature
local buckling failure. A study with respect to ductility in
practical pitched-roof frames has been published in Ref 2. The
results pertinent to the present paper are superimposed in
Fig 1. Domains have been demarcated which are acceptable for
reasons of ductility. Such domains could, for instance, be increased by increasing the design load level. A proposal in this
regard is suggested in Ref 2. Acceptable ductility regions show
as a range of allowable frame span to column height ratio.
The lower ductility boundaries are generally associated with
sway failure modes, whereas the upper ductility limits shown in
Fig 1 are the result of symmetrical total frame or partial
frame collapse modes. There exist more critical modes of failure, however, it is assumed that due to good practical engineering design such modes are avoided in the first place. Full
details of the assumptions made to analyse ductility are given
in Ref 2.

FRAME SWAY

Sway at eaves level of pitched-roof frames must be controlled
to prevent structural functional damage at the serviceability
limit state. Such sway limitations may be responsible that
frames designed by first-order, rigid-plastic hinge theory can
not be executed with the cross-sections derived from plastic
collapse. A simple approximate assessment of the eaves sway is
here undertaken in order to examine whether the member depth
derived from stability considerations or those satisfying sway
limitations are likely to govern. Approximately, the sway, Δ,
at eaves level of a typical external column in a frame may be
expressed as a function of the curvature in such column. The
curvature in a column, in turn, can be expressed by the column
moment, M, and the sectional parameters of such column. Such
approach was already used in Refs 2 and 3. The following
equation is obtained:

$$\Delta = \frac{M}{EI} \frac{H^2}{K} \tag{1}$$

with K=3 and K=4.5 for a pinned and fixed column respectively.

The term EI is the column stiffness and H is the column height. Assuming E=205 000MPa, f_y=300MPa, that at service M equals 0.8 times the yield moment and that the allowable eaves sway of the bare structure must not exceed Δ/H=1/150; gives

$$\frac{S}{d_r} = C \frac{d_c}{d_r} \frac{S}{H} \frac{300}{f_y} \qquad (2)$$

where S is the frame span, d_c is the column depth, d_r is the rafter depth, f_y is the yield strength of the steel and C=8.4 for pinned column bases and C=12.5 for fixed column bases. A typical upper boundary range of Eq 2 is shown in Fig 1 for comparison purposes. It can be seen that sway is always more critical than the stability aspects; except for larger span to height ratios where practical slenderness limits prevail.

CONCLUSION

The object of this comparative investigation has been to identify domains for regular practical pitched-roof steel frames in which in-plane stability, ductility or sway limitations at service govern the design of such frames. The following findings were obtained:

(a) In plastic design, in-plane stability aspects are usually less critical than sway deflection demands on the choice of the sections; except for larger span to height ratios. Hence, the full benefits of plastic design may not be exploited in many practical frames.

(b) Distinct regions are indicated where ductility is a problem. A high ductility demand is indicated for low span to height ratios of multi-bay frameworks. However, the need to provide for a higher than necessary collapse load, to cope with sway displacements, will beneficially reduce the ductility problem in these regions.

A limited rigorous parametric study has so far confirmed point (a) above. A wider study is in progress.

REFERENCES

1. Scholz, H, Stability control of pitched-roof frames by allowable member depth. J. Con. Steel Res., 1991, 19, pp. 253-267

2. Scholz, H, Approximate assessment of ductility in pitched-roof steel frames. In Proc. PSSC 1992, October, Tokyo, to be publ.

3. Scholz, H, A contribution to the assessment of ductility in plastically-designed sway frames. J. Con. Steel Res., to be published

STRUCTURAL TUBE SECTIONS
IN CLEAN DESIGN OF MANUFACTURING FACILITIES

Kasi V. Bendapudi
Manager, Structural Engineering,
Lockwood Greene Engineers Inc.,
4201 Spring Valley Road, Dallas, Texas 75244 U.S.A

W.M.Tepfenhart
Member of the Technical Staff,
AT&T Bell Labs,
480 Redhill Rd., Middletown, NJ 07748 U.S.A

ABSTRACT

Structural tube sections offer unique advantages for clean design. Effective utilization of connection techniques for tube sections can result in optimal design, ease of erection, and shorter construction schedules. Such a technique is described in this paper.

INTRODUCTION

Manufacturing facilities such as food processing and pharmaceutical plants require a high degree of cleanliness. Square and rectangular steel shapes for framing members offer a number of sanitation advantages [1]. Inherent properties of tube sections (torsional stiffness, higher yield strength (Fy), section moduli about both principal axes, and their ability to restrain end rotations) offer significant advantages for structural design. In order to achieve optimal benefits, innovative design methods leveraging these properties need to be utilized.

Structural Design Approach

Over-Design, Under-Design phenomenon: Usual framing design consists of simple beams with non-moment resisting column bases. Columns are designed for kl/r=1.0 (sidesway prevented). In general, tube to tube connections offer a high degree of restraint at beam ends against rotations. The beams will be over-designed if treated as simple beams. On the

other hand, columns experience sidesway and rotation as a
result of under-design with kl/r < 2.0. This is more pronounced when the structure is over-loaded. Unless the frame
is fully braced with due regard to story stiffness [2][3], the
failure mode will be in the form of Sidesway Buckling. Use of
bracing members as a corrective measure is not prudent in any
clean design. Conversely, the columns can be over-designed
and the beams under- designed. In this case, the failure mode
will start in the beams which is a gradual process and is
preferable.

Proposed Structural Design Concept: Under this concept
along main frames, the girders will be designed for fixed end
moments. Type 1 construction (rigid or continuous framing)
[2]. The columns will be designed for the appropriate moments
(see Fig.1). The interior beams will be designed as continuous beams over knife edge supports for gravity loads only
(see Fig.2). The interior beams are not designed to resist
any lateral loads, and thus do not contribute towards any
lateral stability of the frame. Bending moments in the columns
due to beam gravity load do not significantly affect over all
frame stability in the elastic range [3][4]. It will be more
convenient to meet story stiffness requirements under this
design method, because each column can be independently
restrained and need not brace any straining columns. The
moment resisting connections between the columns and girders
require special attention. It is assumed that all sections
are compact, the aspect ratio is < 6 to avoid lateral torsional buckling considerations, and the beams are loaded in
the plane of their minor axes.

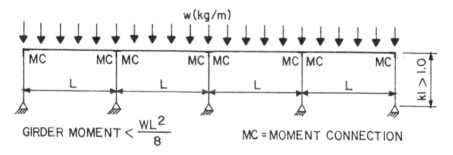

Figure 1. Structural Model for Main Frames

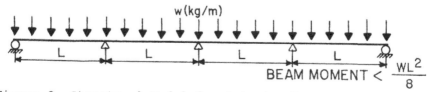

Figure 2. Structural Model for Interior Beams.

618

Connections: Pocket type seated connections (see Fig.3) can be effectively used to resist moments. Seats will be fabricated and welded in the shop to the columns. Erection of beams is relatively simpler and the field welds are designed for the calculated moment. It is assumed that the top and bottom walls plus 1/6 of the depth of the girder resist most of the forces resulting from calculated moment (see Fig. 4a). The remaining cross section is assumed to resist shear force. Universal bending stress and the assumed stress distribution at fixed ends are shown in Fig. 4b and 4c respectively.

The force required to develop the weld is equal to $MD/2I$ and the force required to develop the shear capacity is equal to $wL/2$. For example, at the interior beams for a 3 span condition with all spans equally loaded with uniform load w, the force required for welds at the top and bottom walls will be equal to $(wL)LD/20I$.

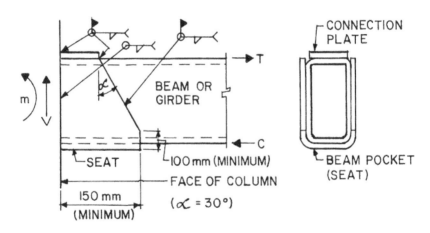

Figure 3. Moment Connection Detail

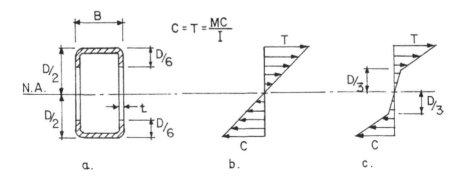

Figure 4. Assumed Stress Distribution at Moment Connection

DISCUSSION

The authors' assumption of the bending stress distribution is reasonable due to the mass distribution in the plane of bending. Determination of exact stress distribution requires analytical studies and laboratory tests. In the context of this design concept, the exact distribution of bending stress at the connection would not impair the strength and stability of the structure. If the weld is continuous at the joint and is designed for combined shear and tensile forces, such an assumption is immaterial. Furthermore, for sanitation purposes it is advantageous to provide continuous welds at these joints.

Pocket type seated connection design is compatible with the assumptions used in the analysis. Experience has shown that connections for TS shapes are cumbersome, and the design engineers are not coordinating adequately with fabricators and erectors to take full advantage of their expertise in order to simplify and optimize the connections. Changing from the traditional simple beam concept to this proposed design approach would result in material savings. Reducing erection costs by utilizing pocket type seated connections is an important consideration as well.

Modifications at the connections would allow structural tube framing design more economical, reduce erection costs, and result in faster construction schedules. The structural design concept illustrated in this paper maximizes the benefits of the unique structural characteristics of the TS shapes.

REFERENCES

1. Imholte, T.J., Engineering for Food Safety and Sanitation, Technical Institute of Food Safety, Minnesota, 1984, pp. 30-44.

2. American Institute of Steel Construction, Inc., Manual of Steel Construction, Allowable Stress Design, In Specification for Structural Steel Buildings, AISC, Chicago, 1989, pp. 5-24, 5-35, 2-308.

3. Yura, J.A., The Effective Length of Columns in Unbraced Frames. AISC Engineering Journal., April, 1971, pp.37-42.

4. Lu, L. W., Stability of Frames Under Primary Bending Moments., Journal of the Structural Division, ASCE, Vol. 89, No. ST 3, June, 1963.

DESIGN OF FRAMES ACCORDING TO EUROCODE 3

T. ABECASIS; D. CAMOTIM; A. REIS
Invited Lecturer , Associate Professor and Professor
Department of Civil Engineering, Instituto Superior Técnico,
Technical University of Lisbon , Av. Rovisco Pais, 1096 Lisboa Codex, Portugal

ABSTRACT

In this paper a systematic approach for the use of the provisions of Eurocode 3 (Part 1) concerning the design of frames is proposed. A sequential diagram is presented which contains the main steps, options and decisions associated with the corresponding safety checking or determination of an ultimate load combination. From the several aspects that must be considered in the establishment of such diagram attention is paid exclusively to the ones involved in ultimate limit states. Finally, some suggestions are made with the aim of clarifying the choices and/or lowering the amount of work required to design frames according to Eurocode 3.

INTRODUCTION

The coming into effect of Eurocode 3 (Part 1:1 General Rules and Rules for Buildings) makes it very important for designers to be familiar with the philosophy, principles and rules contained in this Design Code. Particularly in what concerns the design of building frames the use of Eurocode 3 involves a process of sequential analyses, options and decisions which is not clearly identified.

In this paper a list of the main steps to be followed in the safety checking of building frames is presented. Each step is associated with a number of possible options which are discussed and illustrated by means of a sequential diagram.

One final remark to mention that no reference is made either to serviceability limit states or to ultimate limit states induced by fatigue or involving joints and connections.

CLASSIFICATION OF THE FRAME

Frames can be classified according to (i) their sway-resistance, to (ii) the material and geometrical characteristics of the cross-sections of their members and to (iii) the type of their connections.

Sway-Resistance

A frame is classified as a *braced frame* if the resistance is supplied by a bracing system which is sufficiently stiff to carry all the horizontal loads (to be sufficiently stiff the bracing system must reduce the horizontal displacements of the frame by, at least, 80%) and as *unbraced frame* in all the other cases.

A frame is classified as a *non-sway frame* if the response of the frame to in-plane horizontal forces is sufficiently stiff for it to be acceptable to neglect any additional internal forces or moments arising from horizontal displacements of its nodes, and as a *sway-frame* if the above condition is not met. In the case of a sway-frame the effects of the horizontal displacements of the nodes must be taken in account.

A braced frame is treated as a non-sway frame andthe classification of a frame as sway or non-sway depends on the load case under consideration. For a given load case the frame may be treated as a non-sway frame if it satisfies the criterion involving the critical load ratio $V_{Sd}/V_{cr} \leq 0.1$, where V_{Sd} is the design value of the total vertical load and V_{cr} is its elastic critical value for failure in a sway mode.

Classification of the Cross-Sections

The classification of the cross-sections of the frame members depends on the proportions and yield strength of their compression elements (notice that the number and location of the compression elements is a function of the load case). Cross-sections are classified as *Class 1, Class 2 ,Class 3* or *Class 4* .

When a cross-section has compression elements belonging to different classes, it is normally classified by quoting the highest of such classes and Class 4 cross-sections must be replaced by an *effective cross-section* which is then treated as a cross-section of Class 3.

Types of Connections

Connections are classified according to their bending rigidity as *nominally pinned, semi-rigid* or *rigid* and according to their strength as *nominally pinned, full-strength* or *partial-strength*. Notice the fact that the designation "nominally pinned" can be associated with either the bending rigidity or the strength of the connection.

ALLOWANCE FOR IMPERFECTIONS

Suitable equivalent geometric imperfections are included in the analysis of the frame with values which reflect the possible effects of all types of practical imperfections, i.e., residual stresses and geometric imperfections. Three types of imperfections are considered, namely (i) Frame imperfections, (ii) Bracing system imperfections and (iii) Member imperfections.

Frame Imperfections

The effect of frame imperfections must be included in the global analysis of the frame by means of an equivalent geometric imperfection in the form of an *initial sway* characterized by an angle ϕ. The initial sway may be replaced by a closed system of equivalent horizontal forces and the resulting internal forces and moments are then used for member design.

Bracing System Imperfections

The effects of imperfections must be considered in the design of bracing systems which are required to provide lateral stability to beams and compression members, by means of an equivalent geometric imperfection of the members to be restrained in the form of an initial bow. The initial bow may also be replaced by a set of equivalent stabilizing forces and the resulting internal forces are then used for the design of the bracing system members.

Member Imperfections

The effects of member imperfections may be neglected when carrying out the global analysis of the frame except in the case of compressed members with moment-resisting connections, belonging to a sway frame and such that $\overline{\lambda} > 0.5 \sqrt{(Af_y/N_{Sd})}$, where N_{Sd} is the design value of the compressive force and $\overline{\lambda}$ is the in-plane non-dimensional slenderness for the non-sway mode (flexural buckling of compression members). In these cases an equivalent initial bow imperfection must be included in a second-order analysis of the frame.

Normally the effects of member imperfections on its own design are incorporated in the buckling formulae given in Eurocode 3.

CHOICE OF THE METHOD OF GLOBAL ANALYSIS

Statically determinate frames will obviously be analyzed using *statics* but in the case of statically indeterminate frames an *Elastic Global Analysis* (E.G.A.) or a *Plastic Global Analysis* (P.G.A.) may be used. The P.G.A. can be carried out using either *Rigid-Plastic Methods* or *Elastic-Plastic Methods*. In this paper only rigid-plastic methods are discussed.

An E.G.A. may be used in all cases but the use of a P.G.A. (rigid-plastic) requires the satisfaction of *geometrical* and *material* requirements. The geometrical requirements are: (i) the cross-sections are of Class 1 at all plastic hinge locations where rotations may occur, (ii) the cross-sections have an axis of symmetry in the plane of loading at all plastic hinge locations and (iii) lateral restraint must be provided at all plastic hinge locations. Material requirements are that, for the steel utilised, (i) $f_u/f_y > 1.2$, where f_y and f_u are the yield and ultimate tensile strengths, (ii) the elongation at failure on a gauge length of $5.65 \sqrt{A_0}$ (A_0 is the original cross-section area) is not less that 15% and (iii) $\varepsilon_u \geq 20 \, \varepsilon_y$, where ε_y and ε_u are the strains corresponding to the yield and ultimate tensile strengths.

CALCULATION OF INTERNAL FORCES AND MOMENTS

Internal forces and moments (I.F.M.) may be determined using either an E.G.A. or a P.G.A.. The effects of the imperfections must be taken into account when performing the global analysis (normally only the sway imperfection must be considered at this stage).

Concerning the need for performing a *second-order analysis*, it is stated that it may be avoided if (i) the frame is braced or non-sway or (ii) design methods which make indirect allowances for 2nd order effects, are used.

The structural model ("Type of framing") adopted for the calculation of the I.F.M. and the assumptions made in the design of the members must be consistent with (or conservative in relation to) the method of global analysis chosen and with the anticipated behaviour of the connections.

It is always necessary to consider the appropriate arrangements of the variable loads in order to determine the critical combinations of I.F.M. required to verify the resistance of individual members and connections. In the case of sway frames an additional verification, corresponding to the stability of the frame as a whole, must be performed which takes into account the initial sway imperfections.

Elastic Global Analysis

When a first-order analysis is performed the resulting moments may be redistributed by modifying the moments in any member by up to 15% of the peak moment in that member provided that all members in which the moments are reduced have Class 1 or 2 cross-sections and the I.F.M. remain in equilibrium with the applied loads.

In sway frames the second-order effects associated with the sway mode must always be included. This can be done *directly*, by using a second-order elastic analysis or *indirectly* by using a first-order elastic analysis combined either with (i) amplified sway moments or with (ii) the consideration of sway-mode buckling lengths for the design of the members. However, the use of a first-order analysis is conditioned by the fulfilment of some conditions which involve the calculation of the critical load ratio (V_{Sd}/V_{cr}).

The value of (V_{Sd}/V_{cr}) is obtained, in the general case, by means of an elastic stability analysis of the frame. However, in the case of beam-and-column type frames, an alternative approximate method is proposed where the critical load ratio is given by the largest of the values (one per storey)

$$V_{Sd}/V_{cr} = \max_{i} \delta_i \cdot V_i /(h_i.H_i) \qquad (1),$$

where δ_i is the horizontal relative displacements between the top and bottom of storey i, h_i is the height of storey i, H_i is the total horizontal reaction at the bottom of storey i and V_i is the total vertical reaction at the bottom of the storey i. The values of δ_i correspond to a first-order analysis of the frame subjected to the horizontal and vertical design loads and to the equivalent horizontal forces replacing the initial sway imperfection. This approximate method can also be used to classify beam-and-column frames according to its sway resistance.

Plastic Global Analysis
Only the case of rigid-plastic methods (R.P.M.) is dealt with. The elastic deformations of the members and foundations are neglected and plastic deformations are assumed to be concentrated at plastic hinge locations.

The R.P.M. can only be used for a second-order analysis if a simplified method with indirect allowance for second-order effects is adopted. However, even if such a method adopted a second-order analysis can only be performed if (i) the critical load ratio does not exceed 0.20 and (ii) some additional requirements are met.

CHECKING THE FRAME STABILITY

Non-Sway Frames
It is sufficient to check the frame stability with a first-order global analysis (elastic or plastic). The stability of the frame is assured if the resistance of individual members and connections is verified for the critical combinations of I.F.M..

The in-plane buckling lengths for the non-sway mode must be used for member design. However, in the case of a R.P.A. these buckling lengths have to be determined with due allowance for the effect of plastic hinges. In particular, columns containing plastic hinge locations must be checked using buckling lengths equal to their system lengths.

In a braced frame the bracing system must be design to resist (i) any horizontal load applied to the frames it braces, (ii) any horizontal or vertical loads applied directly to the bracing system and, (iii) the effects of the initial sway imperfections (or equivalent horizontal forces) from all the frames which it braces.

Sway Frames
Sway frames must be analyzed to verify their resistance to failure in non-sway modes as described in the previous section and also to verify the sway stability of the frames as a whole. For this last stability analysis the arrangement of variable loads critical for failure in a sway mode must be considered and the effects of the initial sway imperfections are to be included.

If a second-order elastic analysis is adopted the resistance of the members in the sway-mode is verified for the calculated critical combination of I.F.M. using the in-plane buckling lengths for the non-sway modes. When a first-order elastic analysis is adopted either the in-plane or the sway-mode buckling lengths may be used, depending on the method chosen to account for the second-order effects.

Finally, when the amplified sway moments method is used in a rigid-plastic analysis the buckling lengths for the non-sway mode must be considered, with due allowance for the effects of plastic hinges.

CHECKING THE RESISTANCE OF FRAME MEMBERS

In order to check the resistance of the individual frame members when subjected to the critical combinations of internal forces and moments it is necessary to verify the *resistance of their cross-sections* and also their *buckling resistance* (essentially flexural buckling resistance of compression members and lateral-torsional buckling resistance of beams). A description of the procedures involved in these verifications can be found in a companion paper by the authors (Ref. 2).

CONCLUDING REMARKS

1) A sequential diagram was presented which corresponds to the use of Eurocode 3 in the safety checking or determination of an ultimate load combination of frames, with respect to

ultimate limit states. It contains six main steps which were clearly identified and briefly described. This diagram involves a considerable rearrangement of the provisions of Eurocode 3.

2) Frames are classified according to their sway-resistance, the classification of their members cross-sections and the type of their connections. The first two classifications are dependent on the load case.

3) Three types of equivalent geometric imperfections are mentioned in Eurocode 3. Normally only the effects of an initial sway must be considered in a global analysis of the frame. The design of a bracing system must incorporate the effects of the geometric imperfections on the members of frames which it braces. Both frame and bracing system imperfections can be replaced by sets of equivalent forces.

4) An elastic global analysis of the frame can always be used. A rigid-plastic analysis can only be used if certain requirements are satisfied (elastic-plastic analyses were not dealt with here).

5) The stability of a frame is always assured by the verification of the resistance of the individual members and connections for the critical combinations of internal forces and moments. In braced or non-sway frames only a first-order global analysis (elastic or plastic) is required. In sway frames second-order effects must be considered and sway imperfections must be included. When an elastic global analysis is used second-order effects may be considered either directly (second-order analysis) or indirectly (using the results of a first-order analysis). In a rigid-plastic analysis second-order effects can only be accounted for in special situations and always in a indirect fashion.

6) Depending on the sway-resistance of the frame and on the method used to calculate the internal forces and moments the design of the individual members may be based either on the in-plane buckling lengths for the non-sway mode (most cases) or on the sway-mode buckling lengths. When a rigid-plastic method is used due allowance must be given for the effects of plastic hinges.

7) The practical application of the rules listed in Eurocode 3 to the design of frames will be made significantly easier by the elaboration of design aids aimed at clarifying the choices and/or lowering the amount of work involved. Examples of design aids that it would be convenient to prepare are:

- Design aids for the *determination of the elastic critical value of the vertical load (V_{cr}) for failure in a sway mode* for frames different from the ones already considered in Eurocode 3 (beam-and-column frames).
- Design aids for the *choices involved in considering second-order effects* in terms of the type of structure and the type of global analysis chosen.

REFERENCES

1. Eurocode n°3 - Design of Steel Structures, Part 1.1 General Rules and Rules for Buildings,1992.

2. Abecasis, T.; Camotim, D.; Reis, A. - "Design of Plate Girders According to Eurocode 3"; Proceedings of the Construction Steel Design Conference, Acapulco, 6-9 December, 1992.

THE EXPERIMENTAL TESTING OF FULL SCALE, MULTI-STOREY, THREE-DIMENSIONAL NON-SWAY FRAMES.

CRAIG GIBBONS
Ove Arup & Partners
13 Fitzroy Street, London, W1P 6BQ, UK

DAVID B MOORE
Building Research Establishment
Garston, Watford, WD2 7JR, UK

ABSRACT

As part of an extensive experimental study into the behaviour of semi-rigid connections [1], two full scale, three-dimensional multi-storey frames have been tested. This paper presents a brief overview of the salient features of the experimental set-up and briefly explains how the complexities arising from the three-dimensional nature of the tests were addressed. This paper is a supplementary document to a 'poster presentation' at the First World Conference on Constructional Steel Design, Acapulco, Mexico, December 1992.

INTRODUCTION

Many experimental tests have been performed on isolated semi-rigid connections and flexibly connected subframes. Tests on full-scale three-dimensional frames are somewhat less numerous. However, experimental data on full scale frame behaviour is important. Firstly it enables the effect of column continuity through a loading level to be investigated - a parameter not present in many subframe tests - and secondly; it confirms whether the experimentally observed performance of isolated joints and subframes is indeed representative of their behaviour when they form part of an extensive frame. This latter point is of particular importance if the extensive work on isolated specimens is to be incorporated into universally accepted methods of semi-rigid and partial strength frame design.

The frame tests were carried out in the large structures testing hall at the Building Research Establishment near Watford, England. A brief summary of the large scale testing facilities at the Building Research Establishment is presented in reference [2].

THE EXPERIMENTAL TESTS

General arrangement

Each frame was 2\3 storeys high, two bay, single span of overall dimensions 10m x 9m x 3.8m. Figure 1 shows the general arrangement of the first of these frames, F1.

Initially, it was proposed that the frames should be two storeys high. However, it was soon realised that rotational restraint from the column loading block acting directly at the head of the column would have had an unquantifiable effect on the behaviour of the second storey column segment. This problem was averted by introducing a third storey to part of the frame thus isolating the second storey column from any undesirable restraint effects. The height of the third storey (1.8m) was chosen as half that of the lower storeys. As the positional bracing at the column head provided only a very small degree of rotational restraint (single bolt connection), the stiffness and distribution of moments within the third storey column were similar to that of a column of twice the length bending in double curvature. Therefore, in terms of stiffness and moment distribution, the proposed arrangement for the third storey was representative of a full height storey in a more extensive multi-storey frame. This was subsequently confirmed when the experimental data was appraised.

Flush end plate connections were used throughout each of the frame tests. There were a number of reasons for selecting this particular type of connection. Firstly, cleated connections are very often susceptible to bolt slip at relatively low moment levels. Once bolt slip has occurred, the precise moment rotation characteristics of the connection are usually irreversibly changed. The flush end plate connection however is not affected by bolt slip to the same degree and therefore exhibits similar moment-rotation behaviour under repeated moderate loading. Using such a connection, small levels of load could be applied to the frame without irreversibly deforming either the connections or the frame members. This was an important facility which allowed the loading devices and load control systems to be fully commissioned prior to carrying out a test to failure.

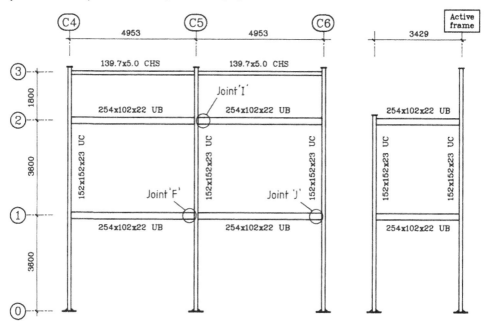

Figure 1: General arrangement of the full-scale test frame

Method of load application.

Loads were applied to the frames by a series of hydraulic rams which were situated beneath, and reacted against, the reinforced concrete test hall floor slab. High strength steel tendons and Macalloy bars, which passed through a regular grid of vertical ducts in the floor slab, transferred load from the rams to the frame via substantial 'saddle' and spreader beams. Each frame member was loaded by a pair of tendons tensioned by a pair of rams operating at the same pressure, thus providing a balanced arrangement of load to the saddle beams.

The actual applied loads were measured by tension load cells coupled into the tendons and Macalloy bars. In the case of the beam loads, two cells were connected in series into each length of tendon. One cell was used to monitor the applied load and the second cell was used as a direct feedback signal with a predefined cut-off threshold to prevent inadvertently overloading the beam. In the case of the Macalloy bars, which applied load to the column head, a single cell was located in each bar to provide a direct measure of the applied load with the measurement of column shortening providing the necessary safety feedback signal. Beam loading was 'load controlled' whilst column loading was 'displacement controlled'.

Deformation measurement

One of the main challenges was to develop a means of measuring the three- dimensional deformation at the mid-height of each of the column segments. Specifically, measurements were required of the linear displacements δ_x, δ_y and δ_z together with the twist rotation ϕ_z. The solution was to use a monitoring system comprising six strategically placed L.V.D.T.'s (Linear Voltage Displacement Transducers) at each measurement location. Using the principles of three-dimensional co-ordinate geometry, the individual components of deformation could be determined from the linear L.V.D.T. readings. The system was effectively a simplified version of a seven L.V.D.T. device capable of measuring all six components of three-dimensional deformation [3] which was used in a recent experimental study of column subassemblages. A total of six column displacement were used in each frame test requiring 36 L.V.D.T.'s.

Rotations at beam ends and column node points were measured using 'hanging dumb-bell' devices. This system comprised a 145mm x 65mm cylindrical steel weight (dumb-bell) connected to an armature on the frame via a thin strip of spring metal. Measurement of the flexural strains in the spring metal as the frame rotated with respect to the dumb-bell (which maintained its orientation due to gravity), gave a direct measure of the absolute rotation. After a load increment had been applied to the frame, it was probable that the hanging dumb-bell rotation devices would tend to oscillate about the vertical gravity reference datum. To improve the accuracy of measurement, 50 consecutive scans were taken for each rotation device and the average value used. The interval between scans was specifically chosen as half the 0.8s natural period of dumb- bell oscillation. At this interval, the alternating positive and negative errors in the measurement of the true rotation were approximately equal and therefore cancelled out when averaged.

Force Measurement

It was important that the column axial force, major and minor axis moments and the warping bi-moment could be monitored accurately up to column collapse. This involved monitoring the steel strains whilst the member was behaving both fully elastic and elastic-plastic with extensive yielding across the section. It is well known that the readings from adhesive fixed foil strain gauges in regions of plasticity are often inaccurate and spurious. This is due to the occurrence of discrete planes of concentrated strain (Luder's lines) rather than even straining over the length of the gauge. The problem was averted by applying the concept of 'redundant gauges'. Ten foil strain gauges were positioned around the cross section at each measurement location - six more than were necessary to satisfy the equilibrium equations under elastic conditions. By post-processing the recorded strain readings, it was possible to identify those gauges which were situated in elastic zones. Least squares error approximation techniques were then applied to generate the three-dimensional, linear strain (and hence stress) profile from these 'elastic gauges' - the readings from gauges in plastic zones were ignored. Using this approach it was possible to clearly identify the plastic zones and thus monitor the spread of yield across the section as the test progressed. A full description of the approach is presented in reference [4]. Beam forces were measured using less sophisticated, conventional approaches. A total of 244 foil gauges were used in each test.

Data logging and experimental control.

Two Solatron Orion Data Loggers were used to provide the 667 data channels necessary to monitor the frame instrumentation. This comprised 488 channels to energise and record the 244 strain gauges; 26 for monitoring the rotation devices, 24 for monitoring the applied loads; 63 for measuring bolt forces and 66 devoted to monitoring L.V.D.T.'s. Load was applied to the frame via two hydraulic pumps. A pair of dedicated servo-control amplifier units controlled the pressures, and hence loads, applied to the individual loading rams and monitored the safety feedback signals. Both the data loggers and the amplifier units were controlled directly from a remote minicomputer. It was therefore possible to specify an applied loading, initiate scanning of the instrumentation and view the recorded data via computer.

CONCLUSIONS

The experimental appraisal of two full-scale, three-dimensional, non-sway frames has been performed. The quality of experimental data which has been collected has justified the effort which was expended in addressing the experimental complexities.

REFERENCES

1. Gibbons, C., 'The strength of biaxially loaded beam-columns in flexibly connected three dimensional steel frames', Ph.D. Thesis, University of Sheffield, 1991.

2. Weeks, G., 'Laboratory Testing of Large Structures', I.A.T.R.L.M.S., Methodology and Testing of Structures No. 3 - tests on full-scale structures, International Colloquium, Bucharest, 1969, pp. 204-222.

3. Gibbons, C., Kirby, P.A. and Nethercot, D.A., 'A mechanical-electrical device for measuring the three-dimensional deformation of steel columns', Proc. I.Mech.E., Journal of Strain Analysis, Vol. 26, No. 1, 1991, pp. 1-8.

4. Gibbons, C., Kirby, P.A. and Nethercot, D.A., 'The experimental assessment of force components within structural steel 'H' sections', Proc. I.Mech.E, Journal of Strain Analysis, Vol. 26, No. 1, pp. 31-37.

Damage Analysis of Steel Building Frames

I. S. Sohal[1] and L. Cai[1]

1. Introduction

In the United States, the steel structures are being designed on the basis of first order or second order elastic analysis[1]. However, recent developments in computer technology have created a potential for integrating the second order inelastic damage analysis with the design practice. This integration will assist in designing the steel structures to avoid a total collapse, but with the desired probabilities of specified levels of damage.

However, in order to integrate the analysis and design, the analysis method must be very efficient. The efficiency of the analysis method will tremendously increase, if each members of the system can be modeled by only one beam-column element. To this end, we present tangent stiffness matrices for members with fully developed, hardening and softening plastic hinges at the ends and intermediate locations, derived by stability function approach.

These developments will be implemented in the computer program for carrying out second order inelastic damage analysis of steel frames. The numerical result for a frame will be compared with the results without damage.

2. Tangent Stiffness Matrix for Beam-column with Fully Developed Plastic Hinges

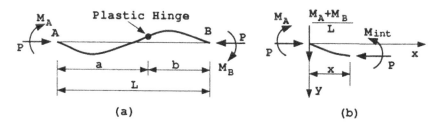

Fig. 1 Beam-Column with an Intermediate Plastic Hinge

For a beam-column with an intermediate plastic hinge as shown by Fig. 1(a), the tangent stiffness matrix is obtained by considering equilibrium of a deformed segment of beam-column and by using boundary conditions as

---

[1]Department of Civil and Environmental Engineering, Rutgers University, Piscataway, NJ 08855-0909

$$\left\{ \begin{array}{c} \dot{M}_A \\ \dot{M}_B \end{array} \right\} = \left[\begin{array}{cc} K^0_{11} & K^0_{12} \\ K^0_{21} & K^0_{22} \end{array} \right] \left\{ \begin{array}{c} \dot{\theta}_A \\ \dot{\theta}_B \end{array} \right\} \tag{1}$$

where

$$K^0_{11} = \frac{EI}{\gamma L} [kL \sin ka(kL \cos kb - \sin kb)]$$

$$K^0_{12} = K^0_{21} = \frac{EI}{\gamma L} (kL \sin ka \cos kb)$$

$$K^0_{22} = \frac{EI}{\gamma L} [kL \sin kb(kL \cos ka - \sin ka)]$$

$$\gamma = \sin kL - kL \cos ka \cos kb$$

where a and b are as shown in Fig. 1(a) and $k^2 = P/EI$. The above stiffness coefficients can be simplified by substituting Taylor Series for sine and cosine terms.

Fig. 2 An Additional Plastic Hinge at End A

When an additional plastic hinge is developed at end A, then $\dot{M}_A = 0$. Thus from first equation of the set (1), $\dot{\theta}_A$ can be obtained in terms of $\dot{\theta}_B$. By substituting $\dot{\theta}_A$ in second equation of the set (1), we obtain

$$\left[K^0_{22} - K^0_{21} \frac{K^0_{12}}{K^0_{11}} \right] \dot{\theta}_B = \dot{M}_B \tag{2}$$

When hinges are developed at both ends, both $\dot{M}_A$ and $\dot{M}_B$ are zero.

3. Tangent Stiffness of Beam-column with Hardening Plastic Hinges

Herein, we assume that the moment-curvature relationship is such that the slope of hardening portion varies from EI at initial yield moment to zero at the fully plastic state[2]. Then ρ_i, a parameter to denote the stage of hardening of hinges, can be expressed as

$$\rho_i = \frac{M_i - M_{yc}}{M_{pc} - M_{yc}} \tag{3}$$

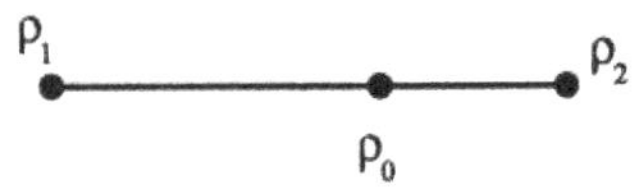

ρ_1 $\bullet$————————$\bullet$————$\bullet$ ρ_2

ρ_0

Fig. 3 Three Plastic Hinges and Their Parameters

Let ρ_0, ρ_1, ρ_2 be the parameters to denote hardening at point 0, 1 and 2. Then modified stiffness coefficients are obtained as

$$K''_{11} = (K'_{11} - K'_{12}\frac{K'_{21}}{K'_{22}}\rho_2)(1 - \rho_1) \tag{4}$$

$$K''_{12} = K''_{21} = K'_{12}(1 - \rho_1)(1 - \rho_2) \tag{5}$$

$$K''_{22} = (K'_{22} - K'_{21}\frac{K'_{12}}{K'_{11}}\rho_1)(1 - \rho_2) \tag{6}$$

where

$$K' = (K^0 - K^e)\rho_0 + K^e \tag{7}$$

in which K^0 is the stiffness matrix of beam-column with a fully developed intermediate hinge and K^e is elastic stiffness matrix of a beam-column.

4. Tangent Stiffness Matrix of Beam-Column with Softening Plastic Hinges

The softening in the plastic hinge may be denoted by ω_i, which may be determined by using post-local-buckling $M - \Phi$ relationship[3] as

$$\omega_i = \frac{M_{max} - M}{M_{max} - M_{min}} \tag{8}$$

where M_{max} is moment at peak of $M - \Phi$ curve and M_{min} is the lowest moment which can be attained in the softening branch, $(EI)_{max}$ is the slope of $M - \Phi$ curve at the point of beginning of the softening branch. Knowing ω_i, the stiffness matrix in the softening branch can be obtained from Eqs. (4–7) by replacing ρ_i with ω_i at softening hinges and replacing EI in K^e by $(EI)_{max}$. In fact, in the softening branch of $M - \Phi$ curve, tangent stiffness matrix is function of hinge length. The development of such stiffness matrix is currently in progress at Rutgers University and will be presented in subsequent paper.

5. Numerical Results

The developed computer program is used to analyze a symmetrical portal frame with tubular sections as shown in Fig. 4. The frame is subjected to constant concentrated vertical joint loads P and an increasing horizontal load at the left top joint. The curve between horizontal load vs. sidesway at the left top joint is obtained without considering local buckling and compared with that obtained with local buckling for $D/t = 96$ and $D/t = 72$. The Fig. 4 shows that there are significant differences between two curves. It means that consideration of local buckling for the large values of D/t is important.

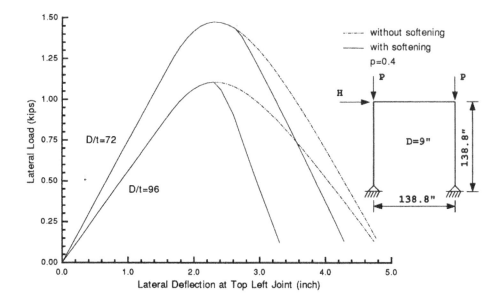

Fig. 4 Load-Deflection Curves of Tubal Portal Frame

References

1. Load and Resistance Factor Design Specification for Structural Steel Buildings, AISC, Chicago, IL, 1986.

2. King, W. S., White, D. W. and Chen, W. F., 'Second-order Inelastic Analysis Methods for Steel-Frame Design', Journal of Structural Engineering, Vol. 118, No. 2, 1992, pp.408–428.

3. Sohal, I. S. and Chen, W. F., 'Local Buckling and Sectional Behavior of Fabricated Tubes', Journal of Structural Engineering, Vol. 113, No. 3, 1987, pp.519–533.

4. Chen, W. F. and Lui, E. M., 'Structural Stability: Theory and Implementation', Elsevier, New York, NY, 1987.

A COMPUTER AIDED DETAILING PROGRAM
FOR HIGH RISE STEEL BUILDINGS

RICHARD J. ALONSO, P.E.
Charles Cohn & Son, Structural Consultants, Inc.
70 East Old Country Road, Hicksville, NY 11801

DEAN KOUTSOUBIS, P.E.
DBK Associates, Inc.
10 Raintree Road, Melville, NY 11747

ABSTRACT

This paper will present CCS-CAD, a computer aided detailing system involved in the design of steel connections and the detailing of steel members for structures of all types with an emphasis on high rise steel buildings. The flexibility of the program and its integration within the engineering process is described.

INTRODUCTION

Due to the highly competitive nature of the construction industry, it has become necessary to improve productivity and reduce costs. As structures becomes more complex, the need to automate the design and detailing process becomes extremely important. Motivated by these considerations, Charles Cohn & Son decided to automate their production of fabrication drawings and documents for steel structures with an emphasis on high rise buildings.

Truly effective automation must be flexible enough to accommodate different building codes, which vary by geographical region, and the different practices and conventions of producers of structural steel. These different practices may include the creation of assembly drawings as well as metric dimensions. In addition, a flexible automated system must be able to incorporate nonstandard connections and details. These nonstandard connection are generally required by increasingly complex structures.

Based on their need for a flexible system, Charles Cohn & Son, with the assistance of DBK Associates, developed CCS-CAD, a computer aided detailing and drafting program. Charles Cohn & Son has been in the detailing business since 1924. They have worked on such structures as the Citicorp building, World Trade Center, 1675 Broadway, and the AT&T building, all in New York City. DBK Associates is a consulting firm specializing in the automation of engineering and design processes.

CCS-CAD was developed with the thought of merging the power of an automated detailing system with the flexibility of a drafting package specifically tailored for the needs of the detailer. Those features of CCS-CAD which contribute to its power and flexibility will be presented.

In-House Development

CCS-CAD was developed in-house to run on a Hewlett Packard series 9000 computer using the UNIX Operating System. The decision to use the UNIX Operating system was made because of its portability, its powerful development environment and its ability to support multiple users. Of prime importance in the decision to develop an in-house program, rather than purchase one, is the need to have the ability to modify the program as specific needs arise. The development of the program is driven by the needs of the projects that are worked on. This can be particularly valuable when the structures vary substantially from the norm, or when a method of presentation is required that does not meet the customary presentation of commercially available programs. As modules are developed, they are used on actual jobs. If a particular feature is required for a given job it is immediately incorporated into the existing program. The development team gets immediate feedback from the users. This feedback is used to improve the system.

Integrated Modules

CCS-CAD is composed of modules that perform various detailing functions including the generation of standards, plans, and details. Modularity is an important feature of CCS-CAD. Each module can be run independently, or they can be integrated such that the output of one program may be used as the input of another program in a manner which is transparent to the user.

For example, the column sizes entered in the column schedule can be transferred to the plan module while the information provided in the plans and column schedule can be extracted by the beam and column detailing module. This information can also be used by the material ordering program. Connections in the beam and column detailing modules may be specified by the user, automatically extracted from the job standards, extracted from a connecting beam or designed within the program.

Drafting Module with Plan Enhancements

Modules are set up to automatically generate complete standards and details with the minimum amount of information. Drawings can be sent directly to the plotter or to a file that can be read by the drafting module. This gives the user the ability to further customize any drawing that could not be completed by the computer. In addition, individual details can be combined together to create more complicated drawings.

The drafting capabilities of CCS-CAD are set up with the structural steel drafter in mind. Graphic entities are defined to optimize the performance with respect to steel detailing and include welds, hole descriptions, sections and bevels.

The drafting module also contains all the features necessary to quickly create erection drawings. These drawings include the dimensions provided by the engineer as well as beam sizes, columns and grids. In addition, this module gives the detailer the ability to create a layout drawing for detailing purposes. The layout drawing contains all dimensions and angles necessary to detail the structural members on the plan.

Access to AISC Database

All routines access the AISC database installed on the computer. By having access to the database, all section properties are immediately available. This cuts down on the amount of input required and also provides a check that the section entered is a valid section. AISC shapes can be generated automatically by the drafting program. These shapes can be shown in cross-section, or longitudinally using either a top or side view.

Engineering Calculations

In all high rise steel buildings a certain amount of the engineering is left up to the detailer. This includes the design of connections, doubler plates and stiffeners. Many of the connections that are designed require engineering calculations. The inclusion of engineering calculations is an important feature of CCS-CAD. By having the computer generate these calculations while generating the drawings, the need to create the calculation manually is eliminated.

Flexibility of Input

CCS-CAD supports a variety of methods to input information including menus, commands, graphics, and batch input. In most cases, the user has the opportunity to select the input method which best meets his or her needs.

Interactive menus give users the opportunity to see all the options at once. This method of input is especially useful for novice users. A command line is available for more experienced users as a short cut to perform functions without traveling through the menus. Applications which are visually oriented allow graphical input and manipulation through the use of a mouse or locator tablet. Interactive graphics allows the user to immediately see the results of the input. Many applications support batch input as well. Input Files are in ASCII format and can be created manually using a text editor or automatically from the output of another module.

Interaction with Other Programs

As the use of computers throughout the construction industry becomes more prevalent, the smooth flow of information between computer systems becomes more important.

With this in mind, CCS-CAD can transfer data to and from other computer systems using neutral formats such as the DXF format. By interfacing with structural analysis software, much of the erection plan can be generated directly from the engineer's structural model. By interfacing with other CAD software, CAD files can be sent directly to the fabricator. And as fabricators automate their procedures with numerical control machines, CCS-CAD has the flexibility to provide information in a format that can be used directly by those machines.

Productivity Enhancements

Every effort is made to improve the productivity of the user. Whenever possible, default values are assigned to required input. The default values are those values which are most frequently used. The user may override the default values at any time. This minimizes the amount of input that must be specified by the user.

CCS-CAD employs extensive error checking techniques. For interactive processing, invalid input is identified immediately as it is entered. In batch processing all the errors are logged in a file and displayed after the program has finished running. The program checks all designs with respect to their geometric and structural integrity.

CCS-CAD is setup in manner that makes it easy for the user to learn and use the program with minimal access to reference material. However, for those instances where assistance is required by the user, a help facility is available. The user can type in a keyword and the program will provide a detailed explanation of a procedure or definition of a term.

Because much of the input involves numerical lengths, users can enter numbers as a decimal or a fraction with feet and inches. Furthermore a calculator is built in to the program which accepts any combination of decimal and fraction.

CONCLUSION

CCS-CAD is continually expanded as additional features are required. Efforts are continually made to further improve the flexibility and power of existing applications. Current developments include the enhancement of CCS-CAD to be used for bridges as well as high-rise buildings.

In conclusion, the automation of the detailing process is necessary to improve productivity and performance as the design of steel structures becomes more complex. The implementation of a custom developed program can provide the power and flexibility required. It also can be altered as changing needs arise. The inclusion of engineering functions within the detailing program contributes to that power and flexibility.

Special Structures

DEVELOPMENT OF A SPACE FRAME CONSISTING OF COLD FORMED SECTIONS

J C CHAPMAN[1] D BUHAGIAR[2] B E CODD[3] R J PRINGLE[3] P J DOWLING[4]

ABSTRACT

The chords of the Conder Harley spaceframe consist of orthogonally intersecting back to back cold formed channels. The diagonals consist of cold formed tubes having flattened and bent ends which are bolted between the intersecting channels. Design development embraced production, construction, research, and design methodology. Design methods must treat various forms of local and overall instability, some of which are discussed in the paper. Data and insight were provided by experiments and numerical analysis, which supported the formulation and verification of design methods. Several structures have now been built to European requirements, and some of these are described.

1. INTRODUCTION

The system described was the invention of an Australian architect, Edwin Codd. He found in his practice a need for a cost effective space frame system, which was not available in Australia. He saw that the need could be satisfied by combining the manufacturing advantages of cold forming with the avoidance of discrete nodal connections. The design concept was to use continuous chord members, with tubular diagonals, the flattened ends of which are bolted between the intersecting chords (Figure 1). The usual practice is to use cold formed channel sections for the chords and cold formed tubes for the diagonals. The advantages of cold forming include limited stockholding of pre-galvanised strip, automated production, high accuracy, on line powder coating and rapid response. Various configurations are possible, the commonest being square on square, but diamond on square also has advantages. When the European licence for the Harley system was purchased by the Conder Group, they approached Imperial College to undertake a programme of research

(1) Chapman & Dowling Associates. Visiting Professor Imperial College, London
(2) Civil Engineering Department, Imperial College. University of Malta.
(3) Conder Group, Winchester, England.
(4) Head of Civil Engineering Department, Imperial College. Chapman & Dowling

leading to the formulation of design rules which would be consistent with British and European practice, and which would provide for European loadings, greater spans, and increased nodal distances. The paper gives an account of the development, which is continuing. At the time of writing, eight structures have been completed in the UK and Europe.

2. FABRICATION

The pre galvanised material currently being used is highly formable and has a yield stress of 390 N/mm^2. Chord thicknesses vary between 1.5 and 4.0mm. For black steel, thicknesses up to 8mm with 460 N/mm^2 yield stress are available. Strip widths up to 1m can be rolled.

The strip stock is first slit to the required width. Hole centres are pre-programmed and punched automatically, before forming the sections. Holes can be punched at any longitudinal position on up to five gauge lines. The tolerance on hole centres at a joint is ±0.5mm and hole diameters are 1mm greater than bolt sizes. Different diameters can be accommodated on a gauge line. The strip stops for about 1 second for punching, but sections are cut to length by flying shears without stopping. If painting is specified, sections then enter the coating line, where a hard chip resistant powder coating is applied. Chords can be produced at the rate of 2000m^2 of space frame per shift.

The diagonals are formed from pre-galvanized strip and resistance-welded in a tube mill. The weld locally destroys the zinc coating but this is restored both inside and outside by zinc spraying. The tube is then flattened, punched, cut to length, and bent in a single operation. Tubes then enter the painting line.

3. ERECTION

Frames are normally assembled at ground level on screw jacks. Chord lengths are typically about 12m and can be man-handled. Joints are made hand-tight and after assembly is completed, bolts are tightened to a standard torque by power wrench.

Most Harley spaceframes have been lifted into position by cranes, but a specially designed synchronised hydraulic hoisting system has now been used on several projects. Alternatively, the spaceframe can, if necessary, be assembled in its final position on scaffolding.

4. DESIGN

The designer must first be aware of the possibilities and limitations of the production system. The section currently being used was designed to resist compression and the eccentricity moment which is inherent in the Harley node, and to achieve a reasonably balanced section, with a stable lip. The width must be such that the flattened ends of the diagonals have adequate end bearing length, and the open width has to be sufficient to allow bolting. The maximum tube diameter is limited by the width of section. Recent designs have used tube diameters ranging between 42 and 70mm.

It is necessary to use the same chord size (but the thickness can vary) throughout a chord layer, and it is usually advantageous to use the same size throughout a structure. In regions of high shear (and hence large joint moments) nodal inserts are used to increase the crippling resistance of the channel webs. If required, the inserts can extend over the full length of a chord member, to increase its axial and bending resistance. Because torsional buckling resistance is less than the flexural buckling resistance, it is sometimes advantageous to cap the sections at intervals along the length. The insert sections are also used as splice plates. Splices are placed away from nodes, at about quarter span.

Washer plates of various thicknesses are used to ensure a satisfactory transfer of the diagonal forces into the channel sections. Various packing devices are required to compensate for changes in tube thickness and diameter, and to fill the void between the diagonal tab and the channel web.

It will be apparent that several design checks are required for every connection and every member. To achieve a design response time which is commensurate with that of construction, it is necessary to computerise the design process, and to link closely the design, costing, scheduling, and production processes. As with hot-rolled sections, detailing of connections is just as important as member design. It will also be apparent that the highly automated production process demands clear and accurate scheduling of components. Equally, construction planning must include clear and accurate instructions for order of production, packaging, delivery to site, and erection procedure.

The paper describes the principal phenomena for which design rules have been formulated or adapted, and which have general relevance to thin walled open sections.

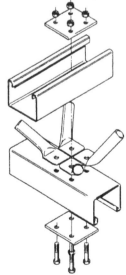

Figure 1 The Conder Harley joint

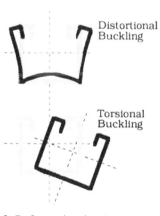

Distortional Buckling

Torsional Buckling

Figure 2 Deformational and torsional buckling

5. LIP BUCKLING

Lip buckling involves deformation of the cross section, unlike flexural and torsional buckling (Figure 2). Lips which are in compression can buckle as struts on an elastic foundation; the stiffness of the foundation is provided by the transverse stiffness of the cross section. If the stiffness is approximated as the stiffness of uncoupled transverse strips subject to disturbing forces applied at the centroid of the effective lip (that is, the lip table, part of the web, and the lip downturn), then the critical lip buckling stress can be readily calculated[1]. The wavelength of buckling (which might typically be from one half to one quarter of the span) is also given by the design formulation. The lip buckling resistance can then be found by introducing an initial deformation, which enables a Perry formulation, similar to that used to calculate the resistance of columns to flexural buckling, to be applied. Also, non-linear elasto-plastic finite element analysis can then be used to verify that the formulation, with its various assumptions and approximations, gives satisfactory results. Figures 3 and 4 show typical comparisons for critical stress and for strength. It can be seen that for the chosen sections, the design formulation, whilst being simple and approximate, provides satisfactory estimates of critical stress and strength. It can also be seen that for slender sections it is essential to take lip buckling into account when the lip is in compression, as a result of axial force or bending moment. This can be done by replacing the yield stress by the lip buckling strength in the column design formulation. The investigation also demonstrated that local buckling of the sides of the channel did not diminish the foundation modulus, and that lip buckling could be treated in the same way, whether caused by axial compression or by bending. Figure 5 shows the development of lip buckling up to and beyond failure. In all the lip buckling analyses, deflection and rotation of the member were prevented.

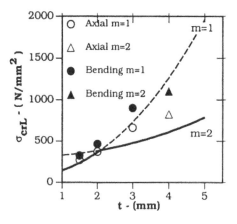

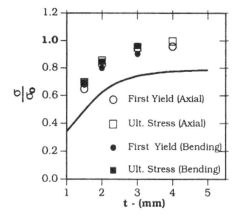

Figure 3 Lip buckling critical stress -
design curve and FE results.
One and two half waves.

Figure 4 Lip buckling strength -
design curve and FE results.
One half wave.

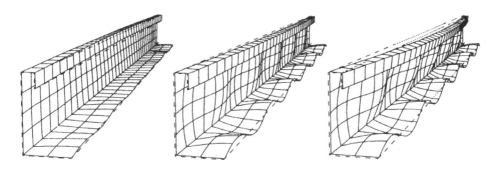

Figure 5 Lip buckling deformations before yield, at first yield (σ_{lip} = 252 N/mm²), and at failure (σ_{lip} = 271 N/mm²). 125 x 100 x 1.5mm section. L = 1.2m. σ_o = 390N/mm²

6. TORSIONAL FLEXURAL BUCKLING

For symmetrical I sections, in which the centroid and the shear centre coincide, torsional buckling can occur without flexure. Torsional buckling might or might not occur at a smaller load than flexural buckling about the minor axis. Torsional flexural buckling occurs in columns of open cross-section when the centroid and the shear centre do not coincide. Twisting then causes the centroid to deflect with respect to both flexural axes, with the result that the axial force exerts moments about both axes. The column therefore rotates and translates. For thin walled sections of uniform thickness there is no significant elastic deformation of the cross-section, unless lip buckling occurs.

For typical dimensions, torsional flexural buckling occurs at a much smaller stress than does flexural buckling, as can be seen in Figure 6. According to current UK design practice the strength is given by the Perry formulation for flexural buckling, but the effective length is increased by the square root of the ratio of the critical flexural buckling stress to the critical torsional flexural buckling stress. In effect, the Euler buckling stress is replaced by the torsional flexural buckling stress, and the imperfection constant, which is a function of slenderness, is also increased. The effect on strength so calculated is shown in Figure 6, which also shows the result of elasto plastic finite element solutions. The stresses at failure at the corners of the section are shown in Figure 7. Figure 8 shows the development of torsional flexural buckling for the same example. Rotations in the central span of a three-span column, shortly before failure, can be seen in Figure 9. Because failure occurs at a smaller axial stress than for flexural buckling (Figure 6), larger deformations can be sustained before failure occurs. For the examples chosen, the modified Perry design method gives a reasonable estimate of failure load. It should be mentioned however that in BS 5950 (1987) the expression for the critical torsional buckling stress, which is used in the expression for torsional flexural buckling, contains a multiplier 2, which implies a degree of warping restraint at the connections corresponding to a reduction in torsional effective length from L to 0.7L. This degree of warping restraint will not necessarily exist in practice and any reduction in effective length should be justified by the particular arrangement. The multiplier has been taken as 1.0 in calculating the strength shown in Figure 6.

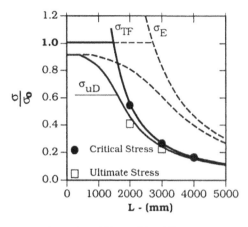

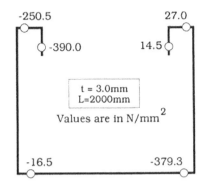

Figure 6 Torsional flexural buckling.
Design curve and FE results

Figure 7 Torsional flexural buckling.
Stresses at failure

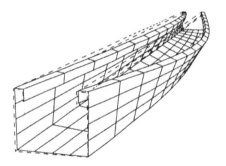

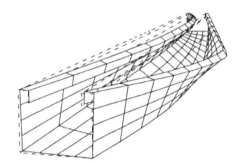

Figure 8 Torsional flexural buckling deformations at failure ($\sigma_u/\sigma_o = 0.41$), and post failure, 125 x 100 x 3mm section. L = 2m.

7. LATERAL TORSIONAL BUCKLING

BS 5950 (1987) states that lateral torsional buckling about the minor axis need not be considered. For the cross-section being used, it seemed that this assumption was not justified. Because in the Harley system the eccentricity moment about the minor axis can be relatively large, and because the lateral torsional buckling mode is similar to the torsional flexural mode, a simple approximate formulation for critical stress was introduced to provide an interim basis for design. When imperfection was taken into account, the estimated minor axis lateral torsional buckling strength for the span being used was about 30% below the yield stress (390 N/mm^2). Consideration of minor axis lateral buckling provided some justification for using a linear interaction between torsional flexural buckling strength, and lateral torsional buckling strength about major and minor axes. Further justification was provided by frame tests, for the particular configuration being used. Recent finite element calculations have shown that the interim formulation for critical stress gives satisfactory results for the given section (Figure 10). The strength for particular spans given by finite element solutions is also compared with strength based on a Perry type strength formulation. The development of buckling is shown in Figure 11. A critical stress treatment by Attard[2] has recently come to our notice. Results from that formulation are also shown in Figure 10. This critical stress formulation should provide a more accurate design basis than the approximate interim formulation referred to above.

Figure 9 3 x 3m chord member.
Incipient compression failure

Figure 10 Lateral torsional buckling
Design curves and FE results

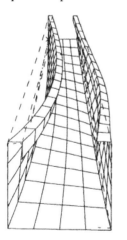
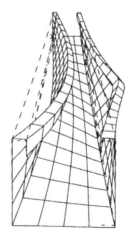

Figure 11 Lateral torsional buckling deformations at failure ($\sigma_u/\sigma_o = 0.84$), and post-failure. 125 x 100 x 2m.

8. FRAME BEHAVIOUR

Chord members are subjected to a combination of compression (or tension) and bending, which is predominantly about the minor axis. For design purposes, an interaction equation of basically linear form is used. Confirmation that the design approach to the individual and combined load effects is satisfactory for the sections and arrangements employed, was provided by a series of component and frame tests (Figure 12). Further experimental and analytical confirmation is required for more general application.

Frame and component tests also provided data to support design formulations for the buckling strength of the diagonals, end bearing strength, bolt forces, web crippling, and bending of the washer plates.

Figure 12 3mm Test Frame   Figure 13 Market development, Derby

9. APPLICATION OF THE CONDER HARLEY SPACE FRAME SYSTEM

The key to any successful construction system lies not only in its production efficiency but also in meeting the needs of the builder and the user. Some of the more important considerations are:

(1) Buildability on sites having varying conditions
(2) Unusual support conditions and layout
(3) Architectural requirements
(4) Cladding and services requirements
(5) Building plan constraints
(6) Reponse time and site programme requirements

The following recently completed projects highlight some of the features of the Harley system which have been used to provide solutions for specific client requirements.

The Eagle Centre Market redevelopment, Derby

The Derby project brought together many of the features of the Harley system in satisfying the constraints set by the client and the construction company. The original market structure was based on a hexagonal and octagonal layout, and was condemned by the Fire Officer. The market place is located above a three storey concrete car park on a floor which slopes several metres from one end to the other. Column positions were constrained to the positions of the columns below. The use of the space frame structure allowed a regular superstructure to be constructed on columns which met that constraint. The components were craned on to the site, and assembly and erection proceeded by various methods. Where uninterrupted public access was required, the spaceframe was erected on scaffolding in its final position. Where obstructions prevented assembly on the ground in its final position, it was assembled where space permitted and translated into position by cranes. Where possible, the space frame was assembled on the ground slab and lifted using the hydraulic system already referred to.

The space frame has satisfied the architectural objective of providing a light and airy public space, whilst facilitating construction under very difficult site conditions (Figure 13).

Buxton canopy

Unloading canopies often require irregular column positions and have difficult site conditions. The modular spaceframe adapts readily to these conditions, and permits irregular column spacings. In the case of the Buxton canopy the Harley system provided a logical and pleasing solution to the problems presented by the layout requirements and the site conditions (Figure 14).

Canopies for the Channel Tunnel loading terminal

At the gateway to Britain from continental Europe, Eurotunnel required stylish and durable but economical canopies for the loading terminal. The canopies were to be erected over existing toll booths. The proximity of the ocean and the exposed position of the structure necessitated a high standard of protection against corrosion.

The polyester powder coated Harley frame satisfies the architectural and functional requirements, whilst meeting the rather tight schedule for design and construction.

Pay kiosk canopy at Liejkenshoek Tunnel, Antwerp

A canopy was required to suit an existing design and layout for the support trees. A high standard of corrosion resistance was specified. The construction schedule was such that the canopy had to be erected over the kiosks, and this was achieved by assembling the canopy at ground level beside the kiosk area and translating it into position by cranes. The colour-coated frame and the elegant trees provide an appropriate and attractive solution to the design specification.

Figure 14 Loading canopy, Buxton

Figure 15 Pay kiosk canopy at
Liejkenshoek Tunnel, Antwerp

10. CONCLUSION

(1) The development embraced considerations of production, conceptual design, construction, experimentation, analysis and design methodology.

(2) It can be foreseen that cold formed members will find an increasing role in engineering structures.

(3) The development of production methods and design concepts must be matched by continuing experimental and analytical research, and development of design methodology.

ACKNOWLEDGEMENTS

The authors would like to acknowledge the support and co-operation of Conder Group plc, the Commonwealth Scholarship Commission, and the University of Malta.

REFERENCES

1. Buhagiar D, Chapman J C, Dowling P J: Design of C-Sections against Deformational Lip Buckling. Eleventh International Specialty Conference on Cold-formed Steel Structures, Missouri, USA, 1992. Submitted for publication.

2. Attard MM: General Non-Dimensional Equations for Lateral Buckling. Thin-Walled Structures 9 (1990) 417-435.

3. Codd B, White SH: Eagle Centre Market refurbishment, Derby. The Structural Engineer. Volume 70, No.5, 1992

ON REPAIR OF FATIGUE AFFECTED BRIDGES

Kentaro YAMADA
Department of Civil Engineering
Nagoya University
Nagoya 464-01, Japan.

Masaru TSUCHIHASHI
Department of Civil Engineering
Nagoya University
Nagoya 464-01, Japan.

Akimasa KONDO
Department of Civil Engineering
Meijo University
Tenpaku-ku, Nagoya, Japan.

ABSTRACT

Fatigue cracks were found at the ends of vertical members of two two-hinged steel arch bridges. Structural analysis was carried out for original structures and for retrofitting structures, such as increasing the stiffness of vertical members, providing diagonal members and/or providing bracing members between the arches. Cumulative fatigue damage analysis was also carried out for details at the ends of vertical members. A comparative study was also carried out to see the effectiveness of rehabilitation scheme. It was found that spandrel type bracing is an effective means for rehabilitation of such bridges.

INTRODUCTION

From fatigue analysis point of view, the live load on highway bridges have been considered to be small as compared to the dead load. For this reason, fatigue design is usually not considered at the design stage for the highway bridges in Japan [1]. In recent years a number of fatigue damages were reported to have occurred as a consequence of ever increasing heavy truck traffic and other reasons[2,3]. For instance, end of suspender of Langer and Lohse bridge were reported to have damaged by bending and torsional vibrations due to wind. Fatigue cracks were also found at weldments of plates connecting the upper flange and the end of lateral bracing and girder . In addition to this, fatigue damage was also reported on upper and lower ends of vertical members of deck type arch bridges.

In most of the cases, secondary members are found to be more affected. Despite the fact that the fatigue damage of such secondary members may not be serious and of direct considerations, the serviceability of the structure is prone to be affected by such damages, especially in view of increasing heavy traffic. The authors feel that fatigue assessment considerations should be given to such members right at the design phase in addition to a proper maintenance scheme. This study signifies the importance of such considerations by citing some real cases.

Earlier Design Considerations : In Japan, until late 60's, design philosophies were primarily dictated by the computation tools available in hand. Calculations were also made by applying simplified assumptions, such as idealized hinged condition at the joints of vertical members of arch bridges, thereby averting bending moments at such joints. In reality there exist bending stresses and such unaccounted bending stresses have caused fatigue cracking at the ends of vertical members. In view of growing traffic, in terms of both number of vehicles and their weights, it has become inevitable that the design specifications should consider flexure induced loading in short vertical members of arch bridges at the design phase.

Schematics of the Study : The analytical work of this study is based on following steps. First, structures are modeled as either plane or space frames according to the nature of structure as described later. For the joints and the details in question, influence lines of stress and displacement are computed by applying Finite Element Method. Computations for stress range and its frequency of occurrence are made by applying a model truck on the bridges and counting them by the Rain-Flow Counting Method. For fatigue damage assessment, modified Miner's Rule, with JSSC Recommendations of Fatigue Design [4], is employed. In order to account for traffic growth, computations for every case are repeated by considering various patterns of increment in traffic.

CASE 1: STUDY OF 2-HINGED STEEL-BOX ARCH BRIDGE

Case Description : The bridge (hereafter referred as bridge-A), as shown in Fig. 1, was constructed in 1963[4]. This is a 2-hinged deck type arch bridge with box type arch ribs. The overall span length is 106.5 m, while the arch span and the rise are 85.0 m and 13.0 m, respectively. The bridge is symmetric with respect to center line and has relatively lesser track width of 6.0 m. After about 20 years of service, fatigue cracks were observed at the upper and/or lower end of the vertical members, as indicated by circles in Fig. 1. It is interesting to note that most of the cracks occurred in the shorter vertical members, and that their distribution was symmetric with respect to both transverse and longitudinal axes of the bridge.

Structure Modeling and the Analysis Scheme : Owing to the symmetry of the box girder the structure is modeled as plane frame as shown in Fig. 2. To elaborate on significance of joint condition on stress distribution, the analysis is carried out in a variety of manner. In addition to this, different cross sections of the vertical members as well as bracing effect is also considered for the proposed rehabilitation scheme. The overall analysis scheme is described below.

1. Model A-1: Considers the original structure assuming the joints of vertical members with arch rib or girder as fixed.
2. Model A-2: Considers the original structure assuming the joints of vertical members with arch rib or girder as hinge.

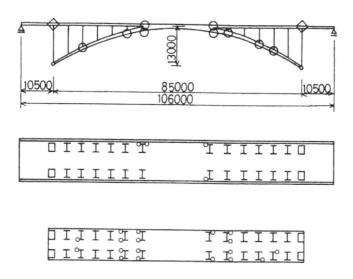

Figure 1. Plan and elevation of Bridge-A.

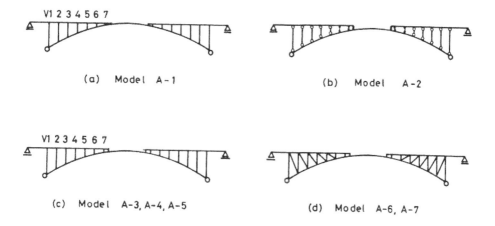

Figure 2. Analytical model of Bridge-A.

3. Model A-3: Considers a revised structure by increasing the cross sections of short vertical members V7.

4. Model A-4: Considers a revised structure by increasing the cross section of vertical members V4 through V7, with the cross section of vertical member V2.

5. Model A-5: Considers a revised structure by increasing the cross section of vertical members V2 through V7, to the same cross section of vertical member V1.
6. Model A-6: Considers a revised structure by providing bracings as diagonal members to the vertical members with end condition assumed as hinged.
7. Model A-7: Considers a revised structure by providing bracings as diagonal members to the vertical members with end condition assumed as fixed.

Results of displacement and stress ranges :   Fig.3 shows deformed shape of analytical Models A-1, A-5 and A-7 and stress wave at the upper ends of the vertical members, V2, and V6, due to 20 $tonf$ truck load. Analytical results show that the increase in the bending stiffness of the short vertical members decreases the deformation and the stress range. For example, Model A-3 shows displacement at 1/4L is about 80 % of Model A-1. This implies that the increase in stiffness of the short vertical members helps to decrease the deformation and the stress ranges, and thus prolong the fatigue life. It is also advisable to use fatigue resistant structural detail at the ends of the vertical members to avoid any further fatigue cracking. Significant reduction in both deformation and the stress ranges is obtained by providing diagonal members, which is demonstrated by Model A-7.

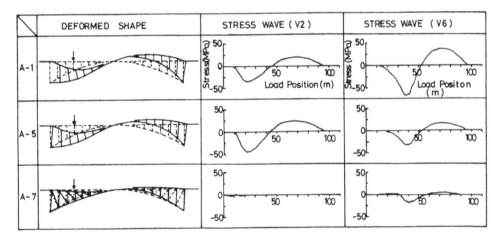

Figure 3. Comparison of deformations and stresses of models A-1, A-5 and A-7.

Calculation of fatigue life :   The ends of the vertical members are fillet welded directly on the upper flange of the arch ribs or girders as shown in Photo 1. Fatigue cracks were formed either in arch rib at the end of weld toe of gussets or in gusset along the filled weld toe. This joint is assumed as load carrying fillet weld, and joint classification E(80) of JSSC [4]is applied assuming that fatigue crack emanates from the fillet weld toe. The bridge is situated in a rural area, and in order to estimate the fatigue life, three types of traffic conditions are assumed, as shown in Fig. 4. They are; (a) The number of daily trucks is originally 500, and increases by 3% annually, (b) it increases constantly by 250 annually, and (c) it is 500, and stays the same. The weight of the trucks is assumed as 20 $tonf$.

Photo 1. Fatigue cracks emanating from the lower end of vertical member of Bridge-A.

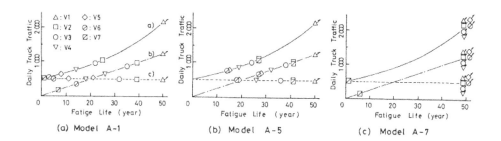

Figure 4. Fatigue life estimation of Bridge-A for various truck traffics.

The computed fatigue lives for each vertical members are plotted on the assumed traffic condition, as shown in Fig. 4. For the original structure, Model A-1, the short vertical members, such as V7, V6 and V5, show relatively short fatigue lives, and that corresponds to the reality. By increasing stiffness of the vertical members, such as Model A-5, the fatigue lives also increase. When the future truck traffic is expected to be less, this rehabilitation measure, with providing fatigue resistant details at the ends of the vertical members, may be sufficient. By applying the diagonal members, all vertical members shows the sufficient fatigue lives, except V7 where diagonal members can not be provided due to insufficient space. This rehabilation measure seems to be the most favorite in order to increase fatigue resistance of the vertical members, of course it requires large cost for repair.

CASE-2: STUDY OF 2-HINGED SKEWED ARCH BRIDGE

Case Description :   The second bridge, hereafter referred as Bridge-B, was constructed in 1963 [5]. The schematic view is shown in Fig. 5. The bridge is situated in one of the earliest modern highway system in Japan, and for the last ten years, the traffic has risen to about 40,000 vehicles per day in one direction. It is a deck type arch bridge and skewed to 62° with longitudinal axis. The overall span, arch span, arch rise and the deck width are 62.2 m, 54.0 m, 11.0m and 9.9 m, respectively. The structural skeleton is composed of two I-shaped arch ribs with 26 vertical members.

A number of cracks were observed at different locations, such as at the upper and the lower end of vertical member, at gusset plate, at end of stringers, at upper end of girder web, etc. The cracks, which we have considered in this study, occurred at locations marked by circle in Fig. 5(b). These cracks can be attributed to various reasons but the dominant factor is supposed to be extremely large number of trucks and occasional overloaded traffic condition.

Originally, the deck was of reinforced concrete. Later, in 1989, owing to serious traffic induced fatigue damage, the deck was replaced with orthotropic steel deck. In addition to this, the arch was stiffened by providing diagonal members, lateral bracing and crossframes.

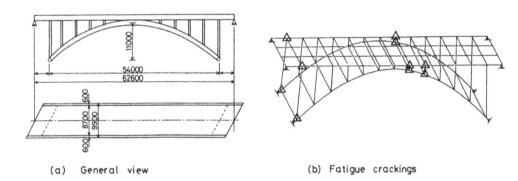

(a)   General view   (b) Fatigue crackings

Figure 5. Plan and elevation of Bridge-B and location of cracks.

Structure Modeling and the Analysis Scheme :   Owing to skew symmetry of the deck and low torsional stiffness, space frame analysis seemed appropriate. In all, seven cases were analyzed, out of which four of them are shown in Fig. 6.

1. Model B-1: The original skew-symmetric arch bridge with lateral bracing. The structure consists of arch rib , girder, stringer and vertical member.
2. Model B-2: Adding diagonal members to the structure of Model B-1.
3. Model B-3: Adding crossframes to Model B-2.
4. Model B-4: Proposed rehabilitation measure with lower lateral bracings, diagonal members and crossframes. It has orthotropic steel deck of 14 mm thickness with

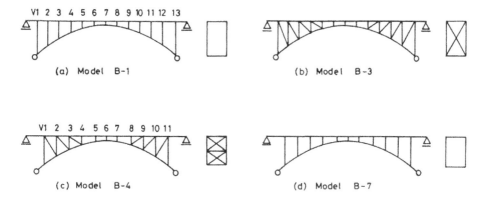

Figure 6. Analytical models of Bridge-B.

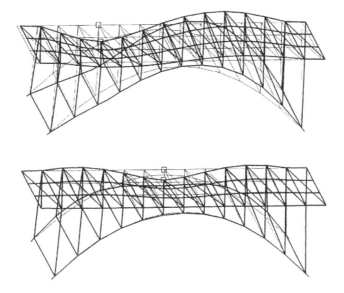

Figure 7. Deformed shape of Model B-1 for Bridge-B, when the load is applied at the quarter point and at the center.

longitudinal stiffener of bulb plates. In addition to this, concrete filled walls are placed and they increase the stiffness of girders.

5. Model B-5: The same as Model B-4, but without diagonal members.
6. Model B-6: Similar to Model B-4, but without crossframes.
7. Model B-7: Similar to Model B-5, considering only presence of lower lateral bracing.

Photo 2. Fatigue crack emanating from lower end of short vertical member of Bridge-B.

Results of stress range : Structural analysis is carried out in the same manner as Bridge-A. Fig. 7 shows the deformed shape of the original structures, Model B-1, when the load is applied at the quarter point (L/4) and at the center (L/2). Because of the skewness of the bridge, the deformation is asymmetric, and thus leads to a significant lateral and vertical deformations.

Due to the asymmetric deformation of the arch ribs, stress ranges at the ends of the vertical members are rather high. It actually causes fatigue cracks at these details, as shown in Photo 2. Fatigue cracks emanates in the vertical member at the fillet weld toe.

Computed fatigue life : Since this bridge is situated in the major highway, the number of traffics is extremely high in recent years, owing to the rapid economic growth in Japan. In 1987 about 40,000 vehicles passed this bridge daily, and about 26 percent of them were trucks. Overloaded trucks were also observed frequently.

In the fatigue analysis of the vertical members of Bridge-B, two traffic conditions are assumed : a) the number of daily truck traffic is linearly increased to 10,400 in 25 years, and b) the number of daily truck traffic remains the same as it was in 1987 i.e. 10,400 trucks per day. The weight of the truck is assumed to be 20 tonf.

The computed fatigue life is plotted in Fig. 8. For Model B-1, the original structure, stress range due to 20 tonf truck are large and thus yields to rather short fatigue life at the end of short vertical members such as V6 and V8.

By applying diagonal members, for example for Model B-2, stress ranges at the short vertical members can be significantly reduced. In this analysis, the stress range at V8 can be lowered by 50 percent, and thus prolong the fatigue life to approximately eight times. Providing diagonal members in arch ribs also reduces the vertical and lateral deformations of this bridge. It seems to help reducing any additional deformation-induced fatigue cracks in the secondary members.

Rehabilation of the bridge : The Bridge-B was retrofitted as follows. First, new vertical members, diagonal members and lateral bracings were placed at the positions.

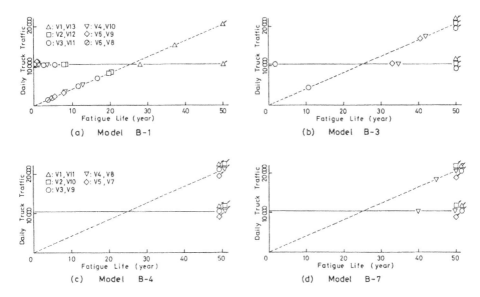

Figure 8. Fatigue life estimation of Bridge-B considering two truck traffic conditions.

Then, the old vertical members, stiffening girders, and the concrete decks were removed, and new orthotropic steel decks were placed. Because of the limitation in space and time, the deck replacement was done in seven blocks, only in thirteen nights. The final shape of the rehabilitated structure can be modeled as Model B-4, where new vertical members, new orthotropic steel decks and diagonal members and bracings are considered.

The rehabilation of this bridge required rather complicated and difficult work at site. However, the Bridge-B is located in one of the major highways and carries a very heavy daily traffic. Suspension of these traffics will greatly affect the economy of this area. Moreover, both ends of the bridge are tunnels, and train and other major highway vehicles run beneath the bridge. All of these geological and economical conditions prevented frequent repair works, and large rehabilation measure such as the one, now feasible.

CONCLUDING REMARKS

Stress analysis and fatigue assessment are carried out on two fatigue affected steel arch bridges. One is located in a rural area, and has arch ribs of relatively rigid box section. The other is located in a heavily loaded major highway, and has a relatively flexible arch rib. Both bridges developed fatigue cracks at the end of the vertical members. High bending stress ranges at these details due to moving trucks are the main reason for these cracks. Various rehabilitation measures are considered in the analysis, and their feasibility largely depends on the daily truck traffic in the future. Providing diagonal members improves the stiffness of the arch, and hence improves the fatigue resistance of the individual members as well.

REFERENCES

1. Japan Load Association, Design specifications for highway bridges, 1990 (in Japanese).

2. Nishikawa, K., Fatigue problems in highway bridges, Bridges and Foundation Engineering, 1983, pp.19-23 (in Japanese).

3. Mizuki, A. et al., Field measurement and repair of road arch bridges which have fatigue crack, Kawata Technical Report, 1985, pp.244-250 (in Japanese).

4. Japan Society of Steel Construction, JSSC Recommendation for Fatigue design, JSSC No.14 1989.11(in Japanese).

5. Expressway Technical Center, Technical Report for Repairing Meishin Expressway Bridges, 1990, pp.76, (in Japanese).

RESIDUAL FATIGUE LIFE OF OLD RAILWAY BRIDGES

F. MANG and Ö. BUCAK
Versuchsanstalt für Stahl, Holz und Steine, Universität Karlsruhe
Testing Center for Steel, Timber and Stone, University of Karlsruhe, Kaiserstr. 12, 7500 Karlsruhe

ABSTRACT

In our days, fatigue steel structures, as for example railway bridges, are designed for a certain service life. Old railway bridges, for example those of the last century, are not calculated for fatigue loadings, because the adequate knowledge was missing. The assessment of the remaining service life is therefore one of the main problems for numerous existing railway bridges.

When assessing the residual service life of old bridge structures, the question arises whether they can be still used after having exceeded the theoretical service life, or if they should be replaced by a new structure after having been newly classified, or if they are still usable after reinforcement or on the basis of shorter service life.

The determination of the loads and the state of the bridge face the difficulty of recording the exact history of the loads and the state of the bridge. Insufficient knowledge about the strength and fatigue behaviour of old steel grades as well as constructions aggravate the above decision.

This paper deals with experimental investigations on five complete bridge structures and on their components of the German Federal Railways, and three bridges from the suburb train in Berlin (suburb train). The results of these component tests are compared with data known from literature as well as with other results of similar investigations performed at the institute, in Karlsruhe.

INTRODUCTION

Steel structures subjected to fatigue loading, such as railway bridges are designed for a given service life.This is due to the fact that fatigue fractures occur after exceeding a critical number of load cycles with a sufficiently high loading level. This calculated failure is covered with a corresponding "safety factor".

For older iron and steel structures, design "for a set time" was not usual. There are numerous structures which have been subjected to fatigue loading for more than 100 years.

After reaching and exceeding the "standard service life", fatigue-loaded structures often cannot be reliably assessed with regard to possible residual life. One reason for this is insufficient knowledge about **the behaviour of the static and fatigue strength of steels** (wrought iron, puddled steel) **and the structures used in the 19th century**. The oldest railway bridge, the

Koblenz/Waldshut bridge between the Federal Republic of Germany and Switzerland, which is in the charge of the "Versuchsanstalt für Stahl, Holz und Steine, dates from 1858 and is still in operation today. This bridge is shown in Fig. 1.

Fig. 1   General view and detail of Koblenz/Waldshut railway bridge

Extensive measurements under traffic load and investigations on the material allow the operation of this bridge to continue at least for the next twenty years. Presently, the gap in single elements due to **rust formation** (Fig. 2) seems to be a problem. One of the main problems in this case is to assess the influence of the corrosion. This mode of corrosion often occurs when the distance between the rivet is large. Once the corrosion process starts, the crack progress gets faster in a short time. A complete removal of the rust in these areas is impossible. In our opinion, the rust should be removed as far as possible and after that the gap is filled with a flexible lute. In this way, this area is protected against rain and aggressive climate. Experiences with this type of protection exist since only a few years, so that a final statement about the long-time behaviour cannot be made at present.

Fig. 2a   Local rust formation   Fig. 2b   After repair with a flexible lute

The other important problem is concerned with the **safety against brittle fracture at low temperature**. It is therefore necessary to examine the ductility of the steel. New testing methods (COD and Ji) are applied in order to obtain a statement about the ductility of these old steels.

Since the design and strength of joints substantially influence the load-bearing behaviour of the total structure, such joints from old steel bridges have to be investigated more precisely. When assessing old steel bridges, material specimens have been taken from the diagonal members or from areas of the construction with lower stresses (for example areas of the supporting points of the construction) for the examination. The data obtained were and are still subjected to very high safety factors since a safe assessment is required. Tests performed in Karlsruhe [5] with preloaded and without preloaded specimen show no influence of the preloading on the fatigue life of the specimens [6, 7, 15, 16 and 17].

The aim of recent investigations in Karlsruhe [6, 7, 17, 16 and 19] is to obtain real residual service life by studies on original joints and structural members or full structures, and compare them with data on simple specimens (small size specimens, i.e. plates with holes) already tested in the institute or which are available from literature. Some characteristics may have to be newly determined.

Knowledge of the residual service life of old bridge structures is required by their operators or by people who need to know how much longer these structures can be operated under present and projected loading conditions. The protection of historical structures and economic considerations are the governing factors.

One of the recent projects was the so-called gun railway (museum railway), which runs in the area of the Wutach Valley in the south of Germany. The other project deals with the bridges for the suburb trains in Berlin. Fig. 3 shows one of these bridge structures of the museum railway and fig. 4 one typical bridge of the suburb train in Berlin.

Fig. 3   A bridge structure of the museum railway

Fig. 4   A bridge structure of the suburb train in Berlin

In order to determine the remaining service life of old railway bridges under load conditions, the following procedure has been applied:

1) Survey of the bridge by visual inspection (state of corrosion). In our experience, this was the determining factor for some of the bridges.

2) If static calculations are available, determination of the critical areas (peak stressed areas). If not, a static calculation of the critical parts of the bridge should be performed and from this, the stress range at the critical positions should be determined.

3) Sampling of less critical but representative areas of the structure.

4) Test investigations, for example chemical analysis, tensile tests and fatigue tests on original material. This is necessary because the material data of this time vary in a wide range.

5) Additional static calculations based on actual material properties.

6) Determination of the previous loading on the bridge (after the new tests, this point can be ignored)

7) Expected future loading (load spectrum) for this bridge.

8) Application of the Miner-rule (linear damage accumulation hypothesis) for the determination of the residual service life as well as the safety factor for this particular bridge.

9) Fracture mechanics investigations (for example COD-tests, J-Integral ...).

10) Measurement of strains on the critical areas of the construction under traffic (application of the rainflow-method to take the load spectrum).

11) Fixing the inspection intervals

Nowadays, sufficient data for material properties have been established by preliminary surveying of various fatigue loaded structures which can be used for initial assessment. Significant investigations need only be done in the final phase for the purpose of confirming the assumptions made. Presently, there is still a gap in knowledge about the behaviour of such bridges as a total system or regarding the load-bearing capacity of complete structural members in the original state.

Questions about

-    the rivet slip
-    the rivet initial load
-    the displacement under shear stress and
-    load distribution on various elements of a structural member

are still unknown. These parameters influence the fatigue behaviour. For this reason, checks on full-scale structural members typical of an actual bridge are needed.

For this purpose, five complete bridges with a length of 4.50, 4.80, 6.80 and 9.5 m dating from 1877-1899, and four main girders of suburb bridges in Berlin with a length of 5.0, 8.5 and 18.5 m dating from 1899-1913 have been investigated in the laboratory of the "Versuchsanstalt für Stahl, Holz und Steine" (testing center of the University), in Karlsruhe.

INVESTIGATIONS ON THE MATERIAL

Tensile tests were carried out on material specimens taken from various bridge structures (more than 60 different types of bridges) and yield strength, ultimate strength and elongation were determined. Figures 5 shows the results for the yield strength of these materials.

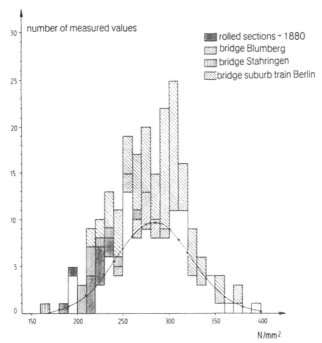

Fig. 5   Results of investigations on the material - distribution of yield strength [1, 18]

The results of the tensile tests carried out on specimens from various parts of the bridges, which were tested in the last 10 years [2, 5, 9, 18 and 19], confirm the material data obtained from other similar structures.

FATIGUE TESTS WITH WROUGHT IRON

The fatigue behaviour of steel materials from 1856 to 1900 was investigated with about 150 specimens. These test specimens were taken out from numerous railway bridges. All test specimens were notched by holes as they exist for riveted constructions. The holes in the test specimens were partly the original holes after removing the rivets carefully. Fig. 15 shows the test results and the lowest line is the line for 97.5% survival probability.

The results can be summarized as follows:

-   The new test results are also covered by the S-N-line evaluated from earlier tests.

-   The dimensions of the test specimens are representative for practical cases, if the specimens are taken from an existing bridge.

-   The influence and differences on the fatigue life between original holes and newly drilled holes exist. But this influence is covered by the given Wöhler-lines.

-   The characteristic stress range ($\Delta\sigma = S_R$) for 97.5% survival probability and for $2 \cdot 10^6$ load cycles is about 100 N/mm^2 for representative test specimens for practical cases.

FATIGUE TESTS ON FULL-SCALE STRUCTURES

Full-scale test specimens were taken from a number of old bridges and fatigue tests were performed on them. One of the tests with a complete bridge was the bridge of the museum railway in the area of Wutach Valley (Blumberg). The bridge was installed into the 50 MN testing machine of the institute. The photograph of fig. 6 shows the installation corresponding to the critical load configuration, load case 1.

Stress results of static tests under both critical load configurations 1 and 2 agree well with calculated values (calculated $\sigma = 135$ N/mm^2, measured 137 N/mm^2 measured).

Calculations carried out with the actual dimensions before testing indicated that load case 1 was the critical load configuration. For this reason, a fatigue test was performed with this load configuration. The test load was twice the maximum load expected with a stress ratio R = +0.35. Under this load, a stress of $\sigma = 137$ N/mm^2 occurred on the outer surface of the main girder. The stresses were calculated with the loads applied and the net cross section.

Flexible supports were used in order to allow for possible irregularities of the bridge. To determine the stress rate on the main girder, critical areas were monitored by strain gauges. With a load cycle of 108.070 the test was terminated, since a correct load application was not possible because of the occurrence of a large crack. The connection of cross-girder/main girder was where the failure occurred. This is shown in fig. 7.

Fig. 6   Bridge specimen in the 50 MN testing machine

Fig. 7   Connection of cross girder/main girder; fracture occurred in web plate

The bridge was cut into small pieces in order to investigate the main girders, longitudinal girders and connections of the cross girder to the main girder.

Fatigue tests on both main girders and longitudinal girders were carried out with a 600 kN resonance testing machine.

The test on the main girders with a span of 3300 mm was performed as a four-point bending test. The distance between points of load application was 800 mm. With this arrangement a constant bending stress occurred in the central area of these girders. Fig. 8 shows the cracks in the main girder I.

Fig. 8   Main girder I after the test. Crack occurred in the web first followed by the angle
L-section) and finally the chord plate. The points of the cracks were different and not
on a line.

Four test specimens were taken from the longitudinal girders and tested in a three-point bending test.

As with the tests on main girder specimens, the cracks start from a rivet and run diagonally through the web. Afterwards, cracks occurred at the L-sections of the same specimens (figs. 9 and 10).

Fig. 9   Longitudinal girder after the test

Fig. 10   Detail of fig. 9

The main girders of other bridges were tested on the same testing conditions with different maximum load. The cracks were initiated at different locations, so that it is difficult to precisely indicate the old crack locations basing on current tests. Fig. 11 shows one of the main girders of the bridge Stahringen after failure.

Fig. 11   Main girder of the bridge Stahringen after failure

The photograph of fig. 12 shows on of the main girders of the Westkreuz bridge in Berlin (suburb train bridge) with a length of 8.5 and a specimen height of 1.80 m in the 50 MN testing machine (fatigue load 25 MN).

Fig. 12   Main girder of the Westkreuz bridge in Berlin in the 50 MN testing machine
(fatigue load 25 MN)

With a nominal stress of 160 N/mm^2 on the outer tension chord, the test was carried out a 3-point bending test with an ultimate stress ratio of R = +0.1. Under this load, the crack occurred on the member of the built-up plate girder, starting from a rivet, for a load cycle 503.180. The test has been continued until the L-profiles and also the outer plate of the tension chord cracked on two different positions. This happened for a load cycle of 543.730. Figure 13 shows the failure on the main girder of the Westkreuz bridge in Berlin.

Fig. 13   Main girder of the Westkreuz bridge in Berlin after failure

After the fatigue test, the cracked main girder was statically loaded with an overload of 35% compared to the fatigue load in order to simulate a fracture. At a temperature of +5°C, the crack on the web plate yawned at first, and on the top of the crack yielding figures became clearly visible. Then, the cracks on the tension chord increased through yielding without the occurrence of a sudden crack.

After the test, the position of the crack has been carefully separated and the rivets have been removed. Figure 14 shows this situation with the cracks discovered additionally, especially those on both segments of the tension chord.

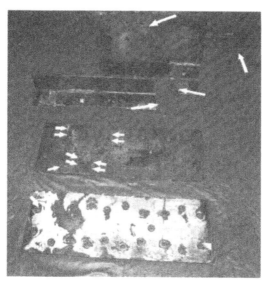

Fig. 14   Additional cracks on the plates of the tension chord of the Westkreuz bridge (they have been discovered after removal of the rivets on the area of the cracks)

The results of these investigations were compared with older test data on puddle steel specimens with drilled holes available in the institute. It is evident that the results from tests on the main and longitudinal girder are within the scatter range of the tests on small specimens (see fig. 15).

WIDE PLATE TESTS

The rivets of the main girder of the bridge Calw were carefully taken off and the flange sheets were exposed. Test specimens of 600 mm x 2500 mm with 4 rows of rivet holes were subjected to fatigue tests in a tensile testing machine of 3 MN with a stress ratio of +0.1. The results demonstrate the critical behaviour of the full-scale tests (wide plate tests) in comparison with that of the normal tests with small-size specimens. Fig. 16 shows one of the test specimens after failure. The cracks occurred on various positions at the same time. The above mentioned difference between the results of wide plate test specimens and small-size test specimens can be seen in fig. 16.

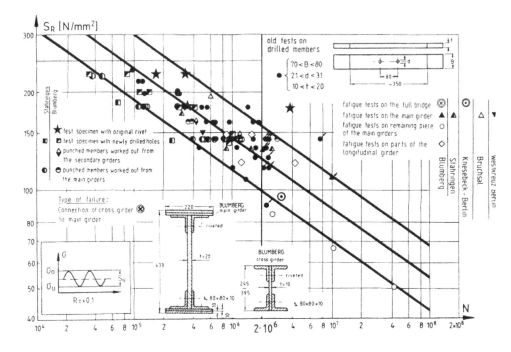

Fig. 15   Comparison of results of the full-scale tests on the Blumberg bridge (museum railway), Stahringen bridge, Bruchsal bridge, suburb train in Berlin and bridge Neustadt with earlier investigations on small test specimens with original and new drilled holes.

Fig. 16   Wide plate test specimens after failure

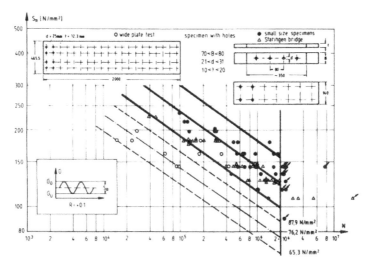

Fig. 17   Results of investigations on wide plate tests in comparison with the results of small specimens

FATIGUE TESTS ON TEST SPECIMENS WITH PRE-LOADED BOLTS

As the fatigue strength of riveted constructions depends also on the pre-load in the rivet (low pre-load in the rivet leads to a high reduction of fatigue strength), a number of fatigue tests were carried out on test specimens with pre-loaded bolts. The materials used were taken from the bridges Koblenz, Stahringen and Calw. the results obtained up to now show an increased life corresponding to the value of the pre-load. The failure cracks shift from the edge of the hole to the full cross section. The location of fracture is shown in fig. 18. It can be seen that, depending on the amount of the pre-load force of the bolts, the allowable load cycle increases. This advantage is especially important for bridges, which have been additionally reinforced with a plate or which have been repaired by means of pre-loaded high-strength bolts and a new segment.

Fig. 18   Test specimen with pre-loaded bolts after failure

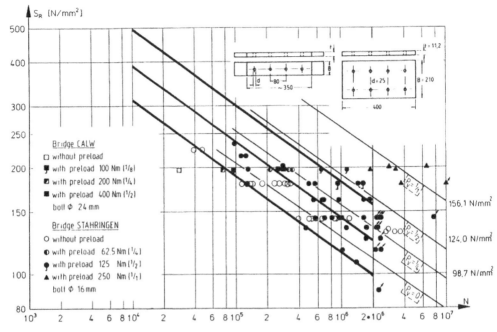

Fig. 19   Comparison of the test results, specimen with and without pre-loading of holes

Previous results of investigations on complete bridge structures and on their large-size structural members are shown in the figures 20 and 21 in comparison with the tests results known from literature.

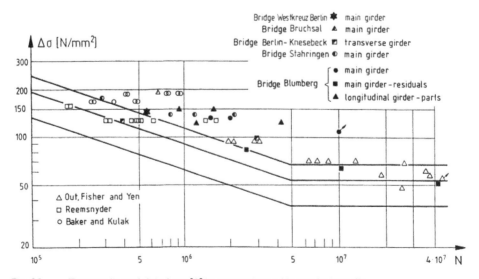

Fig. 20   Test results and data from [3] in comparison with new test results

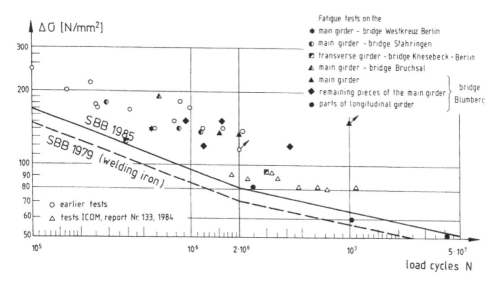

Fig. 21   Test results and data from [4] in comparison with new test results

INVESTIGATIONS ON FRACTURE MECHANICS

In order to obtain information about crack initiation, susceptibility to cracking and crack propagation in wrought iron and early mild steels, investigations on fracture mechanics were carried out. The determination of proper characteristic fracture mechanical values depend on the usually existing restrictions regarding the sizes of the test specimens. Therefore, three-point bending test have been performed in Karlsruhe [10], whereas tests on CT-specimens have been made in the University of Aix-la Chapelle [18] and BAM in Berlin [18]. These tests are still running.

The tests showed that the placement of a suitable initial crack in a test specimen to obtain reliable evidences was diversely problematic. In this respect it can be stated that very large scatter in the results is to be expected for wrought iron in particular.

The fracture mechanical static tests demonstrated that the failure of a cracked test specimen was associated partly with large deformation. This leads to the conclusion that the linear statical fracture mechanics is not applicable to the investigated specimens as could be presumed basing on the existing knowledge. This can also be explained by the fact that the necessary minimum plate thickness to generate a uniplanar elongation in the test specimen is not to be found in the historical steel structures tested in this project (maximum plate thickness of puddle steel $\leq$ 12 mm). According to the knowledge available, the concept of the plastic fracture mechanics (COD, J-Integral) to evaluate the results from the cracked specimens or structural elements can be applied, which, however, will cause a significantly higher expenditure. The reason is that the measurements of crack expansion and elongation have to be made with great care during testing and require increased effort.

On the other hand, the ΔK-concept on the basis of the so-called Paris equation

$$\frac{da}{dN} = C \cdot \Delta K^m$$

ΔK = range of the stress intensity factor
C,m = constants
da/dN = crack growth rate

could be successfully applied also to the test materials in the crack propagation tests carried out simultaneously. This is an important basis for the determination of the required intervals for the inspection of the fatigue loaded structures.

Figures 22 and 23 present some of the da/dN results. As they show, very large scatters are to be expected for wrought iron particularly, which means a careful observation of the steels and their characteristic data.

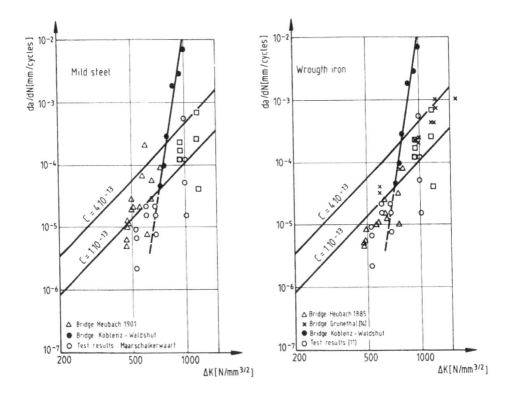

Fig. 22   da/dN-results for test specimens made of mild steel [11, 12, 14, 18]

Fig. 23   da/dN results for test specimens made of wrought iron [11, 12, 14, 18]

On account of this, a method was developed in the University of Aix-la-Chapelle [12, 18], with which the real material values can be determined for very small specimens. A round specimen or 60 mm diameter was taken from the structure by core drilling. The borrow areas, where the samples were taken were the web plates and/or the flange metal sheet of less loaded parts of the bridges. Fig. 24 shows some of these samples.

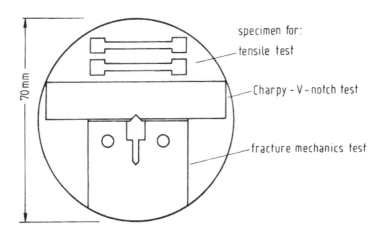

Fig 24   Miniaturized test elements and available tests specimens for various tests [12, 18]

At present, this method is being checked with further test data. Fig. 25 shows a comparison between the samples 1 CT-10 and 1/2 CT-10.

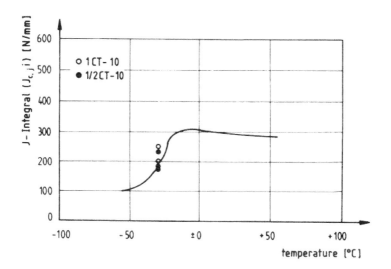

Fig. 25   Comparison between 1 CT-10 and 1/2 CT-10 samples [12]

As recommended beforehand, the J-integral can be used to determine the residual life of a bridge. This can be defined in general as follows.

$$J \quad = \quad \int_T (W \cdot d \, y - T \cdot \frac{du}{dx} \cdot ds)$$

where
W = Energy density
T = Vector of stresses
U = Vector for displacements
T = Integration path around the crack tip
ds = Element of the integration path

The J_{crit} or Ji-values were determined with probes obtained by drilling. For Central Europe these tests were carried out at -30°C. According to the number of the test values, either the minimum or the average value minus two times standard deviation (when sufficient number of samples was available) was taken. The following values are listed as an example.

J [N/mm^2]	s [N/mm^2]	min J [N/mm^2]	Remarks
95	29.	42	Hammer bridge c) Mild steel 1907
92	29.	65	Berlin i) Breisgauer Str.
98	32.	44	Berlin i) Spanische Allee
73	26.	33	Berlin i) Wilhelmsdorfer Str.

The $J_{appl.}$ were determined by Finite Element calculation as a rule.

A more difficult problem is to detect the cracks in a real structure. In order to determine the required inspection intervals,it has to be found out, what is the minimum dimension of a crack in a structure, which can be detected with a sufficient reliability by applying the test or measuring methods at disposal. It can be stated from the experience obtained by the investigation that the cracks equal to or larger than 10 mm can be detected visually. However, this must be taken into consideration that the crack has to crawl out under the rivet head in order to be detected.

The required initial crack length for the fracture mechanics calculation is therefore between 25 and 35 mm depending on the rivet diameter and rivet head dimension.

At present, the investigations are still running

CONCLUSIONS

The results of the fatigue tests on old steel grades as well as on old bridge structures are presented in this paper. The following conclusions can be derived out of them:

- The test results from the total bridge specimens and full-scale tests cannot be compared with the test results from small-size specimens, because the system failed in a different detail.

- The test results from the complete main girder specimens correspond well to those from the tests on small-size tests specimens. However, the values are located near the lower band line for all results.

- Wide plate tests show lower results for fatigue life than those for tests with small-size test specimens.

- A pre-loading of the rivet holes by high-strength bolts results in a higher fatigue strength or higher fatigue life.

- At present, the investigation is still running.

REFERENCES

1. Mang, F. Stahl im Altbau und Wohnungsbau (Steel in Old and Residential Building Construction - Final report). Abschlußbericht zur Forschungsstudie des Landes Nordrhein-Westfalen (VBI-72.02-92/77)

2. Mang, F.; Steidl, G.; Bucak, Ö. (1985) Altstahl im Bauwesen (Old Steel in Construction). Schweißen und Schneiden 1985, No. 1, p. 1-5

3. Brühwiler, E.; Hirt, M.A. (1987) Das Ermüdungsverhalten genieteter Brückenbauteile (Fatigue Behaviour of riveted Bridge Structural Members). Der Stahlbau 1987, No. 1, p. 1-8

4. Herzog, M. (1960) Erwiderung zur Zuschrift von Tschumi, M. auf Herzog, M. Abschätzung der Restlebensdauer älterer genieteter Eisenbahnbrücken (Reply to a letter: Estimation of the Remaining Service Life of Older Riveted Railway Bridges). Der Stahlbau 1986, No. 5, p. 159-160

5. Baehre, R.; Kosteas, D. (1979) Einfluß der Vorbelastung auf die Restnutzungsdauer schweißeiserner Brücken (Influence of the Pre-Load on the Remaining Service Life of Welding Iron Bridges). Bericht Nr. 7496 der Versuchsanstalt für Stahl, Holz und Steine der Universität Karlsruhe, Januar 1979 (not published)

6. Steinhardt, O. (1977) Festigkeitsverhalten von Schweißeisen aus Brückenbauwerken des 19. Jahrhunderts (Fatigue Strength Behaviour of Welding Iron of Bridge Structures of the 19th Century). ETR 26 (1977), No. 6, p. 383-387

7. N.N., Unveröffentlichte Untersuchungen der Versuchsanstalt für Stahl, Holz und Steine der Universität Karlsruhe (Unpublished investigations of the testing laboratory for steel, timber and stones)

8. Stier, W.; Kosteas, D.; Graf, U. (1983) Ermüdungsverhalten von Brücken aus Schweißeisen (Fatigue Behaviour of Bridges Made of Welding Iron). Der Stahlbau 1983, No. 5, p. 136-142

9. Wenzel, F. (1988) Erhalten historisch bedeutsamer Bauwerke (Preservation of Historically Important Structures). Jahrbuch 1987, SFB 315, Universität Karlsruhe (TH), Ernst & Sohn, 1988

10. Baehre, R.; Wittemann, K. (1991) Rißfortschrittsuntersuchungen an Puddelstahl und Flußstahl aus den Jahren 1850 - 1905 (Crack Growth Investigations on Wrought Iron and Mild Steel from the Years 1850 - 1905), Jahrbuch 1991, SFB 315, Universität Karlsruhe (to be published)

11. Maarschalkerwaart, H.M.C.M. van (1990) Determination of Inspection Intervals for Riveted Structures. IABSE Workshop Remaining Fatigue Life of Steel Structures, Lausanne April 1991, p. 335

12. Dahl, W.; Schumann, O.; Sedlacek, G. (1990) Method to Back Decisions on Residual Safety of Bridges. IABSE Workshop Remaining Fatigue Life of Steel Structures, Lausanne April 1991, p. 313

13. Brandes, K. (1991) Crack Growth Tests to Assess the Remaining Fatigue Life of Old Steel Bridges. IABSE Workshop Remaining Fatigue Life of Steel Structures, Lausanne April 1991, p. 103

BRIDGE DECKS CONSTRUCTED USING WELDED STEEL GRID

NEAL H. BETTIGOLE, P.E., F.ASCE
Exodermic Bridge Deck Institute
P.O. Box 374, Westwood, NJ 07675, USA

ABSTRACT

The development and use of a new method of constructing bridge decks using an unfilled, open steel grid composite with a reinforced concrete slab is presented. Starting in 1982, exodermic deck was conceived, tested, installed, and used. In late 1991, a change in the method of construction of this bridge deck type has made it applicable to a much broader range of projects.

INTRODUCTION

Innovation in bridge design faces numerous obstacles. I have learned this first hand through my efforts to bring a novel bridge deck design, which I have named "exodermic", into the mainstream.

Innovation requires a number of important factors for success:
1. Change from existing product or technique must be significant.
2. The innovation must have a firm foundation in theory.
3. Substantial and thorough real world testing is required.
4. Manufacture or construction must be possible without extensive retooling or work force training.
5. Time. Perhaps most important of all, it takes a considerable amount of time to get people to change how they do things.

I would like to illustrate these points through a discussion of the bridge deck concept that I developed.

DISCUSSION

First a bit of history ....

In 1982, after 26 years of bridge design experience, I developed an idea for a new way of building a bridge deck. The concept was straightforward, and I was surprised to learn that no one had tried it before: Take an open steel grid, as widely used on moveable structures, and rather than filling (or half-filling) it with concrete as has been the practice for 60 years, place a <u>reinforced</u> concrete slab <u>on top</u> of the open grid, and connect the two by welding shear connectors to the grid and embedding them in the concrete slab.

An exodermic deck was first installed on the Driscoll Bridge, a 4400 foot (1340 meter) multiple span structure over the Raritan River - and the largest structure on the Garden State Parkway in New Jersey. The project involved the widening of the bridge from 10 to 12 lanes. Pre-cast exodermic modules were used. Prior to this job, and at the sensible suggestion of the Parkway, two exodermic panels were tested at Lehigh University, with excellent results in static and fatigue testing.[1] This installation was opened to traffic on June 1, 1984, and looks today as it did 8 years ago.

Compared to alternative bridge deck types, in the lab tests at Lehigh, and more recently at West Virginia University, exodermic has been found to be far stiffer than alternative decks of the same weight, or, viewed the other way around, far lighter when compared to decks of comparable stiffness.

Thus, the first requirement for a successful innovation -- that it provide meaningful improvement over existing alternatives -- is met.

The research work done at Lehigh and West Virginia, and the subsequent instrumentation and testing (before and after redecking) of a bridge over the New York State Thruway[2] confirmed the validity of the theoretical design basis for exodermic deck. Innovation success factors 2 and 3, which require solid theory and confirmation through testing, could be checked off.

Innovation success factor #4 requires ease of manufacture and installation. Manufacture and construction of exodermic deck requires no significant change in tooling or worker skills. The steel grid is manufactured (currently by L.B. Foster and IKG/Greulich) in the same way it has been constructed for 60 years. Preparing the grid for an exodermic deck involves only the addition of tertiary bars to which vertical studs are welded (together providing the requisite shear connection with the concrete). This addition requires simple punching, cutting, and welding operations as used every day by the manufacturers in the production of their traditional products.

The concrete slab requires nothing unusual - reinforcing steel (generally epoxy coated) and 4000 psi (27 KPa) concrete are all that is required.

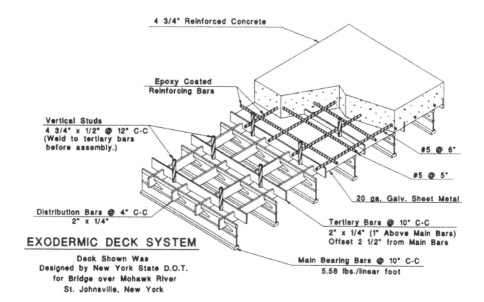

4 3/4" Reinforced Concrete

Epoxy Coated Reinforcing Bars

Vertical Studs
4 3/4" x 1/2" @ 12" C-C
(Weld to tertiary bars
before assembly.)

#5 @ 6"

#5 @ 5"

20 ga. Galv. Sheet Metal

Distribution Bars @ 4" C-C
2" x 1/4"

EXODERMIC DECK SYSTEM

Deck Shown Was
Designed by New York State D.O.T.
for Bridge over Mohawk River
St. Johnsville, New York

Tertiary Bars @ 10" C-C
2" x 1/4" (1" Above Main Bars)
Offset 2 1/2" from Main Bars

Main Bearing Bars @ 10" C-C
5.58 lbs./linear foot

The drawing, above, depicts a typical exodermic deck design. The main bearing bars in the grid can be special rolled shapes, such as the type shown here, or structural T's with depths up to 9 inches (23 cm) or more. By selecting the proper combination of steel grid components and spacings, and concrete type and reinforcement, an engineer can choose a deck design to meet a wide range of weight, span, and strength/stiffness requirements. Exodermic decks are internally composite, and are made composite with a bridge's floor system through the embedment of headed studs.

One aspect of the exodermic concept did slow its adoption until recently. I had always thought of exodermic as a prefabricated module (with the concrete pre-cast on each grid panel) in order to reduce erection time to minimize traffic interruption. In retrospect, it is possible to see how this slowed the acceptance of exodermic -- pre-casting added a complex and costly step to the process. Handling the pre-cast modules, at about 65 lb/sf (318 kg/m²) is significantly more difficult than the 15 lb/sf (73 kg/m²) open grid. Installation on the bridge required new skills from the contractor: grouting joints in both directions, welding studs through openings over 9½ inches (24 cm) deep in the deck modules, and finishing grouted joints to match adjacent precast surface treatment.

Fortunately, other people spotted this difficulty, and found the now obvious solution: cast the reinforced concrete slab in-place once the steel grid has been put down. Placing the steel grid requires only ordinary steel handling skills; the reinforced concrete slab is placed like any other - the only difference being its minimal thickness (3 to 4½ inches - 7.6 to 11.4 cm).

In 1991, 3 bridges were redecked with an exodermic design, where the concrete slab was cast on the bridge.

This bridge, a 600 foot (183 meter) double cantilever with a 30 foot (9 meter) suspended span, located on a major 4 lane route in New York State, spanning a tributary of the Hudson River, was redecked with an exodermic deck beginning in the Autumn of 1991. The main bearing bars of this deck are 3" (7.6 cm) structural T's, and the 4½" (11.4 cm) reinforced concrete slab was cast-in-place.

Selection of the exodermic deck was made under a value engineering clause by the contractor, who estimated that by making the change, one year would be saved in total project time. Cost and time savings were achieved through the simplified construction requirements of an exodermic deck and elimination of the expensive overlay specified in the original design.

Also in 1991, a cast-in-place exodermic deck was specified for a 60 foot (18 meter) pony truss bridge over Metro North's Railroad tracks in Dobbs Ferry, New York. The significance of this deck installation is the span length (11 feet - 3.35 meters), and deck orientation (with traffic, spanning between floor beams).

Preliminary results of static testing with strain gauges indicate considerable stiffening of the whole structure. Stresses in the truss, and its measured deflection under load, indicate that it is composite with the exodermic deck.

The requirement for success with an innovation that is most difficult for an impatient inventor is the fifth one mentioned above, the passage of time. It takes time for people to develop the impression that something new is first worth considering, and then, after more time passes, they believe that it has been tried elsewhere successfully. Time will pass before the sharper decision makers are willing to take a chance with the new product or idea. They must come to feel that not only is the innovative choice the appropriate one for a specific project, but that they gain something (pride in a job well done; to be viewed as a creative thinker by peers and supervisors) in making that choice, while not risking their job or reputation by so doing.

Cast-in-place exodermic decks are catching on quickly. This Spring, in an in-house design, the New York State Department of Transportation specified a cast-in-place exodermic deck for the redecking of a 600 foot (183 meter) stringer-girder bridge in St. Johnsville, NY over the Mohawk River. In this case, an exodermic deck replaced a conventional reinforced concrete slab.

Exodermic decks are currently being seriously considered in bridge rehabilitation projects in 9 states, a number of which involve quite large structures.

CONCLUSION

I believe that after 10 years, and with the recent advent of cast-in-place exodermic, most bridge designers are not only ready to consider this innovative choice in bridge decks, but to go further, and make the commitment.

REFERENCES

[1] Daniels, J. Hartley and R.G. Slutter, Behavior of Modular Unfilled Composite Steel Grid Bridge Deck Panels. Report No. 200.84.795.1, Lehigh University, Fritz Engineering Laboratory, Bethlehem, Pennsylvania, January, 1985.

[2] Darlow, Mark S. and Neal H. Bettigole, Instrumentation and Testing of Bridge Rehabilitated with Exodermic Deck. Journal of Structural Engineering, Vol. 115, No. 10, ASCE, New York, NY, October, 1989.

GROUT INJECTION : A MEANS FOR STRENGTHENING DAMAGED OFFSHORE PLATFORMS

S. PARSANEJAD
Senior Lecturer

A. SALEH
Lecturer

School of Civil Engineering
University of Technology, Sydney
P.O. Box 123, Broadway
N.S.W., AUSTRALIA   2007

ABSTRACT

The effect of grouting on the behaviour of damaged tubular members has been studied both experimentally and analytically. The results have indicated that grouting is an effective means for strengthening damaged tubes with a moderate degree of damage. In this paper, an overview of the past work is first given. Then some results from a recent numerical analysis of fully grouted damaged tubes will be presented.

INTRODUCTION

Damage to tubular members in offshore platforms, caused by collision of supply boats or dropped objects, can reduce the member's strength significantly. [1-9]. In such cases some measure of strengthening is warranted. An economical measure is grouting of the damaged member.

The damage can take the form of a dent, overall bending or a combination of these effects. To study the effect of grouting in enhancing the strength of these damaged members experimentally, twenty-nine small scale specimens have been tested [10-12]. The specimens were subject to various combinations of dent and overall bending. Full grouting as well as partial grouting in the dented region have been included in the experimental program. It has been shown that in fully grouted members, the tube and the grout act compositely throughout the loading and unloading range. In partially grouted tubes, however, the behaviour is unpredictable. In some cases the bond between the tube wall and grout fails prematurely before the attainment of the ultimate load while some specimens maintained their composite action during the test. It was also observed that grouting is fully effective in supporting the tube wall and preventing the local failure in the dented region prior to the attainment of the ultimate capacity of the member.

Analytical expressions, amenable to hand calculations, have also been developed for estimating the strength of both fully and partially grouted damaged tubes [13-15]. At present. a numerical approach is being adopted for the parametric study of these structural elements.

Here a summary of the past experimental and analytical work will be given followed by a brief description of the numerical analysis.

Experimental Work

To date the test results for 29 small scale specimens are available. The diameter to thickness ratio (D/t) of the tubes were within the range of 30 to 60. The depth of dent to diameter ratio (d/D) varied from 0.07 to 0.16. The slendernesses within the imperfection sensitive zone were selected in the range from 60 to 112. The ratio of the maximum induced bending deflection at mid-length (the dent location) to the member length (δ/L) ranged between 0.000 to 0.0052. The length of the grouted segment in partially grouted tubes was also varied. The details of the test procedures, test data and test results are given in detail in References 10 to 12.

The results indicated the following:

1.   The effect of grouting in fully grouted tubes is two fold. Firstly it provides support for the dented zone and prevents the change of cross-sectional shape of the damaged sections before the attainment of the ultimate load. Secondly, it acts compositely with the tube.

2.   Full grouting compensates the loss of strength incurred as a result of a moderate damage (i.e. d/D = 0.15 in combination with δ/L = 0.004) on the original tube.

3.   The effect of grouting in partially grouted tubes is somewhat unpredictable. Some specimens exhibited the two fold effects of the fully grouted tubes. In specimens with short grout lengths (i.e. one-third of the length of the member) the failure was triggered by local buckling at the end of the grouted segment rather than by yielding of the dented zone. In some other specimens the bond between the grout and the tube wall failed prematurely. The grout, however, provided full support for the dent and prevented its local distortion. The failure of the members was initiated as a

result of plastification of the dented zone.

4.   Partial grouting, although not as effective as full grouting, enhances the behaviour of the damaged tubular members pronouncedly.

Analytical Work

Fully grouted members - A simple analytical expression (Equation 1) has been proposed [13] based on simple beam-column analogy, full composite action and first yield collapse criterion. The composite dented section, as shown in Figure 1, is assumed to represent the cross-sectional rigidity of the beam-column.

$$\left(\frac{\sigma_u}{\sigma_y}\right)^2 - \left[\frac{1 + K_{tr}}{\lambda^2_{tr}} + m\right]\left(\frac{\sigma_u}{\sigma_y}\right) + \left(\frac{m}{\lambda^2_{tr}}\right) = 0 \tag{1}$$

Where A_{tr} = transformed cross-sectional area at the dent; A^*_{tr} = transformed cross-sectional area of the undamaged section; E_s = elastic section modulus of steel; $K_{tr} = A_{tr}e_t/Z_{tr}$; P_u = ultimate axial load; Z_{tr} = elastic section modulus of the transformed section at the dent with respect to the dented side; $e_t = e + \delta + e_{tr}$; e = eccentricity of the load with respect to the tube centre; e_{tr} = distance between the centroids of dented and undented transformed cross-sections; L = effective length of the member; $m = A_{tr}/A^*_{tr}$; r_{tr} = transformed radius of gyration at the dent; δ = maximum overall bending; $\lambda_{tr} = (1/\pi)(L/r_{tr})(\sigma_y/E_s)^{1/2}$; σ_y = yield stress of steel; and σ_u = ultimate average axial stress on the transformed undamaged section = P_u/A^*_{tr}.

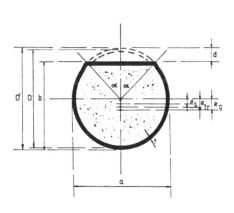

Figure 1. Dented section

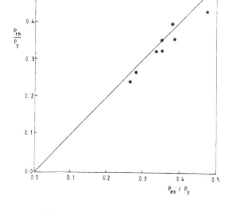

Figure 2. Comparison of theoretical and experimental results

The expression has been compared with experimental results of Reference 10. As it can be seen in Figure 2, it gives a close lower bound estimate of the experimental ultimate capacities. The single point that lies above the line of $P_{ex}/P_y = P_{th}/P_y$ corresponds to a specimen which developed a local buckle at a point other than at the dent location. It is conceivable that an undetected local imperfection caused this local buckle prior to failure of the dented section.

Partially grouted members - Two expressions (Equations 2 and 3) have been proposed for estimating the ultimate strength of partially grout filled damaged tubular members {14 - 15}.

$$\frac{k}{\left[1 - \frac{P_u}{P_y} \lambda^2_e\right]} = \frac{2 \, \text{Sin} \left[\frac{\pi}{2}\left(1 - \frac{P_u}{P_y}\right)\right] - (\text{Sin} \, \alpha - \alpha \, \text{Cos}\alpha)}{2\pi \frac{P_u}{P_y}} \tag{2}$$

$$\left(\frac{\sigma'_u{}^2}{\sigma_y}\right) - \left[\frac{1 + K_s}{\lambda^2_e} + 1\right] \frac{\sigma'_u}{\sigma_y} + \frac{1}{\lambda^2_e} = 0 \tag{3}$$

Where A_s = cross-sectional area of steel; D = mid-thickness diameter; $I = \{I_{tr}L_g + I'_s(L - L_g)\}/L$; I'_s = moment of inertia of the undamaged steel sections; I_{tr} = transformed moment of inertia at the dent; $K_s = A_s e'_t/Z_s$; $P_e = \pi^2 E_s I/L^2$; $P_y = A_s \sigma_y = \pi D t \sigma_y$ = elastic section modulus of the steel section at the dent with respect to the dented side; d = depth of dent; e_s = eccentricity caused by the dent; $e'_t = e + \delta + e_s$; $k = e'_t/D$; L_g = length of the grouted segment; $\alpha = \cos^{-1}(1 - 2d/D)$; λ_e = equivalent slenderness = $(P_y/P_e)^{1/2}$; and $\sigma'_u = P_u/A_s$.

Equations (2) and (3) are based on the assumption that, in the beam column of Figure 3, the grout prevents the change in cross-sectional geometry at the dent and contributes to the stiffness of the grouted segment, but does not take part in the load carrying function. That is the steel section at the dent carries the full load. These assumptions are consistent with the experimental observations pointed out earlier.

Equation (2) expresses the intersection point of the pre-ultimate and post-ultimate load - deflection curves as schematically shown in Figure 4. As shown in Figure 5, Equation (2) gives acceptable predictions of the experimental strengths of ten specimens [12] with a maximum deviation of 14 per cent.

Equation (3) is based on the beam-column analogy normally adopted by codes of practice. It treats the structure of Figure 3 as a beam-column, with a uniform equivalent bending rigidity, which

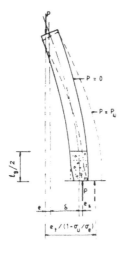

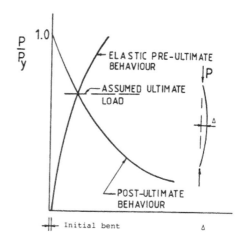

Figure 3. Beam- column model

Figure 4. Pre- and post-ultimate behaviour of a column with initial bent

satisfies the first yield collapse criterion at the dented section. The estimates of strengths by Equation (3) is compared with the ten experimental results of Reference 10 in Figure 6. It is apparent that the estimates are an acceptable lower bound for the experimental ultimate capacities.

Numerical Approach

The work presented in the preceding sections relates to the effect of grouting in enhancement of the

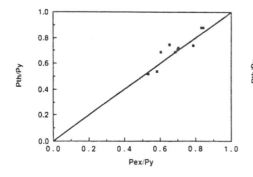

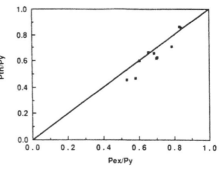

Figure 5. Comparison of theoretical and experimental results

Figure 6. Comparison of theoretical and experimental results

behaviour of damaged tubular members in isolation. To establish the effectiveness of this repair technique in rehabilitation of offshore platforms containing damaged members, a numerical approach is being adopted. An existing method of analysis [16 - 18] capable of incorporating geometric (including large deflection) and material non-linearities is being modified to cater for undamaged empty tubular, undamaged grouted tubular and dented grouted sections. The program is being tested for analysis of grout filled damaged tubular members. It will then be applied to the study of the behaviour of the whole platform in the pre-damaged, post-damaged and post-grouted states.

The method of analysis uses a constant structure stiffness which is assembled from the stiffness matrices of beam elements with assumed linear load-deflection relationships. This matrix is used to generate a series of load and deflection vector pairs which are then utilized in conjunction with out-of-balance forces to iteratively determine the actual response of the structure throughout the non-linear region. A deformation controlled procedure operating on the entire deformation vector enables the load deflection path in the post buckling domain to be found. A detailed description of the method is given in References 16 to 18.

The method is applied to one of the fully grouted test specimens (D1) of Reference 10. The specimen has a mid-thickness diameter of 68.45 mm, a tube wall thickness of 1.398 mm, a length of 1860 mm, a steel yield stress of 494 MPa, an elastic modulus of 199 GPa for steel, a grout cylinder strength of 31 MPa, an elastic modulus of 16.5 GPa for grout, a depth of dent to diameter ratio of $d/D = 0.1425$, and an overall bending to length ratio of $\delta/L = 0.00042$.

The behaviour of the steel is assumed as linear elastic - perfectly plastic. The behaviour of the unconfined grout cylinders up to the crushing load has been experimentally measured (see Figure 7). The post-ultimate behaviour is unknown, however, since the grout is made of a cement-water mix with some expansive agent, its failure is expected to be extremely brittle with little or no post-ultimate strength.

Tests have also been made on grouts confined in tubes where compression was applied to the grout only. Tests indicated a large increase (over 100%) in grout strength followed by a ductile post-crushed behaviour. Two points are noteworthy in relation to these results. The observed increase in strength is expected to be an upper bound for the strength of the grout in tubular members, since the lateral expansion of the tube due to Poisson's effect would reduce the confining pressures. The ductile behaviour observed in the post-crushed range is purely due to the lateral restraint of the tube to the crushed (powdered) grout.

Because of these considerations, three different grout behaviour, labelled "a", "b" and "c" in Figure 7, is assumed. Plots "a" and "b" assume some post-crushed strength after reaching the unconfined strength. Plot "a" assumes a ductile plateau while plot "b" assumes an abrupt but linear unloading to a post-collapse strain of 0.004. An assumption of sudden vertical unloading after crushing has also been tried, but led to numerical instability. Plot "c" assumes an increase in the crushing strength and a post-ultimate plateau due to the confinement of the grout.

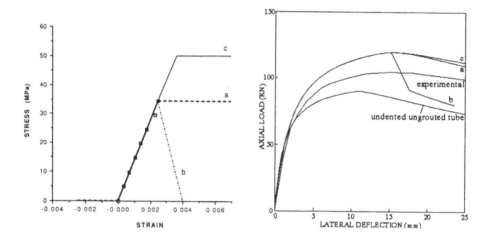

Figure 7. Grout behaviour   Figure 8. Load-deflection behaviour

The numerical results obtained for Specimen D1, based on the grout behaviour "a", are plotted in the form of load versus lateral deflection at the mid-length in Figure 8. The experimental results and the load-deflection plot of the original undamaged and empty tube is also shown. The latter is also obtained using the adopted method of analysis. It can be observed that the numerical results predict the shape of the load-deflection path very accurately, but over estimates the ultimate strength by 14 per cent. It is also observed that the deflection at the ultimate load is predicted almost exactly and the grout has strengthened the damaged tube to a level far beyond the original strength of the empty undamaged tube.

The grout behaviour "b" results in a sudden unloading which is contrary to the experimental behaviour. The grout behaviour "c" gives almost exactly the same pre-ultimate behaviour as the behaviour "a". This is because in this specimen the ultimate load is reached when the compressive strain of the grout is near 0.0025. The change in grout behaviour, which is assumed to occur at strains larger than 0.0025, is therefore affecting the post-ultimate behaviour of the specimen. The grout behaviour "c" seems to predict the post-ultimate rate of unloading more closely than the behaviour "a". It is apparent that the confinement of the grout has a pronounced effect on the behaviour of grout filled damaged tubular members and need to be explored further.

Concluding Remarks

A brief summary of the previous and on going work on the behaviour of grout filled damaged tubular members was given. The work is being extended at present to the study of the effect of grouting of damaged members on the behaviour of the whole platform.

REFERENCES

[1]   Iqbal, S., Sohal, S. and Chen, W.F., Local Buckling and Sectional Behaviour of Fabricated Tubes, J. of Struct. Engr. Vol. 113, No. 3, Mar. 1987.

[2]   Ueda, Y. and Rashed, S.M.H., Behaviour of Damaged Tubular Structural Members, J. of the Energy Resources Technology, Transactions of the A.S.M.E., Vol. 107, September, 1985.

[3]   Ellinas, C.P. and Valsgard, S., Collisions and Damage of Offshore Structures, J. of Energy Resources Technology, Transactions of the A.S.M.E., Vol. 107, September, 1985.

[4]   Ellinas, C.P., Ultimate Strength of Damaged Tubular Bracing Member, J. of the Structural Division, A.S.C.E., Vol. 110, No. 2, February, 1984.

[5]   Ellinas, C.P. and Walker, A.C., Damage on Offshore Tubular Bracing Members, Proceedings of the IABSE Colloquium on Ship Collision with Bridges and Offshore Structures, Preliminary Report, Copenhagen, p. 253-261, 1983.

[6]   Smith, C.S., Strength and Stiffness of Damaged Tubular Beam Columns, Offshore Technology Conference, Houston, May, 1981.

[7]   Smith, C.S., Somerville, W.L. and Swan, J.W., Residual Strength and Stiffness of Damaged Steel Bracing Members, Offshore Technology Conference, Houston, May, 1981.

[8]   Taby, J. and Moan, T., Theoretical and Experimental Study of the Behaviour of Damaged Tubular Members in Offshore Structures, Norwegian Maritime Research, No. 2, 1981.

[9]   Smith, C.S., Kirkwood, W. and Swan, J.W., Buckling Strength and Post-Collapse Behaviour of Tubular Bracing Members Including Damage Effects, Second International Conference on Behaviour of Offshore Structures, Imperial College, London, August, 1979.

[10]   Parsanejad, S., Experimental Strength and Post-Collapse Behaviour of Grout-Filled Damaged

Tubular Members, International Conference on Structural Faults and Repair - 87, University of London, July 1987.

[11]   Parsanejad, S., Tyter, S. and Chin, K.Y., Experimental Investigation of Grout-Filled Damaged Tubular Members, in "Steel Structures - Advances, Design and Construction", Edited by R. Narayanan, Elsevier Applied Science, London and N.Y., 1987.

[12]   Parsanejad, S. and Gusheh, P., Test on Partially Grout-Filled Damaged Tubular Members, Civil Engineering Transactions of I.E. Aust., Vol. CE 30, No. 5, Dec. 1988.

[13]   Parsanejad, S., Strength of Grout-Filled Damaged Tubular Members, J. Struct. Engr. ASCE, Vol. 113, No. 3, March 1987.

[14]   Parsanejad, S. and Gusheh, P., Analytical Expression for Ultimate Strength Analysis of Partially Grout-Filled Damaged Tubular Members, Civil Engr. Trans. of I.E. Aust., Vol. CE 31 No.1, 1990.

[15]   Parsanejad, S. and Gusheh, P., Behaviour of Partially Grout-Filled Damaged Tubular Members, to appear in J. Struc. Engr. - ASCE, 1992.

[16]   Sedlacek, G., Lopetegui, J., Saleh, A., Stutzki, Ch., Ein computerorientiertes Verfahren zur statischen Berechnung raeumlicher Stabwerk unter Beruecksichtigung nichtlinearer Effekte, Der Bauingenieur 60, 297 - 305, 1985.

[17]   Lopetegui, J., Sedlacek, G., Some Applications of Load-Deformation States, Proceedings of the International Conference on Numerical Methods in Engineering; Theory and Applications, NUMETA'87, Swansea, 6-10 July 1987.

[18]   Saleh, A., Lopetegui, J., Nonlinear Analysis using Linear Elements, Proceedings of the Sixth International Conference in Australia on Finite Element Methods, July 8th - 10th, 1991, Sydney, Australia.

THE ULTIMATE STRENGTH AND STIFFNESS OF UNIPLANAR TUBULAR STEEL X-JOINTS LOADED BY IN-PLANE BENDING

G.J. VAN DER VEGTE[*], L.H. LU[*], J. WARDENIER[*], R.S. PUTHLI[*/**]

* Delft University of Technology, Faculty of Civil Engineering,
Stevin Laboratory - Subdivision Steel Structures, Stevinweg 1,
P.O. Box 5049, 2600 GA Delft, The Netherlands.
** T.N.O. Building and Construction Research,
Lange Kleiweg 5, 2600 AA Rijswijk, The Netherlands

ABSTRACT

Little experimental data on X-joints loaded by in-plane bending is available compared with axially loaded X-joints. In order to check the influences of the geometrical parameters on the static strength under in-plane bending as given by several design recommendations, 15 finite element analyses have been performed on X-joints. The influences of the geometrical parameters β and 2γ on the in-plane bending strength are made clear. The experimental results of one X-joint tested to failure, and used for calibrating the model, are also presented.
Finally, the ultimate moments resulting from the numerical analyses have been compared with those obtained from several design codes and recommendations.

NOMENCLATURE

d_0 : outer diameter of the chord

d_1 : outer diameter of the braces

$f_{y,0}$: yield stress of the chord member

$f_{u,0}$: ultimate tensile stress of the chord member

l_0 : length of the chord

t_0 : wall thickness of the chord

t_1 : wall thickness of the braces

M_{ipb} : in-plane bending moment on the braces

Illustrations at the end of the paper.

$M_{u,num}$: numerically determined ultimate (in-plane bending) moment

α : the geometric chord length parameter $2*l_0/d_0$

β : diameter ratio d_1/d_0

γ : chord radius to thickness ratio $d_0/2*t_0$

τ : the wall thickness ratio t_1/t_0

INTRODUCTION

Experimental data on the strength and non-linear behaviour of uniplanar tubular joints under in-plane loading is limited as compared to axially loaded joints [1]-[10]. For X-joints loaded by in-plane bending, only 18 test data are available [8]-[10].

Due to the extensive costs of experiments and the still decreasing computer costs, it is very attractive to create and use reliable numerical models to obtain a better insight in the moment - rotation behaviour of hollow section joints loaded by in-plane bending. Therefore, in this study, non-linear finite element analyses on tubular X-joints loaded by in-plane bending, have been carried out.

The study consists of a total of 15 numerical analyses. The range of β and 2γ values, which are the most influencing parameters on the strength of the joints, covers the whole range as generally applied in tubular structures. Four β values (ratio between brace and chord diameters), varying from 0.25 to 1.0, and four 2γ values (ratio between chord diameter and chord thickness) varying from 14.5 to 50.8, have been considered.

The general purpose finite element computer program MARC has been used for the numerical work. Material and geometrical non-linearity are taken into account.

Finally, the static strength values resulting from the numerical analyses are compared with the values recommended by several design codes (API [11], AWS [12], Eurocode 3 [13] and the formulae recommended by Gibstein [1] and Wardenier [14]).

RESEARCH PROGRAMME

The research programme is summarized in table 1. The configuration of the X-joints is shown in figure 1. A total of 15 finite element analyses on tubular X-joints loaded by in-plane bending have been performed. Four β values and four 2γ values have been considered (β - 0.25, 0.48, 0.73, 1.0 and 2γ - 14.5, 25.4, 36.9, 50.8). For the joints X1 to X8 the wall thickness ratio τ is set to 0.5, for the joints X9 to X16 τ is taken as 1.0. For all joints the geometrical chord length parameter α is set to 12.0. The nominal dimensions of the joints are given in table 2.

The steel grade of all chord members is Fe510 with f_y - 355 N/mm^2 and

f_u - 510 N/mm^2. To avoid premature brace failures, the steel grade used for the braces is StE 690 with a nominal value of the yield strength f_y - 690

N/ mm^2. However, in this research programme, a practical stress-strain

relationship has been used with a yield strength of 745 N/mm^2 [15]. (Joint X1 has not been analysed, since for this combination of β and 2γ the full plastic moment of the braces is achieved before the joint fails, even when the braces are fabricated with a higher steel grade.)

TABLE 1
Research programme

	2γ	14.5	25.4	36.9	50.8
$\beta = 0.25$		X1	X2	X3	X4
$\beta = 0.48$		X5	X6	X7	X8
$\beta = 0.73$		X9	X10	X11	X12
$\beta = 1.0$		X13	X14	X15	X16

TABLE 2
Summary of the geometries investigated

| | Nominal dimensions (mm) | | | | Non-dimensional parameters | | | |
| | Chord | | Braces | | | | | |
Joint	d_0	t_0	d_1	t_1	α	β	2γ	τ
X1	406.4	28.0	101.6	14.2	12.0	0.25	14.5	0.5
X2	406.4	16.0	101.6	8.0	12.0	0.25	25.4	0.5
X3	406.4	11.0	101.6	5.6	12.0	0.25	36.9	0.5
X4	406.4	8.0	101.6	4.0	12.0	0.25	50.8	0.5
X5	406.4	28.0	193.7	14.2	12.0	0.48	14.5	0.5
X6	406.4	16.0	193.7	8.0	12.0	0.48	25.4	0.5
X7	406.4	11.0	193.7	5.4	12.0	0.48	36.9	0.5
X8	406.4	8.0	193.7	4.0	12.0	0.48	50.8	0.5
X9	406.4	28.0	298.5	28.0	12.0	0.73	14.5	1.0
X10	406.4	16.0	298.5	16.0	12.0	0.73	25.4	1.0
X11	406.4	11.0	298.5	11.0	12.0	0.73	36.9	1.0
X12	406.4	8.0	298.5	8.0	12.0	0.73	50.8	1.0
X13	406.4	28.0	406.4	28.0	12.0	1.0	14.5	1.0
X14	406.4	16.0	406.4	16.0	12.0	1.0	25.4	1.0
X15	406.4	11.0	406.4	11.0	12.0	1.0	36.9	1.0
X16	406.4	8.0	406.4	8.0	12.0	1.0	50.8	1.0

FINITE ELEMENT ANALYSES

The main characteristics of the finite element analyses can be summarized as follows :

- The finite element analyses have been performed using the general purpose finite element program MARC which can model the joint behaviour under material and geometrical non-linearity.
- Pre- and post processing has been carried out with use of the program SDRC-IDEAS.
- Eight noded thick shell elements (MARC element type 22) are used to model the joints. Both the corner nodes and the midside nodes of these elements have six degrees of freedom. Quadratic interpolation functions are used for coordinates, displacements and rotations. Transverse shear strains are taken into account.
- Due to symmetry in the joint geometry and loading, only a quarter of each joint has been modelled. The total number of elements used to model the joints is about 420. The finite element meshes used for the different β values are shown in figure 2.
- To exclude the influence of axial forces in the chord, two forces have been applied on the top of the braces, resulting in pure bending moments at the chord-brace intersection (see figure 3).
- The updated Lagrangian procedure has been used to update the displacement field after each iteration.
- The material properties are represented by logarithmic stress-strain curves which have been modelled as step-wise linear relationships (figure 4). These true stress-true strain relationships have been obtained after converting experimentally determined tensile curves [15]. Futhermore, the Von Mises yield criterion and isotropic strain hardening have been used.

Modelling of the welds

For all X-joints except the joints with $\beta = 1.0$, the geometry of the weld has been modelled in accordance with the welding details as recommended by the AWS D1.1 [12]. Shell elements have been used to model the welds, which is shown in figure 5. The dashed lines indicate the mid-planes of the shell elements. The dotted lines indicate how the geometry of the weld has been moved from the outer surfaces to the mid-planes of the circular members. Elements AB (at the crown point) and A'B' (at the saddle point) represent the brace wall, while elements BC and B'C' represent the chord wall. Elements AC and A'C' are used to model the fillet part of the weld i.e. the additional volume exceeding the plate thickness. It can be shown that the throat thickness of the fillet part of the butt weld, after averaging at the saddle and crown point, is about $0.4\ t_1$. Therefore, the wall thickness of the elements AC and A'C', used to model the fillet part of the butt weld, is set to $0.4\ t_1$.

CALIBRATION OF THE NUMERICAL MODELS

In ref. [10] and [16] experimental and numerical research is presented in which the moment-rotation behaviour of one uniplanar and three multiplanar X-joints loaded by in-plane bending is described. The values of the non-dimensional geometrical parameters β and 2γ are 0.6 and 40.0. The numerical and experimentally determined moment-rotation curves for the uniplanar joint are shown in figure 6. The numerical and the experimental moment-rotation curves can be seen to show good agreement, both for the initial stiffness as well as the strength at Yura's deformation limit (Yura's deformation limit : ref. [7]).

NUMERICAL RESULTS

In figure 7, the numerically determined moment-rotation curves are presented for each group of X-joints with the same β-value. The non-dimensionalized moment $M_{ipb}/f_{y,0}* t_0^2* d_1$ has been plotted against the rotation. Furthermore, the deformation limit as suggested by Yura [7] is plotted in all figures. The static strength values as well as the values recommended by the design codes and recommendations considered are summarized in table 3. Ultimate moment is considered to be reached if a maximum in moment is observed on the braces or when the rotation of the braces exceeds a practical deformation limit, which is equal to 80 $f_{y,0}/E$, as suggested by Yura [7].

In figure 8 the non-dimensionalized numerical ultimate moments have been plotted as a function of β, while in figure 9 the non-dimensionalized static strengths have been presented as a function of 2γ. Both in figure 8 as well as in figure 9, curves for the different design codes and recommendations are presented.

TABLE 3
Numerical results for the X-joints loaded by in-plane bending

Joint	$M_{u,num}$	$\dfrac{M_{u,num}}{f_{y,0}*t_0^2*d_1}$	API	AWS	Eurocode 3	(1)	API	AWS	Eurocode 3	(1)
							$M_{u,numerical}/ M_{u,formula}$			
			*1.7	*1.8	char.	mean	*1.7	*1.8	char.	mean
	[kNm]		[kNm]	[kNm]	[kNm]	[kNm]				
X1			184	120	102	111				
X2	45.9	4.97	60	47	44	48	0.763	0.968	1.046	0.954
X3	24.1	5.52	28	24	25	27	0.846	1.000	0.963	0.878
X4	14.3	6.21	15	13	16	17	0.952	1.072	0.924	0.842
X5	344.4	6.39	537	426	369	405	0.641	0.808	0.933	0.850
X6	149.6	8.50	175	153	160	175	0.853	0.975	0.938	0.855
X7	85.1	10.23	83	75	91	100	1.027	1.128	0.936	0.854
X8	50.6	11.51	44	41	56	62	1.155	1.241	0.898	0.819
X9	873.3	10.51	1153	855	877	962	0.757	1.022	0.996	0.908
X10	385.6	14.21	377	315	379	415	1.024	1.222	1.018	0.928
X11	218.5	17.04	178	157	216	237	1.227	1.393	1.012	0.923
X12	136.9	20.18	94	86	134	147	1.453	1.600	1.022	0.932
X13 *	1595.5	14.11	2027	1616	1626	1783	0.787	0.988	0.981	0.895
X14 *	752.8	20.38	662	577	702	770	1.137	1.304	1.072	0.978
X15 *	443.6	25.41	313	283	400	439	1.418	1.567	1.108	1.011
X16 *	283.6	30.72	165	153	248	272	1.714	1.852	1.142	1.042

Remarks :   * = modelled without a weld
            (1) = formula recommended by Wardenier [14]

DISCUSSION OF THE NUMERICAL RESULTS

From the results of the finite element analyses, summarized in the moment-rotation diagrams in figure 7, the ultimate moment-β curves in figure 8 and the ultimate moment-2γ curves in figure 9, the following observations can be made.

For each group of X-joints with the same 2γ-value, it appears that an increasing value of β results in a nearly linear increasing non-dimensionalized ultimate moment which is in agreement with the design codes considered.

For all X-joints, it appears that increasing values of 2γ results in increasing non-dimensionalized ultimate moments. The ultimate moment, however, decreases for larger 2γ values.

For the joints with large β and 2γ values (X12, X15 and X16), it is observed that the rotation at maximum load level is smaller than Yura's deformation limit. For joint X5, the full plastic moment of the brace is reached before the deformation limit is achieved. For the other joints, the rotation capacity is sufficiently large.

COMPARISON WITH SEVERAL DESIGN CODES

The ultimate moment values found for the X-joints can be compared with the predictions of several design codes. The numerically determined static strength values are compared with the following codes and recommendations :

- The API RP2A design code [11] gives lower bound design values for uniplanar joints loaded by in-plane loading. The design values incorporate a safety factor of 1.7.

- The AWS D1.1-88 [12] design code gives also lower bound design values. The design values incorporate a safety factor of 1.8.

- Eurocode 3 [13], based on the formula recommended by Wardenier [14], gives design values for joints loaded by in-plane bending. The characteristic values can be obtained by multiplying the design values by $\gamma_m = 1.1$.

- Gibstein [1] and Wardenier [14] recommend the following formula for the mean strength of uniplanar X-joints under in-plane loading :

$$M = c * \beta * \gamma^{0.5} * f_{y,0} * t_0^2 * d_1 \qquad (1)$$

with: $c = 6.0$ according to Gibstein
       $c = 5.85$ according to Wardenier

The small differences between the values obtained from API $*$ 1.7 versus AWS $*$ 1.8 are caused by the fact that for the API the load capacity formulae are used, whereas for the AWS the punching shear formulae are used. As a consequence, the AWS provides different curves for different τ values.

It should be mentioned that mean values can not be compared with characteristic or lower bound values. Nevertheless, all formulae have been indicated in the figures for comparison. Comparison of the numerical results with the available formulae is presented in table 3. The following observations are made :

For all joints except those with high β and high 2γ values (X15 and X16), the numerical ultimate moments lie below the values determined with the formulae of Gibstein and Wardenier. This may be caused by the smaller dimensions of the test specimens in the experiments for which the weld sizes may be larger. Consequently, the strength increases. The β and 2γ tendency in the formulae of Gibstein and Wardenier (and Eurocode 3) is in good agreement with the results of the finite element analyses.

The numerically determined strengths are in reasonably good agreement with the values of Eurocode 3 multiplied by 1.1 (i.e. characteristic values).

The API does not provide a lower bound curve for all numerical static strengths. For low 2γ values (14.5 and 25.4), some data points fall below the API lower bound line, which was also observed earlier by Mäkeläinen [17]. In addition, the 2γ influence on the strength is not included by the API, which is not correct.

None of the numerically determined ultimate moments lies below the lower bound line of the AWS (except joint X5, which is a brace failure). However, AWS does not give a good representation of the 2γ influence on the static strength of X-joints under in-plane loading.

CONCLUSIONS

Based on the results of the 15 finite element analyses on uniplanar X-joints loaded by in-plane bending, the following conclusions can be drawn :

- An increasing value of β results in a nearly linear increasing value of the non-dimensionalized ultimate moment, which agrees with the design codes and formulae considered.
- An increasing value of 2γ results in an increasing value of the <u>non-dimensionalize</u>d ultimate moment $M_u / f_{y,0} * t_0^2 * d_1$. This effect is more pronounced for larger β values. Note, however, that the ultimate moment decreases for larger 2γ values.
- The API rules do not provide lower bound values of the static strengths. The 2γ influence is not taken into account, which is also not correct.
- In spite of the fact that the AWS provides lower bound static strengths, the 2γ influence is not in line with the ultimate moments.
- For all joints, the numerical ultimate moments are in reasonably good agreement with the formula of Eurocode 3 multiplied by γ_m ($\gamma_m = 1.1$).

- The formula recommended by Gibstein as well as the formula recommended by Wardenier gives, in general, higher ultimate moments than the finite element results. The β and 2γ influences, however, are represented quite well.

REFERENCES

1. Gibstein M.B., The Static Strength of T-Joints Subjected to In-Plane Bending., Det Norske Veritas Report 76-137, April 1976.

2. Stol H.G.A., Puthli R.S., Bijlaard F.S.K., Experimental Research on Tubular T-Joints under Proportionally Applied Combined Static Loading., Boss Conference, Delft, 1985.

3. Toprac A.A., Natarajan M., Erzurumlu H., Kanoo A.L.J., Research in Tubular Joints : Static and Fatigue Loads., OTC Proceedings, Vol. 1, No. 1062, 1967

4. Tebbett I.E., Beckett C.D., Billington C.J., The Punching Shear Strength of Tubular Joints Reinforced with a Grouted Pile., OTC Proceedings, Vol. 2, No. 3463, 1979

5. Sparrow K.D., Ultimate Strength of Welded Joints in Tubular Steel Structures.

6. Akiyama N., Yajima M., Akiyama H., Otake F., Experimental Study on Strength of Joints in Steel Tubular Structures., JSSC Vol. 10, No. 102, 1974 (in Japanese).

7. Yura J.A., Zettlemoyer N., Edwards I.F., Ultimate Capacity Equations for Tubular Joints., OTC Proceedings, Vol 1, No 3690, 1980.

8. Stamenkovic A., Sparrow K.D., Load Interaction in T-Joints of Steel Circular Hollow Sections., Journal of Structural Engineering, ASCE, Vol. 109, No. 9, Sept. 1983.

9. Yura J.A., Swensson K.D., Strength of Double-Tee Tubular Joints : Interaction Effects., Phase 3 Final Report, Vol. 3 of 3, Phil M. Ferguson Structural Engineering Laboratory, The University of Texas at Austin, 1986.

10. van der Vegte G.J., Koning C.H.M. de, Puthli R.S., Wardenier J., The Static Strength and Stiffness of Multiplanar Tubular Steel X-Joints., Internat. Journal Offshore and Polar Eng., ISOPE, vol.1, no.1., March 1991.

11. American Petroleum Institute, Recommended Practice for Planning, Designing and Constructing Fixed Offshore Platforms, API RP2A, 17th Edition, 1987.

12. American Welding Society, Structural Welding Code, AWS D1.1-88, 1988.

13. Eurocode No.3, Design of Steel Structures, Part 1 : General Rules and Rules for Buildings, 1989.

14. Wardenier J., Hollow Section Joints., Delft University Press, 1982.

15. Eurocode No.3, Design of Steel Structures, Part 1 : General Rules and Rules for Buildings, Chapter 3 : Design Against Brittle Fracture, Background Documentation, 1989.

16. van der Vegte G.J., Koning C.H.M. de, Wardenier J., The Static Strength and Stiffness of Multiplanar Tubular Steel X-Joints. Intern. Journal of Offshore and Polar Engineering, ISOPE, vol.1, no.3., Sept 1991.

17. Mäkeläinen P.K., Puthli R.S., Bijlaard F.S.K., Strength, Stiffness and Non-Linear Behaviour of Simple (Multi-Braced) Welded Tubular Joints. T.N.O. Report BI-87-72/63.6.1098, 1988.

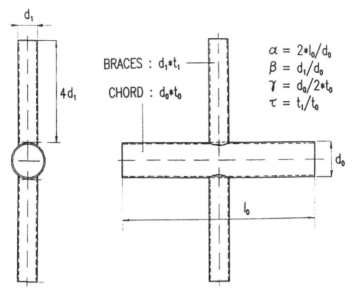

Figure 1 : Configuration of the X-joints.

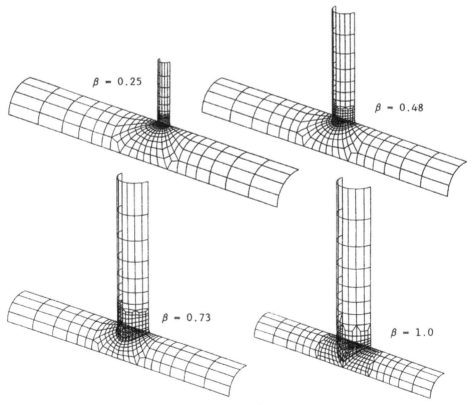

Figure 2 : Finite element meshes.

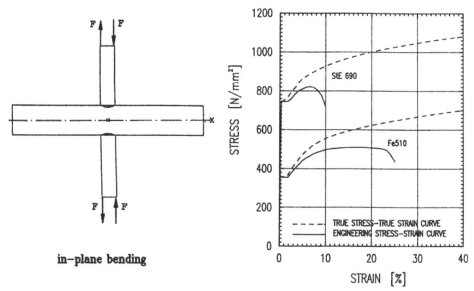

in-plane bending

Figure 3 : Loading of the X-joints.

Figure 4 : Engineering and logarithmic
stress-strain relationships.

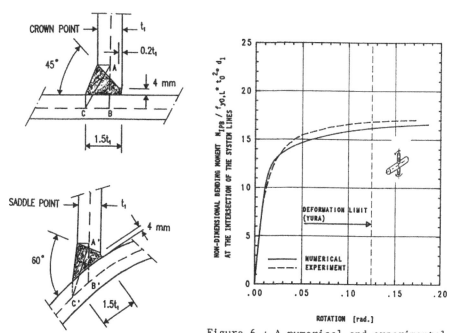

Figure 5 : Welding details for crown
and saddle point.

Figure 6 : A numerical and experimental
moment-rotation curve [16].

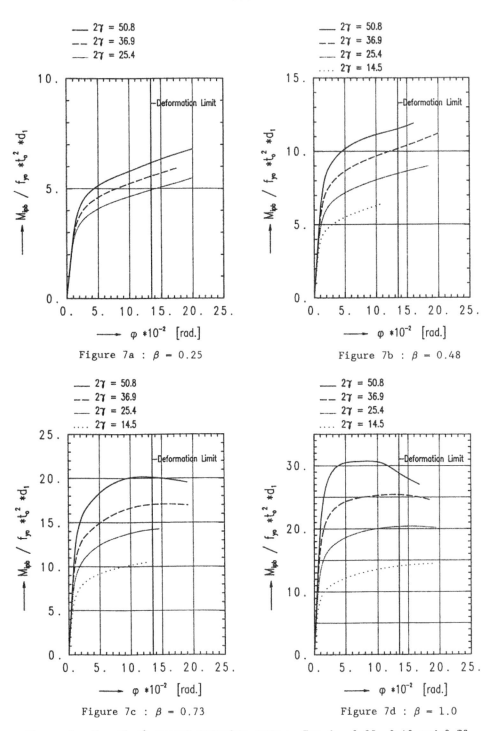

Figure 7a : $\beta = 0.25$

Figure 7b : $\beta = 0.48$

Figure 7c : $\beta = 0.73$

Figure 7d : $\beta = 1.0$

Figure 7 : Numerical moment-rotation curves. For $\beta = 0.25$, 0.48 and 0.73 the results are given for the joints modelled <u>with</u> a weld.

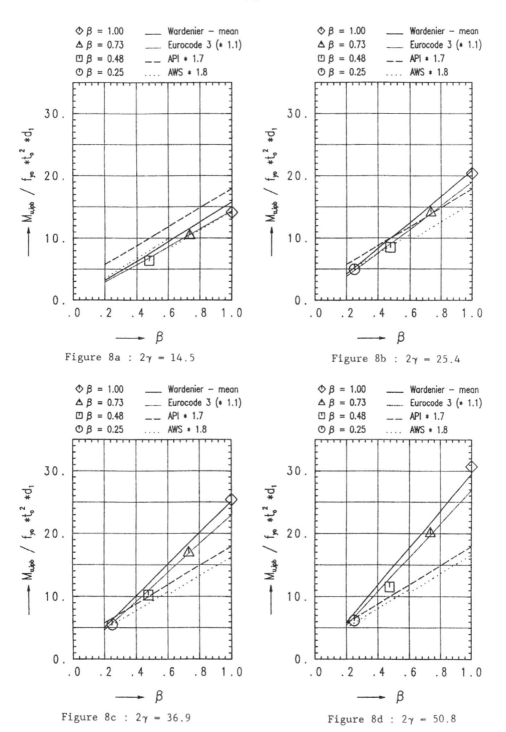

Figure 8a : $2\gamma = 14.5$

Figure 8b : $2\gamma = 25.4$

Figure 8c : $2\gamma = 36.9$

Figure 8d : $2\gamma = 50.8$

Figure 8: Ultimate moment - β curves.

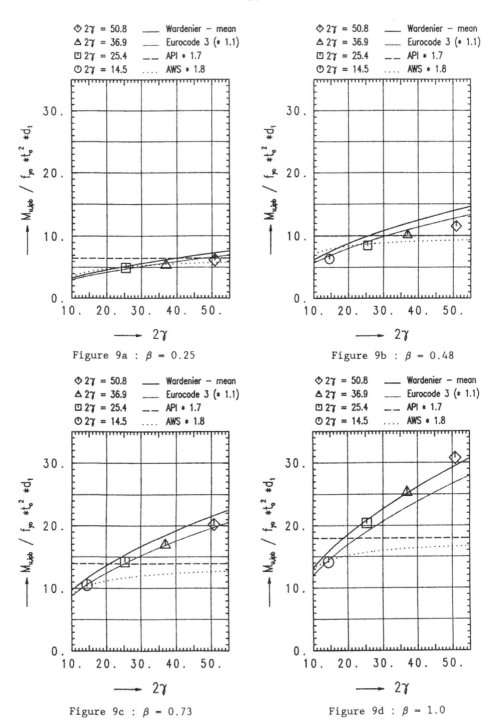

Figure 9a : β = 0.25

Figure 9b : β = 0.48

Figure 9c : β = 0.73

Figure 9d : β = 1.0

Figure 9: Ultimate moment - 2γ curves.

DUCTILE CROSS SECTIONS FOR BRIDGE PIERS

Eiichi WATANABE   Kunitomo SUGIURA
Department of Civil Engineering
Kyoto University
Yoshida-Honmachi, Sakyo-ku, Kyoto 606, JAPAN

Sho-ichi HARIMOTO   Toshiyuki HASEGAWA
Bridge Design Department
Komai Tekko Inc.
2-5-1 Nakajima, Nishiyodogawa-ku, Osaka 555, JAPAN

ABSTRACT

Presented herein is a study on the effective cross sectional shapes superior in strength and ductility. First, the ultimate strength and ductility of several thin tubular cross sections subjected to the constant axial compressive force and monotonic bending are evaluated experimentally. Also made is the comparison of the experimental results with the analytical results by NASTRAN/FEM analyses. Second, the cyclic deterioration process of strength and stiffness is further evaluated experimentally. It is concluded from both experimental and analytical results that the circular section is found to be superior in strength and ductility. Also found is the superiority of the longitudinally stiffened boxes using stiffeners of the Massonnet factor of more than 3 in the ductility.

INTRODUCTION

With the world-wide trend toward a unified limit state design codes of structures, the assessment of the structural reliability is becoming increasingly more important, particularly, the case of the design of steel structures against extremely strong earthquakes. Naturally, it is necessary to study the inelastic behavior of steel structures under complex cyclic loading of the axial force, shear, bending or their combinations. The static strength of structures, of course, is considered to be the most important criterion against small or medium earthquakes; however, against extremely severe earthquakes, the high energy absorption capacity or the ductility may be thought to be the most important.

In the past, for the improvement of the strength and ductility, the followings have been studied extensively: 1)effect of the cross sectional shape; 2)effect of the inelastic characteristics of stuctural steel; 3)composite construction; 4)limitation on the thickness of plate/shell elements and slenderness of beams/columns; and 5)effect of the redundancy of structures. Among these, the present study focuses on selecting the most effective cross section from many alternatives for bridge towers and piers from viewpoint of the strength and the ductility.

For bridge piers, the box section and circular section have been used without recognition of any physical superiority or inferiority, but by the appearance and economical aspect. Thus, the objective of the current study is to provide the physical basis on the selection of the cross sectional shape of steel bridge piers, paying attention to both the strength and ductility. In this study, different cross sections: box with sharp/round corners, stiffened box, and circular section were compared to each other by keeping the sectional area, slenderness parameter and section modulus as nearly equal as possible.

OUTLINE OF EXPERIMENTS AND FEM ANALYSES

Since the plastic mechanism of structures is contributed considerably to the strength and ductility, the plastic hinge behavior of the beam-column segments is tested under monotonic/cyclic bending and constant axial compressive force. Taking into account the survey by Nakai et al.[1] on the proportion of axial compressive stress and compressive bending stress for 88 steel rigid fame bridge piers designed under the 1973 JSHB Code[2], the axial force is determined to be taken as $0.2P_y$ and $0.33P_y$(P_y: the yielding axial force). In order to simulate such a loading condition, the test setup shown in Fig. 1 is utilized, which is consisting of the closed-loop servo-controlled hydraulic actuators and the micro-computer for control and data acquisition. In the case of monotonic bending, the quater section of test specimens is also analyzed by MSC/NASTRAN(large deformation-elastoplastic analysis). Details of this structural testing system and FEM analyses may be referred to Refs. 3 and 4.

The loading condition for each specimen is summarized in Table 1, whose shapes and nominal dimensions are shown in Fig. 2. For the sake of comparison among different cross sectional shapes, it is necessary to unify the sectional properties affecting the strength and ductility, so that (1) cross sectional area is constant considering economical aspect; (2) slenderness parameter is constant for strength of global stability; (3) width-to-thickness ratio is taken to guarantee the yield stress for strength of local stability; and (4) rigidity of stiffeners is three times of the required value by JSHB.

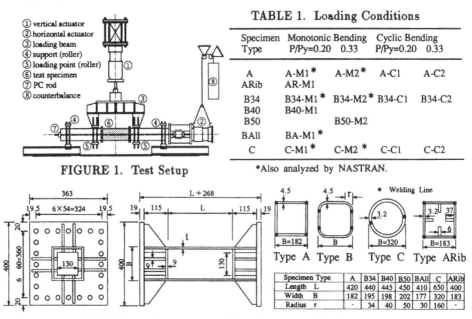

TABLE 1. Loading Conditions

Specimen Type	Monotonic Bending P/Py=0.20	0.33	Cyclic Bending P/Py=0.20	0.33
A	A-M1*	A-M2*	A-C1	A-C2
ARib	AR-M1			
B34	B34-M1*	B34-M2*	B34-C1	B34-C2
B40	B40-M1			
B50		B50-M2		
BAll	BA-M1*			
C	C-M1*	C-M2*	C-C1	C-C2

FIGURE 1. Test Setup   *Also analyzed by NASTRAN.

(1) vertical actuator
(2) horizontal actuator
(3) loading beam
(4) support (roller)
(5) loading point (roller)
(6) test specimen
(7) PC rod
(8) counterbalance

Type A Type B   Type C   Type ARib

Specimen Type	A	B34	B40	B50	BAll	C	ARib
Length L	420	440	445	450	410	650	400
Width B	182	195	198	202	177	320	183
Radius r	-	34	40	50	30	160	-

FIGURE 2. Dimensions of Test Specimens(Nominal, unit: mm)

RESULTS AND DISCUSSIONS

Figs. 3 shows the experimental bending moment(M)-curvature(ϕ) curves(lines) under monotonic bending, where the bending moment and curvature are normalized by the yielding values, M_y and ϕ_y, respectively. Also shown are analytical results(symbols). The moment-curvature relations under cyclic bending are also shown in Fig. 5. The comparative observation can be summarized as follows:

EFFECTIVE CROSS SECTIONAL SHAPE
The strength of C-type is the most highest and the stabilization of hysteresis under cyclic bending is guaranteed with the largest amplitude of curvature. It is concluded that the circular section is superior in strength and ductility according to the current design specification. However, at the same time the cyclic deterioration is also seen to be significant after its peak due to rapid propagation of buckled wave along the circumference.

OPTIMAL RADIUS OF ROUND CORNERS
B34, B40 and B50 are types corresponding to the radius-to-width ratio of $r/B = 1/4$, 1/5 and 1/6, respectively. As this ratio becomes larger, the strength and ductility are improved. It is concluded that the box with round corners is superior, since the progagation of buckled wave of flange plates may be delayed by the existence of round corners and less stress concentration occurs at corners.

EFFECT OF STIFFENING BY ROUND CORNERS
In comparing ARib-type to B34-type, it is found that ARib-type is superior in strength and ductility. No strength deterioration is observed even at $\phi/\phi_y=8$ because the local deformation is prevented by stiffeners. The round corners is not equivalent to the use of stiffener plates with rigidity ratio of less than 3.

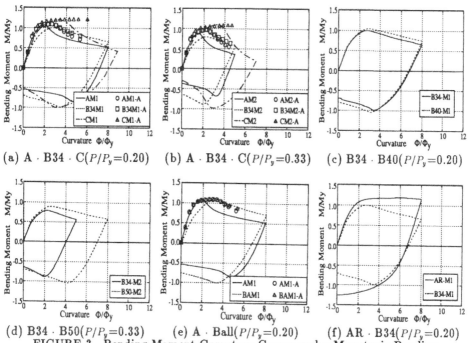

(a) A · B34 · C($P/P_y=0.20$)   (b) A · B34 · C($P/P_y=0.33$)   (c) B34 · B40($P/P_y=0.20$)

(d) B34 · B50($P/P_y=0.33$)   (e) A · Ball($P/P_y=0.20$)   (f) AR · B34($P/P_y=0.20$)

FIGURE 3. Bending Moment-Curvature Curves under Monotonic Bending

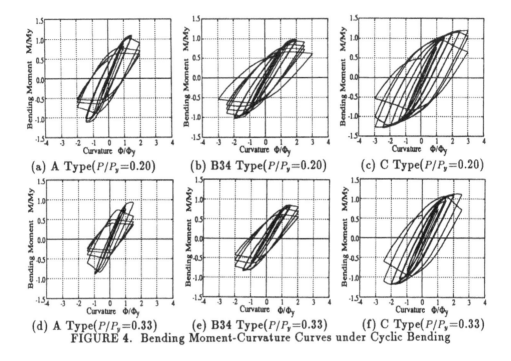

(a) A Type(P/P_y=0.20)   (b) B34 Type(P/P_y=0.20)   (c) C Type(P/P_y=0.20)

(d) A Type(P/P_y=0.33)   (e) B34 Type(P/P_y=0.33)   (f) C Type(P/P_y=0.33)

FIGURE 4. Bending Moment-Curvature Curves under Cyclic Bending

CONCLUSIONS

The monotonic/cyclic loading tests and FEM analyses were carried out in order to investigate the effective cross sectional shape paying attention to both the strength and ductility. Conclusions drawn from the current study can be summarized as follows:

- Circular cross section is superior in strength and ductility as well as aesthetically good appearance if the intensity of axial compression is around 20% of yielding value. The stiffened box is also found to be superior.
- The use of the round at corners of box sections can improve the ductility. In addition, the strength and ductility can be improved by increasing the radius of round corners. However, it should not be expected to be equivalent to fully stiffened box section using stiffeners of the Massonnet factor of more than 3.

REFERENCES

1. Nakai, H., Kawai, A., Yoshikawa, O., Kitada, T. and Miki, T., A Survey on Steel Rigid Frame Bridge Piers: Part 1 & 2, The Bridge and Foundation Engineering, 1982, 16(6-7), pp.35–40, & 43-49(in Japanese).
2. Japan Road Association, Specifications for Highway Bridges(JSHB), Maruzen, Tokyo, Japan, 1973(in Japanese).
3. Watanabe, E., Emi, S., Isami, H. and Yamanouchi, T., An Experimental Study on Strength of Thin-walled Steel Box Beam-columns under Repetitive Bending, Proc. of JSCE, Structural Engng./Earthquake Engng., 1988, 5(1), pp.21–29.
4. Watanabe, E., Sugiura, K., Kanou, M., Takao, M. and Emi, S., Hysteretic Behavior of thin Tubular Beam-Columns with Round Corners, J. of Const. Steel Research, 1991, 18(1), pp.55–69.

PARAMETRIC STUDY FOR THE OPTIMUM DESIGN
OF GUYED TOWERS

Murty K.S. Madugula and Subbarao V. Majety
Department of Civil and Environmental Engineering
University of Windsor, Windsor, Ontario, Canada N9B 3P4

ABSTRACT

A parametric study of a guyed tower is presented. Design was carried out in accordance with Canadian Antenna Tower Standard CAN/CSA-S37-M86. Variation in the mass and the total cost of a guyed tower due to the variation in (a) the number of guy levels, (b) inclination of diagonal web members, (c) radial distance of guy anchor points, (d) height of guy connection points, and (e) size of the mast, are studied.

INTRODUCTION

The demand for tall communication towers is increasing rapidly due to new developments in the telecommunications area. Self supporting towers prove to be satisfactory for low to medium heights, but for heights exceeding 100 m, guyed towers are probably the only economical solution. The increased demand for guyed towers necessitates the search for optimal design criteria for given loading conditions.

Designing an optimum guyed tower is a difficult task because of its nonlinear structural behaviour and the large number of variables that influence the design. As a first step, a parametric study was undertaken to identify the effect of several key variables on the total cost of the guyed tower. In parametric study one design variable is chosen at a time while other design variables are fixed tentatively. The variations in the objective function (i.e., cost of guyed tower for the present study) are studied with respect to the chosen design variable. The process is repeated with all variables to identify key design parameters that influence the optimal value of the objective function.

The optimization problem was formulated as follows:

Objective Function = Cost of guyed tower

The cost of guyed tower is expressed as

(Cost of mast + Cost of guys + Cost of anchorages)

where

Cost of mast = (Total mass of 90^0 angle members x Unit price) + (Total mass of 60^0 angle members x Unit price)

Cost of guys = (Total mass of guys x Unit price) + (Number of connections at ends of guy cables x Unit price)

Cost of anchorages = (Mass of concrete x Unit price) + (Volume of earthwork excavation x Unit price)

The following prices obtained from a local tower fabricator, were used to estimate the cost of the structure:

90^0 angles: \$3.30/kg; 60^0 angles: \$4.50/kg; guys: \$6.60/kg; cable connections at ends: \$600 each; concrete: \$392/m³; and earthwork excavation: \$65/m³

Design Variables:
θ = web inclination with respect to horizontal
r_n = radial distance of guy anchor point for cable "n"
r_{n+1} = radial distance of guy anchor point for cable "n+1"
h_n = height of guy connection point for cable "n"
h_{n+1} = height of guy connection point for cable "n+1"

Geometric Constraints: $r_{n+1} - r_n \geq 0$ and $h_{n+1} - h_n \geq 0$

PARAMETRIC STUDY

A computer program was developed to analyze, design, and estimate the mass and the total cost of a guyed tower [1]. Matrix stiffness method was used for the static analysis of the mast. The non–linearity due to axial loads was accounted for by using the geometric stiffness matrix. An iterative procedure was adopted to satisfy compatibility conditions of guy attachment points.

A 100 m high guyed tower and a 350 m high guyed tower were studied to understand the key parameters contributing to the total cost. The cross–section of the mast was an equilateral triangle. The design variables considered in the parametric study are:

a) number of guy levels; b) inclination of web members to the horizontal; c) radial distance of guy anchor points, d) height of guy connection points, and e) size of the mast.

The design was carried out in accordance with Canadian Standards Association Standard CAN/CSA–S37–M86 "Antennas, Towers and Antenna Supporting Structures" [2]. Structural steel angles were used for mast members; guy strands and bridge strands were considered for guy cables. Guy anchorages were deadman anchorages.

Mass and cost were estimated in the design process and the variations of both mass and cost were studied with respect to the design variables. Finding the minimum cost design was given higher priority than minimum weight.

The study was carried out for the following four loading conditions representing extremes of wind and ice loading in Canada:

1. 850 Pa reference velocity pressure for wind & 50 mm radial ice thickness
2. 850 Pa reference velocity pressure for wind & 10 mm radial ice thickness
3. 450 Pa reference velocity pressure for wind & 10 mm radial ice thickness
4. 450 Pa reference velocity pressure for wind & 50 mm radial ice thickness

Effect of Number of Guy Levels
The objective of the study of the effect of number of guy levels was to understand how the number of guy levels influences the total cost of a guyed tower. In order to determine the optimum number of guy levels it was necessary

to fix other parameters such as mast size, web inclination with the horizontal, and location of guy anchor points. It was assumed that all cable connections are equally spaced on the mast, all cable chords are at 45⁰ inclination, the mast size is 2.0 metres, and the cantilever projection at top is 10 m.

The cost of 100 m high guyed tower decreased initially for high wind conditions and increased steadily for low wind conditions. The minimum cost was observed at five guy connections for high wind conditions and at two guy connections for low wind conditions. From this experiment, two guy connections are adopted for uniformity to study the effect of other parameters on the mass and total cost of a guyed tower.

The cost of the 350 m high guyed tower decreased steadily as the number of guy connections was increased from six to fourteen. However, the reduction in cost is not significant when guy connections are more than seven for low wind conditions. From this experiment, the number of guy connections is taken as seven for uniformity to study the effect of other parameters on the mass and total cost of a guyed tower.

Effect of Inclination of Web Members to the Horizontal

The variation in the mass and the total cost for both experiments are similar. The minimum cost was observed for web inclinations in the range 35⁰ to 50⁰. The effect of web inclination was very significant on the mass and the total cost of a guyed tower for all loading conditions.

As the inclination of web members increased, the following effects were observed:

1. Reduction in the number of panels in the mast.
2. Increase in the effective lengths of web members and leg members resulting in the selection of heavier sections for leg and web members.
3. Reduction in the weight of web members initially because of the reduction in the number of panels; however, as heavier sections are chosen, there is an increase in weight.
4. Reduction in the weight of leg members at small web member inclinations.

Effect of Radial Distance of a Guy Anchor Point

1. At higher inclinations, the cable stands steeper which results in increased tension; however, the length of cable, the mass and the total cost of guyed tower decrease initially.
2. The increase in cable tension results in more axial thrust on the mast. This will result in the selection of heavier sections for leg members, thus increasing the mass and total cost.
3. The forces on the anchor block vary significantly as the cable chord inclination varies. When the cable is steeper, the vertical component of the force on anchor block increases and it results in the increase in weight (cost) of concrete block.
4. An increase in chord inclination results in reduction of axial thrust on the mast. Consequently, the weight of mast members can be reduced.

Effect of Height of a Guy Connection Point

1. As the height of a cable connection is increased, the increase in cable length results in design of heavier cables.
2. The axial thrust on the mast increases as cable stands steeper and it results in the design of heavier sections for leg members.
3. On the other hand, the deflections in the mast are reduced and this results in the design of mast members with less weight.
4. The forces on the anchor block also vary. The vertical component of force increases since cable becomes steeper and it results in the increase

of weight of deadman anchorage.

Effect of Size of the Mast
The effect of mast size on the mass and total cost of a guyed tower is very significant. As the mast size is increased, the following observations can be made:
1. The sectional properties of the mast such as moment of inertia increase resulting in improved stiffness of the structure and, hence, in the design of mast members with less weight.
2. With the increase in stiffness of the structure, deflections in the mast are reduced, resulting in cables of less weight.
3. The lengths of web members increase and the weight component of web members increases.

CONCLUSIONS

From the limited analytical investigations carried out in the present study, the following conclusions are arrived:
1. With an increase in the number of guy levels from 2 to 10, the mass of 100 m guyed tower decreased initially (upto 6 guy levels) and then increased. However, the cost was minimum with 5 guy levels for high wind conditions and 2 guy levels for low wind condition. On the other hand, for the 350 m guyed tower, both the mass and cost decreased with an increase in the number of guy levels from 6 to 14 for all load cases.
2. The effect of inclination of web members was very significant on both the mass and the total cost for all load cases. Minimum mass and cost were observed for web inclinations in the range 35^0 to 50^0.
3. The effect of radial distance of a guy anchor point is pronounced for high wind conditions and is insignificant for low wind conditions. Grouping of cable anchorages for multi-level guyed towers is economical for deadman anchorages.
4. The effect of variation in the height of a guy connection point is significant for all load cases.
5. For a guyed tower of 100 m height, the optimal design is observed at a mast size of 1.5 m. For a guyed tower of 350 m height, the optimal mast size is 2.0 m.

ACKNOWLEDGEMENTS

The financial support provided by the Natural Sciences and Engineering Research Council of Canada for carrying out this study is gratefully acknowledged.

REFERENCES

1. Majety, Subbarao V., "A Parametric Study for the Optimum Design of Guyed Towers", M.A.Sc. Thesis, Department of Civil and Environmental Engineering, University of Windsor, Windsor, Ontario, Canada, 1991.

2. Canadian Standards Association, CAN/CSA-S37-M86, "Antennas, Towers, and Antenna-Supporting Structures", Rexdale, Ontario, Canada, 1986.

CONTRIBUTION OF LIMIT ANALYSIS TO DESIGN OF CYLINDRICAL SHELLS

Y.A. HASSANEAN (*), D.O. LAMBLIN (**), G. GUERLEMENT (**)
(*) Dept of civil engineering, Assiut University, Assiut, Egypt
(**) Service of Mechanics of Materials and Structures, Polytechnic Faculty of Mons,
B 7000 Mons, Belgium.

ABSTRACT

Cylindrical shells reinforced with rings are largely used in mechanical industry. The sollicitation is generally a pressure combined with axial load. Limit analysis axisymetric complete solutions may give precise analytical information about the parameters of the design with very good physical interpretation of their influence. Applications in relation with offshore structures are investigated.

THEORETICAL BACKGROUND

Limit analysis for reinforced cylindrical shells

Let us consider externally reinforced cylindrical shell subjected to internal pressure and constant axial traction strength or internally reinforced cylindrical shell subjected to external presure and constant axial compression strength. Shell and stiffeners are made of rigid perfectly plastic materials with yield stresses respectively equal to σ and σ^*. First it is assumed that the considered shell is sufficiently long so that the end boundary conditions can be neglected. So it becomes possible to consider only one bay of length 2L limited by two planes of symetry (Fig 1). Within the usual hypothesis of limit analysis [1], the flange of the stiffener and the shell itself are treated as sandwich shells with non dimensional generalized variables m, n and $\bar{n}$ in relation with the longitudinal bending moment, the circumferential and axial strength. Non dimensional variables are real values divided by plastic values (n and $\bar{n}$ are positive for traction). Geometries are conveniently caracterized with non dimensional parameters:

$$\alpha_s = L \cdot \left(2/t \cdot R\right)^{0.5} \qquad \alpha_f = l \cdot \left(2/t_f \cdot R_f\right)^{0.5} \qquad (1)$$

In a totally yielded shell (flange of stiffener or shell itself), m and n verify equilibrium equation and belong to some part of a $\bar{n}$ depending yield polygon named plastic profile to wich efficiencies hereafter defined can be associated.

$$\rho_f = \left| \frac{1}{l} \int_0^l n \, dx \right| \qquad \rho_s = \left| \frac{1}{L} \int_0^L n \, dx \right| \qquad (2)$$

These always positive efficiencies, whatever the sign of n, are characteristics easy to use for static study of cylindrical shells. Moreover, due to some properties of yield polygons, the modification of the action of the pressure from internal to external simultaneously with modification of the action of end load from traction to compression doesn't change the numerical values for efficiencies. In the kinematic study, only radial and axial displacement rates combined with circumferential yield lines have to be considered. The normality law applied to yield polygons helps to obtain complete solutions [1].

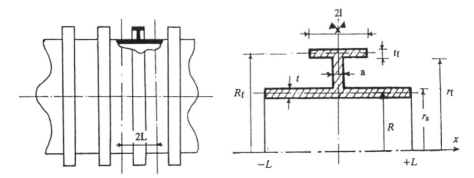

Figure 1 : Current bay with its geometry

Interaction between shell and rings

Such an interaction comes from the hypothesis of a concentraded action of the stiffener and from the study of its web considered as a classical plane stress problem for wich boundary conditions are imposed on radial stresses at the contact circumferences between web, flange and shell. These limit conditions are due to curvature and radial equilibrium of flange and shell. The resulting interaction may be expressed by very surprising simple relations :

$$\bar{p} = \frac{pR}{\sigma t} = \rho_s + S \qquad \text{with S equal to} \tag{3}$$

$$S = \frac{r_s \, l \, t_f \, \sigma^*}{r_f \, L \, t \, \sigma} \, \rho_f \; + \; \frac{a \, r_s \, \sigma^*}{2 \, L \, t \, \sigma} \, \ln \frac{r_f}{r_s} \qquad \text{or} \tag{4}$$

$$S = \frac{l \, t_f \, \sigma^*}{L \, t \, \sigma} \, \rho_f \; + \; \frac{a \, r_s \, \sigma^*}{2 \, L \, t \, \sigma} \left(\frac{r_s}{r_f} - 1 \right) \tag{5}$$

Complete proof of these relations may be found in [2] and [4] for the case with no axial load and in [5] and [6] for other cases. The first definition of S applies to a shell with internal pressure and external stiffeners. The second one applies for external pressure and internal stiffeners. As previously explained, absolute values of efficiencies are independant from pressure's direction. Unfortunately, two definitions for S arise from some unsymetry of the yield polygon (Tresca) for the web of the stiffener. Of course the presented interaction is always related to a failure with a plastic mechanism excluding elastic or elastoplastic instability. For a stiffener with rectangular cross section, one has simply to let $\rho_f = 0$. Efficiencies have to be calculated from choosen plastic profiles for shell and flange of the stiffener. Calculations of ρ_s are depending from the selected value of axial strength modifying the yield polygon of the shell. Calculations of ρ_f are much more simple and may be found in [8] or [2], ρ_f being an one parameter function depending only from α_f. In a first approximation, α_f may be taken equal to 1 for current stiffeners. Preceeding relations show obviously the dependance of the limit pressure versus the shell geometry and the strength parameter S for the stiffener. Definitions of S permit a very good physical interpretation of the geometrical parameters of the stiffener. With $\rho_f = 1$ and $r_f - r_s << r_s$, S may be writed under approximate forms, showing the influence of the area of the stiffener. :

$$S = \frac{\sigma^*}{\sigma} \, \frac{1}{2 \, L \, t} \left\{ a \, (r_s - r_f) + 2 \, l \, t_f \right\} \qquad\qquad S = \frac{\sigma^*}{\sigma} \, \frac{1}{2 \, L \, t} \left\{ a \, (r_f - r_s) + \frac{r_f}{r_s} \, 2 \, l \, t_f \right\} \tag{6}$$

For internal pressure some small reduction in the efficiency of the aera of the stiffener's flange is seen. Generalization of definition of S may be obtained for a I stiffener (external and internal flange) and a sandwich shell [2], [7].

APPLICATION TO OFFSHORE STRUCTURES

Introduction

Primary members of some offshore structures are reinforced cylindrical shells subjected to external pressure and compression. Reinforcement is made with very stiff diaphragms and tee stiffeners. A great number of stiffeners (may be ten or more) are localized between two diaphragms considered as simple or built in supports. A part of the member between two diaphragms may be considered as a long shell and designed with preceeding theory.

Calculation of ρ_s for long shell

Dominating pressure : In this case with axial load smaller in absolute value than half the plastic axial strength of the shell, the yield polygon for the shell is an octogon (8 sides) in the m, n plane. Four possibilities have to be considered for the failure mechanism of the shell and the stiffeners. The two first possibilities are related with failure of the shell and the stiffeners. The two last possibilities are related with failure of the shell between two successive undeformed tee stiffeners. The boundary between the two types of solutions gives minimal values to realize for S if we want to have undeformed tee stiffeners at the collapse of the shell. Complete analytical derivation may be found in [5]. For axial load equal to zero (lateral pressure only), general solutions reduce to solution presented in [2], [3] or [4]. In this case, the yield polygon of the shell becomes a hexagon [1], [4]. Some drastically limited results (due to lack of place) are presented in figure 2 for an axial load equal to thirty percent of the axial plastic load. They give non dimensional pressure $\bar{p}$ against strength parameter S of stiffening rings for different values of α_s. On such a graphic a line parallel to S axis indicates undeformed stiffeners at collapse.

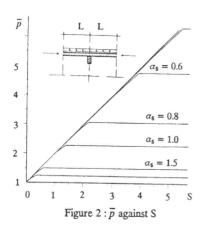

Figure 2 : $\bar{p}$ against S

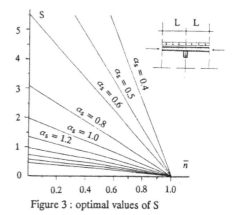

Figure 3 : optimal values of S

Axial load dominated loading : In this case with axial load greater or equal to half the plastic axial strength of the shell, the yield polygon for the shell is a rectangle in the m, n plane. Important simplifications appear giving :

$$\bar{p} = 1 + S = 1 + k\,(1 + \bar{n})/\alpha_s^2 \qquad 0 \le k \le 2 \qquad -1 \le \bar{n} \le -0.5 \quad (7)$$

Relations (7) gives under parametric form (k is a variable parameter) the limit pressure of a long shell with simultaneous collapse of the shell and the stiffeners. For k = 0, the shell without stiffeners is in membrane state for each value of $\bar{n}$. For k = 2, (7) gives the limit pressure of a built in shell of length 2L and the minimum value to realize for S to obtain undeformed stiffeners at collapse.

Optimal design of stiffeners : with presented solutions it is possible to determine optimal value for the strength parameter S. Such a value has to be realized to have only failure of the shell between rigid stiffeners. Figure 3 presents optimal values of S versus $\bar{n}$ for different values of α_s.

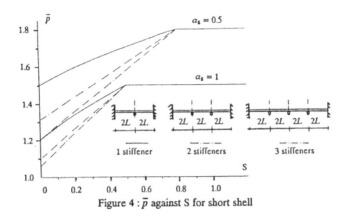

Figure 4 : $\bar{p}$ against S for short shell

Experimental verifications : Actual theory gives good agreement with experimental results published in [9]. Detailed comparaison may be found in [6]. Sometime for experiment with a small number of stiffeners, it is important to consider short shell and real boundary conditions. Some results are presented in figure 4 for a short built in shell under hydrostatic pressure with a variable number of stiffeners. Complete derivation may be found in [6].

ACKNOWLEDGMENTS

Financial support provided by the Egyptian Government to the first writer and by "Fonds National de la Recherche Scientifique" of Belgium to the others is gratefully acknowledged.

REFERENCES

[1] Hodge, P.G., Plastic Analysis of Structures, Mc Graw-Hill, New York, 1959.

[2] Guerlement, G., Ph D thesis, Fac Polytech de Mons, 1976.

[3] Guerlement, G. and Lamblin, D.O., Limit analysis of rings reinforced cylindrical shells, Bull de l'Académie Polonaise des Sciences, 1976, Vol XXIV, 1, 111-118.

[4] Guerlement, G., Lamblin, D.O., Save, M., Limit analysis of cylindrical shell with Reinforced Rings, Eng. Struct, 1987, Vol 9, 145-156.

[5] Hassanean, Y.A., Lamblin, D.O., Guerlement, G., Limit analysis of reinforced shell under pressure dominated loading. To be published.

[6] Hassanean, Y.A., Lamblin, D.O., Guerlement, G., Limit analysis of reinforced shell under pressure and axial load. To be published.

[7] Guerlement, G., Lamblin, D.O., Limit analysis of sandwich shell. To be published.

[8] Lamblin, D.O., Guerlement, G., Interaction curves for bending and axial forces of perfectly plastic curved I beams, J. Struct. Mech., 1972, 1 (2), 187-212.

[9] Tsang, S.K. and Harding, J.E., a plastic mechanism formualtion for the general instability of ring stiffened cynlinders under pressure dominated loadings, Int. J. Mech. Sc, 1985, Vol. 27, N°6, 409-422.

DEVELOPMENTS IN THE STORAGE OF DANGEROUS GASES AND LIQUIDS IN LARGE CAPACITY TANKS

F. MANG

Testing Center for Steel, Timber and Stone, University of Karlsruhe, Kaiserstr. 12, 7500 Karlsruhe

ABSTRACT

In the last ten years, mounded cylindric pressure vessels for the storage of explosive and other dangerous liquids have gained an increasing significance with regard to growing safety require-ments. These tanks are usually placed in a sandbed and covered with earth. Fig. 1 gives an impression of the site, containing 9 vessels in a battery system. In fig. 2 three versions for equip-ment installations are shown. The interactions between operational loadings and soil pressure on the one side, and reactions from elastic foundation on the other side, necessitate an accurate investigation. The nonaxisymmetric loadings create circumferential bending reactions and ring stiffeners have to be installed. With internal pressure, these stiffeners hamper the expansion of the cylindrical shell and cause secondary bending in the longitudinal direction of the vessel. These vessels are usually fabricated on site (see fig. 3). This paper gives a survey of common construction methods and describes the major calculation procedures required for the design.

Fig. 1   9 vessels in a battery system

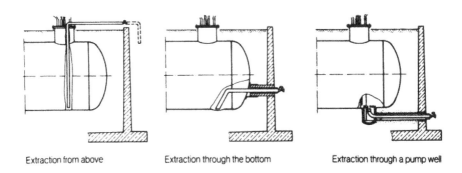

Extraction from above   Extraction through the bottom   Extraction through a pump well

Fig. 2   Versions for equipment installations

Fig. 3   Vessels fabricated on site

INTRODUCTION

For the storage of combustible liquids under atmospheric pressure and liquefied gases under high pressure, the horizontal cylindrical steel vessel with earth cover has been increasingly applied within the last years. The decision for the earth covered type of installation is mainly justified by the safety advantages in respect to external influences on the vessel, such as high temperature in case of fire and dynamic pressure from near by explosion.

The vessels are placed in trenches or above ground and covered with earth. The type of installation as well as storage conditions, structural dimensions and assembly methods have to be considered within the design of the cylindrical steel vessel. As a rule, only single horizontal vessels with capacities up to 400 m^3 and diameters up to 4.0 m are placed in trenches, whereas within an aerial installation under embankment, several vessels are preferably installed in a battery line-up. With this type of installation, vessels with contents up to 4000 m^3, diameters up to 8.5 and lengths up to 100 m are designed at present. Because of the surcharge load together with the relatively large cylinder diameters between 6.0 and 8.5 m, only ring stiffened systems proved to be economical.

MOUNDED STORAGE

Foundation methods

For the foundation of big horizontal storage vessels for liquid gases with earth cover, fig. 4 shows different modes.

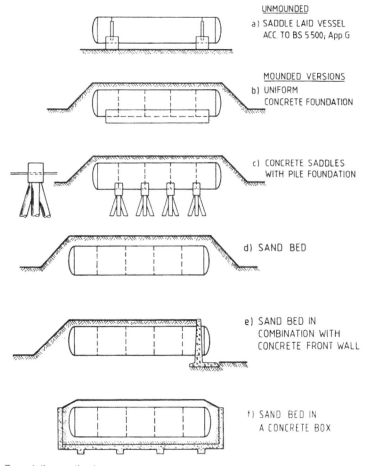

UNMOUNDED
a) SADDLE LAID VESSEL
ACC. TO BS 5500; App. G

MOUNDED VERSIONS
b) UNIFORM
CONCRETE FOUNDATION

c) CONCRETE SADDLES
WITH PILE FOUNDATION

d) SAND BED

e) SAND BED IN
COMBINATION WITH
CONCRETE FRONT WALL

f) SAND BED IN
A CONCRETE BOX

Fig. 4   Foundation methods

The design for a foundation on a continuous rigid reinforced concrete strip footing, to which all loads occurring from uneven settlements can be applied, is only carried out for short vessels. In such cases, a continuous elastic lamination is provided as padding between the reinforced concrete saddle and the steel cylinder. The placement on individual foundations, which in turn may be founded on piles, can only be recommended in exceptional cases, since differential settlements of individual foundations cannot totally be excluded. In addition it is necessary to provide the saddles with sliding bearings, in order avoid horizontal loads on the saddles and pile heads resulting from fluctuations of temperature and pressure.

The continuous storage in a defined sandbed represents the most economic storage method for long storage vessels. This method requires extensive subsoils investigations, since the cylindrical shell is subjected to reactions resulting from differential settlements along the vessel's longitudinal axis.

Most of the tanks of the last years have been placed in a sandbed. With this, variant d) has been used several times, which allows an accessible installation of the piping equipment in the tank's bottom. The accessibility of the tank bottom, however, is a disadvantage with regard to safety aspects and the official approval for this variant is presently refused in various countries; or rather structural measures for the protection of the supply pipes are necessary, since they are exposed to loads from settlements. When arranging the angular retaining wall next to the bottoms, the change of length caused by internal pressure and temperature variation has to be compensated by suitable structural measures (for ex. paddings).

Torispherical or ellipsoidal heads are preferred as bottoms for earth covered cylinders, since their ability of maintaining the cross section is considerably higher compared to hemispherical bottoms, and no additional ring stiffener has to be provided in the transition to the cylinder. Further structural details are developed according to the relevant guidelines (British Standards BS 5500, 1985; AD-Merkblätter, 1989).

In fig. 5, a mounded double-walled vessel for storage is shown. In view of the high storage pressures for liquid gases under service and hydrotesting, partially considerable wall thicknesses are required, depending on the diameter. Since the expenditure for the welding procedure strongly increases with wall thicknesses bigger than 25 mm with regard to preparatory and subsequent treatments, the application of high-strength materials is recommended.

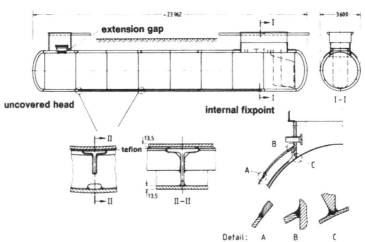

Fig. 5   Mounded double-walled vessel for storage

Dimensioning

On the one hand, earth covered liquid gas storage tanks are pressure tanks for buildings, which have to satisfy the requirements for the relevant pressure vessel codes. On the other hand, they are, owing to their size, constructions which have to be dimensioned according to the structural regulations introduced at present. Stresses from internal pressure are superimposed by stresses from the cylinder's beam action, which is loaded by the soil surcharge, the filling weight, the dead weight and influences from uneven settlements. In addition to stress investigations, checks of the stability are necessary, since axial as well as radial compressive loads occur in the cylinder shell between the stiffening rings. Non-uniform circumferential loads such as earth cover, liquid filling, dead weight and support pressure produce circumferential bending in the tank construction. Depending on the tank's foundation method and stiffness, various distributions of load and soil pressure in the longitudinal direction are the result. The stress level in the stiffening ring is essentially influenced by the size of the foundation's angle of aperture in the cross section of the tank. Basically, the allowable stresses usual for steel structures should be applied. Caused by the hampered expansion of the cylinder shell in plane of the rings, secondary bending stresses occur in the longitudinal direction of the tank. An assessment of these stresses can be done following the British Standards BS 5500, App. A, G.

LOADING CASES

Composition of loading cases

During assemblage and transport of the vessel and installation into the sandbed, but particularly in the hydro testing and operational loading case, a variety of load components are to be considered:

Assemblage condition:	Dead weight, transport
1st pressure test: (without earth cover)	Dead weight, water filling, test pressure differential settlements
Operation:	Dead weight, operational filling, design pressure, earth filling, traffic loads, differential settlements, temperature, friction, passive earth pressure on heads, vacuum, external explosion and earthquake (if applicable)
2nd pressure test: (with earth cover)	such as 1st test plus earth cover

CHARACTERISTICAL LOADING CASES

Earth cover and bearing reactions

The typical problem for mounded cylinders in a sandbed results from the concurrence of the earth surcharge components and the accompanying bearing reactions in the circumferential as well as in the longitudinal direction from elastic foundation.

For this reason, extensive investigations on the subsoil have to precede the design of sand bedded pressure vessels surcharged with earth, in order to determine the external pressure loads for the design of the tanks. The load from earth cover can be formulated as a superposition of sinusoidal radial pressure curves on the upper half. The biggest pressure ordinate at the upper vertex or rather the resulting earth load ensues from the assumptions presented in Fig. 6.

For the load assumption shown in Fig. 6a, the radial component of the hydrostatic soil pressure on the shell has been taken into account. Figures 6b and 6c consider load increasing effects, which can result from settlements in the neighbourhood of saddle laid tanks (6b) or from arc action effects in the soil between tanks arranged closely together (6c).

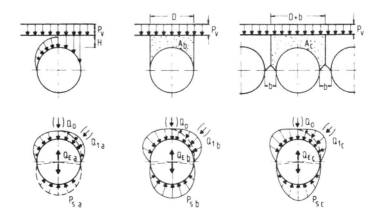

Fig. 6   Load assumptions for earth surcharge

For the load transfer in circumferential direction, the assumption of a bedding angle of 120° - 180° proved to be worthwhile. Up to now, the bearing pressure has been conservatively assumed as sinusoidal radial pressure, where the biggest pressure ordinate on the lower vertex can be calculated from the sum of the individual load components.

Seismic loadings

For the assessment of the behavior of earth covered cylinders upon loads from earth quakes no reliable knowledge is available. Loads in form of horizontal accelerations are often prescribed by the appropriate authority.

The horizontal acceleration occurring in case of earth quakes leads to an increase of the resultant of the individual load components and to a distortion of the direction of effect. According to figure 7, the bearing reaction occurs on a smaller surface, so that the local soil pressure as well as the circumferential reactions increase. For this loading case, however, lower safety is required or rather higher stresses are allowed (for ex. according to BS 5500: + 25%) (British Standards, 1985).

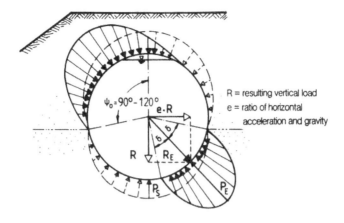

Fig. 7   Assumptions for earth cover load and bearing pressure in case of earth quake

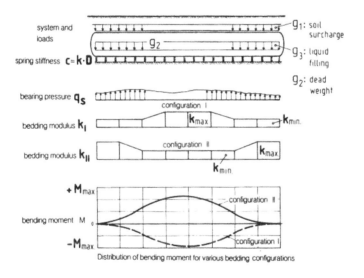

Fig. 8   Distribution of bending moment and bearing pressure for various bedding
configurations
Vessel: 7,10 m diameter and 75 m total length

STRESS INVESTIGATIONS

Elastic foundation

The determination of the reactions related to the tank's longitudinal axis should be realized
according to the theory of beams on elastic foundation for the various loading cases, considering
the characteristic subsoil data, where the most unfavourable distribution of the bedding module

due to differential settlements should be taken into account. Basing on extensive investigations into the subsoil, these predictions are usually prescribed by a soil mechanics expert.

The elastic beam calculation gives the distribution of the bending moment in the longitudinal direction, which is necessary for the investigations on stress and stability, as well as the settlement line from which the bearing reaction per unit of length can be determined, considering the corresponding bedding stiffness. The governing reactions circumferential direction have to be calculated in the cross section of the highest bearing pressure (see Fig. 8).

Superposition of the load components in circumferential direction

The determination of the circumferential forces and moments in the ring stiffened region is based on the following assumptions:

- All bending reactions from earth cover, dead weight and liquid filling within the distance l_r of two rings are assigned to the cross section of the stiffening ring.

- In the area of the so-called "effective width" weld connected rings are additionally loaded by internal pressure as well as by the normal forces from liquid filling, dead weight and earth cover.

Secondary stresses from internal pressure

With internal pressure, the unstiffened shell area widens more compared to the ring stiffened region. The constraint developed in this way, produces bending moments in the tank's longitudinal direction, the maximum of which occurs in plane of the ring stiffener (see Fig. 12).

The extent of the secondary stress considerably depends on the stiffness of the reinforcing ring. To this particular type of stress peaks, reference is made in British Standards 5500, App. A, as so-called "self-limiting" stress, e.g. the stress peak is reduced after yielding takes place. For this reason, BS 5500 allows stresses up to the double yield strength of this particular case.

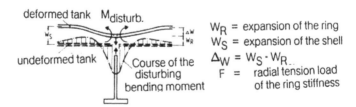

Fig. 9   Effects of internal pressure on ring stiffened shells

Stress investigations

Wall thicknesses of pressure vessels are usually designed for internal pressure according to the relevant codes. However, such a design is not sufficient for an earth covered cylinder, and should be used as a precalculation only. In the unstiffened region of the cylinder barrel, accurate stress investigations are to be carried out, since two-axial stress conditions in superposition with shear stresses are existing. Particularly longitudinal stresses from internal pressure and bending are to be superimposed with the circumferential and shear stresses.

INVESTIGATIONS ON THE STABILITY

Since cylindrical shells are sensitive against compressive loads, investigations on the stability of the cylindrical shell have to be performed for the mounded tanks. The external loadings from earth cover and bearing pressure, acting non-uniformly along the perimeter, have to be taken into account for circumferential direction. Furthermore, it can be required to consider underpressure in the investigation, which might occur due to a possible malfunction under service or human error.

The longitudinal bending loads of the tank system between individual saddles and especially from uneven settlements in a sandbed, as well as axial compression forces from friction and earth pressure on the heads necessitate an investigation for the longitudinal direction.

Investigations on buckling can be carried out according to the relevant pressure vessel codes.

MANUFACTURING AND ASSEMBLING METHODS

Manufacturing and assembling of steel structures

The manufacturing and assembling of horizontal cylindrical tanks with big diameters is usually carried out as a pure site fabrication according to common construction methods for large pipes and pressure vessels. The steel sheets already bent and trimmed by the mill are directly supplied to the site; there they are put together to rings and welded to a pipe section by connecting several rings. The section is then insulated and brought into its final position in the prepared sandbed.

The distribution of sheets, typical for this process of fabrication, is presented in Fig. 10a. For welding, exclusively automatic processes, such as submerged arc welding are applied for the fabrication of the pipe sections. The circumferential seam between the individual assembled sections, which are installed into their final position in the sandbed, is manually arc welded.

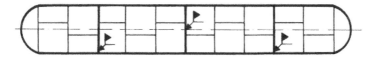

a) conventional pipe construction

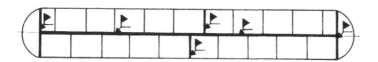

b) centre split pipe method

Fig. 10   Manufacturing methods

In order to reduce the work expenditure on the site in favor of the fabrication in the workshop, the split pipe method has been selected for some projects. Using longitudinal cuts as well as several transversal cuts, the cylinder has been split into semi-shells, the dimensions of which were oriented to the allowable measures of transport. A typical course of the cut is presented in Fig. 13.

Construction of the sandbed and installation of the tank

For the continuous storage in a sandbed, particular attention has to be paid to assembling points of view, since accesses in the sandbed become necessary for fabrication and testing of the final seams.

After testing, these trenches cannot be compacted properly, and thus form softer regions in the bedding. For this reason, possibly big units should be installed in order to limit the number of trenches.

Recently, an increasing number of tanks has been completely prefabricated in assembly station on site or in the workshop and then lifted with heavy-duty cranes into the sandbed. With another procedure, the completed tanks are moved on hydraulic slides into the sandbed. For this purpose, however, two trenches are necessary, crossing the sandbed approximately in the section 1:3. Since fitting problems between a completely prefabricated tank and the sandbed cannot be avoided, the width of the initial bearing sandbed is kept as small as possible, where the angle of aperture can be reduced down to 30°. In order to be able to grant the required compaction of the bearing sandbed in this relatively inaccessible area, a new procedure has been developed by means of which the fine grain sand is flushed with water and compacted in layers with conventional vibrators. However, the compaction gained with this, should be permanently controlled by a soil mechanics expert while filling up the sandbed. The assembly trenches for the hydraulic slides are closed and compacted in a similar way. When calculating the elastic bedding, this region has to be reckoned with a lower bedding stiffness, which should be prescribed by the soil mechanics expert.

For the non-bearing layers of the earth coverage, no special demands are made on the compaction. It should be taken care, however, that only fine grain sands are stored in the intermediate vicinity of the shell. A plan of the courses for the establishment of the earth cover is presented in Fig. 11.

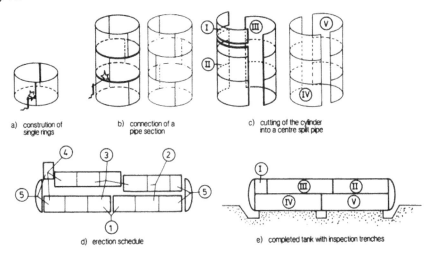

a) constrution of single rings

b) connection of a pipe section

c) cutting of the cylinder into a centre split pipe

d) erection schedule

e) completed tank with inspection trenches

Fig. 11   Assembling course for the split pipe method

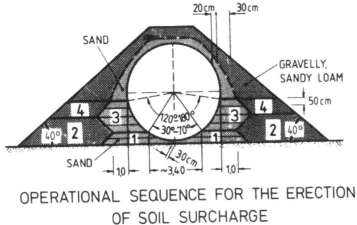

OPERATIONAL SEQUENCE FOR THE ERECTION
OF SOIL SURCHARGE

Fig. 12   Structure of layers for sandbed and earth cover

Quality assessment, performance test and control measures

When dimensioning new tanks a welding factor of 1.0 is taken as a basis almost without exception. For the guarantee of this weld quality an extensive, nondestructive material testing is necessary, where each seam has to be subjected to 100% X-ray or ultrasonic proofs. Usually, these tests are done before the earth cover is installed. After a servicelife of 8-10 years, hydrotesting and weld inspections will be repeated (periodic testing).

Due to the settlement sensitivity of the tanks, which have partly a length of up to 100 m, it is recommended to monitor and record the settlement behavior during hydrotesting and particularly under service in regular intervals.

DOUBLE SKIN COMPOSITE CONSTRUCTION
FOR SUBMERGED TUBE TUNNELS

RAN NARAYANAN and IP LUEN LEE
The Steel Construction Institute,
Silwood Park, Ascot, Berkshire SL5 7QN, UK

ABSTRACT

A new form of sandwich construction, consisting of two relatively thin steel plates with a concrete infill, has been developed for use in the construction of submerged tube tunnels. The paper summarizes the recent investigations into the potential for this form of construction and outlines the advantages of the system. Interim design rules for rapid sizing of double skin composite sections and the planned model test of a complete tunnel cross-section are also briefly described.

1. INTRODUCTION

Double Skin Composite Construction (DSCC) is an innovative form of steel-concrete composite system which has been developed in recent years. The system consists of a layer of un-reinforced concrete sandwiched between two relatively thin steel plates, which are connected to the concrete infill by welded stud connectors. This arrangement was first suggested as an alternative to traditional reinforced concrete systems by Messrs Tomlinson Partnership and Sir Alexander Gibb and Partners, working in association with the University of Wales College of Cardiff for the scheme design of the Conwy Tunnel in North Wales, UK[1,2]. A typical cross-section of submerged tube tunnel using Double Skin Composite Construction is shown in Figure 1. The significant advantages in using DSCC for submerged tube tunnels are outlined below.

- The outer skin in the double skin construction automatically provides water-tightness; the alternative solution using reinforced concrete walls frequently requires a steel skin which does not contribute to load bearing capacity.

- The tunnel could be fabricated in sections at a remote fabrication site and later towed into position, thus ensuring better quality control and speedier construction.

- The weight of the double skin composite elements would be substantially lighter compared with concrete.

- As there is no need to construct a casting yard close to the site, the solution is environmentally more attractive in comparison with other alternatives.

- The outer and inner skins of the double skin composites act as permanent formwork; internal conventional reinforcement would be eliminated, resulting in substantial economies.

- Fire resistance of double skin composites would be high enough for practical purposes, as concrete acts as a heat sink.

- DSCC is distinctly advantageous from the point of view of blast resistance.

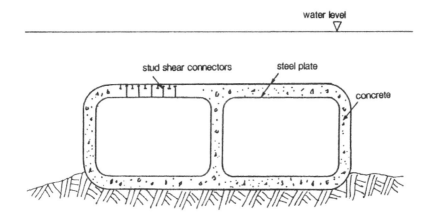

Figure 1. Cross Section of a Double Skin Composite Submerged Tube Tunnel

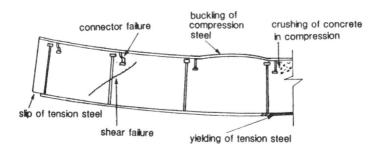

Figure 2. Failure Modes of Double Skin Composite Beam Elements

Experiments on 53 model scale tests and eleven large scale tests conducted at the University of Wales College of Cardiff during 1987-90[3] provided an understanding of the structural behaviour of double skin composite elements subjected to flexural, axial and combined loadings. The tests carried out so far have produced basic information on the behaviour of this structural system and demonstrated that this form of construction is technically feasible. The collapse of a double skin composite (DSC) structure will be triggered by one or more of the failure modes shown in Figure 2.

2. INTERIM DESIGN CRITERIA

Based on the experimental tests and the associated theoretical work, the following interim design criteria have been proposed for rapid sizing of double skin composite sections; these criteria must be regarded as tentative and subject to refinement in the light of investigations currently in progress:

Stud Pullout: All stud connectors must have a length of at least 10 times their diameter. The head diameter should be at least 1.5 times the shank diameter and the head depth should be at least 0.4 times the shank diameter.

Vertical Shear Failure: The design shear strength of concrete traditionally prescribed in the Codes will need to be reduced by around 20%. Stud connectors should overlap for a length of at least 12 times stud diameter in order to resist shear.

Squash Load Capacity for Columns: The squash load may be calculated using a formula similar to the one used for reinforced concrete columns.

Ultimate Moment Capacity: This may be calculated by assuming conventional stress blocks. The concrete stress block should extend to 90% depth of the plastic neutral axis and the maximum compressive force computed from 0.45 times the cube strength multiplied by the stress block area.

Shear Connector Capacity: Connector strength in shear may be taken as 80% of its characteristic resistance for connectors in compression and 50% of the characteristic resistance for connectors in tension. The connectors should have sufficient ductility to accommodate any slip which would occur between steel and concrete.

Stiffness: The stiffness of DSC elements subjected to lateral loading may be calculated assuming elastic behaviour in concrete and steel and by ignoring the concrete below the elastic neutral axis.

Detailing: The stud diameter should not exceed 2½ times the steel plate thickness on the compression face and 2 times the steel plate thickness on the tension face. The maximum connector spacing must be less than (a) four times the stud height and (b) three times the slab thickness. The minimum connector spacing must be greater than (a) 1.5 times the stud diameter plus the maximum aggregate size and (b) the minimum safe welding space. For long studs acting as links, the longitudinal spacing must be less than 0.75 times the thickness of the element.

3. MODEL TEST ON A COMPLETE TUNNEL CROSS-SECTION

A quarter scale model tunnel cross-section will be tested to validate this form of tunnel construction at the University of Wales College of Cardiff during the later part of 1992. An analytical model has been developed for predicting shear forces and bending moments in this test specimen within the elastic stages of incremental loading. This is based on the following assumptions: (a) due to continuity of the steel plating, slip between the concrete and steel will not be significant; (b) the flexural stiffness will not vary significantly around the cross-section. The values of moments, shears, and thrusts predicted by this analysis will be utilised as a basis for comparison with the experimental results.

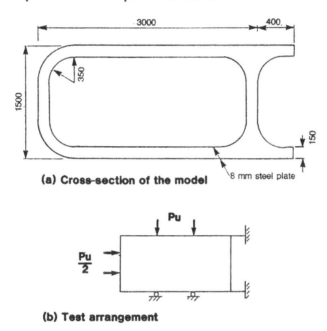

(a) Cross-section of the model

(b) Test arrangement

Figure 3. Model Tunnel Test

REFERENCES

1. Narayanan, R., Wright, H.D., Evans, H.R. and Francis, R.W., *Double Skin Composite Construction for Submerged Tube Tunnels*, Steel Construction Today, Vol 1 (1987), pages 185-189.

2. Tomlinson, M., Tomlinson, A., Chapman, M.L., Jefferson, A.D., and Wright, H.D., *Shell Composite Construction for Shallow Draft Immersed Tube Tunnels*, International Conference on Immersed Tunnel Techniques, Manchester UK, April 1989, Thomas Telford, London, pages 145-156.

3. Narayanan, R. and Roberts, T.M., *Double Skin Composite Construction for Submerged Tube Tunnels*, Engineering Foundation Conference on Composite Construction: Composite Construction II, June 1992, Potosi, Missouri, USA.

RESEARCH ON CRANE BOOMS WITH THE LIGHTEST WEIGHT

Lihua You, Shunpei Yao, Xingming Wei, Xiaobei He
Ist department of Mechanical Engineering of
Chongqing University,China

Lianshan Hong (second author)
Chengdu Seamless Steel tube Works, China

ABSTRACT

In order to manufacture crane booms with the lightest weight, we developed a new kind of low carbon, low-alloy, high quality Structural steel tubes with yield limit 700 MPa. According to different internal forces along booms, we used steel tubes with different wall thickness to different sections and gave calculation of critical forces of this kind of booms. Choosing the lightest weight as optimum object, we studied the optimum design of the booms.

RESEARCH AND MANUFACTURE OF D652 STEEL TUBES

Truck cranes and offshore pedestal cranes have strict demand for their weight. Nowadays, they are trending to be made with the lightest structural weight and big capacity.

Crane booms with big capacity must be made of high strength steel tubes. In recent years, some kinds of steel with high strength have been developed in which WELTEN 80 made in Japan and STE 690 made in West Germany got a wide use in engineering. After studying every kind of low-carbon, low alloy, high strength steel, we developed a new kind of low-carbon, low alloy, high strength steel tubes named D652 according to the raw material in our nation. Their designing chemical composition and mechanical properties are listed in table 1.

In order to study the influence of carbon and various alloys to the rolling, heat treatment and mechanical properties of this new kind of steel tubes, we tested with different chemical composition whose chemical composition and mechanical properties are listed table 2 and table 3 respectively.

Table 1

Chemical composition (%)										
C	Si	Mn	Cr	Ni	Mo	V	B	Al	P	S
<0.16	0.15/ 0.35	0.60/ 1.20	0.60/ 1.20	0.40/ 0.70	0.40/ 0.70	0.04/ 0.08	<0.005	>0.015	<0.025	<0.025
mechanical properties										
σ_s (MPa)		σ_b (MPa)			δ (%)			$(\alpha_{kv})_{-40℃}$ (J)		
>685		>785			>16			> 29		

Table 2

Chemical composition (%)										
C	Si	Mn	Cr	Ni	Mo	V	B	Al	P	S
0.08	0.24	0.80	0.80	0.53	0.49	0.07	0.003	0.03	0.012	0.012
mechanical properties										
σ_s (MPa)		σ_b (MPa)			δ (%)			$(\alpha_{kv})_{-40°C}$ (J)		
688-743		795-852			18-21			> 55		

Table 3

Chemical composition (%)										
C	Si	Mn	Cr	Ni	Mo	V	B	Al	P	S
0.15	0.33	1.02	0.71	1.04	0.57	0.08	0.003	0.03	0.013	0.013
mechanical properties										
σ_s (MPa)		σ_b (MPa)			δ (%)					
835-846		895-914			17					

It can be seen from table 2 that the steel tubes have lower carbon content in the first kind of chemical compositon, but they reach the design demand on their mechanical properties and have better ductility and weldability. In table 3, the yield limit and tensile strength of the steel tubes go up obviously along with the increase of their content of carbon and alloys. In addition, their ductility has little decrease. It indicates that this new kind of steel tubes are suitable for the members of crane booms and other engineering structures for they have good mechanical properties and weldability.

CALCULATION ON CRITICAL FORCES OF BOOMS

The weight of crane booms also have great relations with their design plan. To the lattice booms with tube structures, the previous design commonly uses tubes with same wall thickness. In fact, the internal forces of booms are different along their axis which behave as big at bottom and small at top. So we may use different wall thickness to different sections of booms according to the distributed law of their internal forces. In order to check their stability, we must study computing methods on the critical forces of this kind of booms.

The bottom and top sections of the booms have a varying section area whose inertia moment may be replaced with equivalent moment I_h showed as follows (Note that the explanation on all symbols in this paper have been omitted).

$$I_h = mI_{max} \tag{1}$$

In hoisting plane, booms may be treated as simple supported ladder columns whose computing methods on critical forces have been given in some books.

In swing plane, booms may be treated as ladder columns with bottom clamped and top constrained by the wire ropes. Taking shearing deformation and effect of the wire ropes into account, we may give the differential equations on their deformation as follows.

$$\frac{d^2U_i}{dZ^2} + K_i^2 U_i = K_i^2 \frac{f}{L + \zeta L} Z + K_i^2 \frac{\zeta L}{L + \zeta L} f \tag{2}$$
$$(i = 1, 2, \ldots \ldots)$$

in which $K_i = \sqrt{Ncr}/\sqrt{EI_i(1-KNcr/GA)}$

The general solutions of the above equations are

$$u_i = A_i\cos K_i Z + B_i \sin K_i Z + \frac{f}{L+\zeta L} Z + \frac{\zeta L}{L+\zeta L} f \qquad (3)$$

(i=1,2,......)

The critical force of booms in swing plane may be obtained from expressions (3) according to their boundary and compatibility conditions. For example, the critical force in swing plane of booms consisted of four sections will be determined by the following expressions

$Z=0, \quad U_1=U_1'=0$
$Z=l_1, U_1=U_2, U_1'=U_2'$
$Z=l_2, U_2=U_3, U_2'=U_3'$
$Z=l_3, U_3=U_4, U_3'=U_4'$
$Z=L, \quad U_4=f$

$\qquad (4)$

OPTIMUM DESIGN ON CRANE BOOMS

In this section, we discuss optimum design on crane booms with the lightest weight. Taking offshore pedestal crane booms made of four sections for example, we chose wall thickness of steel tubes and main chord lengths at the first two nodes of every section as design variables

$X=[\delta_1 \ \delta_2 \ \delta_3 \ \delta_4 \ \delta_f \ L_{D1} \ L_{D2} \ L_{D3} \ L_{D4}]^T$
$=[x_1 \ x_2 \ x_3 \ x_4 \ x_5 \ x_6 \ x_7 \ x_8 \ x_9]^T \qquad (5)$

The optimum object may be chosen as the boom weight whose function is

$$F(x)= \pi \ \rho \ \{\sum_{i=1}^{4}[D_i^2-(D_i-2X_i)^2](l_i-l_{i-1})+[D_f^2-(D_f-2x_f)^2]$$

$$\sum_{i=6}^{9} \frac{l_{i-5}-l_{i-6}}{x_i}[\sqrt{(x_i/2)^2+h^2}+\sqrt{(x_i/2)^2+a^2}]\} \qquad (6)$$

in which lo=0 and l_4=L.

The restraints are

(1) global rigidity conditions of booms

$g_1(x)=[\lambda]-\lambda_{hx} > 0$
$g_2(x)=[\lambda]-\lambda_{hy} > 0 \qquad (7)$

(2) rigidity condition of main chords

$g_3(x)=[\lambda]-\lambda > 0 \qquad (8)$

(3) rigidity condition of lattice members

$g_4(x)=[\lambda]-\lambda_f > 0 \qquad (9)$

(4) global stability conditions of booms

$$g_{4+i}(x)=[\sigma]-(\frac{N_i}{A_i\varphi\psi} - \frac{1}{1-\frac{N_i}{0.9N_{EX}}} \cdot \frac{C_{ox}M_{oxi}+C_{Hx}M_{Hxi}}{w_{xi}}$$

$$+ \frac{1}{1-\frac{N_i}{0.9N_{Ey}}} \frac{C_{oy}M_{oyi}+C_{Hy}M_{Hyi}}{W_{yi}}) > 0$$

$$\qquad (10)$$

(i=1,2,3,4)

(5) stability conditions of main chords between the first two modes of every section

$$g_{8+i}(x)=[\,\sigma\,]-\frac{N_{xi}}{A_{xi}\Psi_i}> 0 \qquad (11)$$
$$(i=1,2,3,4)$$

(6) stability condition of lattice members

$$g_{13}(x)=[\,\sigma\,]-\frac{N_{f6}}{A_{f6}\Psi_6}> 0 \qquad (12)$$

(7) restraints on inclination angles of lattice members

$$g_{13+i}(x)=\alpha_{max}-tg^{-1}\frac{2h}{x_{5+i}} \qquad (13)$$
$$(i=1,2,3,4)$$

(8) dimension restraints on wall thickness of main chords

$$g_{17+i}(X)=x_i-(\,\delta_{min})_i> 0 \qquad (14)$$
$$(i=1,2,3,4)$$

(9) dimension restraint on wall thickness of lattice members

$$g_{22}(X)=x_5-\delta'_{min}> 0 \qquad (15)$$

If boom members which meet stability conditions are not weakened locally, they must meet strength conditions. So it is unessential to check their strength. In addition, the fatigue strength on offshore pedestal crane booms is also not calculated for they work discontinuously.

The optimum design on the booms have been given according to the above model. The maximum capacity of boom which is 25.992m in length is 326. 7kN. Assuming that outer diameter of main chords is 73mm and outer diameter of lattice members is 38mm, we may get the following optimum results.

$X_*=[0.0076182, 0.0075759, 0.0063619, 0.0061225, 0.0059987, 1.161, 1.342, 1.365, 1.752]^T$

$F(X_*)=2071.85kg$

The same problem without optimum design will gives the dimensions and weight of the boom as follows under conditions of the same chord lengths between the first two nodes of every section as the optimum plan.

$X=[0.0077, 0.0077, 0.0077, 0.0077, 0.006, 1.161, 1.342, 1.365, 1.752]^T$

$F(X)=2203.91kg$

CONCLUSIONS

1. A new kind of low-carbon, low alloy, high strength structural steel have been developed which have high yield limit, good ductility and weldability. So they are especially suitable for the members of crane booms with the lightest weight.

2. Different wall thickness of steel tubes is used for different sections of booms according to the distribted low of internal forces along the booms. a computing method on critical forces of the booms with bottom claped and top restrained by wire ropes is given.

3. The optimum design of booms with the lightest weight has been studied and the optimum dimensions and the lightest weight of booms have been obtained.

4. It is obvious both that using the new steel tubes given in this paper almost decrease half weight of main chord compared with 16Mn whose yield limit is about 350MPa and that the optimum design decrease 6. 37 % of boom weight relative to normal design.

Regional Developments

EUROCODE 3 - PRESENTATION AND CALIBRATION

COLIN TAYLOR
Manager, Codes and Standards
The Steel Construction Institute
Silwood Park, Ascot, Berks SL5 7QN, UK

ABSTRACT

Several interesting points arise from the preparation of Eurocode 3 'Design of Steel Structures'. The Eurocodes aim at three-way harmonisation between countries, between materials and between industries. Features include a clear distinction between fundamental principles and practical application rules, emphasis on methods of analysis including practical imperfections and the consistent use of self-explanatory symbols, which reduce the need for words in calculations. Avoidance of errors in calculations is assisted in Eurocode 3 by adopting unified formats for member buckling, using slenderness for beams as well as columns, enabling engineers to retain a "feel" for the development of their designs, whilst introducing safe simplifications where justified. Another important feature is the extensive use of statistical calibration against data banks of test results.

INTRODUCTION

Eurocode 3 'Design of Steel Structures' (EC3) is one of the set of nine structural design standards being published by CEN, the organisation for European Standards. The initial version of Part 1 'General Rules and Rules for Buildings' was issued earlier this year with the reference ENV 1993-1-1.

HARMONIZATION

The aim of the Structural Eurocodes is the harmonization of structural design practice throughout the European Economic Area. Whilst the primary motivation was the removal of technical barriers to trade across national boundaries, the opportunity has also been taken to achieve harmonization between the competing structural materials, such as steelwork, reinforced concrete, timber etc, this extending to the use of a consistent safety format (that of Limit States Design using partial safety factors), a consistent level of reliability, consistent terminology, common wording for common rules and a consistent use of major symbols (following ISO conventions).

A third dimension of harmonization is currently being introduced, that between the different industries which use the same construction materials. As has often been said, a beam behaves no differently in a building, a bridge or an offshore platform. The aim is to use similar design expressions, similar terminology and similar symbols for similar structures and structural elements, varying only the loading and the safety factors between applications. Professional structural designers will then be able to adapt to a new sector of industry, without the non-productive familiarisation procedures currently required. Equally important, the processes of cross-fertilization of ideas, and of learning from experience, will not be artificially restricted by arbitrary barriers caused by non-harmonized codes.

FEATURES

Principles and Application Rules

In adopting a harmonized presentation for structural design standards, taking account of the wide differences between current styles in various European countries, but avoiding copying any one of them, part of the solution has been to distinguish clearly between two types of paragraphs, those stating the fundamental requirements ("Principles") and those providing an acceptable means to satisfy these requirements ("Application Rules"). In Eurocode 3 these are printed in different typefaces. Roughly half the paragraphs are of each type, in contrast to Danish codes, which also make the same distinction but consist mainly of Principles. The opposite extreme is represented by British codes (for example) which basically comprise Application Rules. The resulting clarity of expression in the Eurocodes, avoids the need for a separate "commentary" stating the reasons for the rules, though background documents indicating their derivations are still appropriate.

Methods of Analysis

All generally recognised methods of analysis are permitted, subject to necessary limitations, for the global analysis necessary to determine the internal forces and moments in those structures that are not modelled as statically determinate. These methods include:

- elastic analysis
- elastic analysis, with subsequent redistribution
- rigid-plastic (simple plastic)
- elastic — perfectly plastic
- elasto-plastic (plastic zones).

In addition provision is made for the use of:

- first order analysis
- second order analysis
- first order analysis with indirect allowance for second order effects

The various methods are also described in the code, because not all methods are familiar to designers in all the countries concerned.

Allowance for Imperfections

A particular feature of EC3 is the explicit allowance in the calculation procedures for those practical imperfections of construction that have an influence on the resistance of members or structures. These include:

- member imperfections
- frame imperfections
- system imperfections

Member imperfections are introduced in the design of compression members through empirical values of an imperfection factor representing an equivalent lack of straightness. The values of the imperfection factor also account for the effects of typical residual stress patterns and for the beneficial effects of strain hardening. Generally the member buckling resistance is derived from the resistance of the cross-section by applying a reduction factor. This is a function of the relevant imperfection factor and of the slenderness ratio of the compression member (strut slenderness). The only case where this simple procedure cannot be used is in unusually flexible frames, where the interaction between member imperfections and frame imperfections cannot be neglected.

A similar procedure is used for lateral-torsional buckling of unrestrained lengths in beams, using the equivalent slenderness ratio for this form of buckling. More details of this are given below. In this case the member imperfection is related to the beam slenderness of the member, rather than to its strut slenderness.

Frame imperfections are introduced into the analysis of all frames, in the form of an equivalent initial sway or lean. For convenience this can be converted into a closed system of equivalent horizontal forces, except when determining reactions onto foundations. Like the member imperfections, the values of the frame imperfection factor are intended to also account for the possible effects of other forms of imperfection which may affect the stability of frames, such as the effects of lack-of-fit. The basic value of this factor can be reduced in the case of multi-storey and multi-bay frames, to account for the reduced probability of simultaneous occurrence in all bays and in all storeys.

System imperfections are introduced in the design of braced bays and in the design of built-up compression members, such as laced or battened struts. Again these comprise an equivalent lack-of-straightness imperfection. In the case of built-up members the magnitude of the imperfection is intended to also allow for the possible effects of other types of practical imperfection, such as lack-of-fit and residual stresses due to rolling or welding.

In the case of bracing systems, any additional deflections due to the action of the bracing system in resisting externally applied forces (such as wind loads) also have to be taken into account.

Symbols

Symbols have not only been harmonized around the ISO recommendations, the consistent use of descriptive subscripts has been developed to the extent that, once familiarity has been gained, the meaning of any particular symbol in a given expression can generally be determined by inspection, without the need for a complete key listing the symbols which it uses. These self-explanatory symbols do, in effect, approach the concept of "ideograms".

For example, all internal forces and moments are distinguished by the subscript S whilst the corresponding resistances have the subscript R and all design (factored) values also acquire the subscript d. Thus knowing that M represents a moment, the expression:

$$M_{Sd} \leq M_{Rd} \tag{1}$$

represents the requirement that the design value of the internal moment (due to the applied loads) must not exceed the design value of the resistance of the component intended to carry it. In the case of a Class 1 or Class 2 ("Compact") section this can be amplified to:

$$M_{Sd} \leq M_{p\ell.Rd} \tag{2}$$

to clarify that it is the plastic resistance moment that is relevant, and to:

$$M_{y.Sd} \leq M_{p\ell.y.Rd} \tag{3}$$

to clarify that it is the moment about the y-y axis that is of concern.

This system is not completely novel, but is of sufficient interest to deserve mention here. The author first learnt of it from proposals attached to comments from the Netherlands on the draft of EC3, but is uncertain of its origin. After initial reluctance, familiarity seems to convert most users to enthusiasm for the system, because it largely eliminates the need for explanatory words in calculations, such as "factored", "furnished" or "required".

BEAM SLENDERNESS

The slenderness ratio "buckling length"/"radius of gyration" λ which equals ℓ/i is very commonly used in the determination of the buckling resistance of a compression member. It is convenient in that it avoids the need to calculate the theoretical elastic critical resistance N_{cr}, which in the case of a short member may be an abstract value significantly exceeding the "squash load" $N_{p\ell}$. It also has the practical advantage of being a simple ratio, with a physical significance that can readily be grasped and which is unlikely to be incorrectly calculated or copied, whereas an error of a factor of 10 in the value of N_{cr} would not be noticed so easily.

In view of the vital need to prevent the likelihood of avoidable errors in structural calculations, it should be worth more than a passing interest to consider the use of a similar approach when determining the resistance moment of a beam which is subject to lateral-torsional buckling.

Most modern codes define a buckling curve for lateral-torsional buckling on non-dimensional axes such as $M/M_{p\ell}$ against $[M_{p\ell}/M_{cr}]^{0.5}$. These are directly analagous to the axes $N/N_{p\ell}$ and $[N_{p\ell}/N_{cr}]^{0.5}$ used for compression members. Even though most codes use a different expression from that adopted in EC3, the value of $M/M_{p\ell}$ is therefore a function of the beam slenderness λ_{LT}. The relationship between λ_{LT} and M_{cr} is similar to that between λ and N_{cr}. The elastic critical load of a strut N_{cr} is given by $\pi^2 EA/\lambda^2$, where E is modulus of elasticity and A is area. Similarly the elastic critical moment of a beam M_{cr} is given by $\pi^2 EW_{p\ell}/\lambda_{LT}^2$, where $W_{p\ell}$ is plastic section modulus.

Traditionally, designers in most countries have used procedures which require them to determine the value of M_{cr} as an intermediate step. Data for the evaluation of M_{cr} has therefore been included in an informative annex of EC3.

However it is also possible to evaluate λ_{LT} directly, without the intermediate step of evaluating M_{cr}. This is done in current British practice [1]. The two procedures are illustrated by the following example:

EXAMPLE

Calculate the buckling resistance moment of a 356 x 171 x 51 UB in grade Fe 510 steel ($f_y = 355\text{N/mm}^2$) subject to a uniform bending moment over a length of 3 metres between lateral restraints. [Assume the effective length factors k and k_w are both unity].

For a 356 x 171 x 51 UB:

$$W_{p\ell.y} \quad = \quad 895 \text{ cm}^3 \quad \text{from tables}[2]$$

For uniform moment:

$$C_1 \quad = \quad 1.0$$

(a). Using M_{cr} to calculate $\bar{\lambda}_{LT}$:

$$M_{cr} \quad = \quad \frac{\pi^2 E I_z}{L^2} \left[\frac{I_w}{I_z} + \frac{L^2 G I_t}{\pi^2 E I_z} \right]^{0.5} \tag{4}$$

$$I_t \quad = \quad 23.6 \text{ cm}^4 \qquad \left.\begin{array}{l} \\ \\ \\ \\ \\ \end{array}\right\}$$

$$I_w \quad = \quad 0.287 \text{ dm}^6 \qquad \left.\begin{array}{l} \end{array}\right\} \text{ from tables}[2]$$

$$I_z \quad = \quad 968 \text{ cm}^4 \qquad \left.\begin{array}{l} \end{array}\right\}$$

$$E \quad = \quad 210{,}000 \text{ N/mm}^2 \; \left.\begin{array}{l} \\ \\ \end{array}\right\} \text{ EC3}$$

$$G \quad = \quad E/2.6$$

$$M_{cr} = \frac{\pi^2\ 210{,}000 \times 968 \times 10^4}{3000^2 \times 10^6} \left[\frac{0.287 \times 10^{12}}{968 \times 10^4} + \frac{3000^2 \times 23.6}{2.6\ \pi^2 \times 968} \right]^{0.5}$$

$$\bar{\lambda}_{LT} = [W_{p\ell.y}\ f_y\ /\ M_{cr}]^{0.5}$$

$$\bar{\lambda}_{LT} = [\ 895 \times 355\ /\ (435.7 \times 1000)\ ]^{0.5}$$

$$= \underline{0.854}$$

(b). Using λ_{LT} to calculate $\bar{\lambda}_{LT}$:

$$\lambda_{LT} = \frac{L/i_{LT}}{\left[1 + \dfrac{(L/a_{LT})^2}{25.66} \right]^{0.25}} \tag{5}$$

$$i_{LT} = 4.31\ cm \quad \left.\begin{array}{c} \\ \\ \end{array}\right\} \quad \text{from tables}[6]$$

$$a_{LT} = 110\ cm \quad \left.\begin{array}{c} \\ \end{array}\right\}$$

$$\lambda_{LT} = (300/43.1)\ /\ [\ 1 + (3000/1100)^2\ /\ 25.66\ ]^{0.25}$$

$$= 69.6\ /\ 1.066 \quad = \quad 65.3$$

Reference slenderness λ_1 : *)

$$\lambda_1 \quad = \quad \pi\ [\ E/f_y\ ]^{0.5} \quad = \quad \pi\ [\ 210{,}000\ /\ 355\ ]^{0.5}$$

$$= \quad \underline{76.4} \qquad \text{(or from design aids}[3])$$

$$\bar{\lambda}_{LT} = 65.3/76.4 = \underline{0.855} \qquad \text{(cf\ 0.854 in (a) above)}$$

Note 1. This value of $\bar{\lambda}_{LT}$ is actually the more accurate one. The difference arises from using four section properties (I_t , I_w , I_z and $W_{p\ell.y}$), each rounded to three significant figures, to calculate M_{cr} compared to only two (i_{LT} and a_{LT}) for λ_{LT}

Note 2. The strut slenderness $\lambda = L/i_z = 3000/38.7 = 77.5$, compared to 65.3 for λ_{LT}

For either method:

For $\bar{\lambda}_{LT} = 0.855$, using buckling curve a:

χ_{LT} = 0.763 (by calculation or from design aids)

$M_{b.Rd}$ = $\chi_{LT} W_{pl.y} f_y / \gamma_M$

= 0.763 x 895 x 355 / 1.1 / 1000

$M_{b.Rd}$ = 220 kN/m

Alternatively[*] from Concise EC3[3]:

For f_y = 355 N/mm^2 and $\lambda_{LT} = 65.3$:

f_b = 271 N/mm^2

$M_{b.Rd}$ = $W_{pl.y} f_y / \gamma_M$

= 895 x 271 / 1.1 / 1000

$M_{b.Rd}$ = 220 kN.m

[*) The alternative approach used in the "Concise EC3" eliminates the need to calculate λ_1.]

It can be seen from this example that the direct evaluation of λ_{LT} has the advantage of being a simpler calculation, and involves looking up only two parameters, which reduces the risk of error. The value of λ_{LT} is generally 80 - 90% of the value of λ, which facilitates simplifications where the precision of the result is not very significant, but above all enables the designer to retain a feel for the physical significance of the parameter, whereas M_{cr} could equally well be 0.1 or 10 times the value of M_{pl}.

The way the parameters a_{LT} and i_{LT} have been defined helps to minimise the consequences of any errors in referring to section tables. The more influential parameter i_{LT} is approximately equal to a quarter of the flange width b, and therefore accidentally taking the value for a similar section with the same serial size, but a different weight per metre, has little effect. Although the parameter a_{LT} varies more than i_{LT}, it has less influence on λ_{LT}. Even the error introduced by taking the value of a_{LT} for a similar section with a different weight per metre, is less than that caused by the equivalent error in reading the value of even one of the parameters I_z, I_w or especially I_t.

In the expression for M_{cr}, great care has to be exercised when adjusting the powers of 10 for the units used in tables of section properties. Both a_{LT} and i_{LT} have linear dimensions, which reduces the possibility of errors when the expression for λ_{LT} is used.

For all of these reasons, expressions for the direct calculation of λ_{LT} have also been included in the same informative annex of EC3. For the same reasons, the advantages of directly evaluating λ_{LT} are warmly commended to others.

CALIBRATION

Statistical calibration studies have been carried out[4] for many of the rules in Eurocode 3. Details are contained in Background Documents[5]. These studies aimed to ensure that a uniform safety level, defined by a safety index $\beta = 3.8$, was acheived throughout the code. The procedure followed is defined in a joint report[6] from TNO Delft, TU Eindhoven and RWTH Aachen, which forms Background Document 7.01. Together with a small addition, this was also the basis for the initial draft of the proposed additional Annex Z of Eurocode 3 "Determination of Design Resistances from Tests". Subsequently a JCSS Working Document[7] has also proposed essentially the same procedures.

Calibrations carried out in the course of developing and editing Eurocode 3 included amongst others:

- resistance moments of restrained beams

- static strength of bolts

- static strength of welds

- buckling of struts

- lateral-torsional buckling

- failure of columns with bi-axial moments

- effective section properties of cold-rolled sections

- use of high strength steel

- joints between structural hollow sections

- fatigue design rules

In short, every rule for which an adequate data base of test results could be assembled has been submitted to the same procedure. The programme of calibration studies was organised by Prof G Sedlacek of RWTH Aachen in what is believed to be the most extensive exercise of its type so far carried out. The studies themselves were carried out in various centres including Delft, Liege, St Remy and Aachen. Similar procedures were also applied in the development of Eurocode 4 'Composite construction'.

In Eurocode 3, in order to avoid having a large variety of values of the partial safety factor for resistance γ_M, two representative categories were selected as follows:

(1) Cases where failure is related to yield strength, including buckling phenomena

(2) Cases where ultimate tensile strength governs, including net section failure and the strengths of bolts and welds.

Fixed representative values of 1.1 and 1.25, respectively, were selected for these two categories. Where necessary, the coefficients in the design expressions were then adjusted slightly to compensate for variations from the optimum values.

REFERENCES

1. BS 5950, The Structural Use of Steelwork in Building, Part 1. Code of Practice for Design in Simple and Continuous Construction : Hot Rolled Sections, BSI Standards, Second Edition, July 1990.

2. The Steel Construction Institute, Concise Eurocode 3 for the Design of Steel Buildings in the UK, SCI Publication No 116, The Steel Construction Institute, 1992.

3. Taylor, J.C. and Selby, C., Structural Steel Sections : Dimensions and Properties to BS 4 and BS 4848 for use with Eurocode 3, SCI Publication No 117, The Steel Construction Institute, 1992.

4. Dowling, P.J. and Sedlacek, G., Eurocodes — The Current Scene, Paper 11, International Symposium : Building in Steel — The Way Ahead, ECCS/CECM/EKS, September 1989.

5. The Eurocode 3 Editorial Group, Background Documents for Eurocode 3 (numerous), Prepared for the Commission of the European Communities, 1989 - 1991.

6. Bijlaard, F.S.K., Sedlacek, G. and Stark, J.W.B., Procedure for the Determination of Design Resistance from Tests, TNO Report BI-87-112, TNO-IBBC Delft, RWTH Aachen and TU Eindhoven, November 1987, second update November 1988.

7. Kersken-Bradley, M., Maier,W., Rackwitz,R. and Vrouwenvelder,A., Estimation of Structural Properties by Testing for Use in Limit State Design. Working Document, JCSS Joint Committee on Structural Safety, November 1990.

8. ECCS Advisory Committee 5, Essentials of Eurocode 3 : Design Manual for Steel Structures in Building, Publication No 65, ECCS/CECM/EKS, 1991.

CURRENT AND FUTURE TRENDS IN STRUCTURAL STEEL DESIGN IN THE CARIBBEAN

M.W. CHIN PhD,CEng,FICE,FIStructE
Head, Department of Civil Engineering
The University of the West Indies
St. Augustine, Trinidad

ABSTRACT

The paper briefly reviews the historical development of structural steel design in the various islands of the Commonwealth Caribbean and discusses some of the factors which have influenced the use of structural steel in the design of both buildings and bridges over the past decade. Particular attention is given to the development of the Structural Steel Design requirements for buildings in the recently produced Caribbean Uniform Building Code (CUBiC). The paper describes the procedures adopted and the philosophy which led to the selection of the Canadian Steel Standard as the framework on which the CUBiC requirements for structural steelwork is summarised. The design approaches and format adopted are briefly discussed. The need for setting up a procedure for recording experience of its use and for calibration exercises to be carried out is emphasised.

The paper concludes with a discussion on the need for the development of a code for the design of steel bridges in the Caribbean and reflections are offered on possible future developments in steel codes and standards in the Caribbean against the background of international harmonisation and the new generation of European structural codes.

INTRODUCTION

The West Indies, sometimes referred to as the Commonwealth Caribbean, comprises of a series of islands forming an arc along the Eastern end of the Caribbean Sea ranging from Jamaica in the north to Trinidad and Tobago in the south and includes the cooperative Republic of Guyana in South America and Belize in Central America. The territories of the region are classified as LDCs, i.e. the Less Developed Countries such as Grenada, St. Lucia, etc. (i.e. the Leeward and Windward Islands and Belize) and the MDCs, i.e. More Developed Countries comprising of the four independent countries - Barbados, Guyana, Jamaica and Trinidad and Tobago. The LDCs are mostly mini-nations or mini-associated states with very little control over their economies which are basically agricultural. The total population of the region is about five million persons of whom approximately 3.5 million live in Jamaica and Trinidad and Tobago. It is estimated that there are approximately 2,500 engineers currently practising in the region.

Jamaica and Trinidad and Tobago have experienced the longest period of industrial development based on their indigenous mineral resources - bauxite in Jamaica and hydrocarbons in Trinidad and Tobago.

Historically, as British colonies, most of the structural steel design in the Commonwealth Caribbean was based on British codes of practice such as BS449 and BS153 for buildings and bridges respectively. Most of the steel buildings built during the colonial era were low-rise industrial type pitched roof portal buildings and in the case of bridges were of steel trusses. Typical examples are given in Figures 1 and 2. It was not until the mid 1950's that buildings of high-rise (six storeys and over) began to be constructed based on designs to British codes such as BS449 and during the period 1950-1970 there was a steady increase in the number of buildings of between ten to fifteen storeys constructed in both reinforced concrete and structural steel.

During the post colonial era some of the More Developed Countries (MDC's) in the West Indies, notably Jamaica and Trinidad and Tobago embarked on major development plans for roads and highways during the 1960's and 1970's. As a result, during the decade 1968 to 1978 some 60 bridges were built in Jamaica ranging in spans of 12.2 m to 45.73 m utilising composite steel plate girders, reinforced concrete box girders and precast pretensioned AASHTO girders of various cross-sections in spans ranging from 12 m to 24 m. In Trinidad and Tobago a number of major high-way projects resulted in the construction of some 60 bridges during the period 1967 to 1980. Both composite steel girders and prestressed precast AASHTO girders with reinforced concrete decks were used in the majority of bridges.

Building practices in the various islands followed traditional patterns based generally on the use of available materials and technology. Many of these practices were considered inappropriate for the Commonwealth Caribbean when one takes into account the fact that all the territories are subjected to hurricanes and earthquakes. Consequently, the urgent need for a Uniform Building Code for the Caribbean was well recognised by practising engineers in the region. In this paper, therefore, the development of the Caribbean Uniform Building Code (CUBiC) is briefly traced and the procedures adopted and the philosophy which led to the development of the Structural Steel Design requirements are described. The need for calibration exercises to be carried out is emphasised. The paper concludes with a discussion on the need for the development of a Caribbean Uniform Bridge Code and indicates future trends in structural steel design codes for the Caribbean.

FACTORS INFLUENCING THE USE OF STRUCTURAL STEEL

Buildings

It has long been recognised that structural steel is a very versatile material with very favourable strength and ductility properties. Thus the strength of steel permits the use of long, clean spans with a minimum of columns and relatively small members, providing minimum dead loads. Steel's ductility and toughness are particularly important design criteria for building structures subjected to earthquake loads and impact loads such as bridges.

Figure 1. Typical Example of a Low-Rise Industrial Pitched Roof Steel Portal Building

Figure 2. Typical Example of a Truss Type Steel Bridge in the Caribbean

Briggs [1] has examined the relative merits of structural steel for high-rise buildings and with the introduction of a lightweight open web joist flooring system in Trinidad and Tobago in the 1960's, there was an upsurge in the number of steel framed buildings constructed in Trinidad and Tobago especially during the oil boom period of the 1970's when foreign exchange requirements were not a problem and speed of erection became a critical factor. Figure 3 shows such a structure - the National Stadium in West Port-of-Spain, Trinidad.

Bridges

Developments in steel bridge construction over the past decade has enabled structural steel to improve its market share against concrete and as mentioned earlier, a number of composite steel girder bridges have been erected both in Jamaica and Trinidad and Tobago. Figure 4 shows a typical steel plate girder bridge in Trinidad and Tobago. Some of the factors influencing the choice of structural steel for bridges include, inter alia, the following:

- low weight of superstructure resulting in cheaper foundations

- low construction depth resulting in a reduction in costs of abutments and earth works

- improved quality due to prefabrication in factory

DEVELOPMENT OF STRUCTURAL STEEL DESIGN REQUIREMENTS IN CUBiC

The need for the development of a Caribbean Uniform Building Code (CUBiC) has already been mentioned earlier in this paper and the philosophy and methodology adopted in its development have been described elsewhere [2]. However, it is of interest to describe briefly the reasons which led to the selection of the Canadian Steel Standard [3] as the framework around which the CUBiC requirements for structural steelwork were formulated. The short-term Consultant for the structural steel section, Professor Peter Adams summarised the reasons why he and his colleagues Drs. J.L. Kennedy and G.L. Kulak, chose the Canadian Standard as the base code for CUBiC as follows:

1. As evidenced by the activities of the International Standards Organisation (ISO), the definite trend on a worldwide basis was to the limit states design approach. However, since a dual (allowable stress plus limit states) approach was required for the Caribbean, an existing standard that had moved successfully from the allowable stress to the limit states approach was desirable.

2. The Canadian Standard was used as a model by many countries such as France and England in their submissions to the International Standards Organisation.

3. Relative to the American Institute of Steel Construction (AISC) Specification, used in the United States, the Canadian allowable stress standard was comparable while the Canadian limit states standard was both simpler and far superior to the then AISC proposal for a limit states standard.

Figure 3.   An Example of a Steel Structure – the National Stadium in West
            Port-of-Spain, Trinidad

Figure 4.   Typical Steel Plate Girder Bridge in Trinidad

4. The Canadian Standards were clear and well-organised. The limit states standard had been completely reorganised to take advantage of the simplicity of the limit states approach.

5. The Canadian limit states standard was calibrated against the allowable stress standard before adoption. In addition, specific design comparisons were performed before adoption.

6. The limit states standard had been in use for ten years, in parallel with the allowable stress standard. The latter was being withdrawn with designers overwhelmingly supporting the limit states approach.

In summary, the Canadian Standard was one of the few (if not the only) to have successfully moved from an allowable stress to a limit states approach. The limit states version especially is a well-organised, simple, up-to-date document and acknowledged to be so on an international level.

Thus, the Structural Steel Section of CUBiC comprises of three parts viz:-

Section 7A - Limit States Design (LSD)

Section 7B - Working Stress Design (WSD)

Section 7C - Commentary

Sections 7A and 7B were based in large measure on the Canadian Standards Association (CSA) Standard CAN3-S16.1-M84 Steel Structures for Buildings - Limit States Design and on an October 1983 draft of CSA Standard CAN3-S136-M84 Cold Formed Steel Structural Member but Section 7B has been couched, where appropriate, in working stress design format.

Section 7C is a commentary and is intended to clarify the intent of the various provisions given in Sections 7A and 7B.

It should be noted that the decision to have both WSD and LSD available to practising engineers was based on the recognition that Caribbean Structural Designers could not be expected to shift from one design philosophy to another overnight and that the older engineers would need to undergo refresher courses in LSD philosophy before WSD could be withdrawn. Accordingly, it was recommended to the Caribbean governments that both standards be allowed for an initial period of five years. However, in order to be satisfied that the CUBiC provisions for structural steel are safe and do not produce uneconomical structures, it will be necessary to seek funding to undertake a series of calibration/assessment exercises which will embrace the following stages:

Stage 1: Textual examination - the consideration of compatibility, clarity and completeness of text in comparison with other existing codes.

Stage 2: Calculation Exercises - to demonstrate whether a section of the code is workable in the sense that the necessary steps are readily available to the user/designer i.e. user-friendly.

Stage 3: Comparison exercises - to compare design solutions and material/workmanship specifications with other workable codes.

Stage 4: Calibration exercises – to check the overall levels of safety, serviceability and economy of complete assemblages and structures with existing codes.

NEED FOR A CARIBBEAN BRIDGE CODE

A recent questionnaire survey undertaken by the author [4] on past and current practices in the design of short and medium span bridges in the West Indies has confirmed the need for the development of a Bridge Code for use in the Caribbean. However, opinion was divided in respect of which existing bridge code should be used as a base code for the Caribbean. Historically, both the AASHTO and British BS153 Codes have been used in the various West Indian islands. In recent times, however, there has been a strong Canadian influence to use the Ontario Highway Bridge Design Code (OHBDC 1983) [5] in Jamaica due to financing and technical assistance provided by the Canadian International Development Agency (CIDA). In this connection, it is of interest to note some of the differences when designs are made using BS153, AASHTO and OHBDC Codes. Such a comparison was undertaken [4] on a number of representative steel plate girder and prestressed precast AASHTO girder bridges in Trinidad and Tobago utilising a Bridge Design Expert System (BDES) developed by Biswas [6]. The rules of BDES knowledge base in respect of loads, load distribution, allowable stresses and clearances were extracted from the AASHTO specifications. Whilst the designs were generally in agreement, it is pertinent to note that the OHBDC requires the use of limit states design in comparison with the allowable stress design method in BS153 and AASHTO. The OHBDC specifies that bridge evaluations shall be made using the refined methods of analyses required for design and that limit states be used. In the UK BS153 has been replaced by BS5400 which is based on limit states design concepts. However, the UK codes offer little guidance in respect of seismic details. From the results of the questionnaire survey bridge engineers in the Caribbean will need to examine which of these three base codes or combinations thereof will satisfy the relevant aims of standard-isation and the particular conditions in the West Indies in terms of seismic and other loads as follows:

(i)   to simplify bridge design in the region

(ii) to promote overall economy

(iii) to ensure safety, quality and fitness of purpose.

The International Guide on Constructional Steel Design [7] could well provide a suitable basis for such a Caribbean Bridge Code.

FUTURE DEVELOPMENTS IN STEEL CODES IN THE CARIBBEAN

The future work on CUBiC within the context of regional and international harmonisation of codes and standards will be divided between the adoption and/or adaptation of international standards produced by the International Standards Organisation (ISO) to reflect appropriate loads, common performance criteria and general principles of design for buildings and bridges in the Commonwealth Caribbean. In respect of the proposed

Caribbean Bridge Code it is envisaged that a small drafting panel consisting of both practising designers and researchers would be set up to
prepare an initial draft on the basis of an invitation from at least three
firms/organisations similar to what was done for CUBiC.

CONCLUDING REMARKS

On the basis of the foregoing it is appropriate to make the following
concluding remarks:

1. The Caribbean Uniform Building Code (CUBiC) is among the
   first of a series of regional codes which will have a
   profound influence on structural design in the Commonwealth
   Caribbean in the future and will eventually provide a solid
   foundation for moving towards greater harmonisation with
   other international codes such as the Eurocodes.

2. With the recent advances in bridge design codes in the UK
   and North America, there is an urgent need for Caribbean
   engineers to undertake an evaluation of such codes to
   determine their relevance for the development of a Caribbean
   Bridge Code. In this connection it is of interest to note
   that plans are well on the way for a First Caribbean Conference
   on Bridge Engineering to be held in Trinidad in early February
   1993.

3. The opportunity now exists for the practising engineers in
   the Caribbean to collaborate with their counterparts in the
   University of the West Indies and other countries in order
   to play a key role in the development of structural steel
   design codes which will satisfy the essential requirements
   of safety, economy and suitability for use in the Caribbean.

REFERENCES

1. Briggs, M.H.   Structural Steel or Reinforced Concrete.   Proc. 1st
   Caribbean Welding Conference, Fac. of Eng., UWI, Trinidad, Sept.
   1967, pp. 196-201.

2. Chin, M.W.   The Development of a Caribean Uniform Building Code
   (CUBiC) - Some Experiences.   Proc. 10th Triennial Congress of CIB 1986,
   Vol. 5, pp. 1823-1830.

3. Canadian Standard Association - Steel Structures for Buildings - Limit
   States Design, National Standard of Canada Standard CAN3-S16.1-M84
   Canadian Standards Association, Rexdale, Ontario, Canada.

4. Chin, M.W. and Khan-Kernahan, I.K.   Short and Medium Span Bridges in
   the West Indies: An Overview.   Proc. Third International Conference
   on Short and Medium Span Bridges, Toronto, Canada, Aug. 1990, pp. 61 to
   70.

5. Ministry of Transportation and Communications of Ontario.   Ontario
   Highway Bridge Design Code 1983.   Min. of Transportation and
   Communications of Ontario, Canada.

6. Biswas, M. and Welch, J.G.. BDES: A Bridge Design Expert System. Engineering with Computers 2 Springer-Verlag, New York Inc., pp. 125-136.

7. Bjorhovde, R., Dowling, P. and Harding, J. Editors. Constructional Steel Design - An International Guide, Elsevier, 1992.

DEVELOPMENT OF STEEL CONSTRUCTION IN BRAZIL

KHOSROW GHAVAMI
Civil Eng. Dept. PUC-RJ
Pontificia Universidade Catolica do
Rio de Janeiro-Brazil
ILDONY H.BELLEI
Fabrica de Estruturas Metalicas S.A.
JOSE C.D'ABREU
Dept.of Metal& Material Sciences PUC-RJ

ABSTRACT

This article intends to give an outline and a broad review to the development of steel construction design and fabrication in Brazil since the early stages when whole steel structures were imported and assembled on site up to present time when Brazil has become the eighth largest steel producer but still steel construction is not being used commonly yet, although recently several large span steel buildings and box girder bridges have been completely designed, fabricated and constructed completely in Brazil. No attempt has been made to present an exhaustive and critical review of the steel construction industry, norms and research progress in Brazil, but rather to give the state of the art of the steel construction industry for the appreciation of specialists outside Brazil.

INTRODUCTION

A review of the development of steel construction in Brazil is presented which shows the trends in its development and achievements. A short history of the application of metal as a structural material, covering the last two centuries, helps to position the Brazilian steel construction in the world.

Metal as a constructional material started to be used with the industrial production of cast iron in 1759. The first major bridge of 30m arch span was then constructed in England. Quite a few iron bridges followed to be built during the period 1780-1820 which were mostly arch-shaped with main girders consisting of individual pieces forming bars or trusses. In

France cast-iron was more frequently used in the construction of buildings. After 1840 the wrought iron (puddled iron) began to replace cast-iron. Britania Bridge over the Menai Straits in Wales, which was a continous box girder bridge with spans of 70-140-140-70m, was the earliest important example made of wrought iron plates and angles during 1846-1850 [1]. A widespread application of wrought iron occurred around 1880 in the U.S.A., especially in Chicago.

In Brazil the first structure, using imported steel in buildings, was the Santa Isabel Theatre in Recife which was built during the period 1869-71. For its construction 305 tons of cast iron were employed. The first bridge was the Paraiba de Sul in the state of Rio de Janeiro which had 5 spans of 30m and was constructed in 1857. To decrease the importation of steel the Cia Siderugica Belgo Mineira was founded in 1921 to produce steel rods [2]. At this time the steel fabricator FICHET S.A. in Sao Paulo, founded in 1923 with the participation of the French steel industry, started to construct mainly industrial steel buildings.

Before the second World War all main steel structures were shop fabricated and trial assembled in Europe before they were dispatched to Brazil, as there was no skilled labour force available locally. Among several examples the 2km long Teresa Cristina box girder railway bridge in the state of Santa Catarina and the orthotropic suspension bridg in Florianopolis should be mentioned. They were fabricated and constructed by British and German engineers, respectively, between 1920-1927 [3].

The mundial depression in the 30ies and later the second World War caused an interruption to the steel construction development. Consequently the situation had a significant effect in the construction of steel structures in Brazil, as no major steel structure project was executed. However during this time the application of welding, commonly used here for ship repairs, substituted rivetting in Europe and U.S.A. which produced a great advancement in steel construction.

In 1940 the National Steel Company (CSN) was founded in Volta Redonda in the state of Rio de Janeiro and started soon to produce plates, I-beams and rails according to American Standards. Only after the establishment of the steel fabrication plant (FEM) in Volta Redonda in 1953 the local steel industry began to fabricate and build steel structures, using mainly rivetted connections although rivetting was less and less used in USA and Europe. Up to the year 1950 the steel products of CSN were principally used in the construction of industrial buildings and producing mainly truss frame structures using British and American Standards.

A relatively important achievement in national steel construction was the fabrication and assembly of the Guaiba Bridge in Porto Alegre in the 50ies, which was a welded orthotropic steel plate bridge of 58m span, one of the first welded bridge structure to be designed and fabricated by FEM according to British standards. At this period the furnace USIMINAS in the state of Minas Gerais was under construction to produce special steel plates for the construction of long span bridges. In 1970 USIMEC, which is a subsidary of USIMINAS, was constructed with the help of German specialists and started to fabricate 2m long I- and H-section welded beams. These 2 main steel structure fabricators started the development and construction of several important constructions [3],some of which will be described in the following. During the whole of this period very few research programs and studies were carried out into the behaviour and development of steel structures. However one major study into the behaviour of steel box girder bridge during the service had been carried out on the Rio - Niteroi bridge which will be discussed in the following sections.

The lack of substantial research, national steel norms and profound training has lead to have heavier structures as compared with those of their type in industrialized countries. An important step was made in 1983 by issuing the draft of the first Brazilian Steel Bridge Code. This code had been prepared by Chapman and Dowling and is under review by the members of the bridge committee [4]. A further significant contribution was the approval of the first Brazilian Code on Steel Building structure by the Brazilian Association of Technical Norms, ABNT, in 1986.

At present Brazil is the eighth steel producer in the world with 22600000tons/year and is exporting half of its production. Only 1000000tons/year of steel is used in construction. Since 1979 several theoretical and experimental studies into the behaviour of stiffened steel plates, shells and composite structures were carried out at PUC-RIO under the supervision of the first authour.

VARIG AIRCRAFT REPAIR SHOP

Among many industrial steel building recently designed and fabricated in Brazil the VARIG aircraft repair shop next to the international airport of Rio de Janeiro has been chosen to be presented here as it is the largest span steel structure building in South America and is the first one for which several national studies have been carried out, in Brazil, to establish its aerodynamic behaviour [5].

The main structure of the building consisted of two simply supported steel truss frames. The overall dimensions of each hangar are 136m free span and 110m length. Five main truss frames were constructed with spacing equal to 27.5m as shown in Figure 1. Four travelling cranes with the capacity of 10 tons each were suspended from welded girder beams of 55m. These beams were suspended from a transversal truss shed and were connected to the main roof truss at the spacing of 13.75m. The depth of the main truss was 14m which permitted

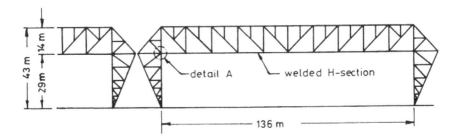

a) Main transversal truss frame

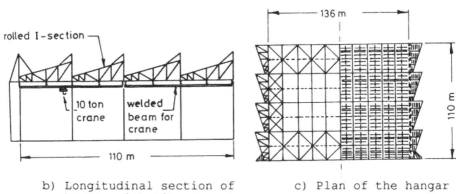

b) Longitudinal section of
hangar with 10tons crane

c) Plan of the hangar

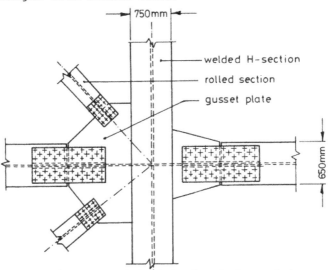

d) Detail A of high friction bolted joints

Figure 1   Overall dimensions and structural detail of VARIG
aircraft repair shop

to limit the maximum deflection of the roof.

The main truss elements were subjected up to 1000tons axial compression force and 800tons tensile force. The slender ratio for compression elements were chosen to be equal in both x and y direction. All the main truss elements were made of welded I- and H-sections of 650mm by 450 to 650mm flange by FEM and assembled on site using high friction bolts. Secondary truss elements were fabricated using laminated profile and assembled by the application of common bolts. The detail of one of the joints is also shown in Fig.1. Although the optimum weight of the structure could be achieved by using both high resistance steel and normal mild steel structure but normal A-36 steel which was produced by CSN according to ASTM was used throughout the project. The total weight of the steel structure of each hangar was 4500tons.

A series of tests were carried out on small models in wind load tunnel to check the validity of the coeffients of forms and suctions. The study showed that for this structure the coeffient of structural form of 1.5 instead of 1.2 and that for suction was 5 times larger than the one recommended in the Brazilian code NB-5.

The design of the structural steel elements was based principally on simple approach using the design rules of American Institute of constructional steel design and erection of steel building, AISC 7th edition, and for the wind load NB-5 as no proper national steel code existed yet in 1970.

BRIDGE CONSTRUCTION

After the second World War engineers in Europe created a more efficient form of construction such as orthotropic plate where steel decks were stiffened longitudinally with open or closed sections and transversally with closely spaced floor beams. The stiffened plate not only served as the deck but also as the top flange of main longitudinal beams [1]. At the later stage box girder bridges were used more frequently. This was because of the great advantages box girder present in relation to other geometrical forms. Box girder structures have larger strength in bending, torsion and consequently have smaller depth of structure, better performance to wind load which leads to better distribution of material and finally to an easy maintenance [6].

The longest span of steel box girder in the world is the President Costa e Silva Bridge between the cities of Rio de Janeiro and Niteroi [1,7]. The three spans of 200-300-200 m over the main shipping channel of the bridge of about 13 km length are of steel twin box girder types.The steel box girders were designed by Howard Needles Tammen and Bergendoff International Inc. of Kansas City. The large fabricted segments were shipped from Great Britain and then assembled, errected on site by the joint venture of two British companies and a Brazilian subcontractor, Montreal Engenharia S.A., during 1969-74. The design and construction of this box girder bridge presented some new engineering challenges, especially at the time when several box girder bridges such as the Danube Bridge in Viena in November 1969, the Milford Haven Bridge in Wales in June 1970 and the Koblenz Bridge in November 1971 had catastrophically collapsed.

For the construction of this box girder bridge a new high strength steel up to 60mm thickness was used for the first time. The welding of these plates and the concern about the accuracy and reliability of design assumptions and methods lead to a conservative design for the unique steel structure. However an extensive on-site measurement of stresses during construction and after completion of the bridge under traffic load were carried out.

The results of the investigation presented new insight into the design of box girder-bridges. For example it was found that the temperature distribution was non-linear and average cross-section stresses in the bridge due to diurnal temperature variation were greater than due to traffic loads; transverse bending stresses due to Poisson's ratio had significant effect, but were not critical [7]. However they need to be studied in detail and be considered in design of other bridges.

Among many box girder bridges designed, constructed and assembled by the Brazilian steel fabricator USIMEC without any foreign participation after the Rio-Niteroi bridge the Victoria box girder bridge of three spans of 175-260-175m in the state of Spirito Santo is of importance technically. This bridge was constructed in 1980 and its principal characteristics are similar to Rio-Niteroi bridge. The first steel box girder of trapezoidal cross-section is the Bertioga Bridge in the port of Santos, which presented several new solutions in the construction of this type of structure .

During the last decade many composite steel railway and roadway bridges were designed , fabricated and assembled in Brazil with their own characteristics. The composite single trapezoidal box girder bridge across the Tocantins River consists of pre-stressed concrete slabs fixed to the top flange of the box girder with shear studs. The studs were embedded in concrete after the completion of the slabs. This is a railway and highway bridge having been constructed while the railway was functioning. Due to the necessity of the iron ore transportation for export the railroad had to be constructed first and later the lateral side highways Therefore at the first stage the bottom flange of the trapezoidal box girder was designed so that the lateral cantilevers could have been constructed later.

One of the longest highway steel bridge structure is the 23 km Linha Vermelha constructed in Rio de Janeiro during 1973-92. The main structure of the bridge consists of composite box girder, truss, truss girder and girder. One of the main innovation in constructing the bridge was the usage of high resistance weather steel COR-TEN B [8]. This was an important step in steel development in Brazil as corrosion is a serious problem. Specially for normal steel the maintenance costs are very high due to low quality control and insufficient standards.

In Figure 2 the longitudinal welded girder with shear stud connectors on the top flange to be embedded in concrete after the installation of the prefabricated concrete slabs and the cross-bracing is shown. The steel structure was supported in most parts either on reinforced concrete columns of R.C. frames as can be seen in Figure 2.

Figure 2   View of the longitudinal girder and the cross-bracing using corrosion resistant steel

STANDARD FOR STRUCTURAL STEELWORK

The major difficulty in the projects, quality control and design of steel structures in Brazil were the lack of norms. The steel norms for buildings (NBR-8800) which are based on the limit state design philosophy was approved and published only in 1986 and since then no revision has been made. Only in 1991 a short manual for the use of the NBR-8800 was published. The Brazilian Code for steel bridges, BCSB, which is under discussion by the committee member of the Associacao Brasileira de Norma Tecnica (ABNT) had been prepared by Chapman and Dowling in 1982, using a new format. In this format the principles of the design are separated from techniques, so that the latter could be up-dated without the need for a complete reissue of the code [4,9,10]. This new concept in code writing facilitates the inclusion of eventual modifications later on.

One of the main factors which delays the approval by the committee members is the lack of relevant experimental results on which design formulas could be soundly understood and checked. For this reason efforts were made by the first author to create an experimental research program at the Civil Engineering Department of Pontificia Universidade Catolica of Rio de Janeiro, which has been going on for more than ten years. The results of which have been reported elsewhere [10-12]. For example based on this studies it was found that the design of stiffened plates according to Brazilian bridges consider most variables and the various modes of buckling collapse but it is a conservative method. The same was true for the design of composite beams. It is hoped that the final version of the code to be discussed by engineers will finally be approved by 1993.

CONCLUDING REMARKS

The brief review concerning the steel construction development in Brazil shows that although the steel production has grown immensely the use of steel construction has not grown proportionally. Nevertheless several technically exceptional steel building and bridges have been constructed based on simple analytical methods during the last two decades.

One of the major problems is the lack of appropriate teaching of steel structures at the Brazilian universities. The lack of research fundings and interaction between universities and interested industries and the dependance on foreign literature e.g. revision of excisting codes contribute to the situation. National experimental research are not available sufficiently which could be used by the committee members to evaluate the formulas of the new code.

However several successful research programs have been carried

out during the last decade which introduces a new stage in steel construction in Brazil. To the knowledge of the authors this situation is worse in other South American countries where the use of steel structures is at a very early stage of development.
During the active participation of the first author in the bridge committee it was suggested that the values of the factors given in the draft bridge code be reviewed after completion of certain design studies using the code but due to lack of funds those were not carried out.

It has come the time where only a governmental agency, with the active participation of industries helped by designers, researchers and fabricators, could develop a more systematic way to further steel construction in Brazil to the benefit of the whole country.

ACKNOWLEDGEMENT

The authors would like to thank Ursula Ghavami for producing the drawings and typing the text. Her unpaid work is still more appreciated at a time when most financial support from governmental agencies and industry has almost stopped completely. Without her help this paper would not have been completed in time to be included in the proceedings of the conference.

REFERENCES

1. Heins,C.P. and Firmage,D.A., Design of Modern Steel Highway Bridges. John Willey & Sons Ltd., 1979, 463pp

2. Bellei, I.H., A Estrutura de Aco no Brasil (Steel structure in Brazil), Internal publication FEM, Rio de Janeiro, 1989, 5pp

3. Mason, J. and Vogel, J., Neuzeitlicher Stahlbau in Brasilien, Der Stahlbau 9-1980, pp261-269

4. BCBS -Brazilian Code for Steel Bridges, Draft, text edited by Chapman and Dowling, Oct.1982

5. Estruturas dos Hangares do VARIG, (Structure of VARIG Hangar), Internal Report FEM, Rio de Janeiro, 1987

6. Lally, A., Steel Box Girder Bridges, AISC-Enginering Journal, 10,4, 1973

7. Ostapenko, A., DePaoli, D.H., Daniels, J.H., O'Brien, J.E., Yen, B.T., Behatti, M.E. and Fisher, J.W., A Study of the President Costa e Silva Bridge during Construction and Service (Steel structure . Final report, submitted to ECEX-Empresa de Engenharia e Construcao de Obras Especiais), Rio de Janeiro,

Fritz Engineering Laboratory Report 397.6, March 1975

8. Bellei, I.H. and Campana, C.A., Elevados da Linha Vermelha, the 6th Seminar on the Use of Steel in Construction and the 3rd Conference of Steel-architecture, Campinas, 1-2 July 1992

9. Chapman, J.C. and Dowling, P.J, A New Design Code for Steel Bridge. Proc. of second International Conference on Short and Medium Span Bridges, Ottawa, Canadian Society of Civil Engineering, August 1986, Vol.2, pp305-319

10. Ghavami, K., The Collapse of Continuously Welded Stiffened Plates Subjected to Iniaxial Compression Load, Proc. Inelastic Behaviour of Plates and Shells. IUTAN Symposium, Rio de Janeiro, edited by Bevilaque, L., Feijoo,R. and Valid, R., Springer, 1986, pp403-415

11. Ghavami, K. and DeLema, L.G., Comportamento em Carga ULtima de Vigas Mistas com Conectores Rigidos e Flexiveis (Ultimate Load Behaviour of Composite Beams with Rigid and Flexible Connectors), Colloquia 83, Santiago, Chile, 1983, pp1-12 (in Portuguese)

12. Ghavami, K., Ultimate Load Behaviour of Stiffened Plates Subjected to Compression Load, Proc.20th Midwestern Mechanics Conf. Purdue University, Indiana, Sept.1987, pp569-574.

ESDEP - A MAJOR INITIATIVE IN STEEL EDUCATION

KEITH EATON, GRAHAM OWENS, CHRISTINE ROSZYKIEWICZ
The Steel Construction Institute, Ascot, UK

JAAP WARDENIER
Technical University of Delft, Delft, The Netherlands

ABSTRACT

In a rapidly changing situation for steel design, with new standards and codes being produced and new technologies being developed, there is an increasing need for training courses, education material and continuing professional development activities to help educate new engineering students as well as retrain established engineers. This need is being addressed throughout Europe, ready to take the steel design profession forward to the 21st century. The Steel Construction Institute and the Technical University of Delft are jointly leading a major project to prepare the European Steel Design Education Programme, ESDEP. This is a comprehensive set of teaching material on steel design and steel construction which will be used throughout Europe. This paper gives details of ESDEP, its contents and its availability.

INTRODUCTION

ESDEP is the most popular acronym today in steel Euro-jargon; it stands for the European Steel Design Education Programme. ESDEP was conceived in the Spring of 1988 when a meeting between several illustrious professors and far-sighted steel producers was convened by Graham Owens (SCI) and Jaap Wardenier (TU Delft) to consider the need to improve the teaching of structural steel design across Europe. There was no doubt in everyone's minds that within the ivory towers of traditional academia, many future designers were discouraged from specifying steel, partly because of unfamiliarity and partly because it was portrayed as yesterday's material with a tired image. With the advent of Eurocode 3 for steel structures, it was time to change the face of structural steel design education and to give steel a new up-to-date image throughout Europe.

ORGANISATION

ESDEP is based on the UK Structural Steel Design teaching programme which was funded by British Steel, and which has educated the educators in British polytechnics and universities over the last six years. The project team consists of Graham Owens (UK), Jaap Wardenier (the Netherlands), Keith Eaton (UK) and Christine Roszykiewicz (UK). Ko de Back (the Netherlands) also made a valuable contribution but had to resign due to ill health. Early in the project, a pan-European Advisory Committee was established under the Chairmanship of Patrick Dowling (UK) and subsequently a Sponsorship Committee was formed under the Chairmanship of Pierre Everard (Luxembourg). In the Autumn of 1988 the UK team visited the major European steel industries and gave slide demonstrations to win the necessary financial backing.

Presently the steel industry throughout the 12 countries of the European Community, together with Austria, Finland and Sweden, is backing ESDEP to the tune of 1,135,500 ECU over four years, and financial support of 835,000 ECU is also coming from the European Commission under the auspices of COMETT (the European Community Action Programme for Education and Training for Technology).

TABLE 1
Countries providing technical input to ESDEP

Austria	Eire	Greece	Norway	Switzerland
Belgium	Finland	Italy	Portugal	United Kingdom
Czechoslovakia	France	Luxembourg	Spain	
Denmark	Germany	Netherlands	Sweden	

The widespread enthusiasm and goodwill from all sides of the European steel spectrum - academia, designers, fabricators and producers - have escalated to such a degree that the experience of working together has resulted in the forging of strong cross-cultural links and has laid the ground for other mutually beneficial working relationships in the new united Europe of the 1990s. Each of the participating nations has something to offer; the objective is to pool this expertise to develop the overall benefits and usage of steel.

DELIVERABLES

ESDEP is producing a comprehensive set of teaching aids, prepared by 18 Working Groups with more than 200 50-minute lectures, plus videos, slides, worked examples and computer software to cover all aspects of structural steel design, ranging from an introduction to steel's role in the European construction industry, through fabrication and fire protection, to offshore structures and seismic design. By making this same material available across all countries in Europe, ESDEP will ensure that future generations of structural engineers will be associated with modern approaches to technology, and the opportunity to design in this modern medium will be unrestricted by national boundaries.

TABLE 2
Lecture syllabus

GROUP NUMBER	WORKING GROUP TITLE	NUMBER OF LECTURES
1A	Steel Construction: Economic & Commercial Factors	4
1B	Steel Construction: Introduction to Design	13
2	Applied Metallurgy	9
3	Construction	15
4A	Protection: Corrosion	5
4B	Protection: Fire	5
5	CAD/CAM/CIM	4
6	General Applied Stability	7
7	Elements	18
8	Plates and Shells	13
9	Thin Walled Constructions	7
10	Composite Construction	15
11	Static Connection Design	13
12	Fatigue Design	15
13	Complete Connection Design	12
14	Structural Systems I - Buildings	18
15	Structural Systems II - Other Structures	24
16	Structural Systems III - Assessment and Repair	4
17	Seismic Design	7
18	Stainless Steel	4
	TOTAL	212

At this stage in the programme, all the lectures and 48 worked examples have undergone a ruthless revision in four technical workshops each of which took place over four days (including a weekend). On each occasion twelve specially appointed experts lead by Jaap Wardenier and Graham Owens agreed to collaborate in a no-fee basis to ensure that this material was up-to-date, technically sound, European in content and in accord with the Eurocodes. The work demanded of participants was very intense but altogether stimulating; it proved to be an ideal opportunity to be away from the fax machine and telephone, and at the end of the day to enjoy discussions and conversation in the company of steel enthusiasts from northern and southern climes.

Apart from the mammoth task of reviewing and editing lectures, 1000 slides are being collated, innovative educational software is being produced and 21 videos have been prepared. More than 300 contributors — advisers, sponsors, working group chairmen, authors, reviewers, etc., from as far afield as Lisbon, Prague, Trondheim and Athens — have all given up valuable professional and personal time to ensure the success of ESDEP. Examples of contributors' enthusiasm must include the Greek who abandoned his sunny island retreat in August to share his expertise with the Computer Committee; the British Steel Manager who made time in a hectic schedule to fly to Paris and lend his thoughts and experience to save a lecture on commercial factors and market forces; and the Czech who planned a special detour from a research trip in Japan to attend an Advisory Committee meeting in London.

FUTURE PLANS

Material is now being produced in its final form and distributed, initially in English, to the project sponsors. They are now beginning the arduous task of translating the material into the principal European languages. The project will be given a grand European launch in 1993, followed by national events in individual countries.

TABLE 3
Project benefits

• Lectures available in 10 European languages
• Wide ranging modular teaching material on steel design and steel construction
• Design rules follow Eurocode 3 and Eurocode 4
• Same information taught throughout Europe
• Three levels of teaching material for undergraduates, post graduate students, and practising engineers.

Everyone is working so hard to steer ESDEP through to the best possible conclusion that one has to consciously stand back to see the longer-term developments that will ensue. What is clear is that ESDEP has already generated a spirit of co-operation and a series of relationships throughout Europe that must be maintained and harnessed in the future. As the European Community strengthens it will want to improve the quality of both its infrastructure and environment. Years of research and development have improved the properties of steel beyond imagination, so that besides its functional properties it can have a strong aesthetic appeal. In an age when environmental issues are prominent across the continent of Europe, future designers will be more acutely aware of their responsibility for the quality of life. Helped by ESDEP, the European steel industry has the opportunity to make a major contribution to our changing urban scene in order to create a strong economic community that is pleasant to live in.

ACKNOWLEDGEMENTS

The authors gratefully acknowledge the many contributions that have made this project possible, including the financial support from the COMETT programme in Brussels, the financial support from the steel industry in 15 European countries, and the contributions-in-kind provided by hundreds of experts in industry and academia throughout Europe.

THE DESIGN OF TALL BUILDING FRAMES
- THE CELLULAR APPROACH

Y S Lau   T S Lok
Nanyang Technological University, Singapore
A R Freeman
Y S Lau, Consulting Engineer, Singapore

ABSTRACT

The paper presents a recursive procedure for the calculation of lateral deflections and joint rotations in tall 2-D building frames in 2 stages, without having to solve simultaneous equations. Starting from a vertical stack of C-cells based on the Grinter-Goldberg model for the frame, average floor joint rotations and storey drifts are calculated from explicit expressions, taking the effects of $P - \Delta$ and axial deformations of the columns into account. In a second stage, individual joint rotations and improved storey drift are calculated from explicit expressions derived for a row of H-cells at each floor. Two cycles at most normally suffice for final design. The method is illustrated by an example of a 19-storey frame.

INTRODUCTION

In the design of structures with many degrees of freedom such as tall building frames, provided that a reasonable degree of precision can be achieved, simplicity and cost effectiveness in computation are high priorities in the choice of method of analysis, particularly in the early stages. The method should preferably be transparent and simple enough for spreadsheet computation, self correcting, and capable with minimal effort of local modifications or refinements to the structure in the final stages. The initial trial values should moreover be sufficiently accurate to be usable without the need for numerous iterations.

THE GRINTER-GOLDBERG FRAME

If there are no setbacks and the base is fixed, a 2-D building frame has 3*N*M degrees of freedom, where N is the number of storeys and M is the number of columns, and a conventional stiffness analysis of the frame under load requires the solution of 3*N*M simultaneous equations.

In the Grinter-Goldberg frame [1,2] axial deformations of members are neglected, and it is also assumed that all the joint rotations in each floor are equal, i.e. all beam midspans are points of inflexion. The degrees of freedom are then reduced by a factor of 3*M/2, to 2*N.

The Grinter-Goldberg frame is equivalent to a single vertical cantilever of stiffness k_i or $\Sigma k_{i,j}$ in storey i, with attached springs at each floor i of rotational stiffness $12E \Sigma k_{bi,j}$ or $12E \Sigma (Ib_{i,j}/L_{i,j})$, as shown in Fig.1, where

   $k_{i,j}$ = stiffness $I_{i,j}/h_i$ of column j in storey i,

   $Ib_{i,j}$ = moment of inertia,

   $L_{i,j}$ = length of the jth girder in floor i at the top of
            storey i.

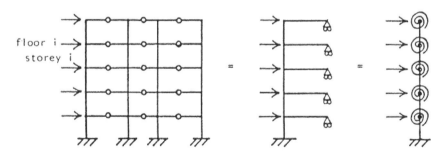

Figure 1.   The Grinter-Goldberg Frame

Stiffness analysis of the Grinter-Goldberg frame requires
(a) partitioning into N*N matrices and condensation followed by the solution of N simultaneous equations, or
(b) the solution of either 2N equations or N equations with 2x2 elements, or
(c) iterative alternating solution of N equations for lateral deflexions and joint rotations.
The partitioned and tridiagonal forms of the stiffness matrix for the Grinter-Goldberg frame are shown in Figure 2.

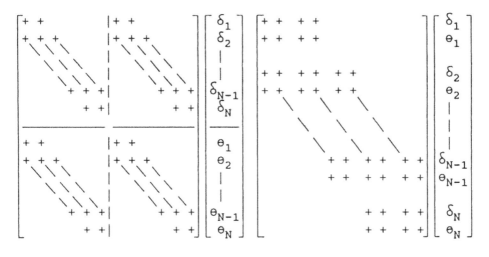

(a) Partitioned form   (b) 2x2 Tridiagonal form

Figure 2. Stiffness Matrix of the Grinter-Goldberg Frame

THE CELLULAR METHOD

The need to solve a large number of simultaneous equations is
avoided in the cellular approach, in which the storeys in a
Grinter-Goldberg frame are initially modelled as overlapping
C-cells. The storey drift and the average of the end rotations
of all the beams in each floor are recursively calculated from
the C-cells in an upward or a downward pass. For drift
control, a single pass is sufficient, unless the total
stiffness of the columns $\Sigma k_{i,j}$ is many times the total
stiffness of the adjacent floor beams $\Sigma kb_{i,j}$.

Each floor is next modelled as a row of overlapping H-cells,
from which the individual joint rotations in the floor and
improved lateral deflexions in the adjacent storeys are
recursively calculated. If modifications to the columns in a
storey or girders in a floor are subsequently needed, only the
H-cell computations for the 2 or 3 floors above and below the
storey or floor under consideration need to be carried out.

THE C-CELL

Adopting the Grinter-Goldberg assumptions and taking $P - \Delta$ into account, the moment equilibrium equations for the floor joints at the top and bottom of storey i may be written as

$$k_{11}\theta_{i-1} + k_{12}\theta_i = R_1 \qquad \ldots (2)$$
$$k_{21}\theta_{i-1} + k_{22}\theta_i = R_2 \qquad \ldots (3)$$

where

$$k_{11} = k_i(4-3k_i/\widetilde{k}_i) + k_{i-1}(4-3k_{i-1}/\widetilde{k}_{i-1}) + 12k_{bi-1}$$
$$k_{22} = k_i(4-3k_i/\widetilde{k}_i) + k_{i+1}(4-3k_{i+1}/\widetilde{k}_{i+1}) + 12k_{bi}$$
$$k_{12} = k_{21} = k_i(2-3k_i/\widetilde{k}_i)$$
$$k_i = \Sigma\, k_{i,j}$$
$$\widetilde{k}_i \triangleq \Sigma\, k_{i,j} - P_i h_i/10E$$

$$R_1 = [F_{i-1}h_{i-1}k_{i-1}/\widetilde{k}_{i-1} + F_i h_i k_i/\widetilde{k}_i]\ /\ 2E$$
$$\quad + [(3k_{i-1}/\widetilde{k}_{i-1})-2]k_{i-1}\theta_{i-2}$$

$$R_2 = [F_i h_i k_i/\widetilde{k}_i + F_{i+1}h_{i+1}k_{i+1}/\widetilde{k}_{i+1}]\ /\ 2E$$
$$\quad + [(3k_{i+1}/\widetilde{k}_{i+1})-2]k_{i+1}\theta_{i+1}$$

Solving,

$$(\ k_{22}R_1 - k_{12}R_2)/D_i = \theta_{i-1} \qquad \ldots (4)$$
$$(-k_{21}R_1 + k_{11}R_2)/D_i = \theta_i \qquad \ldots (5)$$

where

$$D_i = k_{11}\, k_{22} - k_{12}k_{21}$$

Although δ_{fi} can be directly calculated, it is convenient to calculate it from θ_{i-1} and θ_i, as follows:

$$\frac{\delta_{fi}}{h_i} = \frac{F_i h_i}{12Ek_i} + \frac{\overset{m}{\underset{j=1}{\Sigma}}\ k_{i,j}(\theta_i + \theta_{i-1})}{2k_i} \qquad \ldots (1a)$$

since such joint rotations will be subsequently needed, to calculate the individual joint rotations.

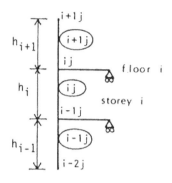

Figure 3. Joint and Member Notation in the storey i C-Cell

Initially Θ_i in storey i is recursively calculated in a downward pass, starting from i = N, neglecting δ_{ai} and assuming

$$\Theta_{i-2} = (F_{i-1}h_{i-1} + F_{i-2}h_{i-2}) / (24\ EI_{bi-1}) \qquad \ldots (6)$$

Alternatively, Θ_{i-1} is recursively calculated in an upward pass starting from i = 2, assuming

$$\Theta_{i+1} = (F_{i+2}h_{i+2} + F_{i+1}h_{i+1}) / (24\ EI_{bi+1}) \qquad \ldots (7)$$

A recursive second pass in the reverse direction yields floor joint rotations from which the storey drifts indistinguishible from those in a Grinter-Goldberg frame are recalculated, using equation (1a). The second pass is unnecessary if the ratio of column stiffness $\Sigma k_{i,j}$ is of the same order of magnitude as or lesser than that of the floor beams $\Sigma k_{bi,j}$.

THE H-CELL

The moment equilibrium equations for joints i,j and i,j+1 in floor i may be written as

$$2k_{11}\Theta_{i,j} + k_{bi,j}\Theta_{i,j+1} = R_1 \qquad \ldots (8)$$
$$k_{bi,j}\Theta_{i,j} + 2k_{22}\Theta_{i,j+1} = R_2 \qquad \ldots (9)$$

where

$$k_{11} = 2(k_{i,j} + k_{i+1,j} + k_{bi,j-1} + k_{bi,j})$$
$$k_{22} = 2(k_{i,j+1} + k_{i+1,j+1} + k_{bi,j} + k_{bi,j+1})$$

$$R_1 = 3(k_{i,j}\delta_{fi}/h_i + k_{i+1,j}\delta_{fi+1}/h_{i+1})$$
$$- k_{bi,j-1}\theta_{i,j-1} - k_{i,j}\theta_{i-1,j} - k_{i+1,j}\theta_{i+1,j}$$

$$R_2 = 3(k_{i,j+1}\delta_{fi}/h_i + k_{i+1,j+1}\delta_{fi+1}/h_{i+1})$$
$$- k_{bi,j+1}\theta_{i,j+2} - k_{i,j+1}\theta_{i-1,j+1} - k_{i+1,j+1}\theta_{i+1,j+1}$$

Solving,

$$\theta_{i,j} = (\ k_{22}R_1 - k_{bi,j}R_2\ )\ /\ D_{i,j} \qquad \cdots (10)$$
$$\theta_{i,j+1} = (\ -k_{bi,j}R_1 + k_{11}R_2\ )\ /\ D_{i,j} \qquad \cdots (11)$$

where

$$D_{i,j} = k_{11}k_{22} - k_{bij}^2$$

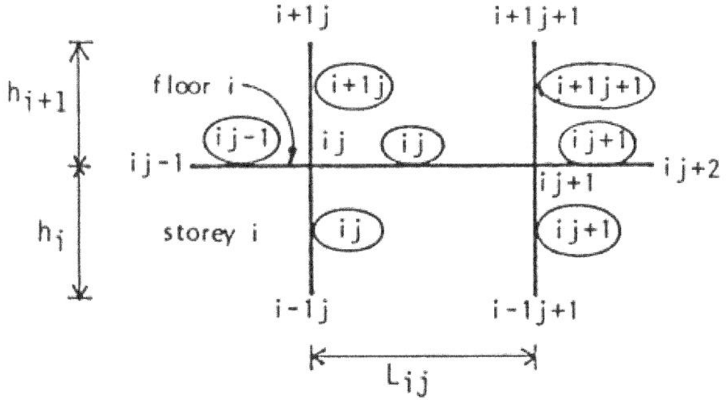

Joint i,j+1 is eliminated in forward pass and joint i,j is eliminated in backward pass.

Figure 4. Joint & Member Notation in the Floor i H-Cell

Initially, $\theta_{i,j}$ is recursively calculated in either an upward pass, starting with $\theta_{i+1,j} = \theta_{i+1}$ from the C-cell analysis, or a downward pass starting with $\theta_{i-1,j} = \theta_{i-1}$, floor by floor. At each floor i in an upward or downward pass, a recursive forward pass from i,1 to i,M using equation (10) is followed by a backward pass from i,M to i,1, using equation (11). The lateral drifts δ_{fi} are calculated after each forward or backward pass from

$$\frac{\delta_{fi}}{h_i} = \frac{F_i h_i}{12E \sum\limits_{j=1}^{m} k_{i,j}} + \frac{\sum\limits_{j=1}^{m} [k_{i,j}(\theta_{i,j} + \theta_{i-1,j})]}{2 \sum\limits_{j=1}^{m} k_{i,j}} \qquad \cdots (1b)$$

A second upward or downward pass floor by floor in the reverse direction to the first normally yields sufficiently accurate joint rotations and storey drifts for all stages of the design. Finally, bending moments and shears in the members are obtained from the member stiffness or slope deflexion equations.

AXIAL DEFORMATIONS OF COLUMNS

If it is assumed that lateral loads are applied at the floors and the joints in a floor remain in a straight line under the loads, the lateral deflexions due to axial deformations of the columns are

$$\frac{\delta_{ai}}{h_i} = \frac{1}{h_i} \int_{z=0}^{z_i} \frac{M_{ok}(z_i - z_k) dz}{EI_{ak}} \qquad \cdots (12)$$

where

$$M_{ok} = \sum_{j=k+1}^{N} F_j (z_j - z_k)$$

$$F_j = \sum_{p=j}^{N} \Delta F_p$$

$$I_{ak} = \sum_{j=1}^{m} A_{k,j} \, x_{k,j}^2$$

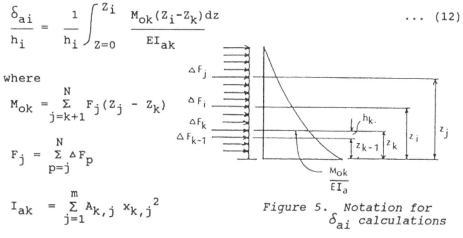

Figure 5. Notation for δ_{ai} calculations

$A_{k,j}$ = area of column k,j in storey k

$x_{k,j}$ = distance of column k,j from centroid

The total lateral drift of storey i, is to a close approximation, $(\delta_{fi} + \delta_{ai})$.

EXAMPLE:

The following example illustrates the application of the cellular method to a 19-storey frame, adapted from [3], under lateral loads.

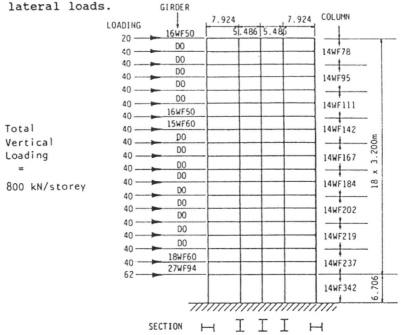

Figure 7. 19-Storey Frame

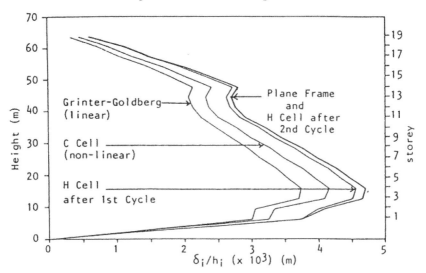

Figure 8. *Non-linear Storey Drifts, Taking into account P - Δ and Axial Deformations in columns.*

REFERENCES

1. Goldberg J.E. (1934), Wind Stresses by Slope Deflexions and Converging Approximations, Transactions ASCE, Vol 99, Paper No. 1878.

2. Grinter L.E. (1937), Theory of Modern Steel Structures, Vol II, Statically Indeterminate Structures and Space Frames, The MacMillan Co., New York.

3. Hurty W.C. & Rubinstein M.F. (1973), Dynamics of Structures, Prentice Hall.

4. Horne M.R. & Merchant W. (1965), The Stability of Frames, Pergamon Press.

APPENDIX 1 − C−CELL DERIVATION

The horizontal shear equilibrium equations for storeys $i-1$, i and $i+1$ and the moment equilibrium equations for floors $i-1$ and i in the Grinter-Goldberg frame, taking into account $P - \Delta$ effects but neglecting axial deformations, may be written as

$$\begin{bmatrix} S_{i-1}^2(1-C_{i-1}^2)k_{i-1} & 0 & 0 \\ 0 & S_i^2(1-C_i^2)k_i & 0 \\ 0 & 0 & S_{i+1}^2(1+C_{i+1}^2)k_{i+1} \\ -S_{i-1}(1+C_{i-1})k_{i-1} & -S_i(1+C_i)k_i & 0 \\ 0 & -S_i(1+C_i)k_i & -S_{i+1}(1+C_{i+1})k_{i+1} \end{bmatrix}$$

$$\begin{bmatrix} -S_{i-1}(1+C_{i-1})k_{i-1} & 0 \\ -S_i(1+C_i)k_i & -S_i(1+C_i)k_i \\ 0 & -S_{i+1}(1+C_{i+1})k_{i+1} \\ S_i k_i + S_{i-1}k_{i-1} + 12k_{bi-1} & S_i C_i k_i \\ S_i C_i k_i & S_i k_i + S_{i+1}k_{i+1} + 12k_{bi} \end{bmatrix} \begin{bmatrix} \delta_{i-1}/h_{i-1} \\ \delta_i/h_i \\ \delta_{i+1}/h_{i+1} \\ \theta_{i-1} \\ \theta_i \end{bmatrix}$$

$$= \begin{bmatrix} F_{i-1}h_{i-1}/E + S_{i-1}(1+C_{i-1})\theta_{i-2} \\ F_i h_i/E \\ F_{i+1}h_{i+1}/E + S_{i+1}(1+C_{i+1})\theta_{i+1} \\ -S_{i-1}C_{i-1}\theta_{i-2} \\ -S_{i+1}C_{i+1}\theta_{i+1} \end{bmatrix} \qquad \dots \text{ (A1)}$$

where $\quad k_i = \Sigma\, k_{i,j}$ and S, C are the Horne-Merchant stability functions [4]

Now $\quad \tilde{k}_i \triangleq \Sigma\, k_{i,j} - P_i h_i/10E$

After eliminating all δ/h's, equation (A1) is approximated by equations (4) and (5).

APPENDIX 2 — H-CELL DERIVATION

The moment equilibrium equations for beam i,j in floor i may be written as

$$\begin{bmatrix} k_{i,j} & 2k_{bi,j} \\ 2k_{bi,j} & k_{i,j+1} \end{bmatrix} \begin{bmatrix} \theta_{i,j} \\ \theta_{i,j+1} \end{bmatrix} = \begin{bmatrix} R_1 \\ R_2 \end{bmatrix} \qquad \ldots (A2)$$

$$\therefore \frac{1}{D_{i,j}} \begin{bmatrix} k_{i,j+1} & -2k_{bi,j} \\ -2k_{bi,j} & k_{i,j} \end{bmatrix} \begin{bmatrix} R_1 \\ R_2 \end{bmatrix} = \begin{bmatrix} \theta_{i,j} \\ \theta_{i,j+1} \end{bmatrix} \qquad \ldots (A3)$$

where $D_{i,j} = k_{i,j}k_{i,j+1} - 4k_{bi,j}^2$

$$k_{i,j} = S_{i,j}k_{i,j} + S_{i+1,j}k_{i+1,j} + 4(k_{bi,j-1} + k_{bi,j})$$

$$R_1 = S_{i,j}(1+C_{i,j})k_{i,j}\,\delta_i/h_i + S_{i+1,j}(1+C_{i+1,j})k_{i+1,j}\,\delta_{i+1}/h_{i+1}$$
$$- S_{i,j}C_{i,j}k_{i,j}\theta_{i-1,j} - S_{i+1,j}C_{i+1,j}k_{i+1,j}\theta_{i+1,j}$$
$$- 2k_{bi,j-1}\theta_{i,j-1}$$

$$R_2 = S_{i,j+1}(1+C_{i,j+1})k_{i,j+1}\,\delta_i/h_i$$
$$+ S_{i+1,j+1}(1+C_{i+1,j+1})k_{i+1,j+1}\,\delta_{i+1}/h_{i+1}$$
$$- S_{i,j+1}C_{i,j+1}k_{i,j+1}\theta_{i-1,j+1}$$
$$- S_{i+1,j+1}C_{i+1,j+1}k_{i+1,j+1}\theta_{i+1,j+1} - 2k_{bi,j+1}\theta_{i,j+2}$$

and
$$\frac{\delta_{fi}}{h_i} = \frac{1}{D_i} \left(\frac{F_i h_i}{E} + [S_{i,j}(1+C_{i,j})k_{i,j}(\theta_{i-1,j} + \theta_{i,j})] \right)$$
$$\ldots (A4)$$

where $D_i = \sum_{j=1}^{m} [S_{i,j}^2(1-C_{i,j}^2)k_{i,j}]$

Equation (A3) is approximated by equations (10) and (11)
Equation (A4) is approximated by equation (1a)

INDEX OF CONTRIBUTORS

Printed and bound by CPI Group (UK) Ltd, Croydon, CR0 4YY

01/11/2024

01782621-0016